Introduction to Electron Holography

Introduction to Electron Holography

Edgar Völkl and
Lawrence F. Allard
Oak Ridge National Laboratory
Oak Ridge, Tennessee

and

David C. Joy
University of Tennessee
Knoxville, Tennessee

SPRINGER SCIENCE+BUSINESS MEDIA, LLC

Library of Congress Cataloging in Publication Data

Introduction to electron holography / Edgar Völkl, Lawrence F. Allard, and David C. Joy.
p. cm.
Includes bibliographical references and index.
ISBN 978-1-4613-7183-0 ISBN 978-1-4615-4817-1 (eBook)
DOI 10.1007/978-1-4615-4817-1
1. Electron holography. I. Völkl, Edgar. II. Allard, Lawrence F. III. Joy, David C., 1943– .
QC449.3.I58 1998 98-42067
502'.8'25—DC21 CIP

ISBN 978-1-4613-7183-0

Originally published by Kluwer Academic / Plenum Publishers, New York in 1999
Softcover reprint of the hardcover 1st edition 1999

10 9 8 7 6 5 4 3 2 1

A C.I.P. record for this book is available from the Library of Congress.

FOREWORD

This book is an authoritative guide to the theory and practice of electron holography and will also serve as a useful guide to the relevant literature. The authors are sure-footed in presenting the whole subject in its essentials and in a concise form. For me, the book also evokes splendid memories of the early encounter with Gabor at the AEI Research Laboratory in Aldermaston, where the Electron Physics Section has been set the task of realizing, if possible, the fantastic vision of "holography" that had jut been revealed to Gabor on the Bank Holiday of Easter 1947, as he sat on a bench awaiting his turn for a game of tennis. He was told in the vision to take an electron micrograph "in phase and amplitude" in a coherent electron beam. When one looked through the developed plate in coherent light one would then see the original object, but afflicted with spherical aberration. This could be corrected optically with a lens of negative spherical aberration! Glaser, among others, thought this was nonsense, since it was not based on theory or experiment. It has taken about fifty years to realize the potential of electron beam holography, but much still remains to be done. The long theoretical and experimental struggle is well described in this book, as are the many applications that did not occur to Gabor himself. This is a fascinating book, which will encourage electron microscopists, theoretical and experimental, to get involved in recent developments.

Tom Mulvey

PREFACE

This book grew out of a joint project in Electron Holography between The High Temperature Materials Laboratory (HTML) of Oak Ridge National Laboratory (ORNL) and the University of Tennessee begun in 1990. It was funded in large part by the Laboratory Directed Research and Development program of ORNL. While we had a clear idea of what we wished to achieve, our initial progress was slow because few other groups were working in this area and so we had to discover for ourselves the special techniques needed to optimize the performance of the microscope when used for coherent beam imaging, and develop the procedures required to reconstruct the holograms that we generated. What we most needed then was a book, similar perhaps to *Introduction to Analytical Electron Microscopy* (Plenum, 1987), which covered all aspects of the technique in sufficient depth and detail to jump-start our efforts.

The aim of this volume, therefore, is to fill that need and to provide electron microscopists, researchers working in the areas of materials science, microelectronics, and the life-sciences, and graduate students with a comprehensive guide to the theory and practice of electron holography. Although the original description of holography by Dennis Gabor in 1947 was aimed at overcoming the problems that then limited the performance of electron microscopes, much of the subsequent development of holography occurred in the fields of optics and microwave engineering because suitable coherent beams of radiation were more readily available in these areas. Within the past ten years, however, the introduction of the field emission electron source has finally provided electron microscopists with a bright and highly coherent beam of radiation suitable for interferometry, and Gabor's vision of the power and utility of this approach is at last being realized. The format of the book is designed to first give the reader an overview of the history, theory, and fundamental concepts underlying holography, then to review the optics of the modern field emission transmission electron microscope and to examine the characteristics of the various key components from which it is constructed. The practice of electron holography is discussed in detail, and, in particular, digital imaging is emphasized as an essential factor in the successful development of quantitative imaging and holography. Finally, through a number of specialized chapters, it provides a practical guide to applying electron holography to a wide variety of problems of current interest. The emphasis thoughout the volume has been to provide sufficient detail and relevant useful information so that any microscopist, with an interest or need in this area and with access to a suitable instrument, could expect to learn how to get this technique working for themselves. The bibliography of the book provides the most comprehensive collection of citations related to electron holography ever assembled and, by itself, represents a significant resource for future research.

The authors wish to thank their colleagues – too many to name individually – who have contributed to this volume by their kindness in allowing us to use materials from their publications, and by their general support. We thank especially Prof. Wilbur

Bigelow for his complete review of the entire book and for his many contributions to revisions of several chapters. Special thanks are also due to the staff of the Central Research Library at Oak Ridge National Laboratories for their invaluable help in tracking references and to the editorial staff of Plenum Publishing for their professional skill in guiding and encouraging this volume. Last, but not least, they are grateful to Dr. Alvin Trivelpiece, Director of Oak Ridge National Laboratory, and the members of the executive committee of the laboratory for their vision in sanctioning the initial project. We are also grateful to Dr. Arvid Pasto, Director of the HTML, and Mr. Ted Nolan, leader of our group, for their guidance and enthusiastic support of this book.

Edgar Völkl
Larry Allard
David Joy

Oak Ridge
June 1998

This book is dedicated to:

Professor Gottfried Möllenstedt

September 14, 1912 — September 11. 1997

Professor Gottfried Möllenstedt was born in Westfalen, Germany in 1912, and died, after a long illness, in September 1997. One of his great scientific contributions was the invention, in 1956, of the Fresnel electron biprism. which forms the basis of modern electron holography. This device was the outcome of his observations of a sample that was charging, but it was his genius that recognized the utility of this effect and allowed him to apply it in a powerful and productive way, to turn Gabor's vision of electron holography into a practical technique. Without the invention of the electrostatic biprism, this book likely would have never been written.

In 1957 Prof. Möllenstedt founded the Institute for Applied Physics at the University of Tübingen, and as its first director he created the environment and the infrastructure that attracted leading researchers from all over the world. This led to a string of discoveries and developments which revolutionized the fields of electron and ion optics. Prof. Möllenstedt also made seminal contributions to the science of electron crystallography, through the development of Kossel-Möllenstedt convergent beam patterns, and to analytical electron microscopy with the fast high resolution electron spectrometer. His many and varied contributions have permanently enriched and empowered the world of electron microscopy. We are proud that he was able to contribute the opening chapter of this book, to help preserve the history of the technique of electron holography for all workers in the field.

Contents

VALID CONVENTIONS THROUGHOUT THE BOOK

$\vec{r}$	...	two-dimensional vector in the object plane or a conjugate plane
$\vec{q}_c$	...	carrier frequency of the holographic interference fringes
$\vec{q}$	...	reciprocal space vector
V_0	...	mean inner potential
U_A	...	Acceleration voltage (usually high tension)
U_f	...	biprism voltage
Φ_m	...	magnetic flux
$\chi(\vec{q})$	...	isoplanatic aberrations
$o(\vec{r})$	...	complex wave leaving object defined in "exit plane"
$a(\vec{r})$	...	amplitude of exit wave of object
$\varphi(\vec{r})$	...	phase of exit wave of object
$A(\vec{r})$	...	image amplitude
$\phi(\vec{r})$	...	image phase
C_S	...	spherical aberration coefficient
C_C	...	chromatic aberration coefficient
Δz	...	defocus value
λ	...	wavelength of electrons
Φ_0	...	phase shift of a single fluxon
$\mathbf{FT}\{\ \}$	...	denotes forward Fourier transform
$\mathbf{FT}^{-1}\{\ \}$	...	denotes inverse Fourier transform
$\otimes$	...	denotes convolution
$:=$	...	equation defining expression on left hand side

LIST OF CONTRIBUTORS

Lawrence F. (Larry) Allard
High Temperature Materials Laboratory
Oak Ridge National Laboratory
Oak Ridge, TN 37831-6064

John E. Bonevich
National Institute of Standards and Technology
Gaithersburg, MD 20809

Altaf H. Carim
Department of Materials Science and Engineering
The Pennsylvania State University
University Park, PA 16802

John M. Cowley
Department of Physics and Astronomy
Arizona State University
Tempe, AZ 85287-1504

W. Johannes (Hans) de Ruijter
EMiSPEC Systems Inc.
2409 S. Rural Rd., Suite D
Tempe, AZ 85282

Bernhard G. Frost
Facility for Electron Microscopy
University of Tennessee
Knoxville, TN 37996-0810

Marija Gajdardziska-Josifovska
Department of Physics and Laboratory for Surface Studies
University of Wisconsin-Milwaukee
Milwaukee, WI 53201

Michael A. Gribelyuk
IBM East Fishkill Analytical Services Group
Hopewell Junction, NY 12533

Rodney A. Herring
Microgravity Sciences Program
Canadian Space Agency
St. Hubert, Quebec, Canada J3Y 8Y9

David C. Joy
Facility for Electron Microscopy
Department of Zoology
University of Tennessee
Knoxville, TN 37996-0810

Michael Lehmann
Institute for Applied Physics and Didactics
Technical University of Dresden
D-01062 Dresden, Germany

Friedrich Lenz
Institute for Applied Physics
University of Tübingen
D-72076 Tübingen, Germany

Hannes Lichte
Institute for Applied Physics and Didactics
Technical University of Dresden
Dresden D-01062, Germany

Georgio Matteucci
Department of Physics and National Institute for Physics and Materials
University of Bologna
I-40126 Bologna, Italy

Martha R. (Molly) McCartney
Center for Solid State Science
Arizona State University
Tempe, AZ 85287-1504

Gottfried Möllenstedt (deceased)
Institute for Applied Physics
University of Tübingen
D-72076 Tübingen, Germany

Giulio Pozzi
Department of Physics and National Institute for Physics and Materials
University of Bologna
I-40126 Bologna, Italy

Wolf D. Rau
Institute for Semiconductor Physics
D-15230 Frankfurt (Oder), Germany

David J. Smith
Center for Solid State Science and Department of Physics and Astronomy
Arizona State University
Tempe, AZ 85287-1504

John C.H. Spence
Department of Physics and Astronomy
Arizona State University
Tempe, AZ 85287-1504

Jürgen Sum
Gatan GmbH
Ingolstädter Strasse 40
D-80807 München, Germany

Akira Tonomura
Hitachi Advanced Research Laboratory
Hatoyama, Saitama 350-03, Japan

Edgar Völkl
High Temperature Materials Laboratory
Oak Ridge National Laboratory
Oak Ridge, TN 37831-6064

JonKarl (JK) Weiss
EMiSPEC Systems Inc.
2409 S. Rural Rd., Suite D
Tempe, AZ 85282

THE HISTORY OF THE ELECTRON BIPRISM

G. Möllenstedt

Institute for Applied Physics, University of Tübingen,
D-72076 Tübingen, Germany

1. The beginning

In 1939, the physicists from Danzig and Königsberg arranged a seminar meeting at the School of Technology at Danzig. It was at that occasion when I, as a young student of physics, heard the term "electron interferometry" for the first time. At that seminar, Walter Franz, then lecturer at the University of Königsberg, spoke on quantum mechanics of the free electron, especially on the behavior of electron waves in magnetic fields. The essence of what he said is documented.[115,116] Later, in 1964 he stated in an article[117] that "In presence of an electromagnetic field, the momentum of a particle with charge e is known to be

$$\vec{p} = m\vec{v} + e\vec{A} \tag{1.1}$$

where $\vec{A}$ is the vector potential from which the magnetic field strength is determined as $\vec{B} = \nabla \times \vec{A}$. The phase difference $\Delta\varphi$ between two rays (a) and (b) connecting two points 1 and 2 is determined by

$$\Delta\varphi = \int_{1(a)}^{2(a)} \vec{p}\,\mathrm{d}\vec{r} - \int_{1(b)}^{2(b)} \vec{p}\,\mathrm{d}\vec{r} = \oint \vec{p}\,\mathrm{d}\vec{r} \tag{1.2}$$

"Introducing the expression for $\vec{p}$ from above, the term $m\vec{v}$ yields the same path difference as in absence of a magnetic field whereas, according to Stokes' theorem, the loop integral over $\vec{A}$ may be transformed to a surface integral over $\nabla \times \vec{A}$, i.e., the magnetic flux Φ_m, yielding

$$\Delta\varphi = \oint m\vec{v}\,\mathrm{d}\vec{r} + e\,\Phi_m \tag{1.3}$$

"This simple relation which should be the first thing taught in a lecture on wave mechanics for beginners after introducing the magnetic field (strangely enough I could not find it in any lecture notes except my own) shows clearly that the phase difference between electron rays depends on the magnetic flux included between the rays, even if the rays do not run in a magnetic field." In 1939, after a lecture by Walter Franz on the consequences of the above equation, Walther Kossel discussed the possibility of an experimental proof, but he came to the conclusion that, at that time, an experimental proof was not feasible.

I (G.M.) remember W. Kossel saying, "I hear the message but I lack an electron interferometer."

2. The first interferometer

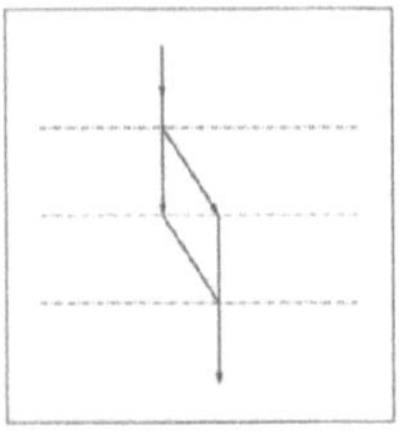

Figure 1.1. Marton's interferometer.[283]

In 1953 in Tübingen, we had with great interest taken notice of L. Marton's paper on an electron interferometer.[283] He used a thin monocrystal (Fig. 1.1) as a beam splitter separating the incoming electron beam into two coherent partial beams which were then superimposed again onto each other using a second and third monocrystal to produce interference phenomena. Fig. 1.2 shows, among others, L. Marton and his wife Claire at a meeting in Tübingen in 1952. The following statements by Marton raised our special attention: 'Because of the short wavelength, the source size, the size of the slits and their separation, the separation of the fringes becomes so small that a major experimental effort may be needed for coping with them.'

Figure 1.2. Ladislaus Marton (1) and Claire Marton (2), behind them W. Kossel (3), H. Mahl (4), B.v. Borries (5) and G. Möllenstedt (6). This photo was taken at the occasion of the 4th annual meeting of the German Society for Electron Microscopy at Tübingen, June 6 to 9, 1952.

In spite of such difficulties we thought that this interferometer was so ingenious that we initially intended to construct this type of interferometer at Tübingen. From my work with convergent beam electron diffraction I had some experience with the preparation of large area thin single crystals. But at last just these experiences induced us not to attempt to build Marton's interferometer. One reason was that the life-time of such single crystals was limited. However, the final decision was due to the experiences by G. Ruthemann with the electron spectrometer and mine with convergent beam diffractograms which showed that an initially monoenergetic electron beam loses its narrow energy distribution by inelastic scattering processes.

Shortly after that we were much impressed when A. Simpson succeeded in obtaining an evaluable interference pattern using gold monocrystals in spite of all such difficulties.[281]

Notwithstanding Simpson's positive results and Otto Rang's beautiful distant interferences using naturally grown thin single crystal lamellae we still decided not to use single crystals as coherent beam splitters for an electron interferometer.[350]

3. The biprism idea

In 1950, at the Süddeutsche Laboratorien (SDL) in Mosbach/Baden, when working with dark field micrographs in the electrostatic electron microscope, I had observed that a thin tungsten wire used to intercept the primary beam in the focal plane of the electrostatic objective lens for dark field electron microscopy had charged up (Fig. 1.3). I also observed that this charging led to double images on the final image screen. At that time I thought that such double images were of no physical interest unless they were coherent.

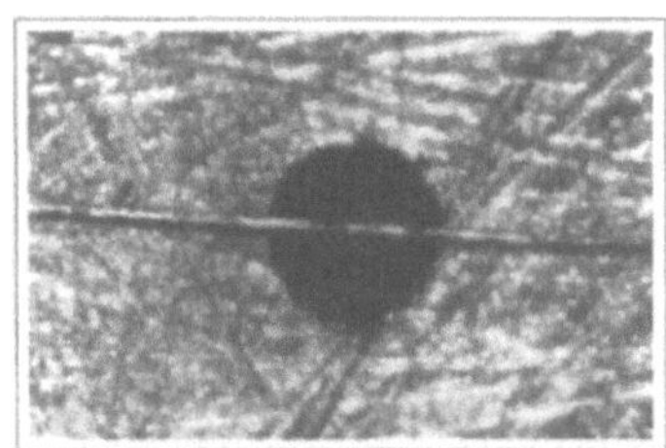

Figure 1.3. Tungsten wire which led to the invention of the biprism.

What would happen if a voltage was applied to an insulated thin filament? May 1952, I discussed this question with Heinrich Düker (Fig. 1.4) who at that time was looking for a subject for a doctoral thesis. He agreed with pleasure to tackle this problem, i.e., the design, construction and testing of an electrostatic biprism. Since he had much technical experience he was especially well suited for making the necessary experiments. Since he died early, I should like to give a short review of his extraordinary life: Born February 27, 1923, in Hechingen, Baden-Württemberg, Germany, he attended primary and secondary school at Tübingen. In 1940 he joined the German navy, became a navy engineer and served until 1945 as a managing engineer on submarines. After several years in post-war prison he studied physics at Tübingen University, passing his diploma examination in 1953, and graduating to Ph.D. (Dr. rer. nat.) in 1956. From 1957 he worked at the Max Planck Institute for Metal Research, Special Metals Division, Stuttgart. After 1964 he worked for the Bosch company, establishing a new research institute in Berlin. Since 1969 he was scientific consultant to the management of the Bosch company. Finally, he became the director of the central research establishment of the Bosch company on the Schiller Height at Gerlingen near Stuttgart, Germany, where he died June 25, 1985.

Figure 1.4. H. Düker, who constructed the first biprism.

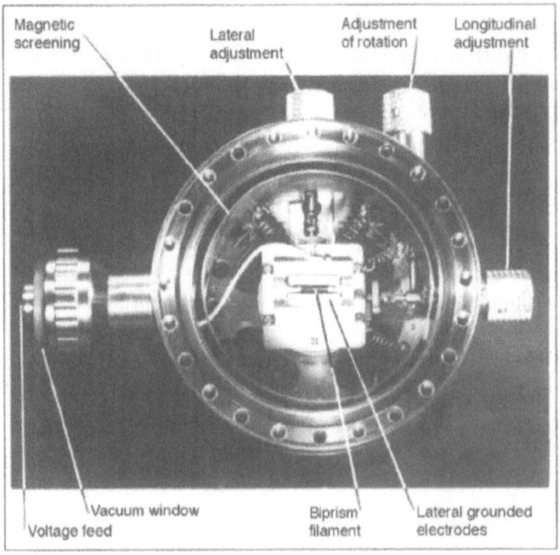

Figure 1.5. Design of the first electron-optical biprism.

4. Biprism design and 1st experiments

In order to make ourselves familiar with biprism interferences we first studied the interferences of visible light using a glass prism. We learned that narrow slits, magnifying lenses, and strictly monochromatic light were needed to succeed. Since at that time lasers did not exist, we had great difficulties to attain sufficient intensities. Nonetheless, these preliminary experiments with visible light were very valuable, because they taught us what difficulties we would have to surmount with electron biprism interferences.

4.1. The first experimental stage for the production of electron biprism experiments

The first experimental stage consisted of the following structural elements:

Electron source: Tungsten cathode with Wehnelt electrode, as used with the electrostatic electron microscope at the SDL in Mosbach.

Lenses for the demagnification of the beam crossover: Axially symmetric, electrostatic lenses as designed by Mahl.

Lens for magnification of the interference pattern: Axially symmetric magnetic lenses.

The design of the electron-optical biprism is shown in Fig. 1.5. A quartz filament $3\,\mu$m in diameter, covered with a thin gold layer by vacuum evaporation, was placed between two grounded electrodes: The distance between the filament and the grounded electrodes was 2 mm. The arrangement of filament and grounded electrodes could be adjusted and rotated from outside the vacuum using adjusting screws. The high voltage supply stabilized to $2 \cdot 10^{-5}$ was designed by S. Panzer at the SDL in Mosbach. It supplied a voltage of about 50 kV, which was completely sufficient, as the limited breakdown strength of the electrostatic lenses did not allow us to work with higher voltages anyway.

4.2. Discovery of the first electron biprism interferences with the first experimental stage

After careful adjustment of the ray paths and after production of a demagnified electron probe for the coherent irradiation of the biprism filament we saw on the final image screen a shadow image of the filament with Fresnel fringes on both sides. But when the filament voltage was increased to superimpose the Fresnel fringes from both edges, we could not see any biprism interferences although we were using powerful optics. But, after photographic recording with an exposure of 30 sec, we were able to see fine interference fringes when we looked at the photographic record using a highly magnifying optical microscope. Fig. 1.6 shows the first electron biprism interference patterns observed in 1954. We published a short note titled 'Fresnel interference experiment using a biprism for electron waves' in the *Naturwissenschaften*.[311,312] We obtained several appreciative letters but we were reprimanded by the well known experimental physicist and textbook author R.W. Pohl from Göttingen who stated that Thomas Young had produced optical biprism interferences ten years before Fresnel.

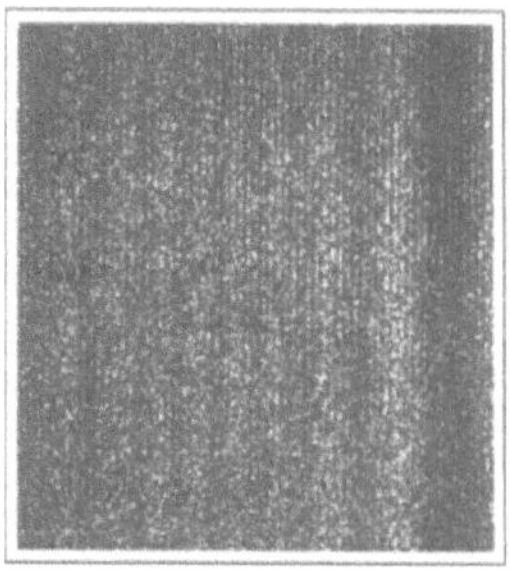

Figure 1.6. The first electron biprism interference pattern (1954) using axially symmetric lenses. Exposure time 30 sec.

4.3. The second technical version of an intensified electron interferometer

After some benevolent encouragement by D. Gabor we decided to start a completely new design of the interferometer.[132] The first experimental results had shown that the intensity of the interference patterns was insufficient. For this reason the axially symmetric lenses were replaced by cylindrical lenses. Thus the previous point shaped electron probe was replaced by a more intensive line shaped probe. Fig. 1.7 shows the design of an electrostatic cylindrical lens.

Fig. 1.8 shows the schematic ray paths and the technical realization of the improved interferometer. The cathode and its adjustment remained unchanged but the circular

Figure 1.7. Design of an electrostatic cylinder lens. The upper grounded electrode is removed. The voltage feed on the right supplies the high voltage. On the bottom left there is an adjusting screw allowing adjustment of the rotation of the stage carrying two cylindrical rods. The rod diameter was 3 mm, the mutual distance between the rod axes 8 mm, and the distance between the rods and the grounded electrodes 4 mm. The focal length of the cylinder lens was 3.6 mm.

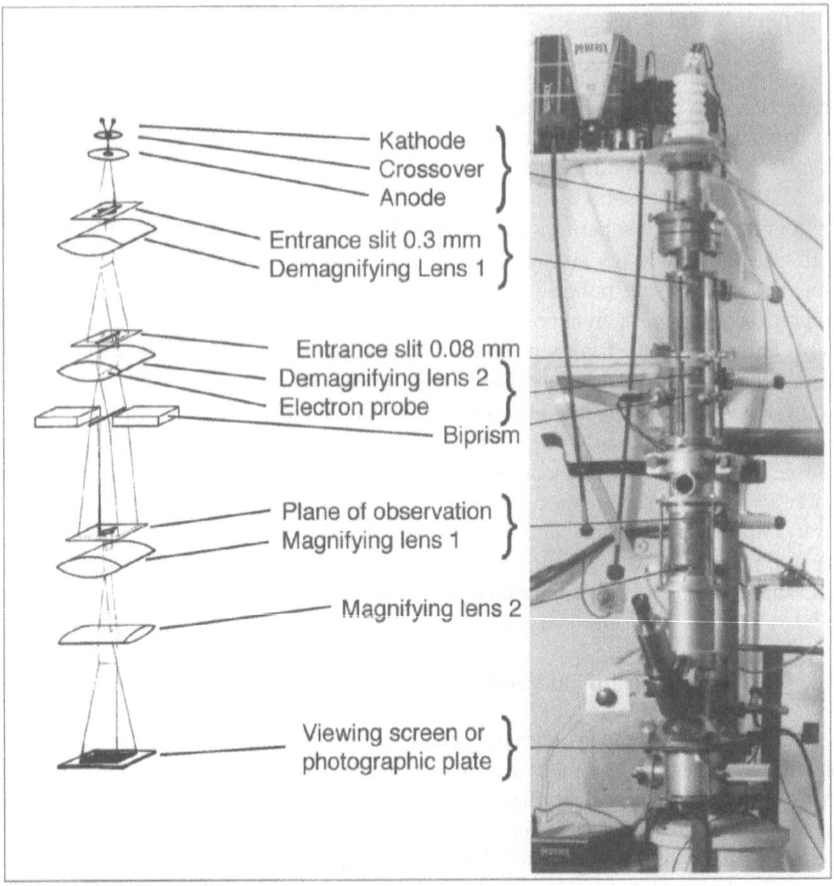

Figure 1.8. Ray paths and technical realization of the improved electron interferometer using cylinder lenses. The two demagnifying lenses had fixed focal lengths of 3.2 mm (lens I) and 5.2 mm (lens II), respectively. The first magnifying lens had a focal length of 3.2 mm whereas the focal length of the second magnifying lens could be varied from 5.2 mm to infinity by varying its lens voltage using a potentiometer. The distance between the electron probe and the biprism filament was 39.5 mm, between the filament and the plane of observation 293.5 mm.

entrance aperture was replaced by a slit 0.3 mm in width. The axially symmetric lenses were replaced by cylindrical lenses.

4.4. Results obtained with the new experimental design

The Fresnel fringes in the shadow image of the biprism filament on the viewing screen appeared much brighter than before, and after optimum adjustment of the slits and the cylindrical lenses many maxima and minima of the fringe pattern became observable. Only a few seconds were needed to record the pattern on a photographic plate. It was even possible to observe interference fringes in the geometrical shadow of the filament. Their fringe period was precisely measured and found to be in agreement with its theoretically expected value.

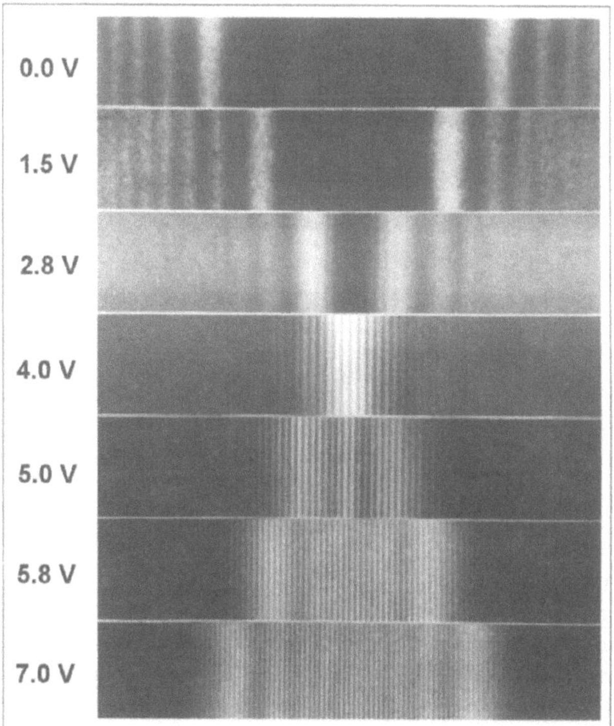

Figure 1.9. Biprism interferences for different values of the positive filament potential U_f. The top pattern is obtained for $U_f = 0$ (shadow image). With increasing filament potential (up to 7 V) the two coherent parts of the beam having passed on both sides of the biprism filament overlap and form interference fringes, the period of which decreases with increasing filament potential.

Fig. 1.9 gives an impression of the brilliance of the biprism interferences for different values of the filament potential. This interferometer allowed to verify de Broglie's relation $\lambda = h/(m\,v)$ using no other data but measurable geometrical distances and the accelerating voltage. We wrote a letter to Louis de Broglie (Fig. 1.10) informing him of our results, and we received the following answering letter:

> Paris, 19 June 1956
>
> Monsieur and dear Colleague,
>
> I was extremely pleased to receive your kind letter and to learn of your beautiful experiments in which you have obtained electron interference by a method analogous to Fresnel's biprism. It was, of course, a great pleasure to see that you have obtained a new and particularly brilliant proof of the formula $\lambda = h/(mv)$, and I shall not fail to make known your experiments to my students. Thanking you most gratefully for your communication, I beg you to accept, Monsieur and dear colleague, the expression of my devoted sentiments,
>
> Louis de Broglie

Figure 1.10. Louis Victor Pierre Raymond Prince de Broglie, 1892-1987

5. Further Tübingen experiments

Very soon, the coherent beam splitter for electron waves was applied in many countries, especially in France, Japan, Italy, the UK, the US, and the USSR. I shall, however, not review any details of such applications in this history of the first steps of electron interferometry.

Finally, I shall shortly mention a few highlights of interferometric experiments made in Tübingen:

1. Max Keller: Measurement of mean inner potentials of solids, and comparison with theory (Fig. 1.11).[224]

2. Rainer Buhl: Design and applications of an electron interference microscope (Figs. 1.12 and 1.13).[35]

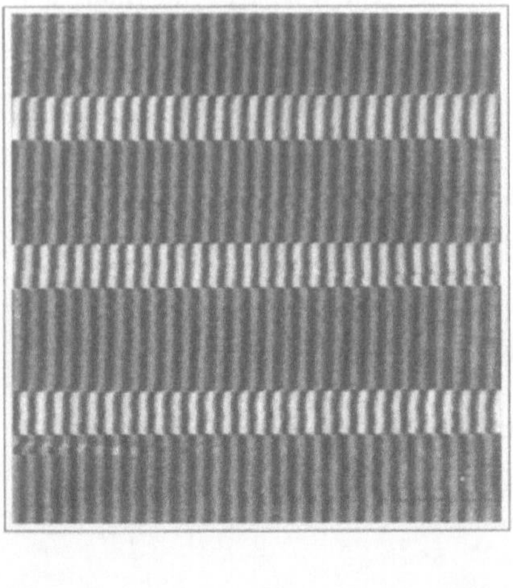

Figure 1.11. Electron interferogram with steps of different thickness. The measured values of the mean inner potential of vacuum evaporated layers of carbon, Al, Cu, Ag, and Au did not depend on the layer thickness, the accelerating voltage, the intensity and the exposure time but in the case of Ag and Au they depended on the rate of evaporation.

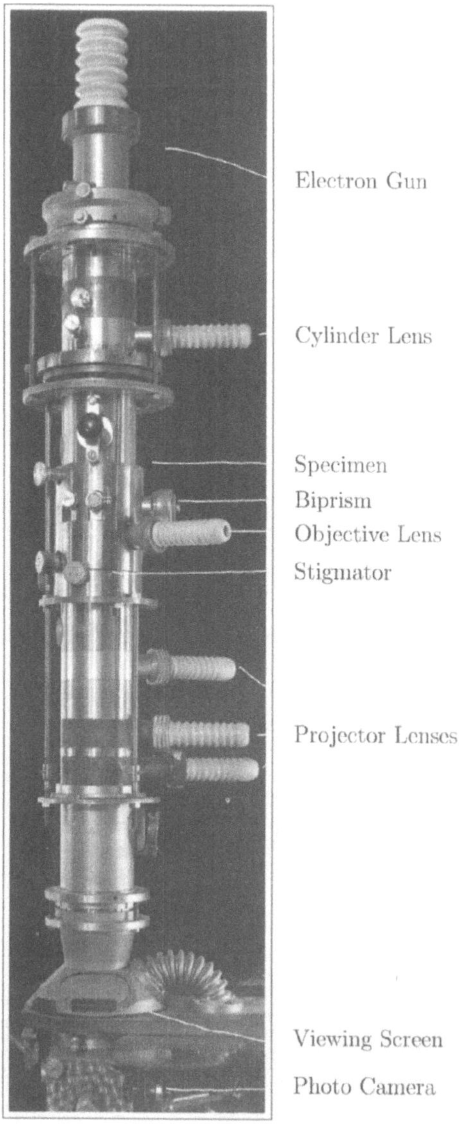

Figure 1.12. Buhl's electron interference microscope.

3. Werner Bayh: Realization of the experiment suggested by W. Ehrenberg and R.E. Siday demonstrating the phase shift of electron waves by the magnetic flux included between two coherently split partial beams.[101] The phase shift is made visible as a fringe shift in the interference pattern (Figs. 1.14 and 1.15).

   Ehrenberg's list of publications treats subjects such as X-ray spectroscopy, diffraction of low energy electrons, electrical contacts, radar and television, nuclear physics, excitation by neutrons, and problems of quantum mechanics. He has written a book titled *Dice of the Gods, Causality, Necessity, and Chance.*

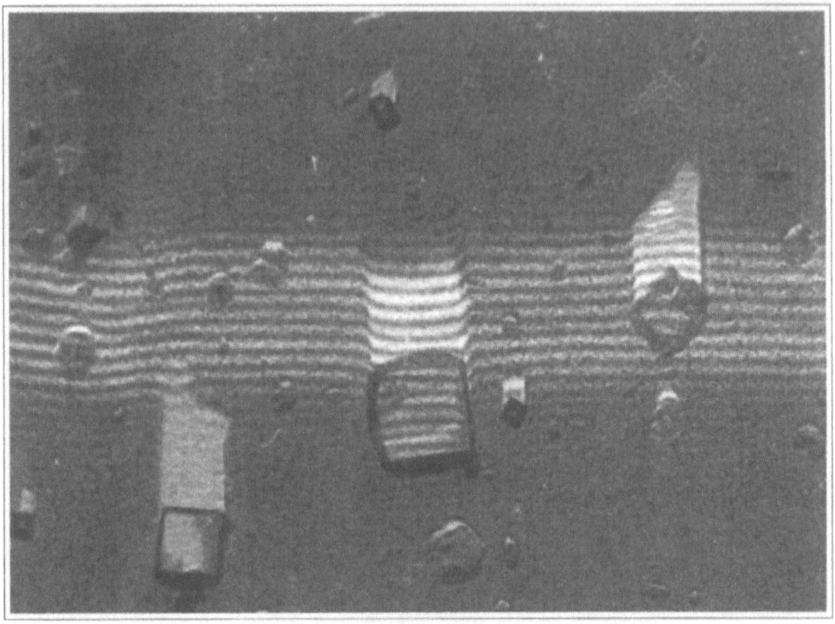

Figure 1.13. Micrograph of an aluminum object.[35] The phase shift becomes visible as a local displacement of interference fringes.

Figure 1.14. Werner Ehrenberg, 1901-1975

Figure 1.15. R.E. Siday, 1911-1957

During his last years at EMI, W. Ehrenberg was in close friendship and exchanged ideas with R.E. Siday. Publications by R.E. Siday treated subjects such as X-ray transitions, magnetic lenses for spectrometers, magnetic prisms, refraction index in electron optics, and beta-spectrometry.

Curriculum Vitae of Werner Ehrenberg:

20th July 1901	Born in Berlin
1907-1919	Attended Mommsen Gymnasium in Charlottenburg
1919-1924	Studies Philosophy, Physics, and Mathematics at the Universities of Berlin and Heidelberg
1924	Graduated Ph.D. Heidelberg
1924-1929	Under different research grants at the Kaiser Wilhelm Institut in Berlin-Dahlem, at the Physical Institut of the University of Berlin, and at the X-ray Laboratory of the Technische Hochschule in Stuttgart
1930-1933	On the staff ('Assistant') of the Technische Hochschule in Stuttgart, at the Physical Institute
1933-1936	At Birkbeck College (Prof. Blackett), under a grant from the Academic Assistance Council
1936-1945	At Electric and Musical Industries (EMI), Hayes, Middlesex
1945-1951	At Birkbeck College (Prof. Bernal) under a grant from the Nuffield Foundation
1949	Awarded M.Sc. (London)
1950	Awarded D.Sc. (London)
1951	Visiting Professor at Purdue University, Lafayette, IN
1951-1962	Reader in Physics at Birkbeck College
1962-1968	Professor in Physics at Birkbeck College

Fig. 1.16 shows the ray paths as suggested by Ehrenberg and Siday.[101] Fig. 1.17 shows the ray paths realized in Bayh's experiment.[15] Fig. 1.18 shows the fringe shift due to the continuous variation of the vector potential.[15] The magnetic flux included between two coherently split partial waves gives rise to a phase shift proportional to the flux included. A phase shift of 2π corresponds to a flux $\Phi_m = h/e = 4.135 \cdot 10^{-15}$ V sec. Bayh's experiment confirmed this relation with an experimental uncertainty of only 4%.

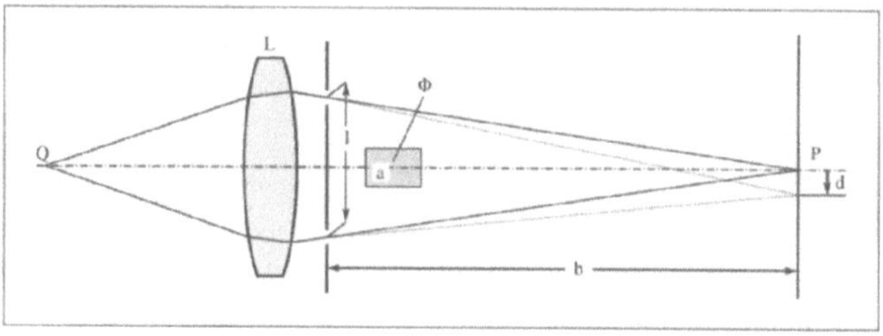

Figure 1.16. Ray paths as suggested by Ehrenberg and Siday.

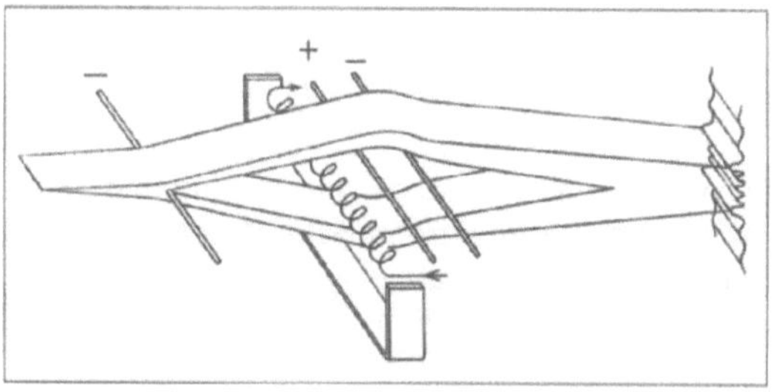

Figure 1.17. Ray paths as realized by Werner Bayh.

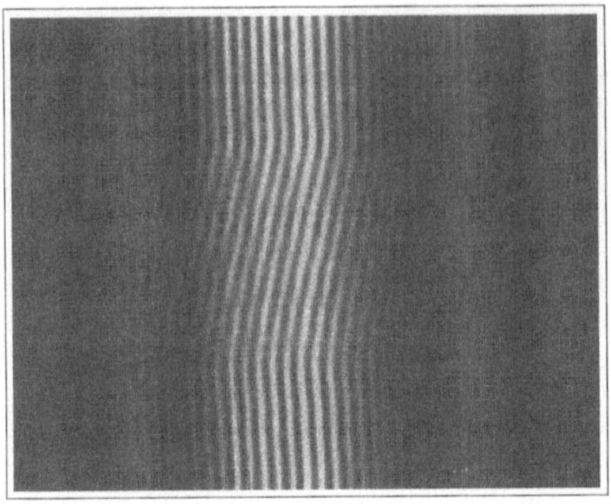

Figure 1.18. Fringe shift due to the continuous variation of the vector potential.[15] Time is plotted in vertical direction. In the center of the fringe pattern the magnetic flux is increased linearly with time.

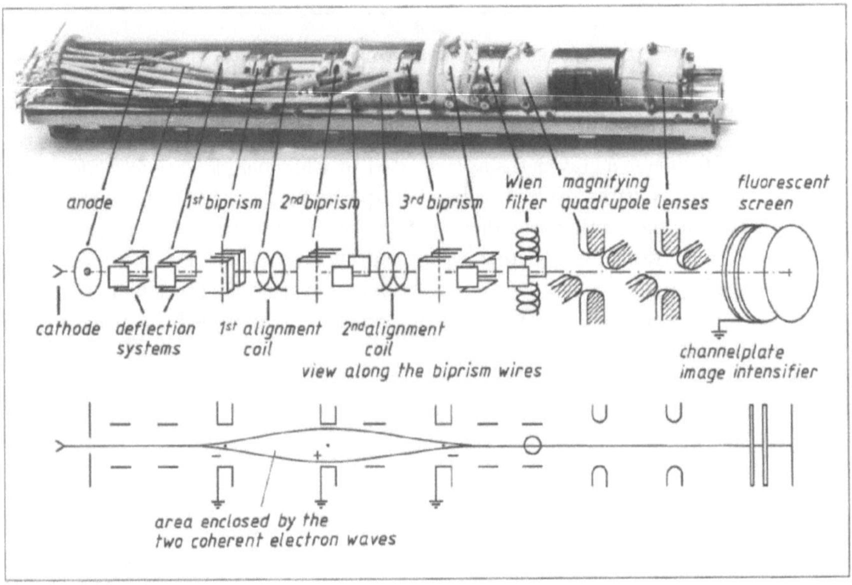

Figure 1.19. Miniature-interferometer for the Sagnac experiment, and its electron-optical elements. The overall-length of the interferometer was 30 cm.

4. Franz Hasselbach and Marc Nicklaus: The Sagnac experiment with electron waves.

   A 'miniature-interferometer' was designed,[170] which could be rotated about a vertical axis in order to realize a Sagnac experiment with electrons.[171, 324] The experimentally observed phase shift in the rotating system was in agreement with theoretical expectations. Fig. 1.19 shows the miniature-interferometer for the Sagnac experiment.

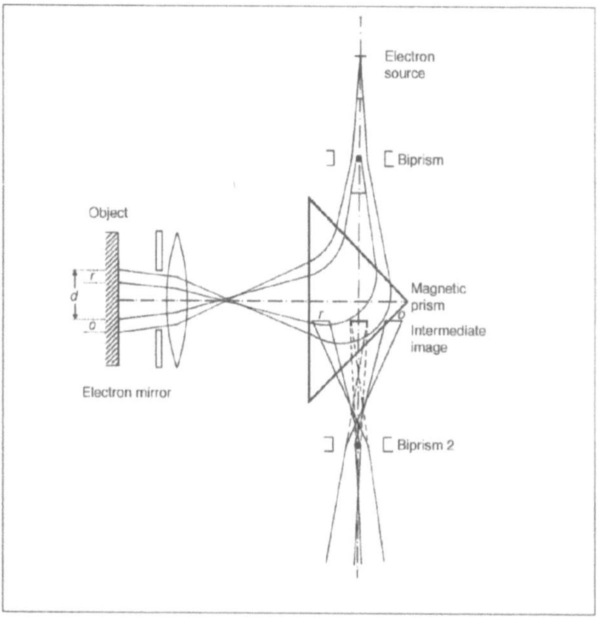

Figure 1.20. Setup of the electron mirror interference microscope.

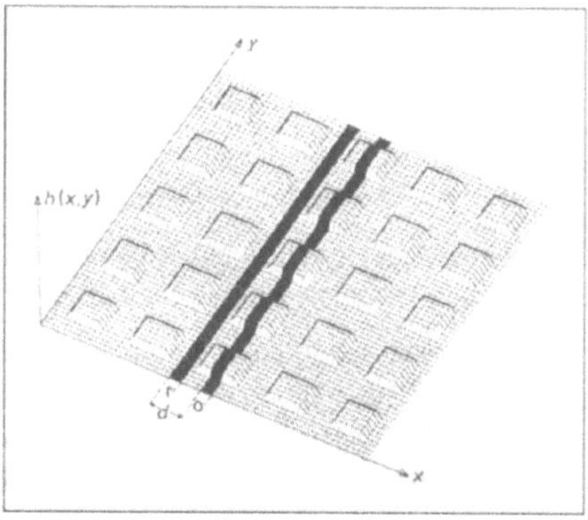

Figure 1.21. An artificial specimen with gold squares 2.5 nm in height, and a lateral spacing of 12 μm.

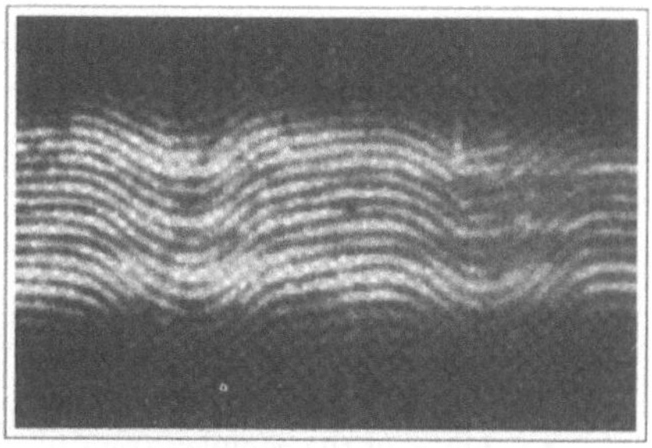

Figure 1.22. Displacement of fringes due to path length differences, i.e., phase shifts in the mirror image.

5. Hannes Lichte and Gottfried Möllenstedt: Measurement of the roughness of supersmooth surfaces using an electron mirror interference microscope.[257]

   The electron wave is split up into two coherent partial waves using a biprism, then magnetically deflected by 90° and reflected in front of the specimen surface. The partial waves are then deflected again to form two virtual intermediate images which are superimposed using another biprism to form an interference pattern. The local fringe shift depends on the local field in the vacuum immediately in front of the specimen surface (Fig. 1.20), and thus contains information on the surface relief. Fig. 1.21 shows a test object, and Fig. 1.22 the interference pattern with local fringe shifts.

⟸

Figure 1.23. Interference pattern with a localized phase shift by π where the value of the quantized flux makes a jump by $h/(2e)$.

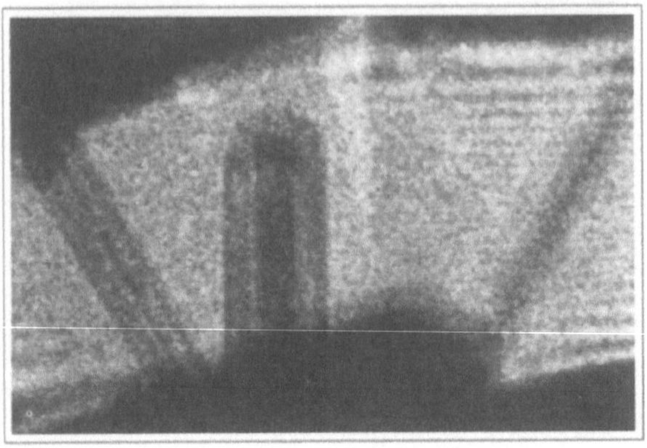

Figure 1.24. Image plane hologram of ZnO needles.

6. Herbert Wahl: Electron-interferometric measurement of quantized magnetic flux in superconducting hollow cylinders[484] and off-axis image plane holography.[485]

   Fig. 1.23 shows the phase jump in the interference pattern of a superconducting hollow cylinder including quantized magnetic flux.

   In Wahl's image plane holography the electron wave is recorded in an image plane hologram. In a first exposure the reconstructed light-optical wave is recorded directly. In a second exposure it is recorded after superimposing a plane wave. The two intensities allow the unequivocal determination of both the modulus and the phase of the complex transmission.

   Fig. 1.24 shows an image plane hologram of ZnO needles, Fig. 1.25 the common registration of two holograms.

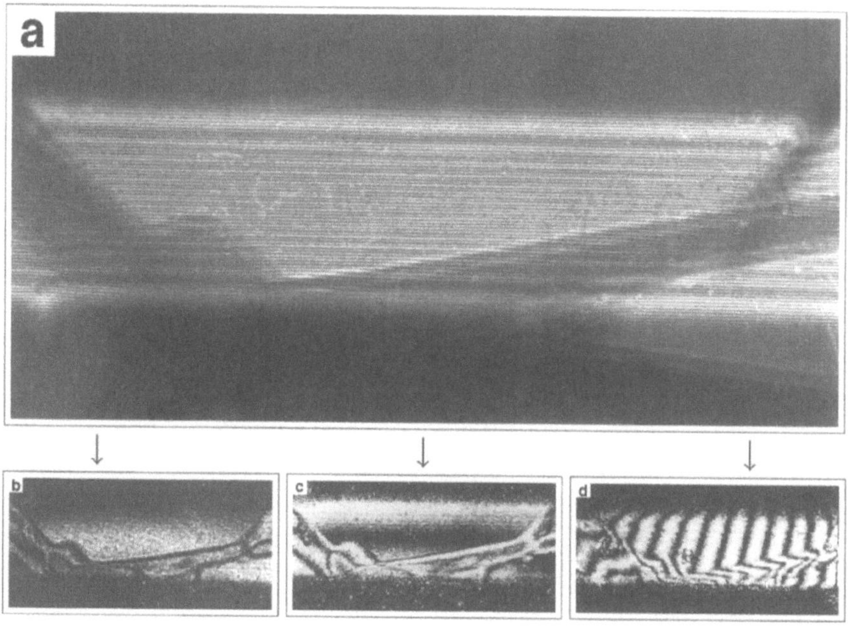

Figure 1.25. a) hologram, b) phase contrast, c) dark field, d) holographic interferometry. Figures b), c), and d) are optical reconstructions of the electron hologram a) using plane reference waves. The width of the interference field is 80 nm.

Acknowledgments:

For over 30 years Prof. Friedrich Lenz (University of Tübingen) has contributed to, and enriched the topics discussed here. For numerous discussions over the years and for translating this text into English I am deeply thankful to him. I am indebted to my former graduate students and coworkers for their advice and for the permission to publish some of the figures. I should further like to express my special gratitude to Dr. B.J. Hiley and J.C.E. Jennings, both Readers at Birkbeck College, University of London, England. I have taken the biographical data on W. Ehrenberg and R.E. Siday from a letter they have sent me.

PRINCIPLES AND THEORY OF ELECTRON HOLOGRAPHY

J.M. Cowley and J.C.H. Spence

Department of Physics and Astronomy, Arizona State University,
Tempe, AZ 85287-1504

1. Introduction

Electron holography was conceived as a means for overcoming the limitation of the resolution of electron microscopes due to the unavoidable presence of aberrations of electromagnetic lenses.[129,130] The essential means for doing this was the formation of an interference pattern between the electron wave scattered by the specimen and a reference wave of known form. The practical realization of this aim had to await the technological developments of smaller, brighter electron sources to give greatly improved coherence of the illumination and more effective, quantitative recording media. With the advent of these aids, rapid progress is now being made toward realization of the potential of electron holography for providing important advances in the power and range of applications of electron microscopy.

Of the many possible forms of electron holography, several have been developed extensively and others are under investigation. In addition to the possibility for the enhancement of resolution, other important applications of the holographic method arise from the fact that an image can be obtained showing directly the changes of phase of the electron wave as it passes through a specimen. In this way electric and magnetic fields existing in and around thin films and small particles may be mapped and measured with high precision and high spatial resolution.

In this initial chapter we first review the principles of the various methods of electron holography and then provide a theoretical basis for their description and discussion. Finally we describe in some detail a few areas of application of the methods which are of particular interest and may not, in some cases, be covered by other contributions to this volume.

1.1. Phase and amplitude images

For electron beams in the energy range of 10^3 to 10^7 eV, the principal interaction of the electron wave with matter is the change in wavelength associated with the acceleration of the electrons by the electrostatic potential field. The wavelength change leads to a phase change of the wave, relative to the phase of a wave in vacuum. Because the electron wavelength is much smaller than the dimensions of the scattering objects, the individual atoms, the scattering of the electrons takes place through only very small angles. It is then a very good approximation to assume that, for very thin specimens, the phase change of the electron wave is proportional to the integral of the potential along a straight-line path through the specimen. The thin object is, in effect, a two-dimensional phase-object for electrons, giving a phase change varying in the two dimensions perpendicular to the incident beam direction, but no change in the wave amplitude. If such an object were to be imaged in an ideally perfect electron microscope system, having no aberrations, defocus or aperture limitations, the contrast would be zero because the image intensity depends only on the square of the amplitude distribution and not on the phase. In practice, there are some amplitude-variation terms that arise from inelastic scattering processes such as the excitations of phonons, plasmons and excited states of the electrons of the solid but, in general, these amplitude variations are at least an order-of-magnitude smaller than the phase variations for thin objects. It would obviously be advantageous to obtain an image having contrast depending directly on the distribution of the phase variations.

The strength of the interaction of the electron wave with a specimen is such that, as the specimen thickness increases (for example, to about 1-10 nm for 100 keV electrons) there are elastic scattering effects which may have the effect of introducing amplitude variations and also the simple weak phase-object approximation fails because multiple-scattering effects become dominant. For even greater thicknesses, the inelastic scattering processes accumulate and produce greater amplitude variations. However, an imaging system giving contrast which is dependent on the phase changes (and so approximately proportional to the thickness and linear in the electrostatic potential) would still be preferable.

To a limited extent, such an imaging system is achieved in normal bright-field electron microscopy by use of phase-contrast imaging techniques such as, most commonly, the defocusing of the microscope. When images are taken out-of-focus, a phase shift is introduced which is proportional to the square of the scattering angle so that, for radiation scattered to particular angles, the waves become in-phase with the incident, transmitted beam and so can give the effect of amplitude contrast, proportional to the phase change. But since this applies only for particular scattering angles, the representation of the phase-change distribution of the object is very poor. This situation is further complicated by the unavoidable presence of lens aberrations, notably the third-order spherical aberration which adds a phase change proportional to the fourth power of the scattering angle.

If images could be obtained which showed both the phase distribution and the amplitude distribution for the wave exiting from a specimen, it would be possible to make corrections for the phase perturbations due to defocus and lens aberrations and so obtain a direct representation of the interaction of the wave with the specimen. This is the objective of electron holography.

1.2. Use of a reference wave

The important concept introduced by Gabor[129,130] was that the relative phase distribution in a wave may be determined by observing the interference pattern produced by superimposing a reference wave of known phase distribution. Gabor's idea was that, for a weakly-scattering thin object, the wave directly transmitted though the object ('unscattered') would be very little perturbed by the object and could be used as a reference wave so that the relative phases of the waves scattered from the object could be derived from the interference effects.

In order that the phase distribution of the reference wave should be well defined, it should be coherent, coming preferably from a point source. Since ideal point sources

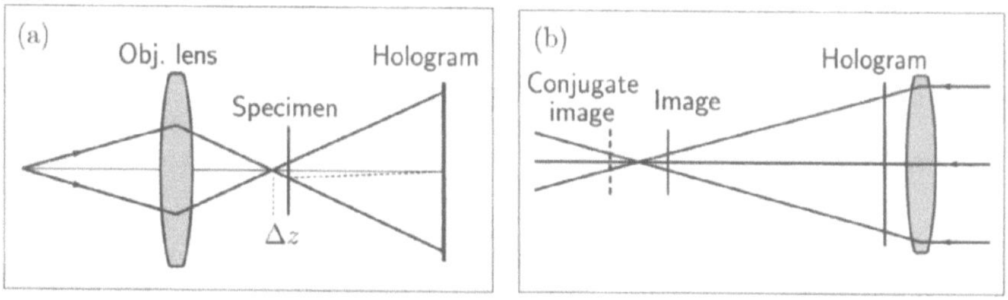

Figure 2.1. (a) Scheme for recording an in-line, point-projection hologram, as suggested by Gabor. (b) Reconstruction from a Gabor-type hologram, showing the formation of the desired image and its conjugate.

do not exist, Gabor proposed the use of a cross-over formed by a strong electron lens demagnifying a very small electron source as suggested in Fig. 2.1(a). If this crossover is placed very close to the specimen, a greatly magnified point-projection ('shadow') image is formed on a distant screen. This enlarged image he called a 'hologram' implying that the image contained the whole of the information about the specimen through the interference effects of the transmitted and scattered waves. The phases of these waves were perturbed by the aberrations and the defocus of the electron lens used to form the small electron probe. However, because the relative phases of the waves were recorded in the hologram, a reconstruction process could be carried out to correct the phases for the defocus and aberration effects. In simple terms, Gabor proposed a method for bringing a shadow image back into focus. The focus error is the distance Δz in Fig. 2.1(a). We will see later that the point-projection shadow images are almost identical with the conventional defocused images.

Because of the large magnification involved in forming the point-projection image, Gabor proposed that the reconstruction could be made using visible light and glass lenses designed to have the aberrations which would compensate for the aberrations of the electron lens (Fig. 2.1(b)). It was known at that time from the work of Scherzer[391], that the resolution normally attainable with an electron microscope could not be improved beyond a certain limit set by the irreducible aberrations (and, in particular, by the third-order spherical aberration) of the objective lens. It appeared that the processes of forming a hologram and forming a reconstructed image, collectively known as 'holography,' offered a means for overcoming this limitation on the resolution.

1.3. In-line holography

At the time when Gabor made his proposal for holography with electrons, no sufficiently bright, small source of electrons was available to test the idea experimentally. Gabor tested the idea by forming a hologram and performing the reconstruction with a light-optical system.[131] For an object consisting of isolated thin lines (e.g., lettering) the reconstruction from a greatly defocused hologram showed a clear recreation of the original except for a rather confusing diffuse background. Fig. 14.4 on page 317 shows an example of such a mesh image reconstructed from a hologram. The diffuse background emphasized a fundamental limitation of this holographic process. Because it is the hologram intensity which is recorded, there is no way of distinguishing between the complex wave distribution and its complex conjugate. Both the original wave and the complex conjugate contribute to the reconstruction.

For example, if the reference wave has the amplitude unity and the wave scattered by the object is represented by the small complex function, $\epsilon(x, y)$ or $\epsilon(\vec{r})$, the intensity recorded is

$$I(\vec{r}) = |1 + \epsilon(\vec{r})|^2 \tag{2.1}$$

Neglecting the term of second order in the small quantity, ϵ, this gives

$$I(\vec{r}) = 1 + \epsilon(\vec{r}) + \epsilon^*(\vec{r}) \tag{2.2}$$

If the phase terms in $\epsilon(\vec{r})$ include the phase shifts due to defocus and lens aberrations, the complex conjugate $\epsilon^*(\vec{r})$ contains phase shifts of the opposite sign. If the reconstruction process subtracts these phase shifts from $\epsilon(\vec{r})$, it adds negative phase shifts to $\epsilon^*(\vec{r})$ so that, in addition to the desired, corrected image, there is an unwanted, conjugate image having minus-two times the original defocus and aberrations. For an object consisting of isolated points or lines, the conjugate image is diffuse and so constitutes a clearly distinguishable background, as seen in Fig. 14.4 on page 317. For the more general type of object with continuous distributions of intensity, the conjugate object cannot be separated out so easily. Means that have been suggested for making this separation will be discussed below.

The effective experimental realization of the Gabor holographic scheme, with electrons used to form the hologram, had to wait about 20 years for the availability of the field-emission gun (FEG) which gave an increase in source brightness by three of four orders of magnitude relative to the previously-used thermionic emission guns. Such high-brightness FEG sources became available with the introduction of scanning transmission electron microscopy (STEM) instruments by Crewe et al.[77] and the later commercial STEM instrument production by VG microscopes, UK. At about the same time, the availability of digital recording of hologram intensities allowed the process of reconstruction from the hologram, recorded with electrons in a STEM instrument, to be carried out by manipulation of the image hologram intensity data in a computer.[57] Experimental tests of this scheme gave reconstructions showing a significant improvement of resolution over that attainable by direct bright-field STEM imaging.[272]

1.4. In-line TEM holography

In the meantime, it had been recognized that, instead of the near-spherical coherent wave from a small cross-over formed by an electron lens, the near-planar wave used to illuminate the specimen in a normal transmission electron microscopy (TEM) instrument could serve to form the hologram.[156,157] If, as suggested by Fig. 2.2, the objective lens of the microscope is defocused so that it forms the image of the wave front occurring some distance after the specimen plane (or the virtual wave front on a plane some distance before the specimen), the magnified image formed in the microscope reflects the effects of interference between the wave transmitted through a thin object and the waves scattered by the object and so constitutes a hologram. The relative phases of the transmitted and scattered waves are modified by the defocus and the aberrations of the objective lens but these phase perturbations may be removed in the reconstruction process, as in the Gabor scheme.

The use of the conventional TEM configuration appeared much more approachable than the point-source configuration. In order to obtain the interference effects in the hologram it is necessary to have sufficiently coherent, near-plane-wave illumination but

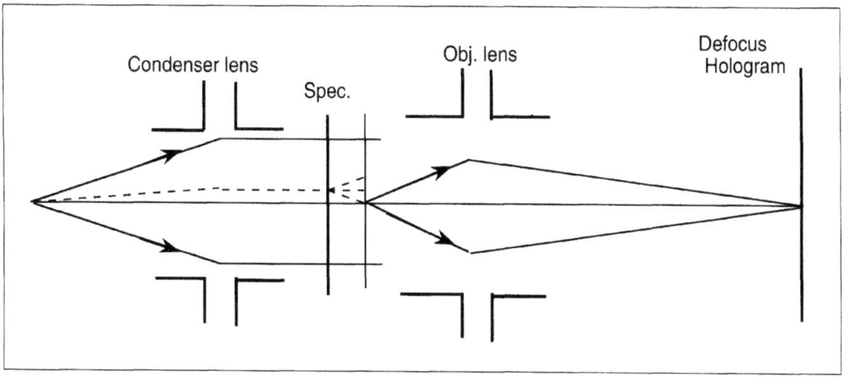

Figure 2.2. In-line TEM holography. Interference between the transmitted and scattered waves occurs on an out-of-focus plane.

this could be achieved, at the expense of intensity, by using a very small source or a source demagnified by a strong condenser lens even before the small, bright FEG sources became available. Experiments were made using optical reconstruction schemes and specimens consisting of isolated small particles.[157,322,449]

For the so-called 'in-line TEM' mode of electron holography, the conjugate-image problem of the Gabor scheme is still present. For the images of small isolated particles, the conjugate image forms a diffuse background that can be disregarded but for more general types of objects it can cause considerable confusion. Various methods have been applied to minimize its influence.

If the defocus used for recording the hologram is very large, the hologram intensity distribution can be described more appropriately in terms of Fraunhofer diffraction, rather than Fresnel diffraction theory. Then the conjugate image formed in the reconstruction process is so very far out of focus that it has little influence on the reconstructed image. Such large defocus values may complicate the practical reconstruction procedures if high resolution is sought, but not in the cases where relatively poor resolution is sufficient, as in the case of reconstructions showing the distributions of phase-change resulting from local variations of the magnetic or electrical fields around the specimen.[278,451]

For relatively high-resolution reconstructions, the effects of the conjugate image may be reduced by varying one of the experimental parameters so that, for a series of recorded holograms, the reconstruction of the desired image should be the same but the reconstruction of the conjugate image will be different. For the in-line TEM mode of holography, the experimental parameter that may be varied most conveniently is the defocus. Through-focus series of exposures are made and the correlation and comparison of image intensities for the series can be used to derive the ideal image phase distribution. Initial experiments on thin amorphous carbon films, considered as thin weak-phase objects, were made by Thon and Willasch[429] and appeared to show a resolution improvement although, because the structure of the specimen could not be known, it was difficult to judge whether there was any increase in the information about the specimen structure. Later experiments using a thin crystalline specimen of known structure were more satisfactory in this respect, but even for thin crystals, viewed in principal orientations, the atoms are usually aligned in rows parallel to the incident beam direction and it can not be assumed that the phase changes are very small. Hence more sophisticated methods for multiple-image correlation are required, such as those of Schiske,[394] Kirkland et al.[227,228] and the 'focus-variation' methods.[327,386]

For the point-projection, Gabor mode of holography, the convenient experimental variable that may be used to reduce the effect of the conjugate image is the translation of the beam over the specimen.[272,487] If the specimen is thought of as moving in one direction relative to the beam then the conjugate image moves in the opposite direction. Thus if the various reconstructions of the desired image are superimposed, the corresponding reconstructions of the conjugate image are smeared out.

The in-line modes of electron holography have been overshadowed by the more popular off-axis modes, to be described below, but are still important for many purposes and have potential for future development. As indicated, they can be very effective for the study of the distributions of electric and magnetics fields such as those in thin films of ferroelectric or ferromagnetic materials. They have been used to create striking images of the distributions of flux lines in superconducting thin films. They are essential for the understanding of images formed with very low-energy (about 100 eV) electrons emitted from very small field-emission tips (see pages 311 ff.). They have great potential application when the interference effects are observed in the diffraction plane rather than near the image plane of an electron microscope.

1.5. Off-axis holography

A very large expansion in the range and impact of holographic techniques came from the realization by Leith and Upatnieks,[251] in developing laser optical techniques, that the reference wave need not be one transmitted through the object but could be a wave passing through the air, outside the object. To achieve this, the incoming wave must have sufficient lateral coherence to allow waves passing through the object and around the object to interfere, or else a beam-splitter must be used so that the incoming wave is split into two coherent parts which can be separated by a considerable lateral distance before being recombined to give interference effects.

For visible light, beams of considerable lateral coherence are easily produced by lasers. Half-silvered mirrors or systems of mirrors and/or prisms may be used to send mutually coherent beams along paths having arbitrarily great separations. Thus many variations of holography were developed with visible light and are described, e.g., in Ref. 49. In the case of visible light, the image signal usually comes from light scattered or reflected from the surfaces of objects. Because the angles between the scattered waves and the reference wave may be quite large, the hologram may contain interference effects on the scale of the wavelength of the light but the recording media available have adequate resolution to record the interference fringes directly without need for magnification. The large angles allow the recording of information on the three-dimensional structure of the object and the production of 3-D reconstructions. A diagram of a typical optical holography system is shown in Fig. 2.3.

If Z_{ref} is the contribution from the 'direct' beam via the mirror, i.e., $\overline{\mathrm{SMD}}$, while Z_{obj} is the (weaker) contribution due to diffuse scattering via path $\overline{\mathrm{SPD}}$ from an object point such as P, then the intensity at D is

$$I = |Z_{ref} + Z_{obj}|^2 = Z_{ref}\, Z^*_{ref} + Z_{obj}\, Z^*_{obj} + Z_{obj}\, Z^*_{ref} + Z^*_{obj}\, Z_{ref} \qquad (2.3)$$

if the difference in path-lengths does not exceed either the temporal or lateral coherence limitations of the laser. The film is developed under special conditions such that its amplitude transmittance is proportional to I. For the reconstruction, the object is removed, the processed film (the hologram) is placed in the same position as the original

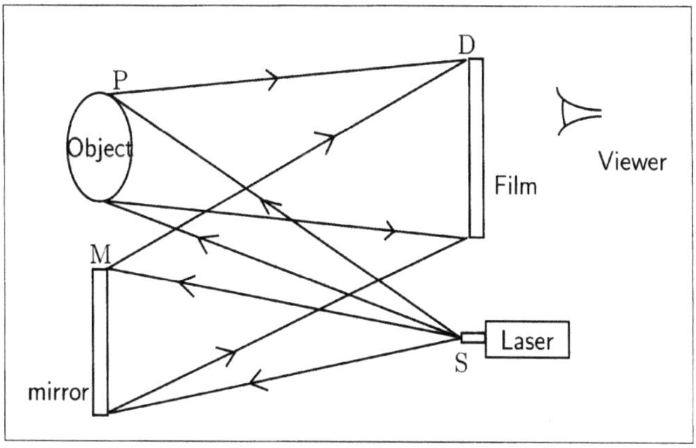

Figure 2.3. Typical arrangement for three-dimensional optical holography.

unexposed film, and the laser and mirror are retained in their original positions. A viewer observing through the processed film then sees a virtual image of the object apparently in its original position, because the amplitude of light at the film plane will now be reproduced exactly as it was during the recording. For a film transmittance proportional to I, the wavefield leaving the processed hologram during reconstruction is obtained by multiplying the above equation by Z_{ref}, giving

$$R = (Z_{ref} + Z_{obj}) + Z_{ref}\, Z_{ref}\, Z^*_{obj} \tag{2.4}$$

if $Z_{obj} \ll Z_{ref}$. We identify the term in parenthesis as a reconstruction of the complex, three-dimensional wavefield in the original image, producing a virtual image at P. The second term is the twin image. The viewer, by control of eye muscles may then focus on different planes at different depths behind the hologram, producing a three-dimensional impression. In principle, the viewer cannot therefore distinguish the view through the hologram from a view of the original object (including Z_{obj}). This is quite unlike the experience of viewing a photographic print, in which the eye can only focus on the plane of the print. A fuller analysis shows that the real and virtual twin images have been spatially separated in this arrangement (if the angle $\sphericalangle$ PDM is sufficiently large), and that a real image can be obtained at P by illuminating the hologram from the viewer's side along $\overline{\text{DM}}$.

Optical holography differs from electron holography in at least three important regards — firstly, optical methods provide a three-dimensional reconstruction, so that the surfaces of opaque objects may be examined at different depths in the reconstruction, and the interior of transparent objects may be examined. Electron holography allows images to be reconstructed at different focus settings; however, it does not allow us to look inside samples — within a sample the electron wave field does not propagate according to the vacuum-propagation law required for holographic reconstruction, and the 'projection approximation' holds to high accuracy. Modern holographic reconstructions are made using computers, which provide all the flexibility and power of modern image processing techniques, however, new problems then arise with the three-dimensional viewing of the reconstruction. In Fig. 2.3, the viewer may choose any plane in depth to focus on and parallax effects occur a computer display always requires the viewer to focus on the screen. Sections and projections may be presented by computer, but the display of three-dimensional information on a two-dimensional computer screen is still an area of investigation.

In Gabor's original scheme, he proposed that holograms could be recorded using electrons. Then the hologram could be magnified and the reconstructions from the holograms could be made using light. In magnifying the hologram in an electron microscope, however, the lateral and longitudinal magnifications may differ. Perfect three-dimensional reconstruction is only possible if the hologram magnification is equal to the ratio of the wavelengths.

A second difference between optical and electron holography is that electron holograms may be recorded in the near or far-field of an object. The far-field condition requires defocus $\Delta z > d^2/\lambda$. Then, for a small object of size d, the virtual image is sufficiently far out of focus to avoid interference with the real image. Optical holography of macroscopic objects does not normally satisfy this far-field condition (e.g., $\Delta z =$ 800 m for a 2 cm size object illuminated by a He-Ne laser), and is thus mostly near-field holography. However, for electrons, a focus defect of only $\Delta z = 300$ nm is sufficient to obtain far-field conditions for electron holography of, e.g., a Buckyball molecule (C_{60}) at 100 kV. In addition, the coherence lengths of optical lasers are of centimeter or meter dimensions, compared to a few hundred nanometers for kilovolt electron beams. Finally, the most important difference between the two methods is that, because of the large wavelength and modest scattering angles involved in optical holography, the holographic interference fringes may be directly recorded on film without magnification, whereas a high magnification of the interference fringes is usually required for electron holography. (The point-projection in-line geometry does, however, allow lensless electron holography if a 'nanotip' field-emission source is used). Thus, in Fig. 2.3, the roughly sinusoidal fringes which are formed due to interference between the direct ray $\overline{\text{SMD}}$ and the scattered ray $\overline{\text{SPD}}$ have a periodicity on the film plane which is larger than the grain size of the film — in the usual electron case the corresponding fringes would be of nanometer dimensions, and not suitable for available detectors.

In optical holography the hologram is formed from the diffuse scattering arising at each point on the surface of a three-dimensional opaque object. It is of interest to ask whether such holograms could be formed with electrons. The process of hologram formation and reconstruction can then be understood most simply in the in-line geometry if we treat the hologram as the sum of a set of Fresnel zone plate patterns, as shown in Fig. 2.4. We recall that the wave-function solutions of the Schroedinger equation may be linearly superimposed, so that the total hologram will be given by summing the complex amplitude distributions of the holograms generated by each point on the object's surface (such as P in Fig. 2.4). Again, if Z_{ref} is the contribution from the direct path $\overline{\text{SD}}$, while Z_{obj} is the (weaker) contribution via path $\overline{\text{SPD}}$ from an object point such as P, then the intensity at D is given by Eq. (2.3). For a point object, the resulting pattern forms a Fresnel Zone plate pattern, consisting of concentric rings with maxima at radii

$$r_n = (2\,f\,n\,\lambda)^{1/2}$$

where $f^{-1} = z_2^{-1} - (z_1 + z_2)^{-1}$. Such a zone-plate has the properties of both a concave and a convex lens with focal length f, and these properties are used to provide the reconstructed images. The complete hologram for an extended, weakly scattering object consists of a sum of such patterns from each object point. (A more realistic analysis for electron holograms must include the effects of any phase shift on scattering, and any phase shifts in the reference beam as a function of angle.) For an opaque, thin

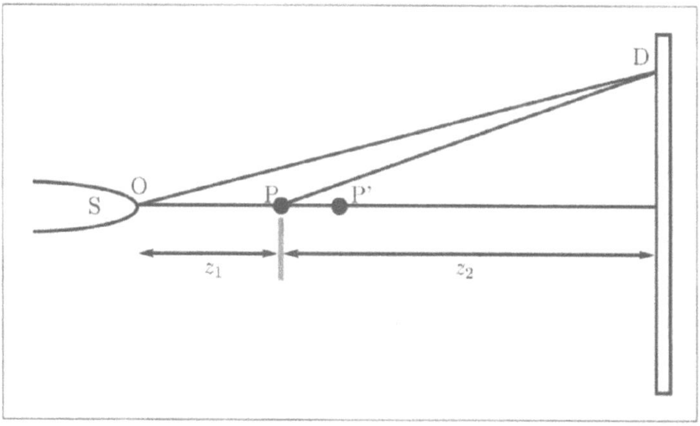

Figure 2.4. Diagram suggesting the formation of a hologram as the sum of Fresnel zone plate patterns due to scattering at points P from a point source, S.

object terminating with a straight edge within the beam (shown in grey in Fig. 2.4), the pattern is a set of parallel Fresnel fringes, whose maxima occur at $x_n = M(2\,z_1\,n\,\lambda)^{1/2}$ (with $M = z_2/z_1$), if the scattering phase is ignored. Thus we can think of the Fresnel fringes as a kind of one-dimensional zone-plate. These fringes extend a distance from the edge approximately equal to the lateral coherence width $X_C = 2\,z_1\,\lambda/d_S$, where d_S is the source size. The fringes appear magnified by $M = z_2/z_1$ at the screen. For an object extended along the axis from, e.g., P to P', interference effects will be observed in the hologram if $\Delta\hat{z}$, the distance between P and P', is less than the temporal coherence length $L_C = 2\,\lambda\,(E_0/\Delta E)$ with E_0 the beam energy and ΔE the energy spread. At 200 kV, with $\Delta E = 1\,\mathrm{eV}$, $L_C = 1\,\mu\mathrm{m}$. (We note that L_C is proportional to $\sqrt{E_0}$ ($\lambda \propto 1/\sqrt{E_0}$), so that low voltage instruments (such as the LEEM and point projection microscope) have relatively poor chromatic coherence. The chromatic aberration of lenses is proportional to $C_C\,\Delta E/E_0$, where C_C is the chromatic aberration constant.)

The holographic reconstruction is again given by Eq. (2.4) and is shown in Fig. 2.5. The first term in parenthesis is a reconstruction of the complex, three-dimensional wavefield in the original image, producing the virtual image P at V, as shown in Fig. 2.5. The viewer again sees a three-dimensional impression. The second term produces light in the reconstruction which converges to a real image P' at R. This light confuses an observer attempting to focus on the virtual image. Conversely, a screen situated at the focus of the real image will contain a background from the virtual image, as discussed previously. The (point-) resolution d_r of the reconstructed image can be understood by treating the hologram (zone-plate) as a lens, whose diameter is approximately equal to the transverse coherence width $M\,X_C$ at the screen. Then $d_r = 1.6\,\lambda\,z_2/(M\,X_C)$.

In summary, three-dimensional electron holography of surfaces in this simple point-projection geometry will be limited both by the twin image problem and by the longitudinal coherence length L_C. Other methods for reflection holography are discussed elsewhere in this book, e.g., on pages 311 ff.

Off-axis holographic schemes, analogous to those used for visible light, may be used for other radiations including acoustic waves[321,402] and radio waves.[404] For X-rays some in-line holography experiments have been described using X-rays from synchrotron radiation sources.[229]

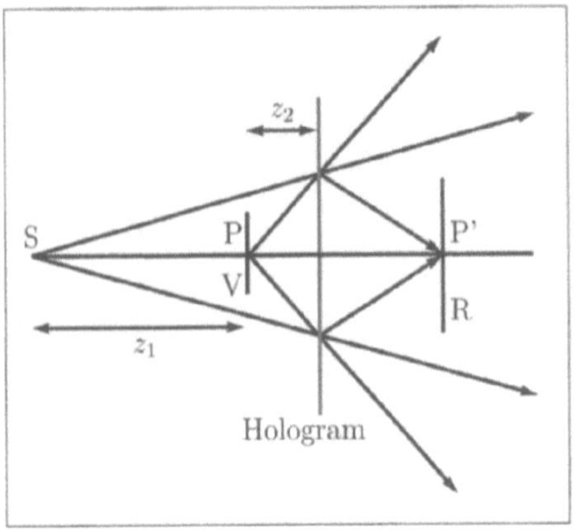

Figure 2.5. Reconstruction of virtual (V) and real (R) images from a hologram using a point source, S.

1.6. Off-axis electron holography

In early attempts to find suitable beam-splitters for high-energy electrons, thin single-crystal films were used in an orientation such that the transmitted beam and one of the diffracted beams had approximately equal intensity.[283,306] Recently, a useful practical form of electron holography has been developed on the basis of such a beam-splitter under the name of 'amplitude division' holography,[288,306,376] but the initial experiments entailed the use of a second thin crystalline film set parallel to the first so that the split beams could be brought together by a second diffraction process to produce an interference pattern depending on the differences in path length involved.

The most commonly-used methods for off-axis electron holography, however, are based on the use of the electrostatic biprism developed by Möllenstedt and Düker.[312] A very thin conducting wire is inserted in the microscope perpendicular to the beam. A positive voltage (or negative voltage depending on the electron optical set-up) on the wire serves to deflect the electron waves on the two sides so that they overlap and produce interference fringes having a spacing inversely proportional to the angle between the directions of propagation. To a very good approximation, the deflection of the waves on each side of the wire is constant, i.e., independent of the distance from the wire. Hence, the interference fringes may be of constant spacing over a large area.

In early observations, interference patterns were obtained with only a relatively small number of fringes because the available sources did not give very good lateral coherence. However, clear evidence was given that perturbations of the fringes resulted from local variations of the path lengths of the waves on the two sides of the biprism arising when the waves traversed different parts of a specimen.[104,194,314] The biprism wires need to be very thin and smooth. They may be made by pulling molten silica to form very fine amorphous fibers which are then coated with a thin film of, e.g., Au or Pt (see also page 92).

The most commonly used configuration for electron holography employing an electrostatic biprism, referred to as the 'off-axis TEM holography' mode is illustrated in Fig. 2.6(a). The biprism is inserted close to the image plane of the objective lens in a TEM instrument, represented in the diagram with only the minimum number of essential components.

Ideally, the condenser lens focuses electrons from a point source to form a plane parallel incident wave at the specimen. Inserting the biprism wire creates a black shadow, surrounded by Fresnel fringes, across the image in the objective lens image plane. As an increasing positive voltage is applied to the biprism wire, the two fields of the image on the two sides of the wire shadow are seen to move together and then overlap. The region of overlap is bright and crossed by interference fringes parallel to the wire. The fringe spacing is inversely proportional to the voltage applied to the wire and the extent of the overlap region (plus the initial width of the wire shadow) is directly proportional to the voltage. Hence the total number of fringes is roughly proportional to the square of the applied voltage.

Limitations on the width of the useful region of the image overlap. i.e., the number of interference fringes which may be observed and the fineness of the fringe pattern, come from the coherence of the incident beam on the specimen and from the resolution of the recording medium. The contrast of the interference fringes is given ideally by the lateral coherence at the specimen level. The lateral coherence function is given, according to the Van Cittert-Zernike theorem, in terms of the Fourier transform of the source intensity function. For an ideally incoherent source (which is a good approximation for an electron-emitting gun), having an intensity distribution $I_0(r)$ at a distance R from the specimen plane, the coherence function at the specimen. measuring the contrast of interference fringes formed when ideal thin slits are placed a distance r apart, is given by (see, e.g., Ref. 71 or 408)

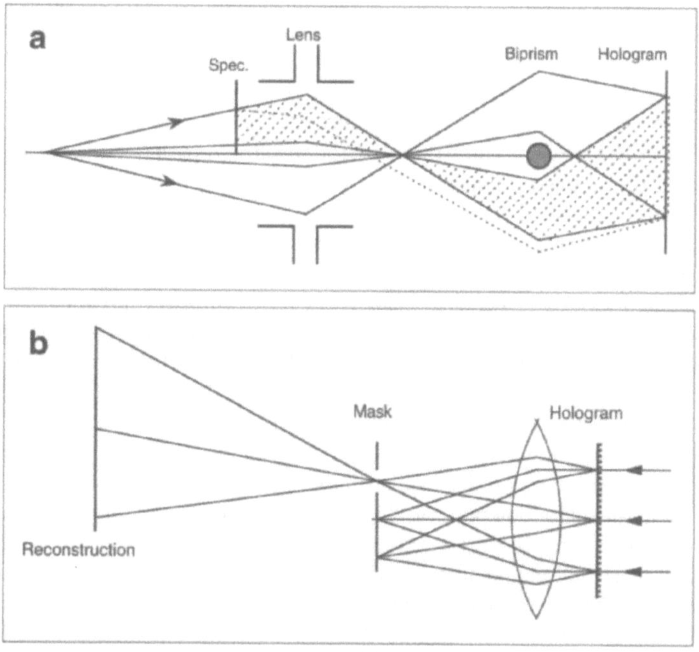

Figure 2.6. (a) Formation of an off-axis hologram in the image plane of the objective lens of a TEM instrument by using an electrostatic biprism to give overlapping of the beams passing through, and outside, the specimen. (b) Reconstruction from the hologram formed in (a) by choosing one side-band of the diffraction pattern in the back-focal plane of a lens (c.f., Fig. 2.7(b)).

$$\Gamma(r) = \frac{\int I_0(r_0) \cdot \cos\{2\pi[2\,r/(R\,\lambda)]\,r_0\}\,\mathrm{d}r_0}{\int I_0(r_0)\,\mathrm{d}r_0} \tag{2.5}$$

Thus for a circular disk source of diameter D, for which the Fourier transform of $I_0(\vec{r}_0)$ is of the form $J_1(\pi\,D\,q)/(\pi\,D\,q)$, with $q = 2\,r/R\,\lambda$ and J_1 being the first order Bessel function, the coherence function is a maximum of unity when the separation of the reference points in the specimen plane, r, is zero and goes to zero for a separation equal to the 'coherence width' of $r = 0.61R\,\lambda/D$, or $r = 0.61\,\lambda/\beta$, where β is the angle subtended by the source at the specimen.

In order to obtain an extensive pattern of fringes in the hologram, therefore, it is necessary to use a very small bright source, preferably that given by a field-emission gun. For example, with a thermionic source with an effective diameter of about 5 μm at a distance of 20 cm, the coherence width is 0.2 μm, while for a FEG the effective source size is about 5 nm so that the coherence width can be 200 μm.

When two points of the image are brought to coincide by the deflection of the biprism, the intensity distribution produced depends on the relative phases of the two overlapping waves and so on the differences of the corresponding path-lengths. Apart from the linear variations of phase due to the geometric paths, giving rise to the pattern of regular fringes, differences of phase may arise from interaction with the specimen. If it can be arranged that one of the beam paths passes through vacuum to provide a reference wave and the other passes through a specimen area, then information is obtained about the phase and amplitude changes due to that part of the specimen. The change of phase is indicated by a lateral shift of the interference fringes. A change of amplitude is indicated by a variation of the intensity. Hence the hologram contains the information concerning both the phase and the amplitude function, i.e., the complete complex wave function, for the wave transmitted by the specimen.

The process of reconstruction, by which the phase function and the amplitude function may be derived is illustrated by the idealized optical system of Fig. 2.6(b). In practice, the reconstruction process is now more usually carried out by computer manipulation of digitized data from the recorded hologram, but the diagram of the optical method conveys the principles of the method.

The hologram recorded on a photographic plate is illuminated by a plane parallel light wave and viewed in an ideal lens system. For no object present, the uniform, cosine-squared, fringe pattern would give three sharp points in the diffraction pattern in the back-focal plane of the lens: a central beam and two diffracted beams with separations inversely proportional to the fringe spacing. Selecting any one of these three diffraction spots by use of an aperture would give uniform intensity in the image plane. If an object is inserted in one of the two interfering specimen regions, the perturbations of the interference fringes produce a distribution of scattered beams around each of the three spots in the diffraction pattern. The intensity distribution around the central spot there corresponds directly to the normal diffractogram of the inserted object. Around one of the sharp diffraction maxima there is a distribution of diffracted amplitude information, including the corresponding phase information, which results from the first-order perturbation of the fringe pattern by the object. Around the other sharp diffraction maximum there is the distribution of diffracted amplitude, including the corresponding phase, which corresponds to the complex conjugate of the distribution around the first. If the regions around the two sharp diffraction spots are separately allowed to contribute to the image, by selecting them alternately with the aid of an aperture, the desired direct image and the unwanted conjugate image may be recorded separately. Thus the major difficulty of in-line holography, the overlapping of these two images, can be avoided.

For Gabor's in-line holography, the correction of the image for defocus and aberration effects may be carried out by adjustment of the focus and aberrations of the optical lens used. For the digital reconstruction these corrections can be made as part of the image reconstruction procedure and the advantage is obtained that the amplitude and phase functions, or the real and imaginary parts of the complex wave function, may be deduced separately.

The reconstruction procedure for off-axis holography is illustrated in Fig. 2.7.[472] Fig. 2.7(a) is the hologram recorded when the wave passing through the thin edge of a crystal is superimposed on the wave passing through vacuum, outside the crystal. Fig. 2.7(b) shows the distribution of intensity in the Fourier transform of the hologram, equivalent to the intensity distribution in the back-focal plane of a reconstructing lens system. The three separate diffraction patterns are clearly visible. When one of the side-band diffraction patterns is chosen, an inverse Fourier transform gives the phase and amplitude distribution, Figs. 2.7(c) and 2.7(d), separately.

The sensitivity of the reconstructed image to the incoherent resolution-limiting factors is evident from the nature of the reconstruction process. Mechanical specimen vibration or drift has the effect of damping the central diffraction pattern from the hologram and the two side-bands to the same extent. For reconstruction from one side-band, the resolution achievable is limited by such incoherent effects in the same way as for normal TEM imaging of equivalent resolution. To avoid overlap of the diffraction patterns, the separation of the center of the side-band from the center of the central diffraction pattern must be about three times the radius of the side-band diffraction pattern; which is equivalent to specifying that the fringe periodicity in the hologram should be less than one third of the least resolvable distance desired in the reconstructed image.[471] Hence, the achievable resolution is limited by any mechanical, electrical or magnetic instabilities which affect the formation or magnification of fringes on this scale.

The various, more detailed considerations of the methods and requirements for off-axis TEM holography, discussions of applications of this mode for various purposes and considerations of the experimental factors and practical limitations have been given in a number of review articles,[73,159,261] in the book edited by Tonomura et al.[453] and in the further chapters of the present volume. Particular mention should be made of the programs designed to enhance the resolution attainable in electron microscopy of thin specimens to achieve the original goal of Gabor,[129,130] taking the resolution limit for intermediate voltage microscopes (operating in the range of 200 to 400 keV) from the present value of about 0.16 nm to approach 0.1 nm.[268] Other programs do not require resolution enhancement but seek accurate measurement of specimen parameters or of the electrical field distributions in specimens[298] and extensive programs make use of the off-axis TEM mode to explore magnetic field distributions, with special modifications to allow observations of time-dependent phenomena.[195]

1.7. Alternative off-axis modes

Many different forms of electron holography appear to be feasible. A few years ago a listing was given of twenty possible modes distinguishable in terms of differences in theoretical basis or experimental approach,[66] and since that time several new modes have been suggested and/or demonstrated. The multiplicity of modes comes in part from the possibility of using either TEM or STEM approaches. The principle of reciprocity[53] implies that for every possible TEM mode there is an equivalent STEM

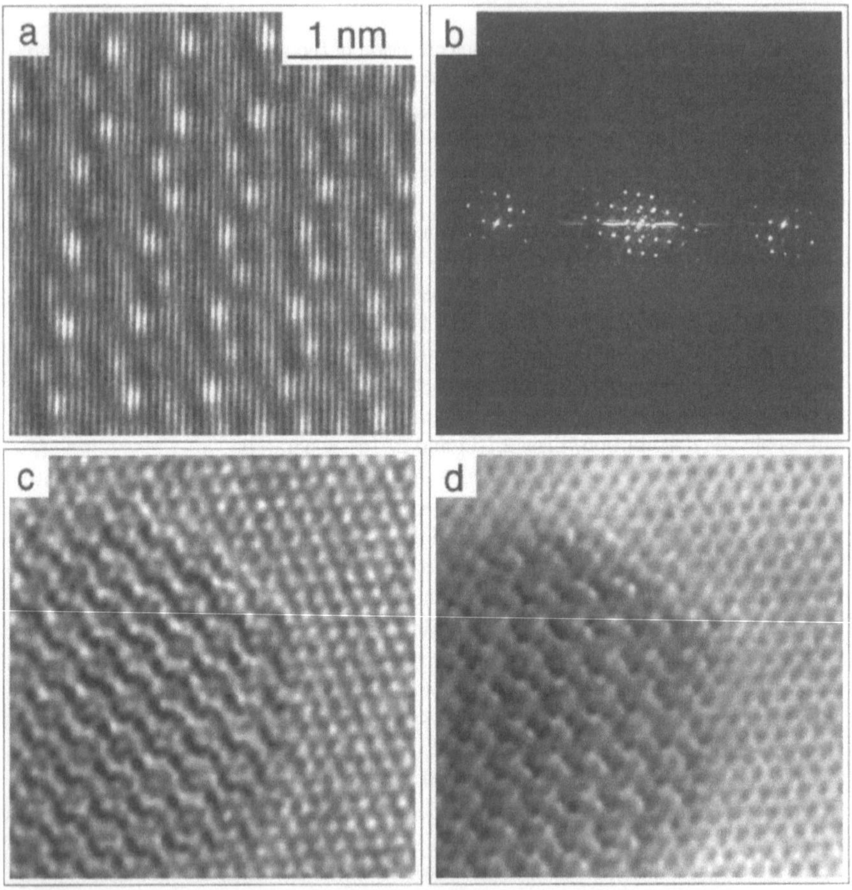

Figure 2.7. Reconstruction from the off-axis TEM hologram of a thin crystal. (a) Portion of the hologram crossed by fine interference fringes. (b) The diffraction pattern of the hologram showing the central pattern and the two side- band patterns. (c) The phase-image, reconstructed from one of the side-bands of (b). (d) The corresponding amplitude image. Courtesy of Dr. E. Völkl.[472]

mode. Also for each in-line mode there is usually a corresponding off-axis mode and also there are various modes for which the hologram is formed in the diffraction plane rather than the image plane of the objective lens of the imaging system.

For example if, in Fig. 2.6(a), the electrons are considered to pass from right to left, rather than from left to right as in TEM, an equivalent off-axis STEM mode of holography is obtained. The electron source point would be made to effectively scan over a two-dimensional area by means of deflector coils placed near to the objective lens. The image, serial in time, would be recorded by use of a small detector at the position of the FEG gun of the TEM system. So far, this scheme has not been attempted because it is obviously very inefficient in its use of the incident electron beam. The specimen would be irradiated with electrons during the whole scan time and the percentage of the transmitted electrons actually collected to form the image would be very small.

In the alternative off-axis STEM holography scheme suggested in Fig. 2.8(a),[63,69,151]

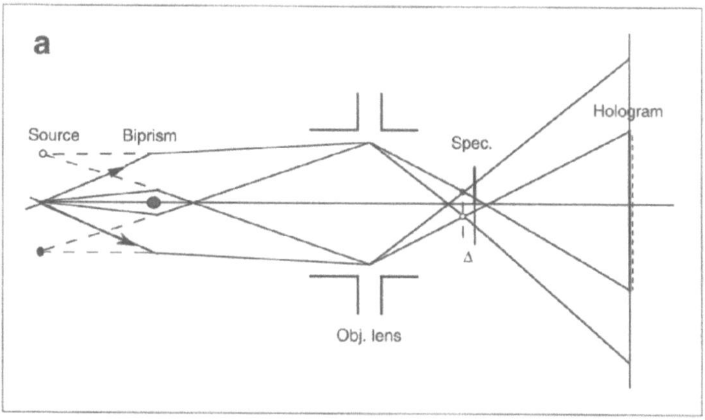

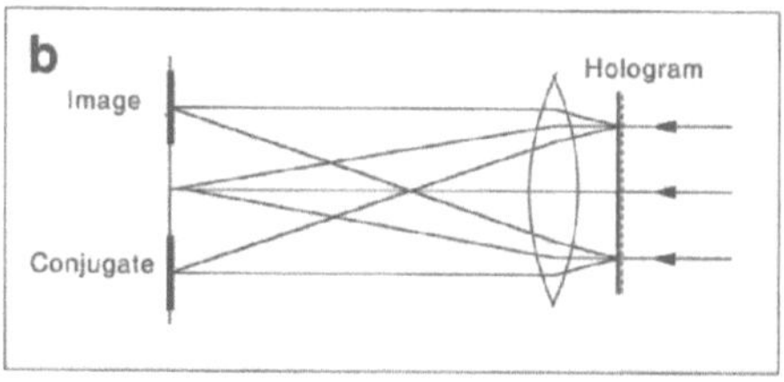

Figure 2.8. (a) The formation of an off-axis hologram in a STEM instrument with an electrostatic biprism in the illuminating system. (b) Reconstruction from the hologram formed in (a) by formation of the image and its conjugate, separated, in the back-focal plane of a lens.

the biprism is again placed close to the effective point source so that two fine probes, representing mutually-coherent, near-spherical convergent waves, are formed in the specimen plane. Interference of these waves at a large distance from the specimen gives a fringe pattern. If one of the focused probes passes through the specimen and one passes through vacuum, the hologram consists of the Fraunhofer diffraction pattern of the portion of the object illuminated by the incident beam, plus a reference wave from the other probe, and thus contains the information on the relative phases and amplitudes of the wave passing through the object. The reconstruction process, suggested in Fig. 2.8(b), then involves a single Fourier transform process (rather than the two Fourier transforms needed for TEM). However, this reconstruction process must be repeated for each pixel of the image since only one image pixel area is illuminated by the beam at any one time. It has been shown by computer simulation that the resolution attainable, the electron irradiation of the specimen and the dependence on experimental parameters for this mode are much the same as for off-axis TEM.[69]

In the form of off-axis TEM holography, which is related by the reciprocity relationship to this second mode of off-axis STEM holography, the angle of incidence of the beam illuminating the specimen is systematically changed so that the relative phases of the waves at any two points in the specimen plane are varied. This is the principle of the so-called 'phase-shifting' mode of TEM off-axis holography.[373]

In the alternative,'amplitude division' mode of TEM off-axis holography,[342,379] a thin crystal is used as a beam-splitter in place of the biprism, as in the early attempts at interference microscopy described on pages 26 ff. It may be considered that, instead of imposing a periodic wave field in the image plane, a periodic modulation of the wave incident on the specimen is used, originating from the interference of the waves transmitted and diffracted by the thin crystal placed close to the specimen. Again, the diffraction pattern formed in the reconstruction process consists of the three sharp points due to the ideal cosinusoidal periodic wave-field surrounded by the first and second order diffraction distributions arising from the modulations due to the specimen. The equivalent STEM mode is similarly feasible.

1.8. Alternative in-line and other modes

For STEM imaging, the flexibility of the detector system makes possible a number of different bright-field and dark-field imaging modes.[67,368] In addition to the annular detector originally used by Crewe et al.,[77] the high-angle annular detectors and thin annular detectors have been shown to be valuable, and split, or multiple-segment detectors have proved effective for special purposes. It was proposed by Veneklasen[468] that for thin, weakly-scattering objects, a specially-configured detector might have the effect of correcting for the aberrations of the objective lens and so serve as a reconstruction device when applied to the in-line hologram.

The detector plane of a STEM instrument contains the intensity distribution of the point-projection shadow image and contains the effects of the interference of the incident transmitted wave with all the scattered waves. The relative phases of the waves vary with angle because of the angular variation of the phases with the defocus and aberration terms generated by the probe-forming lens, and the intensities are modified as a function of angle accordingly. It was proposed that the effects of these modifications could be corrected by forming an image of the detection plane on a mask which imposed a correcting radial intensity distribution and using the intensity transmitted through the mask as the image signal. This technique would have the great advantage that the reconstructed, resolution-enhanced image could be produced on-line, as for a normal STEM image, with no separate off-line computer based reconstruction procedure. The method suffers from the presence of the conjugate images but these may be largely suppressed.[66] However, experimental considerations have so far prevented its successful operation.

A related, less ambitious but more successful use of a special detector system for on-line STEM is that of Leuthner et al..[255] With a biprism in the illumination system of a STEM instrument, the hologram in the detector plane is crossed by interference fringes. For a small region in the middle of the detector area, the fringes are uniform and parallel. But when the one probe which passes through the specimen is changed in phase these fringes uniformly shift laterally and any change in the amplitude of the probe gives them a change of intensity. Hence if this fringe pattern is imaged on a mask consisting of a complementary set of fringes, e.g., a set of opaque and transparent parallel lines, the transmitted signal gives an intensity depending on the phase shift of the probe. The intensity integrated over the area of the fringe set gives the amplitude variation. Thus as the probe through the specimen is scanned over the object, the phase image and amplitude image may be collected on-line, separately. This technique has shown some useful results[255] but has not been pushed to give any effective resolution enhancement.

The in-line, point projection holographic method of Gabor[129,130] has been discussed in terms of the possibility of resolution enhancement which involves the recording of the hologram with the electron cross-over placed close to the specimen and a large degree of magnification. But when resolution enhancement is not a consideration, large amounts of defocus and much smaller magnifications may be used. Then the holograms take on a quite different appearance and their treatment is different.

When the objective lens is greatly defocused, positive or negative, the hologram takes on the appearance of a magnified, point-projection image or 'shadow image' of the specimen. It appears more like a TEM image than a diffraction pattern. The difference from a TEM image, apart from the interference effects which are not usually

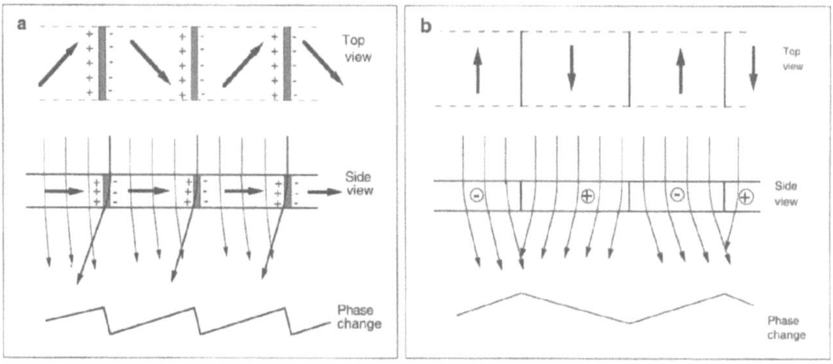

Figure 2.9. The deflection of electron beams and the corresponding phase shifts for transmission of an electron beam through thin films of (a), a ferroelectric with 90° domain boundaries, and (b) a ferromagnetic material with 180° domain boundaries.

very evident, is that the geometry of the image is distorted by the aberrations of the lens. These distortions are gross when the defocus is not very great,[55] but for large amounts of the defocus the paraxial image is very little distorted and may be used as a basis for holographic reconstruction using either in-line or off-axis methods.[75] Both of these methods are valuable for the imaging of magnetic field distributions in thin films or around small particles.

The in-line, greatly-defocused point-projection technique is familiar as the Fresnel mode of imaging ferromagnetic domain configurations in thin films.[41] The magnetic field in one domain may deflect the electrons in one direction and the magnetic field in an adjacent domain may deflect the electrons in the opposite direction (see Fig. 2.9(b)). At the magnetic domain boundary, the electrons from the two domains may be made to overlap, giving a bright line, or they may diverge, giving a dark line. so that a striking image of the domain configuration, such as Fig. 2.10, may be produced. Attempts at quantitative measurements of magnetic fields using this technique, while possible in principle, have been very limited. The off-axis, greatly defocused, point-projection holographic method has been extensively developed for exploring the form of the magnetic fields in thin films or around small particles and for making quantitative measurements of magnetic field distributions.[278,280] The recording and reconstruction techniques are essentially the same as for TEM off-axis holography. Corrections can be made for the distortions of the point-projection image due to the aberrations of the objective lens, but these are not required when the resolution sought is no better than about 1 nm for 100 keV electrons.

The STEM case is matched by the TEM case for in-line or off-axis holography with large values of defocus. When a sufficiently coherent source is used, greatly defocused TEM images of thin films of ferromagnetic materials show the same Fresnel contrast of domain boundaries as in Fig. 2.10. The off-axis mode of TEM holography has been used with great effect, particularly by Tonomura and associates.[451.453] Methods developed for making contour maps of the magnetic field intensities (see Fig. 2.11) have been highly effective in revealing details of fields in small particles. in thins films such as those used in recording media, in the space surrounding magnetic materials and in the regions of flux vortices in superconductors. Similarly the electric fields in and around solids have been determined and displayed as contour maps.[110]

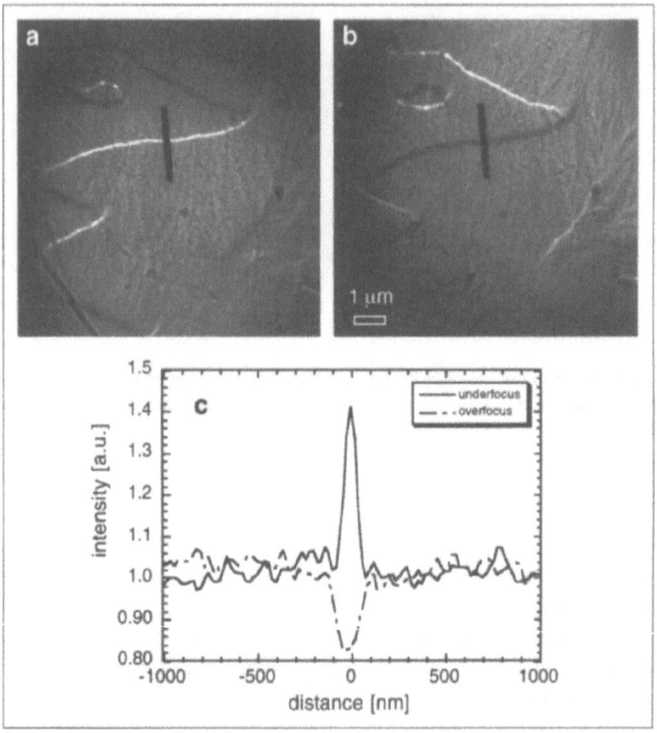

Figure 2.10. In-line, far-out-of-focus holograms (otherwise known as a Fresnel images) of a Co/Cu multilayer thin film showing (a) the underfocus image, (b) the over-focus image, and (c) the intensity variations along the lines indicated in (a) and (b).

1.9. More general approaches

The methods of electron holography which aim at the enhancement of image resolution have shown considerable promise but appear to be limited by various experimental factors so that for intermediate voltages (300 - 400 kV) it appears that progress beyond the resolution limit of about 0.1 nm will become increasingly more difficult. So far, thin crystals have been the object of investigation almost exclusively for the obvious reason that it is relatively easy to verify the results of reconstruction processes when the objects are of known structure. However, the imaging of periodic objects must always be considered as a special case and the ultimate aim must be to improve the resolution limit for interpretable images of general non-periodic objects.

On the other hand, it is known that the Fraunhofer diffraction patterns of thin specimens show elastic scattering to angles corresponding to real-space periodicities as small as 0.05 nm or less before the intensities become very small and submerged in thermal diffuse scattering. The question then arises as to whether the information contained in diffraction patterns with coherent incident beams can be combined with that from the images to extend the effective range of the imaging processes.

For perfectly periodic objects such as thin crystals, this possibility has been explored with some success under the heading of electron crystallography[95,122] using the methods of phase-extension similar to those employed in X-ray diffraction. In general,

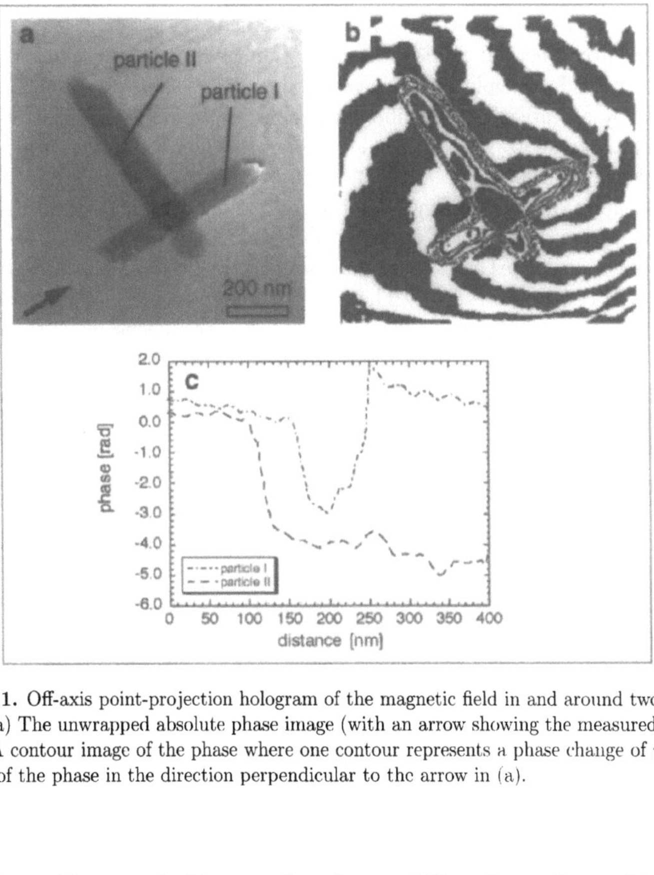

Figure 2.11. Off-axis point-projection hologram of the magnetic field in and around two CrO_2 particles. (a) The unwrapped absolute phase image (with an arrow showing the measured component of B). (b) A contour image of the phase where one contour represents a phase change of $\pi/11$. (c) Line-scans of the phase in the direction perpendicular to the arrow in (a).

however, it would seem valuable to explore the possibility of recording and interpreting the maximum amount of information obtainable from the scattering of electrons by an object using coherent scattering conditions, for example by obtaining a diffraction pattern from every pixel of an image or by obtaining an image with every part of the diffraction pattern of an object. Approaches have been made from various angles toward this goal.

When the incident beam of a STEM instrument is stopped at any one point on the specimen, a coherent convergent beam diffraction pattern is formed on any distant plane of observation and may be observed and recorded by use of a suitable two-dimensional recording system.[64,67] In the case of a thin single crystal, the pattern is a regularly-spaced array of disk-like spots and, if the spots overlap because the probe size on the specimen is smaller than the crystal periodicity, interference effects are visible in the areas of overlap reflecting the phase differences between the adjacent diffracted beams. Hence, in principle, all the relative phases of the diffraction pattern can be determined and, in the simple kinematical approximation, the structure of the crystal may be determined unambiguously with all the accuracy of an X-ray structure analysis.[201] In practice the situation is complicated by dynamical diffraction effects and other experimental considerations[405,406] but useful indications on crystal structure and symmetry can be obtained immediately in some cases.[434,469] An attractive aspect of this approach for single crystals is that, in order to obtain the relative phases for the whole of the diffraction pattern, it is necessary to achieve only that amount of lateral coherence necessary for the interference of adjacent diffracted beams or, in effect, a STEM resolution limit equal to the unit cell periodicity.

An alternative method for the determination of relative phases of a diffraction pattern is to use off-axis holography in reciprocal space. The deflection of the diffraction pattern by a biprism causes different regions of the diffraction pattern, such as adjacent diffraction spots, to overlap and interfere.[341] Such an approach is not limited to crystalline specimens.

The possibilities for recording, correlating and interpreting series of diffraction patterns taken from each pixel of an image area have been considered and explored by means of optical analogues and, to a limited extent, by actual STEM observations by Rodenburg and collaborators.[366,367] The intensities are observed as a function in four dimensions, the two dimensions of the image and the two dimensions of the diffraction plane. Appropriate sections or projections of the four dimensional distribution may be selected to aid interpretations, and an image with resolution enhanced by a factor of two or more may be derived directly.

In an alternate, related approach,[230] the diffraction patterns from each pixel of an image of small area were recorded at TV rates as the beam in a STEM instrument, having a diameter of about 0.3 nm, was scanned over a thin specimen. Each diffraction pattern was Fourier transformed to give the auto-correlation function, or Patterson function, of the illuminated area of the pixel. Correlation of the Patterson functions, which contain information about all the interatomic vectors in the sample area, then allowed a mapping of the scattering matter and the derivation of an 'image' of the structure with an apparent resolution of better than 0.1 nm. As for most resolution-enhancement procedures, the initial experiment was done for a thin single crystal of known structure (Si in [110] orientation) but the extension of the method to non-periodic objects has not yet been made. The combination of this technique with the off-axis scheme using a biprism gives the second off-axis STEM holography mode described in Section 1.7.

2. Imaging theory

For most cases of electron microscope imaging, the coherence width of the incident radiation is such that the imaging process may be assumed, as a first approximation, to be coherent, so that the imaging is described in terms of the modification of wave-functions and the image intensities are found from the modulus squared of the image wave-function. The deviations from perfect coherence are introduced by adding intensities for a small range of incident beam orientations and/or for a small range of wavelengths of the radiation. This contrasts with the familiar, incoherent case of light optics for which the modifications of intensity distributions by the optical system must be considered.

In a coherent imaging process, the wave function at the exit face of the object, $o(\vec{r}) = a(\vec{r})\exp[i\,\varphi(\vec{r})]$ where $\vec{r}$ is the two-dimensional vector with components x, y, which gives rise to the wave-function in the image plane, $O(\vec{r}) = A(\vec{r})\exp[i\,\phi(\vec{r})]$ and the resulting intensity distribution, $I(\vec{r}) = |A(\vec{r})|^2$.

In the simple Abbe description of the imaging process which is appropriate for electron microscopy, it is considered that, for an ideally perfect thin lens, the Fraunhofer diffraction pattern of the object wave appears in the back-focal plane of the objective lens with a distribution $H(\vec{q})$ given by the Fourier transform, $\mathbf{FT}\{o(\vec{r})\}$, where $\vec{q}$ is the two-dimensional vector in reciprocal space with coordinates u, v, which have magnitudes $(2\sin\theta_x)/\lambda$ and $(2\sin\theta_y)/\lambda$, where θ_x and θ_y are the half-angles of scattering in the x-

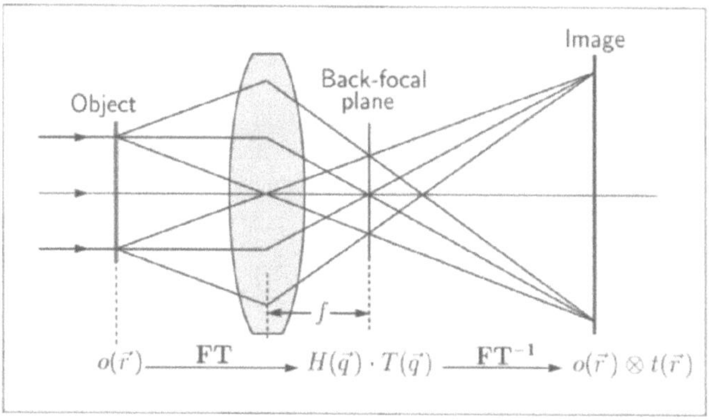

Figure 2.12. Diagram representing the Abbe imaging scheme in which a Fraunhofer diffraction pattern is formed in the back-focal plane of the lens.

and y- directions and the wavelength is λ.

$$H(\vec{q}) := \mathbf{FT}\{\, o(\vec{r})\} = \int o(\vec{r}) \cdot e^{2\pi i \vec{q}\vec{r}} \, \mathrm{d}\vec{r} \tag{2.6}$$

For high voltage electrons it is usually possible to make the small-angle approximation, referring only to paraxial rays, since the angles of scattering rarely exceed 10^{-1} radians. Then the coordinates in the back-focal plane are conveniently written $u = x/f\lambda$ and $v = y/f\lambda$, where f is the focal length (see Fig. 2.12).

The transfer from the back-focal plane to the image plane involves an inverse Fourier transform process, denoted by $\mathbf{FT}^{-1}\{\,\}$, so that, for the ideally perfect thin lens, the complex image wave is $\mathbf{FT}^{-1}\{H(\vec{q})\} = o(-\vec{r}/M)$, where M is the magnification factor. Normally the magnification factor and any image rotation is neglected so that the image is referred to the dimensions of the object and the ideal image is written simply as $o(\vec{r})$.

Real lenses, of course, do not give perfect reproductions of the object wave. They have a finite aperture which is normally introduced by multiplying the distribution of amplitude in the back-focal plane by an aperture function, $B(\vec{q})$, such that

$$B(\vec{q}) = \begin{cases} 1 & \text{for } q \leq q_0 \\ 0 & \text{for } q > q_0 \end{cases} \tag{2.7}$$

The aberrations of the lens, including defocus, give the effect of modifying the phases of the wave in the back-focal plane, so that the wave function there is multiplied by a phase function, $\exp[i\,\chi(\vec{q})]$. The defocus gives a phase term proportional to $|\vec{q}|^2$ and the third-order spherical aberration gives a term proportional to $|\vec{q}|^4$. Other terms of higher order and terms not having radial symmetry will be discussed in some of the following chapters. Then the amplitude and phase modifications of the wave in the back-focal plane are represented by multiplication by the transfer function

$$T(\vec{q}) = B(\vec{q}) \cdot \exp[i\,\chi(\vec{q})] \tag{2.8}$$

where

$$\chi(q) = \pi\,\Delta z\,\lambda\,q^2 + \tfrac{\pi}{2}\,C_S\,\lambda^3\,q^4 \tag{2.9}$$

with Δz as defocus and C_S the spherical aberration coefficient.

Then the complex image distribution in the image plane is given by the inverse Fourier transform of the back-focal plane wave-function

$$\begin{aligned} O(\vec{r}) &= \mathbf{FT}^{-1}\{\mathbf{FT}\{o(\vec{r})\}\cdot T(\vec{q})\} \\ &= o(\vec{r}) \otimes t(\vec{r}) \end{aligned} \tag{2.10}$$

where the inverse transform of $T(\vec{q})$ is the 'spread-function' $t(\vec{r})$, and the $\otimes$ symbol represents the convolution operation, defined by the integral over a dummy variable, R,

$$o(\vec{r}) \otimes t(\vec{r}) \equiv \int o(\vec{R})\cdot t(\vec{r}-\vec{R})\,\mathrm{d}\vec{R} \tag{2.11}$$

and we have made use of the relation that the Fourier transform of a product of two functions is the convolution of their Fourier transforms. Then the convolution of the object wave with the spread functions represents the smearing out of the wave to give the loss of resolution associated with the lens imperfections. The intensity distribution in the image plane is then

$$I(\vec{r}) = |o(\vec{r}) \otimes t(\vec{r})\,|^2 \tag{2.12}$$

Since both $o(\vec{r})$ and $t(\vec{r})$ are, in general, complex, the form of the intensity function is not necessarily related directly to the object structure.

The refractive index, n, for electrons in the electrostatic potential field, $V(\vec{r}, z)$, of a solid is slightly greater than unity and is given by $(1 + V(\vec{r}, z)/U_A)^{1/2}$, where U_A is the accelerating voltage for the electrons, which is usually so large compared with the value of V that the approximation may be made.

$$n(\vec{r}, z) - 1 = \frac{V(\vec{r}, z)}{2\,U_A} \tag{2.13}$$

For a very thin object, with high energy electrons, it may be assumed that the phase change of an electron wave passing through the object in the z direction is given by integrating $2\pi\,(n-1)/\lambda$ along z, so that for an incident plane wave of magnitude unity, the wave leaving the object is

$$o(r) = e^{-i\sigma\,V_p(\vec{r})} \tag{2.14}$$

where σ is the 'interaction constant' of magnitude $\pi/\lambda\,U_A$, and $V_p(\vec{r})$ is the projection of the potential distribution. The Eq. (2.14) represents the so-called 'phase-object approximation' (POA). It is sometimes convenient to add an absorption function, $-\mu_p(\vec{r})$, to the exponent to represent the effect of the change in amplitude of the transmitted wave resulting from inelastic scattering processes, although this term is usually so small that it may be neglected as a first approximation.

The phase object approximation fails when the thickness of the object is increased by an amount which depends on the wavelength of the electron beam and on the scale of the detail which is being considered. This approximation neglects the spread of the wave by Fresnel diffraction effects within the object which can be represented by convolution of the wave with the propagation function $p(r)$. For small-angle scattering, $p(r) \propto \exp(\pi\, i\, r^2/(R\,\lambda))$ for propagation over a distance R. An indication of the amount of the spread is given by putting the exponent equal to π, for which $r = (R\,\lambda)^{1/2}$ so that there is a spread of 0.1 nm for $R = 2.5$ nm and a spread of 1 nm for $R = 250$ nm for 100 keV electrons, and these R values suggest the possible range of thicknesses for which the POA can be used for imaging with the corresponding resolutions.

A simple relation exists between the bright-field image intensity and the projected potential distribution for the special case that the phase object approximation (POA) is valid and also the phase change in the object is very small, as e.g., in the case of a very thin film of light-atom material. Then the expression Eq. (2.14) may be approximated,

$$o(\vec{r}) = 1 - i\,\sigma\,V_p(\vec{r}) \tag{2.15}$$

This is the 'weak phase-object' approximation (WPOA). Inserting this in Eq. (2.12) and writing $t(\vec{r}) = c(\vec{r}) + i\,s(\vec{r})$, in terms of its real and imaginary parts, which, from Eq. (2.8), is the Fourier transforms of $B(\vec{q}) \cdot \cos\chi(\vec{q})$ and $B(\vec{q}) \cdot \sin\chi(\vec{q})$,

$$\begin{aligned} I(\vec{r}) &= |\,[1 - i\,\sigma\,\phi(\vec{r})\,] \otimes [\,c(\vec{r}) + i\,s(\vec{r})\,]\,|^2 \\ &\approx 1 + 2\,\sigma\,\phi(\vec{r}) \otimes s(\vec{r}) \end{aligned} \tag{2.16}$$

Thus the image contrast is given directly by the projection of the potential distribution of the object, smeared out by the spread function, $s(\vec{r})$. Clearly, the optimum imaging of the object is given when this spread function is made to resemble a single sharp, positive or negative peak of minimum possible width. Following Scherzer[392] and Eq. (2.9), this occurs for a defocus such that $\sin\chi(\vec{q})$ is a function which is close to ± 1 for as large a range of $|\vec{q}| = q$ as possible. Plotting the values of $\sin\chi(\vec{q})$ against q for representative values of U_A and C_S as in Fig. 2.13 shows that this condition is satisfied for a negative defocus value (under-focus, or a weakening of the objective lens) of $\Delta z = -(4\,C_S\,\lambda/3)^{1/2}$ such that the value of χ first decreases to about $-2\pi/3$ and then increases, becoming zero for $q_m = 1.51\,(C_S\,\lambda^3)^{-1/4}$, and then increases more and more rapidly. The $\sin\chi$ value is then close to -1 for a wide range of q values which is decreased for either an increase or a decrease in defocus. Then $s(\vec{r})$ takes the form of a narrow negative peak. For values of q greater than the q_m value, the rapid rise of χ associated with the fourth-power C_S term causes $\sin\chi$ to oscillate rapidly, positive and negative. On the assumption that these rapid oscillations imply that information from these high q-values can not contribute interpretable image detail, it may be assumed that the objective aperture can be set at $q = q_m$ and the effective resolution limit of the image may be given by the inverse of q_m. Thus the 'Scherzer resolution' is given at

$$\Delta z_s = 0.66(C_S\,\lambda^3)^{1/4} \tag{2.17}$$

This Scherzer resolution figure has been widely used as a basis for discussion of electron microscope performance and possible means for its improvement. For example it is seen that the resolution may be improved by an increase in the accelerating voltage and a decrease of the wavelength or, less effectively, by a decrease of the spherical

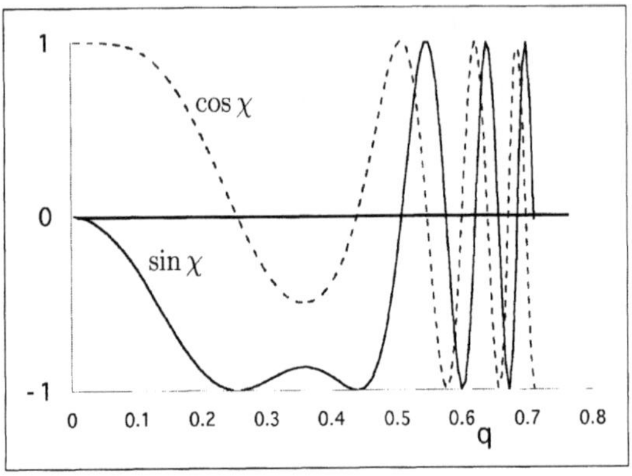

Figure 2.13. The form of the imaginary part, $\sin\chi$, and real part, $\cos\chi$, of the transfer function of a lens for 300 kV, $C_S = 1\,\text{mm}$.

aberration constant. It is a convenient measure of resolution since the value of q_m and the form of the $\sin\chi$ curve may be derived from a Fourier transform of the intensity distribution of the image from a thin amorphous film.[60,71,408] However, it must be emphasized that this whole concept of resolution evaluation is based on the use of the WPOA approximation and most electron microscope specimens are not weak phase objects.

For very thin films, the WPOA approximation may be inadequate for specimens containing elements of even moderate atomic number, especially if the material is a single crystal in a principal orientation so that the electrons pass along rows of atoms. For the POA approximation, Eq. (2.16) becomes much more complicated with terms including both $s(\vec{r})$ and $c(\vec{r})$, and no simple interpretation of the image in terms of the object structure, such as Eq. (2.16), is possible.

For specimens that are too thick for a reasonable application of the POA, it is possible to calculate the form of the object wave, $o(\vec{r})$, by taking into account the spread of the wave in the crystal by Fresnel diffraction effects and the phase changes corresponding to transmission through matter with multiple elastic scattering. For crystalline samples these calculations are commonly made by one of the computer programs based on many-wave dynamical diffraction theory and extensions of these methods have been made for non-periodic objects.[71,234,408]

A further point to be noted is that the simple WPOA theory of imaging given above applies only for a plane incident wave of unit amplitude. In general, if the wave incident on the specimen has a wavefunction $\psi_0(\vec{r})$, it can be assumed that the object wave, $o(\vec{r})$, is given by the product of $\psi_0(\vec{r})$ by a two dimensional transmission function such as Eq. (2.14), provided that the object is sufficiently thin. For thicker objects the calculation of the object wave may be much more complicated.

3. In-line point-projection holography

In the system represented by Fig. 2.1, the electron source may be assumed to be a delta-function so that in the back-focal plane of the objective lens the wave has amplitude unity multiplied by the transfer function of the lens, $T(\vec{q})$. The beam at the specimen is then given by an inverse Fourier transform of this as $t(\vec{r})$ and the beam transmitted through the specimen is $t(\vec{r}) \cdot o(\vec{r})$, where $o(\vec{r})$ in this case is the object wave for a plane-wave incident. In the distant plane of observation, the intensity distribution is then

$$I(\vec{q}) = |\,T(\vec{q}) \otimes H(\vec{q})\,|^2 \tag{2.18}$$

In the in-line holographic scheme of Gabor, the directly-transmitted beam is distinguished as the reference wave so it is appropriate to write $o(\vec{r})$ as $1 - p(\vec{r})$ where $p(\vec{r})$ is assumed to be small. Then, if $P(\vec{q})$ is the Fourier transform of $p(\vec{r})$, Eq. (2.18) becomes

$$\begin{aligned} I(\vec{q}) = {} & |\,T(\vec{q})\,|^2 + |\,T(\vec{q}) \otimes P(\vec{q})\,|^2 - \\ & T^*(\vec{q}) \cdot [\,T(\vec{q}) \otimes P(\vec{q})\,] - T(\vec{q}) \cdot [T^*(\vec{q}) \otimes P^*(\vec{q})] \end{aligned} \tag{2.19}$$

In the absence of an aperture, $|T(\vec{q})|^2 = 1$, and the second, second-order term in P may be neglected. Then the third and fourth terms represent the hologram for the desired image and its complex conjugate, associated with the conjugate image.

If, in the reconstruction process, a lens is used which has a transfer function $T(\vec{q})$, this transfer function is multiplied by Eq. (2.19) to give

$$I(\vec{q}) \cdot T(\vec{q}) = T(\vec{q}) \otimes [\delta(\vec{q}) - P(\vec{q})\,] - T^2(\vec{q}) \cdot [\,T^*(\vec{q}) \otimes P^*(\vec{q})\,] \tag{2.20}$$

and the resulting image formed by an inverse Fourier transform is

$$I_h(\vec{r}) = t(\vec{r}) \cdot o(\vec{r}) - t_2(\vec{r}) \otimes [\,t^*(-\vec{r}) \cdot p^*(-\vec{r})\,] \tag{2.21}$$

Since it is assumed that $t(\vec{r})$, like $T(\vec{q})$, is a known function, this intensity distribution may be multiplied by $t^*(\vec{r})$. The function $|t(\vec{r})|^2$ is just the intensity distribution of the beam at the specimen. Thus the first term on the right side of Eq. (2.21) gives the object function for the part of the specimen illuminated by the incident beam. The second term is the reconstruction of the conjugate image which is smeared out by $t_2(\vec{r})$, the inverse Fourier transform of $T^2(\vec{q})$. From Eq. (2.9) it follows that this is the spread function for twice the defocus and twice the spherical aberration of the lens. Hence the conjugate image has relatively poor resolution and forms a diffuse background.

Various schemes have been proposed for reducing the relative contribution of the conjugate image. It is shown, following Eq. (14.10) from page 316, that the conjugate image is not too distracting if the Fresnel number $N(d) = \pi\, d^2/\lambda\, \Delta f$ is small, where d is the object size. Recursive mathematical techniques such as the Gerchberg-Saxton algorithm[382] may be applied when some characteristic of the image is known, e.g., when it is known that some area of the image should have zero contrast. Experimentally the conjugate image may be suppressed, at the expense of increased exposure of the specimen to electron-beam irradiation, by taking series of holograms for which the reconstruction of the wanted image should be the same but the reconstruction of the conjugate image should be different. The series may, for example, be through-focus series or they could be series with lateral displacements.[272,487]

For the Gabor reconstruction scheme, it is usually assumed that the amount of defocus is small and the diameter of the reconstructed image region is small, comparable with the resolution of a TEM instrument using the same objective lens. Then the image area to be constructed for each incident beam position is so small that only relatively few pixels are required in the calculations for each incident beam position, but it is necessary to make many reconstructions from contiguous beam positions in order to get a useful total image area.

For many purposes, and especially when resolution enhancement is not required, large defocus values may be used so that $|t(\vec{r})|^2$ covers a much larger area, comparable to that of a normal TEM image. Then the convolution in Eq. (2.18) may be written,

$$\begin{aligned} H(\vec{q}) \otimes T(\vec{q}) &= \int H(\vec{Q}) \cdot e^{i\chi(\vec{q}-\vec{Q})}\, \mathrm{d}\vec{Q} \\ &= e^{i\chi(\vec{q})} \cdot \int H(\vec{Q}) \cdot e^{i\chi(\vec{Q})} \cdot e^{-2\pi i(\Delta z\,\lambda + 2C_S\,\lambda^3)\vec{q}\,\vec{Q}}\, \mathrm{d}\vec{Q} \end{aligned} \tag{2.22}$$

where we have neglected terms in C_S and higher orders of $\vec{Q}$ in the exponent since we will be considering only the paraxial region where the defocus term dominates. In terms of the variable $\vec{\rho} = \vec{q}(\Delta z\,\lambda + 2\,C_S\,\lambda^3\,q^2)$, or, since in the small angle approximation, $\vec{q} = \vec{r}/(\Delta z\,\lambda)$, $\vec{\rho} = \vec{r}(1 + 2\,C_S\,r^2/\Delta z^3)$, the integral in this expression is a Fourier transform of the product,

$$H(\vec{q}) \otimes T(\vec{q}) \approx e^{i\chi(\vec{q})} \cdot [o(\vec{\rho}) \otimes t(\vec{\rho})] \tag{2.23}$$

and

$$I(\vec{q}) = |\, o(\vec{\rho}) \otimes t(\vec{\rho})\,|^2 \tag{2.24}$$

which is identical with Eq. (2.12), the intensity of the bright-field TEM image apart from the distortion of the image implied by the substitution of $\vec{\rho}$ for $\vec{r}$. This distortion is seen to be small for the paraxial region of small r and for large values of the defocus, Δz. Thus point-projection shadow images are obtained which appear to give a direct imaging of the object. They have the same resolution as the corresponding TEM images and can be recorded and reconstructed holographically in the same way.[75]

4. In-line TEM holography

When a thin specimen is imaged out-of-focus in a TEM system with a coherent plane-wave illumination, the image intensity is given by Eq. (2.12) with the appropriate spread function, $t(\vec{r})$. The convolution with this spread function involves the interaction of the transmitted and scattered waves. If, as in the previous section, the transmitted wave is separated out by putting $o(\vec{r}) = 1 - p(\vec{r})$, the image intensity becomes

$$\begin{aligned} I(\vec{r}) &= |\,[1 - p(\vec{r})] \otimes t(\vec{r})\,|^2 \\ &= 1 - p(\vec{r}) \otimes t(\vec{r}) - p^*(\vec{r}) \otimes t^*(\vec{r}) + \ldots \end{aligned}$$

for which the Fourier transform is

$$\mathbf{FT}\{I(\vec{r})\} = \delta(\vec{q}) - P(\vec{q}) \cdot T(\vec{q}) - P^*(-\vec{q}) \cdot T^*(-\vec{q}) + \ldots \tag{2.25}$$

If this is multiplied by $T^*(\vec{q})$ and inverse Fourier transformed, one obtains

$$I_r(\vec{r}) = o(\vec{r}) - t_2^*(\vec{r}) \otimes [t^*(\vec{r}) \cdot p^*(\vec{r})] + \ldots \tag{2.26}$$

which represents the wanted image (the phase and amplitude images) plus the conjugate image.

The most common method for removing the ambiguities arising from the conjugate image is to use through-focus series of images. The applications that have been made of reconstructions from TEM images, however, have mostly related to images of thin crystals, viewed in principal orientations so that the simple WPOA assumptions used to derive Eq. (2.26) are not valid and much more sophisticated approaches have been used. Kirkland et al.[227,228,394] started with images obtained at 500 kV of copper hexadecachloro-phthalocyanine crystals and made use of the maximum a-posteriori recursive algorithm for image reconstruction. Van Dyck and Op de Beeck[327] used a 'focus variation' method in which large series of images are taken with small increments of defocus to make use of expressions obtained by differentiation of the image amplitude with respect to defocus.

5. Off-axis TEM holography

In the off-axis scheme of Fig. 2.6(a), it may be considered that, if the reference wave travels through vacuum, the hologram wavefunction is given by adding a plane wave with a tilt corresponding to $\vec{q} = \vec{q}_c$ to the object wave in the image plane, so that the hologram intensity becomes

$$\begin{aligned} I_h(\vec{r}) &= |\, o(\vec{r}) \otimes t(\vec{r}) + e^{2\pi i \vec{q}_c \vec{r}} \,|^2 \qquad (2.27) \\ &= 1 + e^{-2\pi i \vec{q}_c \vec{r}} \cdot [\, o(\vec{r}) \otimes t(\vec{r}) \,] + e^{2\pi i \vec{q}_c \vec{r}} \cdot [\, o^*(\vec{r}) \otimes t^*(\vec{r}) \,] + |\, o(\vec{r}) \otimes t(\vec{r}) \,|^2 \end{aligned}$$

This may be expressed in the form

$$I_h(\vec{r}) = 1 + |\, o(\vec{r}) \otimes t(\vec{r}) \,|^2 + 2 \,|\, o(\vec{r}) \otimes t(\vec{r}) \,| \cdot \cos(2\pi \, \vec{q}_c . \vec{r} + \phi(\vec{r})) \qquad (2.28)$$

i.e., as the sums of the intensities for the two images separately, plus modulated cosinusoidal fringes with phase shifts ϕ representing the phase of the image and an amplitude representing the amplitude of the image.

In the reconstruction process, illustrated by Fig. 2.6(b), the distribution in the back-focal plane of the lens is given by Fourier transform of Eq. (2.27) as

$$\begin{aligned} \mathbf{FT}\{I(\vec{r})\} = \delta(\vec{q}) + [H(\vec{q}) \cdot T(\vec{q}) \otimes H^*(\vec{q}) \cdot T^*(\vec{q})] \, + \qquad (2.29) \\ \delta(\vec{q} + \vec{q}_c) \otimes H(\vec{q}) \cdot T(\vec{q}) + \delta(\vec{q} - \vec{q}_c) \otimes H^*(\vec{q}) \cdot T^*(\vec{q}) \end{aligned}$$

where $H(\vec{q})$ is the Fourier transform of $o(\vec{r})$. Thus the diffraction pattern consists of three parts: the diffraction pattern of the intensity distribution of the normal TEM image, the Fourier transform of the wanted image wavefunction centered on the delta-function at $\vec{q} = -\vec{q}_c$, and the Fourier transform of the conjugate image, centered on the delta function at $\vec{q} = +\vec{q}_c$. The contributions of the conjugate image and the intensity distribution may be excluded by using an aperture to select the wanted image part provided that $\vec{q}_c$ is made large enough. If the reconstructing lens, or its mathematical analogue, can be made to correct the aberrations by multiplying the third term of Eq. (2.29) by $T^*(\vec{q})$, the result is $H(\vec{q})$ and an inverse Fourier transform gives $o(\vec{r})$ directly. In this way the conjugate image problem is avoided and, if the reconstruction process is done in the computer, the phase and amplitude components, or the real and imaginary parts, are obtained separately.

6. Off-axis holography in a STEM instrument

When a biprism is inserted in the illumination system of a STEM instrument, as suggested in Fig. 2.8(a), the effect is to produce two effective, mutually coherent sources and hence two coherent small probes in the specimen plane. Interference of the waves from these two probes gives the pattern of interference fringes in the distant plane of observation, the plane of the diffraction pattern. If one probe passes through the specimen and the other through vacuum, the modulation of the interference fringes reflects the amplitude and phase changes in the specimen.

For large amounts of defocus, as discussed on pages 42 ff, the point-projection image of the specimen resembles a TEM image and the hologram formed can be interpreted and reconstructed as for an off-axis TEM hologram.[280]

For the off-axis form of STEM holography discussed on pages 29 ff, and illustrated in Fig. 2.8(a,b), it may be assumed that the wave function at the specimen level for a separation, e.g., $\vec{a}$, of the incident probes is

$$\psi(\vec{r}) = o(\vec{r}) \cdot t(\vec{r}) + t(\vec{r} - \vec{a}) \tag{2.30}$$

so that the intensity in the hologram is

$$\begin{aligned} I_h(\vec{q}) &= \left|H(\vec{q}) \otimes T(\vec{q}) + T(\vec{q}) \cdot e^{2\pi i \vec{a}\vec{q}}\right|^2 \\ &= 1 + |H(\vec{q}) \otimes T(\vec{q})|^2 + \\ &\quad T^*(\vec{q}) \cdot e^{-2\pi i \vec{a}\vec{q}} \cdot [H(\vec{q}) \otimes T(\vec{q})] + \\ &\quad T(\vec{q}) \cdot e^{2\pi i \vec{a}\vec{q}} \cdot [H^*(\vec{q}) \otimes T^*(\vec{q})] \end{aligned} \tag{2.31}$$

In the reconstruction process, the reconstructing lens, or its mathematical equivalent, gives the inverse Fourier transform of Eq. (2.31) in its back-focal plane as

$$\begin{aligned} I_r(\vec{r}) = \delta(\vec{r}) \otimes [\delta(\vec{r}) + o(\vec{r}) \cdot t(\vec{r}) \otimes o^*(\vec{r})\, t^*(\vec{r})] + \\ \delta(\vec{r} + \vec{a}) \otimes t^*(-\vec{r}) \otimes o(\vec{r}) \cdot t(\vec{r}) + \\ \delta(\vec{r} - a) \otimes t(\vec{r}) \otimes o^*(\vec{r}) \cdot t^*(\vec{r}) \end{aligned} \tag{2.32}$$

Thus, around the central delta-function there is the auto-correlation function of the imaged region. The wanted image and its conjugate image are formed separately around the two delta functions representing the diffraction pattern from the interference fringes (see Fig. 2.6(b)). If, in the reconstruction process, the intensity distribution Eq. (2.31) is multiplied by $T(\vec{q})$, the third term of Eq. (2.32) becomes just $o(\vec{r}) \cdot t(\vec{r})$, or the wavefunction of the illuminated part of the specimen, well separated from the conjugate image.

7. Theoretical formulations for other modes

Similarly, it is possible to write relatively simple mathematical formulations for the various other modes of electron holography which have been postulated or shown to be feasible. For example, for the in-line modes proposed for on-line readout of the corrected image, such as those of Veneklasen[468] and Leuthner,[255] the intensity of the hologram, Eq. (2.19), is multiplied by a masking function. In the Veneklasen scheme, the mask

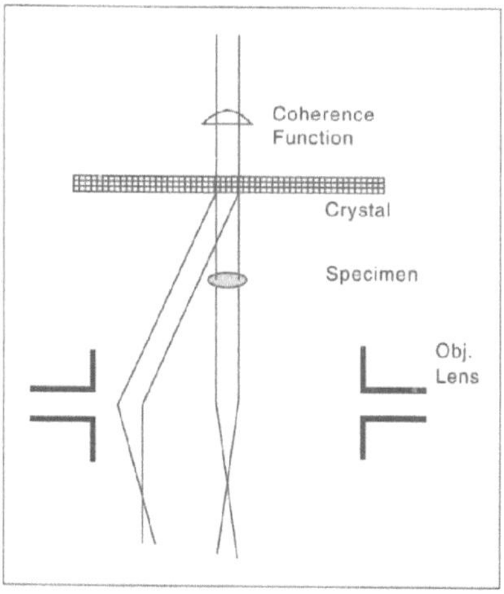

Figure 2.14. Arrangement of a crystal placed before the specimen for amplitude-division, off-axis electron holography.

function would ideally be $T(\vec{q})$ so that the third term of Eq. (2.19) becomes simply $T(\vec{q}) \otimes P(\vec{q})$ for which the Fourier transform gives directly $p(\vec{r}) \cdot t(\vec{r})$. The fourth term of Eq. (2.19) then gives a greatly defocused and aberrated conjugate image. However, it is experimentally difficult to make a mask having the complex transmission function $T(\vec{q})$. It was shown by Cowley[66] that it would be equally effective to multiply the intensity distribution by a real mask function $2\,B(\vec{q})\,\cos[\chi(\vec{q})] = B(\vec{q}) \cdot [T(\vec{q}) + T^*(\vec{q})]$, or $B(\vec{q})\,\sin[\chi(\vec{q})]$.

For the amplitude-division off-axis holography, the beam-splitting device is a thin single crystal placed just before the specimen, rather than a biprism,[71,342,379] as illustrated in Fig. 2.14. If the beam incident on the crystal is considered to be limited by an amplitude function, $f(\vec{r})$, corresponding to the coherence width of the incident radiation, and the crystal gives rise to two strong beams, the transmitted beam and the $\vec{h}$ reflection having amplitude $A_{\vec{h}}$, the wave function at the specimen level may be written as

$$\psi(\vec{r}) = o(\vec{r}) \cdot f(\vec{r}) + A_{\vec{h}} \cdot f(\vec{r} - \vec{a}) \cdot e^{2\pi i \vec{h} \vec{r}} \tag{2.33}$$

where $|\vec{a}| = (2\,\Delta z\, \sin\theta_h)/\lambda$ and θ_h is the Bragg angle. The hologram intensity distribution is then given by convoluting Eq. (2.33) with $t(\vec{r})$ and multiplying by the complex conjugate to give

$$\begin{aligned} I_{\vec{h}} = {} & |[o(\vec{r}) \cdot f(\vec{r})] \otimes t(\vec{r})|^2 + \left| \left[f(\vec{r}) \cdot A_{\vec{h}}\, e^{2\pi i \vec{h}\,(\vec{r}+\vec{a})} \right] \otimes t(\vec{r} + \vec{a}) \right|^2 + \\ & \{[o(\vec{r}) \cdot f(\vec{r})] \otimes t(\vec{r})\} \cdot \left\{ \left[f(\vec{r}) \cdot A^*_{\vec{h}}\, e^{-2\pi i \vec{h}\,(\vec{r}+\vec{a})} \right] \otimes t^*(\vec{r} + \vec{a}) \right\} + \ldots \end{aligned} \tag{2.34}$$

On the assumption that $f(\vec{r})$ is sufficiently wide compared with the desired image area, the Fourier transform of $f(\vec{r})$ may be assumed to be a delta function so that in the reconstruction process, as represented by Fig. 2.6(b), the distribution in the back-focal plane of the reconstructing lens is given by an intensity distribution around the delta function at the origin and two side bands. that around the delta function at $\vec{q} = -\vec{h}$ being

$$\delta(\vec{q}+\vec{h}) \otimes \left[H(\vec{q}) \cdot T(\vec{q}) \cdot A_{\vec{h}} \cdot T^*(\vec{h})\, e^{-2\pi i \vec{h}\vec{a}}\right] \tag{2.35}$$

and if the reconstructing lens has a transfer function $T^*(\vec{q})$, an inverse Fourier transform gives $o(\vec{r})$ multiplied by a constant term.

The requirement for the coherence of the incident radiation in this case is much less than for the more conventional off-axis TEM holography. Because the objective lens is focused on the beam-splitting crystal it is not necessary that the coherence width should be large enough to cover both the object and the reference wave at the specimen level. The coherence patch, $f(\vec{r})$, need only cover the area of the specimen which is imaged coherently, i.e., the pixel size for the defocused image given approximately by $(\Delta z\, \lambda)^{1/2}$. Such a coherence width is readily attainable even in a microscope with a conventional thermionic source.

8. Interference in diffraction patterns and holographic diffraction

When a strong objective lens is used to focus the electron beam from a small bright source, such as a FEG, which may be approximated by an ideal point source, on a very thin single-crystal specimen which has a transmission function given by the Fourier series, $o(\vec{r}) = \sum_{\vec{h}} F_{\vec{h}}\, e^{2\pi i \vec{h}\vec{r}}$, the wave leaving the specimen is $o(\vec{r}) \cdot t(\vec{r})$ and the corresponding diffraction pattern intensity is

$$\begin{aligned} I(\vec{q}) &= \left| \sum_{\vec{h}} \left[F_{\vec{h}} \cdot \delta(\vec{q}-\vec{h})\right] \otimes T(\vec{q}) \right|^2 \\ &= \sum_{\vec{h}} \sum_{\vec{k}} F_{\vec{h}} \cdot F^*_{\vec{k}} \cdot T(\vec{q}-\vec{h}) \cdot T^*(\vec{q}-\vec{k}) \end{aligned} \tag{2.36}$$

where $\vec{h}$ and $\vec{k}$ are reciprocal lattice vectors representing the positions of the diffraction spots having separations proportional to $1/a$ and $1/b$ for a unit cell with dimensions a and b as seen in the direction of the incident beam.

The transfer function $T(\vec{q})$ is limited by the aperture function $B(\vec{q})$, as described in Eq. (2.8). If the diameter of $B(\vec{q})$ is less than the distance between diffraction spots, the functions $T(\vec{q}-\vec{h})$ and $T^*(\vec{q}-\vec{k})$ will not overlap unless $\vec{h} = \vec{k}$. Then Eq. (2.36) reduces to

$$I(\vec{q}) = \sum_{\vec{h}} F_{\vec{h}} \cdot F^*_{\vec{h}} \cdot B(\vec{q}-\vec{h}) \tag{2.37}$$

i.e., the diffraction pattern consists of a set of uniform circular disks having intensities proportional to the squares of the structure amplitudes, $F_{\vec{h}}$. In practice, the single crystal is often not sufficiently thin to give uniform intensity of the disks. If a projection approximation such as the POA, does not hold, the disks are modulated by dark and light bands corresponding to the variations of transmitted or diffracted intensity as a function of angle of incidence.

For the ideal thin-crystal case represented by Eq. (2.36), it is seen that for larger angles of convergence of the incident beam such that the diffraction disks do overlap, interference effects occur in the region of overlap. For example, in a one-dimensional case, if $\vec{h}$ and $\vec{k}$ are the adjacent reflections with indices h and $h+1$, the intensity in the area of overlap is given as

$$I(q) = F_h \cdot F^*_{h+1} \cdot e^{i\chi(h-q)} \cdot e^{-i\chi(h+1-q)} + \dots$$

or, if the center point of the overlap region is taken as origin so that $q - h = e + 1/2$ and $q - (h+1) = e - 1/2$,

$$I(\epsilon) = 2|F_h| \cdot |F_h + 1| \cdot \cos\left[2\pi\epsilon\left(\Delta z\,\lambda + C_S\,\lambda^3/4\right) + C_S\,\lambda^3\,\epsilon^3 + \alpha\right] \tag{2.38}$$

where α is the relative phase of the h and $h+1$ reflections. Thus, as a good approximation for small ϵ, the area of overlap is crossed by sinusoidal fringes with periodicity depending on the defocus. It may be noted that if the center points, $\epsilon = 0$, for all the regions of overlap between adjacent disks are considered, it should be possible to deduce the relative phases of the structure amplitudes F_h for all reflections in the diffraction pattern and hence to perform a structure analysis of the crystal without the 'phase problem' which is the major difficulty of X-ray diffraction structure analysis.[201,405,406] Practical considerations of obtaining sufficiently thin single crystals have limited the experimental realization of this scheme to date. However, clear demonstrations of such fringes and some inferences concerning crystal symmetry have recently been demonstrated in TEM instruments equipped with FEGs.[434,469]

It should be noted that the phases of the structure amplitudes, $F_{\vec{h}}$, are not absolute quantities. A translation by a vector, $\vec{R}$, in real space is represented by convolution of the transmitted wave-function by $\delta(\vec{r} - \vec{R})$ so that the diffraction pattern amplitude is multiplied by $\exp(2\pi i\,\vec{q}\vec{R})$. Then the adjacent h and $h+1$ reflections in the one-dimensional example differ in phase by $\exp(2\pi\,i\,R/a)$ where a is the unit cell periodicity, and the intensity Eq. (2.38) varies sinusoidally as the incident beam is translated across the unit cell. The relative phases of the various reflections may be deduced from the relative phases of this sinusoidal variation. The other deduction of interest is that, if a small detector is placed in the region of overlap of any two reflection disks, the intensity detected, when used as the signal for a STEM display, gives a set of parallel fringes on the display screen.[405,406] This description of the origin of lattice-fringe images in STEM allows an immediate appreciation of the factors such as detector size, detector position, defocus and structure amplitudes which affect the fringe visibility. It may be noted that, for a point detector at the midpoint between two diffracted orders, the STEM image is unaffected by any aberrations (such as C_S) which are even functions of q.

If the objective aperture size is further increased so that the diffraction spot disks have a diameter much greater than the separations of the spots, many diffraction disks may overlap at any point of the diffraction pattern. Then the intensity includes many terms of the double summation Eq. (2.36) and the fluctuation of the intensity at any point varies in a complicated manner with the incident beam position. The STEM image obtained for any position of a small detector gives a two-dimensionally periodic structure image for the appropriate conditions of defocus, etc. This situation corresponds to the situation that the incident beam diameter is smaller than the unit cell dimensions and the diffraction pattern shows no evidence of the periodicity but reflects the structure and symmetry of only that part of the unit cell which is illuminated by the specimen.[58,74]

The possibility that a series of such diffraction patterns from adjacent points within the unit cell may be used as the basis for deriving the structure of the crystal with greatly improved resolution, relative to the normal bright-field STEM image, has been explored by Konnert et al.[230]

The more formal approach to this possibility due to Rodenburg et al.[366,367] may be expressed as follows. The intensity may be recorded as a function of the probe position $\vec{R}$ and the diffraction vector $\vec{q}$ and so can be written as the four-dimensional function, $I(\vec{q}, \vec{R})$, which can be Fourier-transformed with respect to $\vec{R}$ to give the function $G(\vec{q}, \vec{R}')$. If it is assumed that the WPOA holds and absorption effects are neglected, it is possible to write

$$\begin{aligned} G(\vec{q}, \vec{R}') = {} & |T(\vec{q})|^2 \cdot \delta(\vec{q}) + \\ & T(\vec{q}) \cdot T^*(\vec{q} + \vec{R}') \cdot \Psi_s^*(-\vec{R}') + \\ & T^*(\vec{q}) \cdot T(\vec{q} - \vec{R}') \cdot \Psi_s(\vec{R}') \end{aligned}$$

where the wavefunction $\Psi_s(\vec{q})$ representing the scattered wave in the diffraction plane is, in this case, proportional to the Fourier transform of $-i\sigma V(\vec{r})$.

From this it follows that

$$\Psi_s(\vec{R}') = G(\vec{R}'/2, \vec{R}') \tag{2.39}$$

so that the Fourier component of the scattered wave, $\Psi_s(\vec{q})$, is found by taking the Fourier transform of $I(\vec{q}, \vec{R})$ for the reciprocal vector $\vec{q}_2 = \vec{q}/2$.

The availability of the four-dimensional data provides redundancy with respect to the number of unknowns so that the ambiguity in the phase determination is removed and the solution is unique. The resolution of the reconstructed image may be doubled with respect to the bright-field STEM image since, if the objective lens aperture cuts off the information at $|\vec{q}| = q^0$, the scattered wave can be reconstructed for all vectors out to $2q^0$. This has been demonstrated for a case of relatively poor resolution.[367]

To conclude this section, we note that the idea of using a reference wave as a means for determining relative phases is not unique to holography. The use of reference waves to determine relative phases of diffracted waves from a crystal has a long history in X-ray diffraction structure analysis. In his classical paper of 1929,[32] Bragg showed that if there is one heavy-atom peak of scattering amplitude placed at the origin of the projection of a unit cell, all diffraction spot phases could be taken as positive with respect to the scattering from the origin peak. Then the Fourier series $\sum_{\vec{h}} F_{\vec{h}} \exp(2\pi i \vec{h} \vec{r})$ could be summed to give the projected electron density distribution $\rho(\vec{r})$ uniquely. If the summation is done by superimposing the intensity distributions for a number of masks having sinusoidal transmission functions for visible light, a greatly enlarged 'image' of the projection of the structure could be produced. Bragg referred to this process as 'two-wavelength microscopy.' Needless to say, this is not an imaging process in the normal sense of the term since the diffraction pattern is produced by a large volume of crystal containing many millions of unit cells, and the 'image' produced is a periodic averaging over all these cells.

Later developments in X-ray crystallography introduced phase-determining procedures which are more complicated than Bragg's initial heavy-atom method, involving the use of reference waves from a number of heavy atoms or from atoms for which the scattering factor may be varied by using X-ray wavelengths close to an absorption edge. A related valuable technique, that of isomorphous substitution, makes use of a series of crystal structures for which the sites of the many light atoms are the same but heavy atoms of various, widely different atomic numbers are substituted at particular sites.

In recent years some powerful, analogous techniques have been applied to the study of the structures of surface layers of crystals. The diffraction effects are observed for low-energy electrons emitted, or scattered, from particular atoms in the surface layers. The electrons used may be those emitted from the atoms in photoemission processes or those inelastically scattered by the atoms so that they are incoherent with the incident low-energy beam and observed as diffuse scattering in the background of LEED patterns or in Kikuchi-patterns.[179,439,491] The electron waves coming from particular atoms or atom sites are considered to act as reference waves and are used to determine the relative phases of the waves scattered from other atoms. The use of a reference wave in this way has suggested the analogy with holographic methods and the techniques have unfortunately been described as 'electron holography.' Since, however, it is the diffraction patterns, rather than images, which are observed in these techniques, as in the case of the X-ray diffraction methods, this is a misnomer. It has been suggested[68] that a more appropriate term would be 'holographic diffraction.' Then it may be possible to avoid confusion with the low-energy-electron holographic methods which are described on pages 311 ff.

9. Application to the study of magnetic and electric fields

The phase of an electron wave passing through a thin specimen is modified by the potential field due to the ions and electrons in the material, but may also be affected by superimposed electric or magnetic fields either applied externally or resulting from local magnetization of polarization of the material as in the case of ferromagnetic or ferroelectric specimens. The magnetic or electric fields may extend into the surrounding vacuum and so be of considerable extent. The phase changes are then given by integrating along the electron paths. For an electrical field, V, and a magnetic vector potential $\vec{A}$, the phase shift can be written as

$$\varphi(x, y) = \frac{e}{\hbar} \oint V(\vec{r}, z)\, \mathrm{d}z - \frac{e}{\hbar} \iint \vec{B}(x, y)\, \mathrm{d}\vec{S} \tag{2.40}$$

If the extent of the fields in the beam direction is not too great in relation to the fineness of the detail to be resolved (e.g., about $0.3\,\mu$m for 1 nm resolution) the projection approximation can be made and the field can be considered as a phase object with a transmission function $o(\vec{r}) = \exp[-i\,\tau\,\varphi(x, y)]$.

Simple diagrams suggesting the form of the phase shift and the electron paths in a geometric-optics picture are given in Fig. 2.9, page 33, for typical cases of the electric fields in alternating domains of a ferroelectric structure with 90° domain boundaries and the magnetic fields in alternating domains with 180° domain boundaries in a ferromagnetic structure. For the electrical field case it is seen that the electrons are deflected by the component of the electrical field at right-angles to the beam path. Components of the field parallel to the beam path give no phase change if the specimens are at zero potential because the negative and positive contributions must integrate out to zero.[419,509] For the magnetic case, the beam deflections are perpendicular to both the beam direction and the magnetic field direction so that a deflection in the plane of the paper is given by a magnetic field in or out of the paper. For the in-line modes of holography, the Fresnel images of magnetic or electric fields, such as Fig. 2.10, are readily described by considering the beam deflections illustrated in Fig. 2.9. For the off-axis modes, the distributions of the phase shifts suggested in Fig. 2.9 should be considered.

In the off-axis, point-projection or TEM modes, the phase shift for ferroelectric domain boundaries, illustrated in Fig. 2.9(a), gives sudden jumps at the domain boundaries and the interference fringes in the hologram are deflected sideways as illustrated in the images obtained by Zhang et al.[509] The electric fields in the vacuum surrounding the area of a p-n junction in a semiconductor have been visualized by Pozzi et al.[110] and the quantitative studies of the electrical fields in the semiconductor at such junctions have been made by McCartney et al.[298]

In the so-called 'absolute' mode of off-axis electron holography for which one wave passes through the vacuum and the other through the specimen, the phase difference deduced from the hologram represents the integration in Eq. (2.40) carried out around the closed path made by the two paths which may be assumed to lie in the x-z plane. For a thin film of thickness t with an in-plane magnetization, the leakage of the field into the vacuum is negligible so that the phase change is given by

$$\Delta\varphi = C_0 \cdot V_0(x,y) \cdot t(x,y) - \frac{e}{\hbar} \iint B_n(x,y)\, \mathrm{d}x\, \mathrm{d}z \qquad (2.41)$$

where C_0 is a constant and B_n is the component of the magnetic field normal to the plane defined by the beam paths, i.e., in the y-direction. For a specimen of uniform composition and with negligible thickness variation, the first term of Eq. (2.41) may be neglected. Then the x-gradient of the phase is given by

$$\frac{\partial\, \Delta\varphi(x,y)}{\partial x} = -\frac{e}{\hbar} B_n(x,y) \cdot t(x,y) \qquad (2.42)$$

Thus the phase gradient gives a direct measure of the local variation of the normal in-plane component of the magnetic flux density, $B_n(x,y)$, integrated through the specimen thickness. This provides the basis for the quantitative magnetometry of thin films and small particles.

The phase changes due to magnetic fields in thin films of ferromagnetic materials 2-10 nm thick may be quite large, of the order of $10\,\pi$. From the expression Eq. (2.28) it is seen that a phase difference of $2\,\pi$ makes no difference to the intensity of the hologram. The phase map produced by the reconstruction is then 'wrapped' with repetitions of the intensity for each $2\,\pi$ phase change, as is evident, for example in Fig. 2.11. In order to make smaller variations of phase more obvious, or to detect small total phase changes, Tonomura's group has devised a method to obtain a wrapping of the image to give the effect of contouring the phase image with much smaller phase intervals.[451] An optical reconstruction is made from the hologram with a system incorporating a Mach-Zehnder-type interferometer. With a non-linear photographic reproduction of the hologram interference fringes, the diffraction pattern in the back-focal plane of the reconstruction lens contains a series of diffraction orders on each side of the central spot instead of just one order on each side as given by cosinusoidal fringes. The higher-order reflections are correspondingly more sensitive to phase changes and may be chosen to give contouring at smaller intervals of the phase change. Using this approach it is sometimes possible to detect phase changes as small as $\pi/100$. If the reconstruction procedure is carried out by computer manipulation of digitized hologram intensities, rather than by optical methods, contour maps may be produced easily with comparable sensitivity to phase change provided that the intensities are represented on a sufficiently fine scale.

The contouring of the phase distribution allows a ready appreciation of the nature of the magnetic field distributions. The phase difference between any two points vanishes when they are located on an equipotential line of an electric field ($\Delta V = 0$) or on a magnetic line of force ($\int \vec{B}\,\mathrm{d}\vec{S} = 0$). Hence the contour lines indicate the equipotential lines or the magnetic lines of force (see also chapters 7 and 8). Also the phase differences of 2π indicate $\Delta V\, t = \hbar/e$ for a thickness t, or ($\int \vec{B}\,\mathrm{d}\vec{S} = \hbar/e$), so that the electric or magnetic fields may be measured on an absolute basis.

Measurements of magnetic flux made using the off-axis point-projection holography in a STEM machine have given values within 1 or 2 % of the theoretical values, derived from the known magnetic parameters and thicknesses of thin films, with a spatial resolution approaching 1 nm.[280]

An alternative mode of off-axis holography, for use with either the TEM or the point-projection mode in a STEM instrument, is the so-called differential mode in which both of the waves interfering in the hologram plane pass through the specimen. Then the hologram records the relative phase of the two waves. For the off-axis point-projection configuration, for example, the wave leaving the specimen may be written

$$\psi(\vec{r}) = o(\vec{r}) \cdot t(\vec{r}) + o(\vec{r}) \cdot t(\vec{r} - \vec{a}) \tag{2.43}$$

Then the intensity distribution in the hologram is

$$\begin{aligned} I(\vec{q}) = {} & |H(\vec{q}) \otimes T(\vec{q})|^2 + \left|H(\vec{q}) \otimes T(\vec{q})\, e^{2\pi i \vec{a}\vec{q}}\right|^2 + \\ & [H(\vec{q}) \otimes T(\vec{q})] \cdot [H^*(\vec{q})\, e^{2\pi i \vec{a}\vec{q}} \otimes T^*(\vec{q})]\, e^{-2\pi i \vec{a}\vec{q}} + \mathrm{c.c.} \end{aligned} \tag{2.44}$$

As in the absolute mode, the $\exp(2\pi i\, \vec{a}\, \vec{q})$ terms give interference fringes of periodicity $q = 1/a$, but the modulation is a complicated function and useful reconstruction is not possible unless the function $o(\vec{r})$ is known exactly for one of the beam positions.

For the greatly-defocused form of holography used for magnetic imaging and resolutions no better than about 1 nm, it is possible to use the relationship Eq. (2.23) in the form $H(\vec{q}) \otimes T(\vec{q}) = T(\vec{q}) \cdot [o(\vec{r}) \otimes t(\vec{r})]$ so that aberration-correction of the reconstructed side-band gives

$$\psi_r(\vec{r}) = [o(\vec{r} + \vec{a}) \otimes t(\vec{r})] \cdot [o^*(\vec{r}) \otimes t^*(\vec{r})] \tag{2.45}$$

For the relatively poor resolution of most applications, the spread function $t(\vec{r})$ can be approximated by a delta function so that Eq. (2.45) becomes $o(\vec{r} + \vec{a}) \cdot o^*(\vec{r})$. For a phase difference which varies slowly over a distance comparable with the beam separation, a Taylor series expansion gives $\Delta\varphi(\vec{r} + \vec{a}) = \Delta\varphi(\vec{r}) + \vec{a} \cdot \Delta\varphi'(\vec{r}) + (a^2/2) \cdot \Delta\varphi''(\vec{r}) + \ldots$. Then, keeping only the first two terms of the expansion, it is seen that the reconstructed wave becomes $\psi_r(\vec{r}) = \exp[-i\,\vec{a}\,\Delta\varphi'(\vec{r})]$ so that the phase function is proportional to the gradient of the phase difference and gives a direct measure of the projected in-plane component of the local electric field or the magnetic flux density.

The resolution of the mapping in this differential mode can not be better than the separation of the probes on the specimen and so is not as good as for the absolute mode but may still be useful for slowly varying fields.[280] However, in special circumstances it may give relatively high resolution information. Thus, for a transmission function of the specimen which is almost constant apart from a small local perturbation, the phase difference between the two probe positions may be written $\varphi(\vec{r} + \vec{a}) - \varphi(\vec{r})$. Then the reconstruction shows the small local perturbation and its inverse separated by $\vec{a}$, and the form of the perturbation is immediately apparent. This may apply, for example, in the case of an isolated domain boundary between two domains within which the field is uniform.

10. Studies of surface structure

Applications of electron microscopy to the study of surfaces have been valuable in providing information with high spatial resolution on the morphology, crystal structure, defects and composition of the surface layers of atoms and their relation to the bulk structures.[59,503] The techniques are complementary to the STM and AFM techniques, more recently developed, in that they have comparable spatial resolution, give diffraction-based data on crystallinity and crystal defects, relate the surface structure to that of the bulk and are more amenable to quantitative interpretation. The use of holographic techniques in conjunction with the electron microscope imaging of surfaces provides the possibility for even more precise measurements of surface features or parameters.

The inherent difficulty in the application of electron microscopy techniques for surface studies is that, for most surfaces, the vacuum in standard commercial microscopes is not adequate to allow the preparation or observation of clean surfaces. A number of special microscopes have been built with ultra-high vacuum (UHV) environments for the specimen and UHV specimen preparation chambers attached, but such instruments are complicated and expensive and not commonly available. Few attempts have been made to add the further complication of a provision for holography to such an instrument. However, the principles for surface electron holography have been demonstrated in instruments that do not have UHV specimen environments, using surfaces which are not particularly sensitive to the vacuum level, such as those of noble metals or some oxides and semiconductors.

The transmission off-axis TEM mode of holography has been applied to thin crystals which have almost perfect, planar surfaces. Steps on the surfaces of such crystals give sudden changes of thickness and so of the phase changes of transmitted electrons. For example, Tonomura et al.[443] detected the small phase changes associated with surface steps on the surfaces of thin molybdenite crystals by using off-axis TEM holography with the interferometer technique and phase-shift amplification. With a 24-times phase amplification they were able to image surface steps less than 1 nm in height. The lateral resolution in directions parallel to the surface was, in this case, a few tens of nanometers. Tanji and Ishizuka[435] have extended the transmission off-axis approach to the method of imaging the edge profiles of thin crystals with atomic resolution developed by Marks and Smith.[281]

Greatly increased sensitivity to most types of surface structure or defects is achieved, at the expense of some loss of resolution, by using the electron microscope in the reflection mode with the incident beam striking the surface at a grazing angle of incidence. Reflection electron microscopy is analogous to dark-field transmission microscopy in that a strong diffracted beam is directed down the axis of the objective lens and used to form the magnified image. In the reflection case, the diffracted beam is chosen from the reflection high-energy electron diffraction (RHEED) pattern. The angles that the incident and diffracted beams make with the surface are normally 2 or 3 times 10^{-2} radians for 100 keV incident electron energy so that the image of an almost flat surface tends to be foreshortened by a factor of 30 or 50 and the resolution is worse in the beam direction by a corresponding factor. The resolution in the direction at right-angles to the beam in the surface, however, is comparable with that for normal TEM imaging and has been reported to be as good as 0.3 nm.[503] The sensitivity of REM, or the corresponding scanning mode, SREM, obtained in a STEM instrument,[59] to surface structure is illustrated by the fact that single-atom-high steps on the surface are readily visible as strong dark, bright or dark-light lines and changes in the structure or

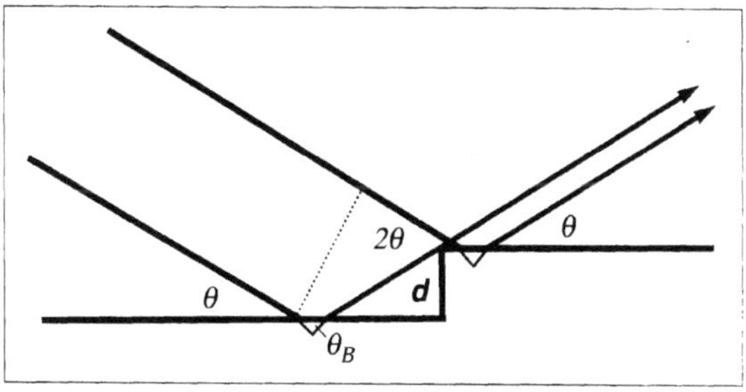

Figure 2.15. Geometry of ray-paths for deriving the phase difference of waves diffracted from the two sides of surface step in the reflection electron microscopy geometry.

composition of the topmost layer of atoms on the surface can give strong contrast.

The difference in phase between waves reflected from the two sides of a surface step can be derived readily from the diagram, Fig. 2.15. The step on the surface has a height, d. The angles made by the incident and reflected beam with the surface are assumed to both be θ. From the geometry of the figure it is seen that the phase difference of the waves reflected before and after the step is

$$d \cdot \frac{2\pi}{\lambda} \cdot \frac{1 - \cos(2\theta)}{\sin\theta} \approx 4\pi\, d \cdot \frac{\theta}{\lambda} \tag{2.46}$$

If θ is a multiple of the Bragg angle for diffraction from planes of spacing d parallel to the surface, then, in the small angle approximation, $\theta = m\,\lambda/2d$, where m is an integer. Then Eq. (2.46) gives a phase change of $2\pi\, m$, and the image contrast is zero. However, for diffraction at the Bragg angle, θ_B, of waves within the crystal, the angle of incidence may be much smaller than θ_B because of refraction at the surface. Also the step height d may be different from the interplanar spacing in the crystal because there is an expansion or contraction of the planar spacing at the surface. Also the reflection may be made with some deviation from the Bragg angle. Hence the phase difference Eq. (2.46) may be large and may vary from a multiple of 2π by any amount.

In order to detect and measure the phase difference due to surface steps and other features on a crystal surface by holographic methods, it is not usually feasible to use a reference wave which passes through the vacuum outside the crystal and so makes an angle of greater than θ with the reflected wave. Instead the separation of the image and reference waves is made to lie in the plane of the crystal face and it is arranged that the reference wave is reflected from a portion of the crystal face which is perfect and featureless. Then the only phase differences are those which arise from the surface detail within the imaged area.

Fig. 2.16 was obtained in this manner, using an interferometer to contour the phase difference distribution, by Osakabe et al.[333] The image shows the strain field around the point of emergence of a screw dislocation on the (110) face of a crystal of GaAs. The displacement of a fringe by the distance between fringes corresponds to a change of surface height by 0.05 nm. On the top left side there is a surface step of height 0.2 nm ending in the dislocation core. On the right-hand side, the bending of the fringes indicates the gradual change of height associated with the strain field which is not uniform but is concentrated along the [1,-1,2] direction. The total phase shift at the step is 3.8 λ and it is estimated that the vertical displacements are measured

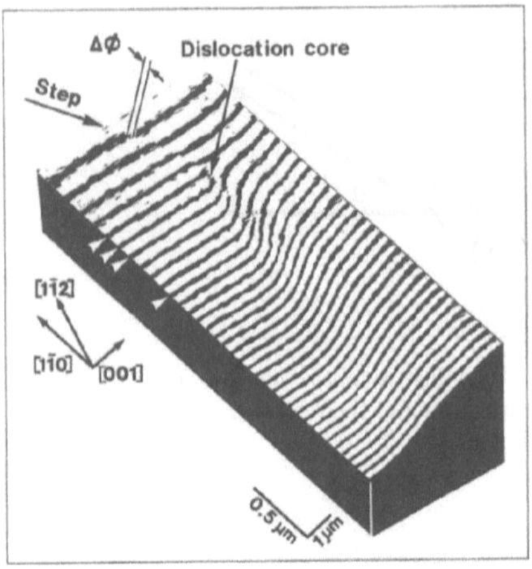

Figure 2.16. Contoured off-axis hologram by reflection from the (110) face of a GaAs crystal showing the strain field around the emergence of a screw dislocation (Osakabe et al.[333]).

with an accuracy of 10^{-3} nm. Similar methods have been used by Banzhof et al.[14] to measure step heights on metal single crystals. A discussion of reflection holography at low voltages can be found on pages 311 ff.

11. Discussion

In this brief review, we have attempted to provide a background and a basis for understanding for the more detailed and specialized articles which follow. We have also mentioned a few topics which are not given a detailed treatment elsewhere in this book but might be of interest for some particular readers. The subsequent chapters will usually provide detailed discussions of the difficulties experienced in the practical realization of the various holographic techniques and also indicate their potential value for various fields of research. However, it may be appropriate here to mention a few of the factors which have general significance in that they influence the possibilities and limitations of several, or all, of the approaches to electron holography.

For all the holographic modes that we have mentioned, the reconstruction process requires a knowledge of the instrumental parameters of the system used to generate the hologram. If it can be assumed that the astigmatism of the microscope has been adequately corrected and that the effects of non-axial and higher-order aberrations can be neglected, the important parameters to be determined are the defocus and the third-order spherical aberration coefficient (although this coefficient may be of sub-nanometer dimensions for the virtual source inside a nanotip field-emitter, it must still be considered, as discussed on pages 311 ff.). The precision with which field strengths can be measured and the extent to which resolution can be enhanced depend on the accuracy with which these parameters can be determined. In particular any appreciable enhancement of resolution depends on the determination of these parameters with a much greater accuracy than is required for normal high-resolution imaging and the standard methods used for deriving values for the parameters are no longer adequate. It has been suggested by Lehmann and Lichte,[249] for example, that to achieve 0.1 nm resolution with a 300 keV microscope having $C_S = 1$ mm, it is necessary to know the value of C_S with an accuracy of a small fraction of 1% and the defocus must be known to within a few nm.

Various methods have been suggested whereby the required precision may be attained. Methods involving the use of high-angle diffracted beams are suspect in that they may be affected by higher-order aberrations. Refinements of the technique involving the Fourier transform of images of thin amorphous phase objects may give adequate results[84] but depend on the presence of amorphous material on or around the specimen and this may not be desirable in practice. Alternative approaches have been suggested which rely on some foreknowledge of the form of the specimen. e.g.. that it is a single crystal viewed in a principal orientation so that the image should consist of isolated atomic peaks with low-potential areas between them.[153]

A further important limitation to the achievement of resolution enhancement may come from the practical consideration of the area and number of pixels of the recording medium. On this basis, Lichte[268] suggests that with the CCD recording systems now available or expected to be developed in the near future, the attainment of 0.1 nm resolution is impossible for 100 keV electrons but should be feasible for 300 keV electrons. One interesting aspect of off-axis holography is that it may act as a filtering process to eliminate the contributions of inelastically scattered electrons to TEM images. Since electrons that have undergone an inelastic scattering process are incoherent with the elastically scattered electrons or with each other, they can not contribute to the interference effects which modulate the hologram fringe pattern but provide only a diffuse background to the intensity distribution. Thus the side-bands in the diffraction pattern formed in the reconstruction process and the reconstructed phase and amplitude images come from the elastically scattered electrons only. This filtering effect may be of importance for improving the contrast of images of thick specimens and is of value even for thin specimens since the theoretical simulation of image intensities is much more easily done if only elastic scattering need be considered. It has been suggested that the filtering of inelastically scattered electrons may apply even for energy losses of much less than 1 eV[268] and may be effective for removing the contributions of thermal-diffuse scattered electrons.[489] However, arguments have been advanced[70] which suggest that the filtering can not be effective for energy losses which are less than the value of the energy spread of the incident beam.

Because of the simplicity of the mathematical description, there is a tendency to discuss the recording of holograms and the reconstruction processes in terms of the weak-phase-object approximation (WPOA). However, it must be recognized that the WPOA is applicable for very few of the experimentally interesting specimens. In particular, when the aim is the enhancement of resolution, the preferred specimens are thin single crystals of known structure held in a principal orientation such that the incident electrons are directed parallel to rows of atoms. Then the reconstructed image should give a recognizable projection of the structure and the success of the reconstruction procedure can be immediately apparent. But crystals which are sufficiently thin and are also resistant to radiation damage by the incident beam almost invariably contain elements of moderately large atomic number and the WPOA fails badly, especially for the principal orientations. What can be deduced in such cases is the exit wave-function for the specimen and this is strongly affected by dynamical diffraction effects so that it has no simple relationship to the crystal structure. The exit wave-function for a crystal can, of course, be calculated with high accuracy for a known structure by making use of any of the computer programs based on various formulations of the many-beam dynamical diffraction theory.[71,408] Arguments can be made that, for the case of crystals in principal orientations, the incident electrons tend to be channeled to flow along rows of closely-spaced atoms of moderate or high atomic number so that, to a reasonable approximation, the exit wave-function gives a non-linear representation of the projection of the structure.[461] This may be a very useful result in some cases. In general, however, there is no method yet devised for inversion of the dynamical diffraction process so that the crystal structure may be derived from the exit wave-function. The method of holographic reconstruction offers the advantage over the use of normal HREM images, apart from any improvement of resolution, that more experimental information is provided for comparison with theoretical results. Both the phase and amplitude images are provided instead of just the image intensities.

Interpretation of the wave-function must rely on a trial-and-error process of fitting the experimentally derived function with calculations for model structures made, preferably, for a range of values of the relevant parameters. Apart from the parameters describing the relative positions, ionicity and thermal vibrations of the atoms in the crystal, the most significant parameters for which accurate values are needed include the crystal thickness, the deviations from the exact axial orientation and the alignment of the incident beam with respect to the symmetry axis of the objective lens. On the other hand, as pointed out by Lichte,[261] the application of holographic methods for crystals may be of value for the investigation of dynamical diffraction effects. Verification of the phase variations predicted by dynamical diffraction theory may be of value in confirming features of the specimen such as thickness variations, or the appropriateness of the assumptions regarding the diffraction conditions such as, for example, the question of whether a relatively simple two-beam diffraction theory is applicable in a particular case. Also, from the exit wave-function for a specimen it is possible to derive the diffraction pattern which would be given by any chosen area of the specimen, no matter how small. Thus nanodiffraction patterns can be derived for parallel-beam incidence rather than for the convergent-beam conditions which are experimentally feasible with a STEM instrument.[74]

11.1. Conclusions

It will be clear from our discussions, and even more from the following chapters, that the methods of electron holography have shown considerable promise. Many aspects of established techniques and many possibilities for new techniques have barely been explored and a great deal remains to be done before the full potential of the methods can be realized. The applications to the study of magnetic field distributions are probably the most thoroughly developed, being almost at the stage of direct application to production-line problems of the magnetic recording industry. The attempts at resolution enhancement are beginning to show real progress. The recent development of methods for the use of very low-energy electrons appears to be rather more complicated than originally imagined but with improved theoretical analysis and further experimental experience, may have an exciting future. Results of this technique in the reflection geometry are just beginning to appear. Other proposed techniques for use of fast electrons have yet to be explored experimentally. The subject of electron holography appears to be alive and well.

Acknowledgments

Supported by NSF award DMR9526100. We are grateful to Dr. E. Völkl for the images of Fig. 2.7, to Drs. M. Mankos and M. Scheinfein for Figs. 2.10 and 2.11 and to Dr. Osakabe for Fig. 2.16. We are grateful to Drs. M. Mankos and M. Scheinfein for Figs. 2.10 and 2.11 and to Dr. Osakabe for Fig. 2.16.

OPTICAL CHARACTERISTICS OF AN HOLOGRAPHY ELECTRON MICROSCOPE

L.F. Allard and E. Völkl

High Temperature Materials Laboratory, Oak Ridge National Laboratory, Oak Ridge, TN 37831-6064

1. Introduction

The growth in popularity of the technique of electron holography is primarily a result of the availability of a new generation of electron microscopes that operate with field emission electron sources of various types. While many important results were obtained by early workers in the field[224,311] utilizing standard thermionic emitters or pointed filaments to obtain a measure of increased brightness and coherence, the field emission source provided a large jump in capability for making interferograms with high contrast fringes, necessary for useful reconstructions of amplitude and phase components of an image. A significant number of TEM or STEM instruments, operating at 200-300 kV with either cold field emitters or field-assisted thermal emitters (Schottky emitters), are now extant. All of these instruments have the potential, with the installation of a simple biprism device, to make functional holograms that can provide information about a sample that is not available with any other technique.

The theory of electron holography, a description of many of the holography techniques currently being developed in the field, and numerous practical examples of the application of these techniques for materials characterization are covered in detail in this textbook. Also, a comprehensive description of the practical aspects of making electron holograms in both TEM and STEM instruments is given in the following Chapter 4. In the present chapter, we will discuss how the optical characteristics of the field emission microscope differ from the conventional thermionic emitter TEM, so that a beginning holographer can better understand his or her instrument and utilize it to produce the best and most reliable holography results. Because of the authors' experience, details of the electron optics and operation of the Hitachi HF-2000 cold field emission TEM will be used as a representative instrument. While there are a number of important differences between cold field emitters and Schottky emitters that may affect general operational results from a field emission microscope, a relative few of these are of importance in electron holography. Consequently, a comparison of the advantages and disadvantages of these emitter types is not relevant to this treatment; only in the event that there is a particular advantage of one over the other for making electron holograms will these differences be mentioned.

1.1. Topics to be covered

In the following Sections, we review details of the electron optical characteristics of the field emission electron microscope that are most important to the acquisition of high quality electron holograms. These include brightness, coherence effects, illumination geometry, isoplanatism and coma, contrast transfer, and finally electron optical tests for quantifying the capabilities of the microscope to make holograms. It is our intent to provide a general understanding of the optics, performance and operational aspects of the microscope that will be of most immediate use. Full mathematical treatments of most of the topics have been published; they will be only briefly reviewed here. This chapter will therefore provide the basis for the practical acquisition of holograms, described in Chapter 4.

2. Brightness

One of the most prominent differences between field emitters and thermionic sources is the characteristic called 'brightness,' β. For a given accelerating voltage, brightness is most directly controlled by source size; that is, the effective size of the crossover of the electron beam at the gun for thermionic emitters, or the size of the 'virtual' source in the case of cold field emitters.[173,496] Brightness increases with a decrease in the size of the source, and the size of the source is related to the size of the emitting area.

Field emitters that emit electrons strictly from the action of a strong electric field are called 'cold' field emitters. These are generally formed from single crystals of tungsten etched to create a tip with a radius of $0.1\,\mu$m or less, having a $\langle 310 \rangle$ axial orientation. In comparison, thermionic emitters such as hairpin tungsten filaments have radii of $100\,\mu$m, LaB_6 emitters have emitting suface flats of $6\,\mu$m, and Schottky emitters have tip radii on the order of $0.5\,\mu$m.[177,432] Thus the physical size of the field emitter plays a large role in the inherent brightness of this gun type. The sharpness of the cold field emitter tip allows electrons to be pulled out of the tip simply by the strong field effect generated by a positive voltage placed on an extraction electrode. This type of gun is effectively approximated by a diode geometry.[175] The strength of the field is on the order of $5 \cdot 10^7$ V/cm.[361,412] In comparison, a thermionic emitter relies on the extra energy input typically by ohmic heating to overcome the work function of the emitter material so that electrons are emitted. The first electrode is negatively charged with respect to the emitter to control the flow of electrons.[174] The Schottky emitter, with its larger tip radius of curvature, sees about one-tenth the field strength as the cold field emitter and thus requires heating to 1800 K for electron emission. It is therefore considered to be a "field-assisted" thermionic emitter.

Taken to an extreme, the ultimate geometry for a field emitter is the so-called 'nanotip' emitter, where a single atom from which all the electrons are emitted resides at the tip. Such nanotip emitters are not yet commercially available, but several studies have shown that they can be successfully produced, and that an increase in brightness of two or more orders of magnitude over conventional field emitters can be expected.[107,298,347]

Figure 3.1 shows two equivalent magnification SEM images, to illustrate the differences in geometry between the 3 most common emitter types. A standard hairpin tungsten filament (Fig. 3.1a) serves as the support for the etched tungsten single crystal field emitter and shows the several orders of magnitude difference between the radii of curvature at the tips (and therefore the emitting areas) of these two emitter types. The

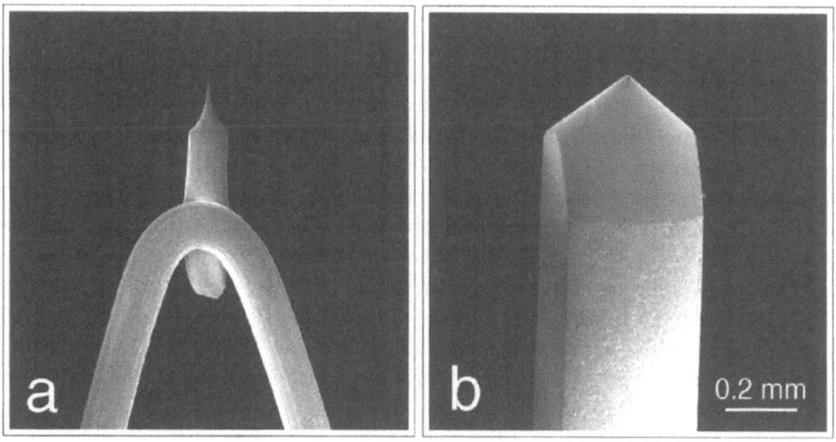

Figure 3.1. a) Typical field emitter, comprising an etched single crystal tungsten wire spot-welded onto a standard tungsten hairpin filament such as is used in a conventional microscope. b) One type of LaB_6 emitter, having a 60° cone angle with a flat surface 5 μm in diameter ground on the tip. (Figure courtesy D.W. Coffey, Oak Ridge National Laboratory)

primary emission surface of a typical LaB_6 emitter (Fig. 3.1b) is the small flat ground on the emitter tip. Properly conditioned Schottky emitters also form facets at the tip, as illustrated at high magnification in Ref. 432.

Brightness is strictly defined as the current density per unit solid angle of the source. For a thermionic gun, the source can be considered to be the real crossover in front of the emitter tip. Relative to the crossover of diameter d_0, if the electrons are emitted in a solid angle defined by the semi-angle α_0, and the emission current is i_e, then the brightness β is

$$\beta = \frac{4\, i_e}{(\pi\, d_0\, \alpha_0)^2} \tag{3.1}$$

with SI units of $Am^{-2}\,sr$, where sr = steradians.

For a constant emission current and at a constant accelerating voltage, the brightness is a constant, and, neglecting aberration effects, does not vary at conjugate image planes through the column.[99,360,467] However, because it is difficult to measure absolute brightness, even in devices designed for the task,[176,399] it is practically useful to consider an effective brightness equation related to Eq. (3.1) that has parameters directly measureable in the TEM

$$\beta_{eff} = \frac{4\, i_p}{(\pi\, d_p\, \alpha_p)^2} \tag{3.2}$$

Here i_p, d_p and α_p are, respectively, the probe current at the specimen, the diameter of the focussed probe (typically taken as the FWHM diameter) and the convergence semi-angle of the probe as defined by the condenser aperture. β_{eff} is always less than β, but, as is the case in TEMs, when the aperture is far from the specimen, $\alpha_p \sim \alpha_0$ and β_{eff} is controlled by aberration effects, which may be small for small α_p. Hence, it is typical to compute the brightness for a given electron microscope setup using the factors in Eq. (3.1), measured by methods detailed, e.g., in Ref. 497.

When the Hitachi HF-2000 cold FE-TEM is operated in normal imaging mode (versus the fine probe or analysis mode) at 200 kV and 20-30 μA emission current, a probe current of $5 \cdot 10^{-9}$ A is typically obtained using a 300 μm condenser aperture (6 mrad convergence semi-angle). The beam diameter (FWHM) obtained with these conditions is 2.3 nm. Substituting these values in Eq. (3.2) gives an effective brightness for the HF-2000 in imaging mode of $1.1 \cdot 10^{13}\,\mathrm{Am^{-2}sr^{-1}}$. With a nominal 100 μm aperture inserted that gives a 2 mrad convergence angle, the current in the probe drops to 0.6 nA. Because the spot size at the specimen remains unchanged at 1 nm, the effective brightness of is essentially the same, as expected.

The numbers are significantly different for a thermionic emitter. As an example, in the JEOL 4000EX TEM using a LaB_6 emitter (a microscope for which we have careful measurements), a probe diameter of 0.2 μm (Spot Size 3) is obtained using a condenser aperture that gives 1.4 mrad convergence semi-angle, at a probe current of $1.95 \cdot 10^{-9}$ A, which yields an effective brightness of $1 \cdot 10^{10}\,\mathrm{Am^{-2}sr^{-1}}$, three orders of magnitude lower than the cold field emission gun.

The brightness of an electron beam is also related directly to operating voltage,[360] so that as the voltage is increased, the brightness is increased. Therefore, there is an advantage in moving from a 100 kV instrument to the 200-300 kV or even higher voltage instruments, to achieve higher brightness operation. This is most easily seen in the derivation of Langmuir's relation.[146,244] If the beam energy is represented by components v_L and v_P normal to and parallel to the electron optical axis, then the lateral component is given by $m\,v_L^2/2 = 3\,k\,T/2$, and the parallel component satisfies $m\,v_p^2/2 = e\,U_A$. The beam angle $\alpha_0 = v_L/v_p = (k\,T/e\,U_A)^{1/2}$, and the solid angle of the beam exiting the gun is $\pi\,\alpha_0^2$. Since β is the current density/unit solid angle, we get

$$\beta = (J/\pi\,k\,T)(e\,U_A) \tag{3.3}$$

showing the direct relationship between brightness and accelerating voltage. In fact, at higher accelerating voltages, brightness is affected significantly by relativistic effects, and U_A in Eq. (3.3) must be replaced by $U = U(1 + e\,U/2\,m_0\,c^2)$. We note that some publications have implied that a brightness increase with voltage is true only for thermionic emitters; however, it is actually independent of the type of electron source.

Brightness of the electron source is also an important instrumental factor controlling fringe contrast in electron holography. As pointed out by Lichte,[258] and discussed in Chapter 9, the contrast of interference fringes in an electron hologram is directly related to the coherent current provided by the illuminating system, which is controlled directly by the brightness of the gun

$$I(\mu) = -\beta\,\ln(\mu)/(2\pi/\lambda)^2 \tag{3.4}$$

where μ is the contrast and β the brightness. Thus, when operating with an illuminating system with a given set of parameters, a gun with higher brightness will yield a higher coherent current at a given level of fringe contrast, and thus exposure times for the hologram can be reduced. Similarly, if the illuminating system is manipulated so that the exposure time stays constant, higher fringe contrast will be obtained in the hologram. Since the CFE gun at present routinely provides the highest brightness of commercially available sources, in the absence of other factors influencing the recorded fringe contrast, it is expected that higher ultimate fringe contrast can be obtained with this emitter type. It must be emphasized, however, that many factors combine to affect the contrast of holograms fringes in any given instrument. These are discussed in Section 7 of this chapter, where a standard method is proposed for quantifying the contrast of fringes that can be recorded. This method makes it possible to monitor the performance over time of a given microscope for hologram recording, or to compare the performance of two or more microscopes for electron holography.

As a final note, it is useful to address the differences in operating geometries between the thermionic electron gun and the field emission electron gun. The negatively biased Wehnelt cap in a standard thermionic triode electron gun produces a physical (real) crossover of the electron beam at a point in front of the emitter tip, as illustrated in Fig. 3.2.[198] A virtual source of the beam is formed at a point between the tip and the real source; the position of the virtual source is controlled by the bias on the wehnelt. As bias is decreased, the virtual source moves toward the real source. At higher bias, the virtual source moves toward the tip of the filament and the emitting area at the tip decreases, until a bias voltage is reached at which the emitting area reaches zero and the beam is effectively cut off. The bias voltage at cut-off is determined by the height of the filament within the wehnelt, and the size of the hole in the wehnelt. For a given geometry, a certain combination of gun bias and filament height produces a maximum gun brightness.

For a CFE, however, because of the positive bias of the diode geometry, the electrons can be considered to be emitted from a virtual source at some point behind the actual tip of the emitter,[176] as shown in Fig. 3.3. The location of this source can be controlled by appropriately choosing the values of the primary extraction voltage (V1) on the extractor electrode, and the voltage (V2) on a focusing electrode. The combination of two voltages acts as an electrostatic lens, so that the ratio V2/V1 defines the operating conditions for the illumination. A set of electrodes after the electrodes that extract the beam and define the source point serve to accelerate the beam to the desired operating voltage. A typical value for the V1 voltage is 4 kV, and the V2/V1 ratio can be varied over a useful range of 4.5 to 7.0 or so. The higher ratio effectively moves the virtual source farther from the emitter tip and provides more beam intensity at the expense of a slight degradation of energy spread. For experiments such as energy filtered imaging, beam currents as high as 40 nA at the sample can be achieved, at V2/V1 ratios in the 6.8-7.0 range. The best conditions for electron holography require, however, choice of a ratio that optimizes beam brightness and intensity so that high contrast fringes can be obtained. In our experience, V2/V1 ratios in the range of 5.5-6.0 give beam currents of 2-5 nA at the specimen and allow holograms to be recorded at 1000 kX magnification with 5-10 sec exposures that give fringe contrast in the 20-30% range. Methods outlined in Section 7 can be used to determine the effects of varying operating parameters on the quality of the holograms that can be recorded.

3. Coherence and contrast transfer

The second factor that gives a field emission microscope a major advantage over one with a thermionic emitter is the 'coherence' of the electron beam. Since the beam coherence plays an important role in the formation of electron holograms, it is useful to review here some of the basic concepts of coherence in order to better understand the factors that control the quality of the hologram. Coherence also plays a strong role in the contrast transfer characteristics of the microscope, and thus in the degree to which holography can be used to improve microscope resolution, as we shall see later. For a full treatment of the concepts of coherence in the electron microscope, see Spence, Chapter 4.[410]

Coherence can be described in terms of two important concepts: 'spatial coherence'

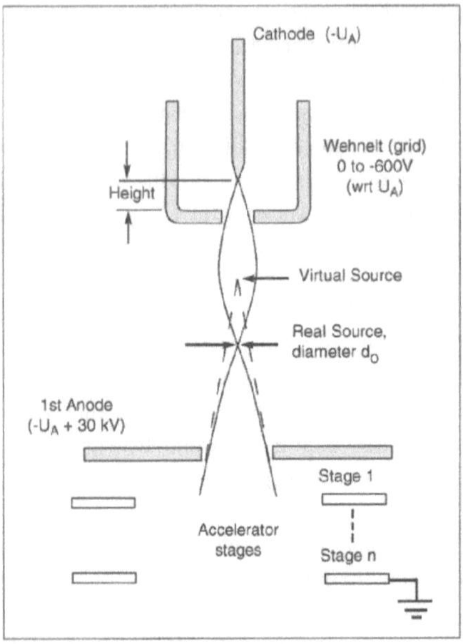

Figure 3.2. Schematic of the geometry of the beam of the triode electron gun of the thermionic emitter. The cathode or emitter is held at $-U_A$. The Wehnelt is biased negatively with respect to the cathode, and the first anode is typically 20-30 kV above $-U_A$. A real crossover, or source, is formed in front of the Wehnelt, giving a virtual source, also in front of the Wehnelt.

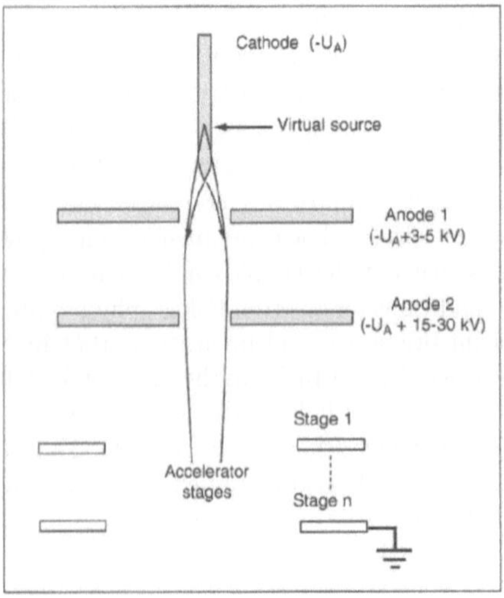

Figure 3.3. Schematic of the geometry of the diode gun of the field emission microscope. The extractor electrode (anode 1) is biased strongly positive with respect to the cathode tip, and the strong field at the tip causes electron emission. The effective source of electrons is a virtual source positioned behind the cathode tip. Control of the voltage on the focusing electrode (anode 2) versus the voltage on the extractor electrode governs the position of the virtual source.

and 'temporal coherence.' Spatial coherence is associated with the size of the electron source, while temporal coherence is associated with instrumental effects such as energy spread in the beam from the electron emitter, and instabilities in the high voltage and objective lens current which lead to deviation from monochromaticity of the beam. The illumination condition can range from incoherent, to partially coherent, to coherent, dependent upon the physical geometry of the emitter, the size of the illumination aperture, and the instrument instabilities.

In considering coherence effects, it is useful to define an 'effective' source of illumination in the electron microscope. This source can be considered to be the illuminating

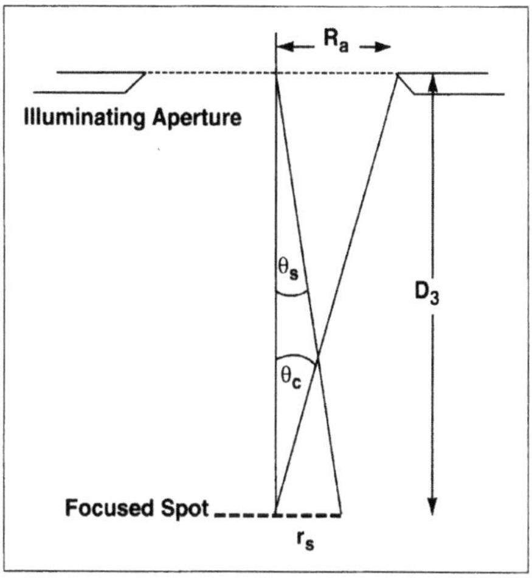

Figure 3.4. Illumination geometry in the transmission electron microscope, with parameters defined that permit calculation of the degree of coherence in the illuminating aperture. (Adapted from Spence.[410])

aperture of the condenser lens, within which each point represents a point source of electrons.[410] A 'coherence width' X_c can be defined that is related to the size of the condenser aperture.

$$X_c = \lambda/(2\pi\,\theta_C) \tag{3.5}$$

Here θ_C is the beam convergence semi-angle, that is, the half-angle subtended by the condenser aperture at the specimen, which is controlled by the aperture size, as shown in Fig. 3.4. The physical concept is that X_c represents a distance in the specimen plane over which the incident illumination can be considered to be perfectly coherent. In like fashion, an 'incoherence width' X_i can be defined as a distance for which larger periodicities are incoherently illuminated and their scattered beams will not interfere to produce a periodic image.

$$X_i = \lambda/(2\pi\,\theta_C) \tag{3.6}$$

This defines the range of periodicities in the sample for which the high resolution image must be interpreted by the theory of partial coherence.[410] In the case of incoherent illumination, there is no correspondence between the collection of source points at the aperture and points in the illuminated area on the specimen.

Note that Eq. (3.5) assumes that the condenser aperture is incoherently illuminated. To determine if this is true for real microscope operating conditions, we consider the semi-angle θ_S subtended by the focussed spot at the illuminating aperture position and, as illustrated in Fig. 3.4 (neglecting OL prefield effects)[410] and define a coherence width X_a in the illuminating aperture as

$$X_a = \lambda/(2\pi\,\theta_S) \tag{3.7}$$

Now we look first at the effects of the illumination system on the degree of beam coherence at the sample, and then at how the illumination system controls the degree to which the condenser aperture is coherently illuminated. We compare the results from the JEOL 4000EX having a LaB_6 emitter with the HF-2000 FE-TEM, as typical examples.

For the JEOL 4000EX operating at 400kV ($\lambda = 0.00167\,\text{nm}$) with a $200\,\mu\text{m}$ condenser aperture, $\theta_c = 1.4\,\text{mrad}$, so from Eq. (3.5), $X_c = 0.19\,\text{nm}$, and $X_i = 1.19\,\text{nm}$. We see that the distance in the sample over which the illumination is coherent barely exceeds the point resolution limit of 0.17 nm for the 4000EX, and so periodicities exceeding 1.19 nm are incoherently illuminated and can not produce periodic contrast in the image.

How well does the $200\,\mu\text{m}$ condenser aperture in the 4000EX approximate an incoherent illumination source? To calculate X_a, we need θ_S. The distance D_3 from the condenser aperture to the specimen on the 4000EX is 16 cm, and for a focused spot radius of $0.1\,\mu\text{m}$, θ_S is 0.00063 mrad. This gives $X_a = 0.43\,\mu\text{m}$, a value very much smaller than the $200\,\mu\text{m}$ diameter of the aperture. So the aperture is incoherently illuminated, and thus the thermionic source of the 4000EX provides incoherent illumination at the specimen. Contrast for images over the range of periodicities between 0.19 nm and 1.19 nm must be described by the theory of partial coherence.

We can calculate similar numbers for the illumination geometry of the HF-2000 to compare the degree of coherence at its illuminating aperture with that of the 4000EX. On the HF-2000, $D_3 = 25.4\,\text{cm}$, $r_s = 1.15\,\text{nm}$ in normal imaging mode using a $300\,\mu\text{m}$ aperture giving $\theta_S = 4.53 \cdot 10^{-6}\,\text{mrad}$, and $X_a = 88\,\mu\text{m}$. This is much larger than the value of X_a for the 4000EX, but still smaller than the diameter of the condenser aperture, so the illumination at the aperture plane is not fully coherent. However, when the $100\,\mu\text{m}$ condenser aperture is used, $r_s = 0.5\,\text{nm}$, giving $\theta_s = 1.97 \cdot 10^{-6}\,\text{mrad}$, and the coherence width $X_a = 203\,\mu\text{m}$. So in this case the condenser aperture is coherently illuminated, and even the fully focused spot at the specimen is fully coherent, at least in terms of spatial coherence. However, because of energy spread considerations, temporal coherence, as characterized by the coherence length, still plays an important role in the extent to which contrast is transferred by the objective lens.

From the above, we see that it is reasonable to describe the way contrast is transferred in both thermionic and field emission microscopes using the theory of partial coherence, which means that the effects of both partial spatial coherence and partial temporal coherence on the contrast transfer function must be considered.

3.1. Contrast transfer considerations

Since in this chapter we are interested only in the optical characteristics of the microscope and how the objective lens transfers contrast from the specimen to the image, we will be concerned only with the behavior of the electron wave after it has interacted with a very thin specimen, or so-called 'weak phase object.' For completeness, we review briefly the pertinent equations from Chapter 2 that lead to the computations of the contrast transfer function of the microscope. Recall that the objective lens of the microscope first transfers the wave that exits from the thin sample first to the back focal plane (BFP) of the lens, where the amplitudes are given by the Fourier transform of the exit wave. To obtain the image wave, in the case of a perfect lens with no apertures or aberrations, the wave function is given simply by an inverse Fourier transform of the wave in the BFP. However, aberrations must be accounted for in the theory, and they are handled by multiplication of the wave in the BFP by a phase factor $\exp[i\chi(\vec{q})]$, where (from Eq. 2.9, page 38)

$$\chi(\vec{q}) = \pi\,\Delta z\,\lambda\,q^2 + \pi\,C_S\,\lambda^3\,q^4/2 \tag{3.8}$$

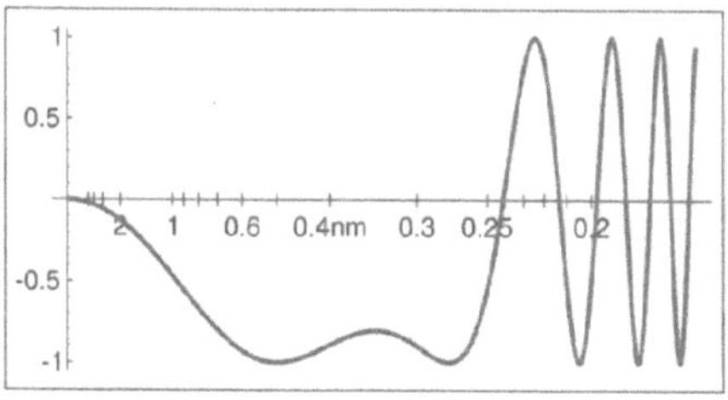

Figure 3.5. Contrast transfer function (CTF) computed for a 200 kV operating voltage, with C_S = 1.2 mm, C_C = 1.4 mm, Scherzer focus Δz = -65 nm. Assumes no aperture in OL and no damping by spatial or temporal coherence effects. The point resolution is nominally given as the first crossing, or 'first zero' of the transfer function, in this case at 0.24 nm.

and q represents distances in Fourier space, or reciprocal dimensions in diffraction space, and Δz and C_S are defocus and spherical aberration parameters, respectively.

In it simplest form (i.e., when no additional aberration effects are included), the 'transfer function' of the objective lens is

$$T(\vec{q}) = B(\vec{q})\, e^{i\,\chi(\vec{q})} = B(\vec{q})\,\{\cos[\chi(\vec{q})] + i\,\sin[\chi(\vec{q})]\,\} \qquad (3.9)$$

where $B(\vec{q})$ is a function included to account for the presence of an aperture in the BFP, if one is used. The aperture function is simply 1 for values of q less than the reciprocal dimension subtended by the objective aperture (or if no OL aperture is used), and 0 for values beyond the edge of the aperture.

To determine experimentally the contrast transfer characteristics of the microscope, a very thin specimen must be used which satisfies the 'weak phase object approximation' (WPOA), as nearly as possible. With the WPOA, only the sine of the transfer function is involved in producing image intensity (see Eq. 2.16, page 39), so the transfer function of the objective lens reduces to

$$T(\vec{q}) = B(\vec{q})\,\sin[\chi(\vec{q})] \qquad (3.10)$$

The function $\sin[\chi(\vec{q})]$ accounts for effects of spherical aberration, but ignores effects of astigmatism, which are assumed to be correctable, and the effects of higher order aberrations such as axial coma, distortion, etc. The important aberrations influencing imaging in the FE-TEM are discussed in more detail later in Section 5, and in Chapter 9.

The optimum transfer of information between the object and the image is obtained when the focus is set to a particular value (the Scherzer focus) such that the sine of the function $\chi(\vec{q})$ has a value near 1 over as large a range of values of q as possible.[61]

Figure 3.5 shows a plot of the contrast transfer function at Scherzer focus ($\Delta z = -1.2\sqrt{C_S\,\lambda}$) for a standard 200 kV microscope, but with the caveat that there is no contribution to contrast transfer from either spatial or temporal coherence effects; that is, it is assumed that perfectly coherent illumination is used. For these calculations, C_S was chosen to be 1.2 mm, the same as for the analytical pole piece of the HF-2000. The horizontal axis is typically labelled in reciprocal units (q, or spatial frequency units), but in this and the remainder of the CTF plots in this chapter, it will be labelled in corresponding real space dimensions (nm), for ease in relating the CTF features to actual distances or spacings in the specimen. In this illustration, it was assumed that no objective aperture was used, so the aperture function was taken as 1 over the full range of spatial frequencies.

This transfer function plot shows the broad transfer band up to the 'first-zero' crossing of the q-axis at about 0.24 nm, which is considered the Scherzer[392] or point resolution of the microscope at $r_{Sch} = 0.66\,C_S^{1/4}\,\lambda^{3/4}$. Within the range of spacings from about 1 nm to the Scherzer resolution, the contrast transferred has a nearly constant negative value. That is, for a weak phase object, diffracted beams from crystal spacings that fall within this 'Scherzer passband' (i.e., from spacings greater than 0.24 nm) will yield 'black atom' contrast in the image. At spatial frequencies positioned near the crossover points at and beyond the Scherzer limit, no information will be transferred, and the corresponding spacings will not be visible in the image. Diffracted beams from crystal spacings within the first band beyond the Scherzer crossover (i.e., spacings from about 0.24 to 0.21 nm) where the transfer function is positive will yield 'white atom' contrast. This contrast reversal occurs alternately for each crossover of the CTF, at increasingly smaller spacings.[362] In this range where the transfer function oscillates so frequently (which, as we shall see, is typical in field emission microscopes), the spacings producing a given diffracted beam may coincide with either a maximum or a minimum in the transfer function, and so it is extremely difficult to interpret an image which shows contrast from spacings below the Scherzer resolution limit.

As indicated earlier, the effects of limited spatial and temporal coherence, which are present in real microscopes, are incorporated into contrast transfer calculations by the introduction of the concept of 'envelope functions.' Spatial and temporal envelope functions serve to dampen the transfer function at higher spatial frequencies.[111] The spatial envelope function $K_S(q)$ (also called the 'source' envelope) is related to the effective source size and beam convergence, the latter of which is governed by the choice of aperture in the condenser system for typical thermionic emitter microscopes (in which the illumination can be considered to originate from all points in a fully-filled condenser aperture).[410] The temporal envelope function $K_T(q)$ (also called the 'chromatic' or 'instrument' envelope) is controlled by the instabilities in high voltage and lens power supplies, and to a lesser extent by external mechanical vibrations and electromagnetic fields.

The Temporal Envelope An expression for the temporal envelope function $K_T(q)$ is given by Fejes[106]

$$K_T(q) = \exp[-(\pi\,\Delta\,\lambda\,q^2)^2/2] \tag{3.11}$$

where the term Δ is defined as the 'defocus spread' given by

$$\Delta = C_C \cdot f_r \cdot \sqrt{\left(\frac{\Delta E}{e\,U_A}\right)^2 + \left(\frac{\Delta U_A}{U_A}\right)^2 + \left(2\,\frac{\Delta I_0}{I_0}\right)^2} \tag{3.12}$$

and $f_r = (1 + (e\,U_A)/E_0)/[1 + (e\,U_A)/(2\,E_0)]$ is a relativistic correction coefficient[363] ($E_0 = 511\,\text{keV}$, the rest energy of an electron), C_C is the chromatic aberration coefficient inherent to the objective lens, U_A is the high tension, ΔE is the root mean square (RMS) value of the energy spread from the emitter, given by $\Delta E = \Delta E_{FWHM}/2.355$, $\Delta U_A/U_A$ is the instability of the high tension and is given as the RMS value of the HT ripple, and similarily, $\Delta I/I_0$ is the RMS value of the objective lens current ripple. The defocus spread Δ is the variation in defocus value related to the effective change in the focal length of the objective lens caused by the various instabilities defined above, and has typical values of 3-10 nm. The C_C of the lens is directly related to focal length, and is typically slightly greater than the C_S of the lens. While C_C is a calculated value provided by the manufacturer, and contains a relativistic correction itself, the additional factor f_r is required to account for relativistic effects on the instability parameters. Note that this factor has sometimes been omitted in the literature from the defocus spread computation, but is important because it can have a significant effect, particularly at higher accelerating voltages. At 100 kV, the relativistic correction in Δ is only 9%, but this increases to 16, 23 and 28%, respectively, at 200, 300 and 400 kV.

Typical values of high voltage and lens current instabilities in modern microscopes are quoted as in the $1 \cdot 10^{-6}$ (RMS) range. These 'ripples' are important only over the time of recording of the high resolution micrograph, and may in fact be routinely lower than the quoted values, which are sometimes given over longer times. In all calculations made here (and likely by virtually all others in the literature), the specification values from the manufacturer are used for CTF calculations.

As indicated in Eq. (3.10), the relative effect of the instability of the objective lens is 4 times that of the high tension; we can therefore calculate the importance of these various factors on the total defocus spread. For a FE-TEM such as the HF-2000 with a nominal RMS HT ripple of $1 \cdot 10^{-6}$ ($= \Delta U_A/U_A$), RMS OL ripple of $1 \cdot 10^{-6}$ ($= \Delta I/I_0$), and an energy spread from the emitter of 0.5 eV at 200 kV, which yields an RMS value of $0.5/(2.355 \cdot 2 \cdot 10^5) = 1.06 \cdot 10^{-6}$, the defocus spread Δ becomes $\Delta = C_C \cdot 2.87 \cdot 10^{-6}$.

Now suppose the high tension supply stability is improved to $0.5 \cdot 10^{-6}$ (a significant amount, and perhaps difficult to achieve in practice) without changing the stability of the lens power supply. Then the defocus spread becomes $\Delta = C_C \cdot 2.74 \cdot 10^{-6}$ or a small improvement of 6.3%. Suppose rather the objective lens stability is improved to $0.5 \cdot 10^{-6}$ without changing the high voltage stability. The defocus spread then becomes $\Delta = C_C \cdot 2.05 \cdot 10^{-6}$ corresponding to a significantly larger improvement of 28.6%.

Thus it is clear that for the FEG microscope, it is important to be concerned with lens stability to a greater extent than the HT stability, in order to push the information limit achievable by the microscope to the highest spatial frequency possible. The importance of these effects on the extent of contrast transfer in both present and future generation microscopes will be detailed in Section 6.

The effect of the instrumental instabilities on the image forming characteristics of the microscope is shown by multiplying the basic CTF by the temporal envelope function. As shown in Fig. 3.6a, this dampens the CTF so that spacings smaller than about 0.13 nm are not transferred by the OL, and thus no information about spacings this size or smaller will appear in an image. However, there is still a significant oscillation in the CTF below the Scherzer limit of 0.24 nm, so that, in the absence of an aperture to completely eliminate contrast transfer in this range, and in the absence of spatial coherence effects, the CTF oscillations would complicate the recorded image and make it difficult to interpret. A different result is obtained when spatial coherence is included in CTF calculations, as shown next.

The Spatial Envelope The most commonly used expression for the spatial damping envelope is given by[111,236]

$$K_S(q) = \exp\{[\pi\, \alpha_0\, (\Delta z\, q + C_S\, \lambda^2\, q^3\,)\,]^2 / \ln 2\} \tag{3.13}$$

This equation is acceptable for illumination that is defocussed and therefore has a Gaussian distribution of illumination angles, where α_0 is the half-angle of the distribution, and the remaining parameters are as previously defined.

For a typical 200 kV microscope with a thermionic emitter that has a high voltage ripple of $1 \cdot 10^{-6}$, an OL current stability of $1 \cdot 10^{-6}$, an energy spread of 2 eV from the

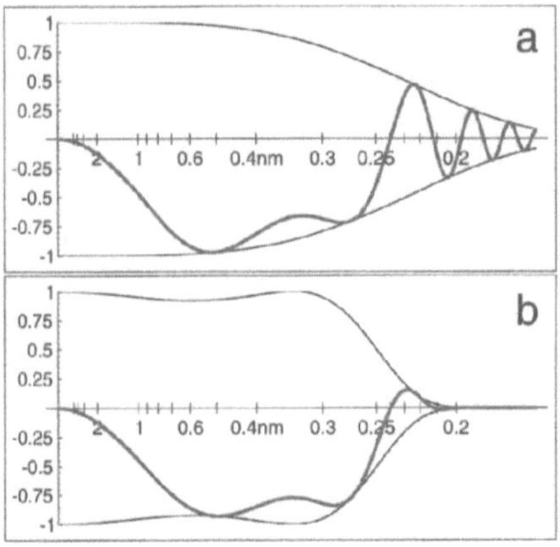

Figure 3.6. a) CTF for a 200 kV thermionic emitter microscope, with parameters as given in the text, showing damping from temporal envelope only; b) CTF showing damping from spatial envelope only.

emitter, and a chromatic coefficient C_C of 1.4 mm, the resultant defocus spread Δ is 7.8 nm. Considering just spatial coherence effects as controlled by illumination angle, for an illumination semi-angle of 1 mrad, the spatial damping envelope at Scherzer focus attenuates the transfer function as shown in Fig. 3.6b. The strong damping caused by spatial coherence effects allows spatial frequencies only above 0.2 nm in spacing to pass, and acts nearly like an objective aperture with a diameter to subtend a reciprocal distance in the BFP of 5 nm^{-1}. So it is clear that under the conditions stated above, the spatial damping controls contrast transfer at the higher spatial frequencies, and the stabilities of the various instrumental elements are 'good enough.'

As a final observation, the temporal coherence envelope does not change with defocus, while the spatial coherence envelope contains Δz and thus varies with defocus. This allows the possibility that a total transfer function (including both temporal and spatial envelopes) might be generated in a field emission microscope at a focus other than Scherzer which would allow for a significant improvement in the information limit of the microscope. This 'optimum focus' condition, described by Lichte,[262] is important for electron holography, since holography offers the potential for correction of the spherical aberration in the microscope, and therefore elimination of the high degree of oscillation of the transfer function up to the extended information limit. For a microscope equipped with a hardware-implemented C_S corrector, the optimum focus condition is also likely to be useful to achieve ultimate point resolution. Details of the benefits of the optimum focus for electron holography are given in Chapter 9.

4. Illumination geometry

In this section, the illumination geometry for the field emission microscope is described, since it differs considerably from that of typical thermionic emitters, and is an important contributor to the extent of contrast transfer in the FE-TEM.

In a conventional TEM with a thermionic emitter, using an objective lens pole piece designed for high resolution imaging, and working at a typical magnification of 500 kX, the beam is usually set near crossover to give short exposure times that minimize the effect of specimen drift on the recorded image. The spot size at crossover is typically on the order of 0.2-0.5 μm. It is usual, then, to measure the convergence semi-angle of the beam, and to enter this value for that of the illumination angle in

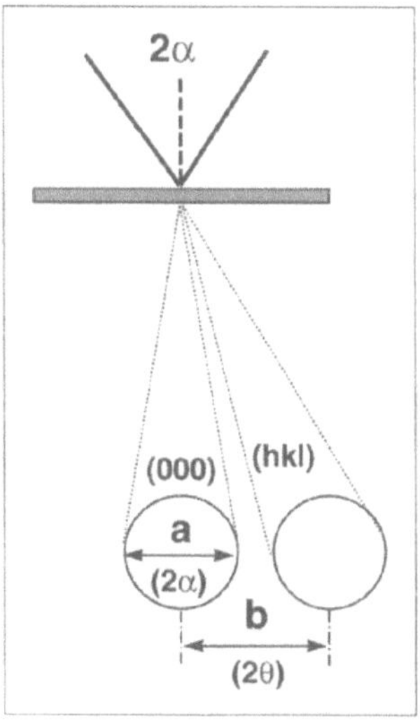

Figure 3.7. Convergence angle geometry for thermionic emitter microscope. The semi-angle of illumination $\alpha = a\,\lambda/(2\,b\,d)$, where λ is the wavelength in nm and d is the interplanar spacing for the (hkl) reflection used.

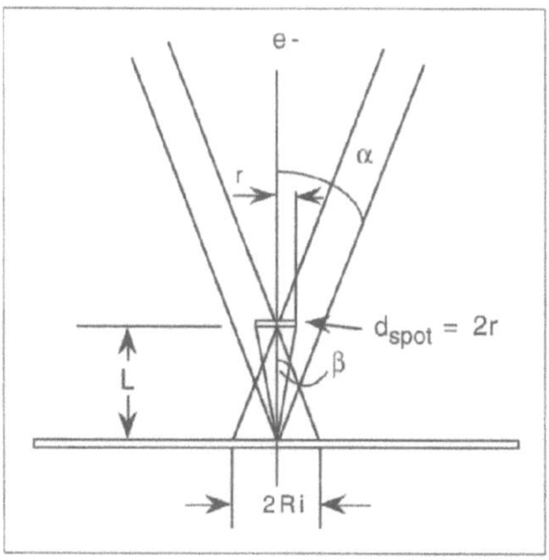

Figure 3.8. Basic illumination geometry for a field emission electron microscope, used for the calculation of the effective illumination angle α.

the computation of the contrast transfer function. Figure 3.7 illustrates this geometry. Typical values of this angle for microscopes with thermionic emitters that are equipped for high resolution imaging range from 0.7 to 1.5 mrad.

For a field emission microscope, however, the ultra-fine spot size at illumination crossover produces a different illumination geometry, and yields an illumination angle very much smaller than the conventional TEM, even with large convergence semi-angles. The small illumination angle results from the need to defocus the incident beam as appropriate to record a high resolution image of a useful area. In order to compute the illumination angle, it is necessary to first measure the convergence angle, and then to be able to measure both the current in the beam, and the current on the viewing screen (as is typically possible with any TEM). This geometry is shown in detail in Fig. 3.8 for an overfocus illumination geometry, but all arguments apply to the underfocus condition as well.

To determine the effective illumination angle for a field emission microscope, we define the following terms: R_i = radius of illumination, R_s = radius of viewing screen, M = Magnification, i_p = total probe current, i_s = current at screen, α = convergence angle, r = radius of spot, β = illumination angle.

This geometry yields the following relationships:

$$R_i = \sqrt{i_p/i_s} \cdot R_s/M \tag{3.14}$$
$$L = R_i/\alpha$$
$$\beta = r/L$$

As an example, using the geometry of the HF-2000, for a 300 μm condenser aperture, $\alpha = 6.3$ mrad. The diameter of the focussed probe, with a current of 10^{-9} A, is 2.3 nm. The screen radius = 10 cm. At a magnification of $M = 500$ kX, with the beam spread beyond the edge of the viewing screen for recording the high resolution image, the screen current is about $5 \cdot 10^{-8}$ A. Then: $L = 0.283\,\mu\text{m} \;/\; 6.3 \cdot 10^{-3}\,\text{rad} = 45\,\mu\text{m}$. Therefore, $\beta = 0.00115\,\mu\text{m} \;/\; 45\,\mu\text{m} = 0.026$ [mrad].

The illumination angle that is used in a field emission microscope for high resolution imaging is thus on the order of 50 times smaller than for thermionic emission microscopes. The effects of this small illumination angle on the CTF for the field emission microscope will be illustrated in Section 6.

5. Isoplanatism and coma

The full image forming capabilities of an objective lens are generally characterized by plotting the function χ relative to changing focus values, for particular values of the spherical aberration constant C_S. However, χ embodies not only spherical aberration and axial astigmatism which is assumed to be correctable, but several other aberrations as well. These include chromatic aberration, and the third-order aberrations distortion, field curvature, and coma (which got its name from the word comet, which describes the shape of its aberration figure).[172] Of these, chromatic aberration and coma are most important for high resolution imaging and holography. Chromatic aberration effects can only be minimized by reducing instabilities in power supplies when the instrument is designed, and by elimination of environmental effects such as stray fields. Coma, however, which results from a misalignment of the illuminating beam with respect to the axis of the objective lens, can be virtually eliminated by aligning the illumination tilt so that the axis of the beam is accurately parallel to the axis of the objective lens, giving the so-called 'coma-free' illuminating condition.[400,508] This is important in field emission microscopes, because it gives optimum positioning and extent of the 'isoplanatic patch.'[161] This term refers to the image area over which individual image regions away from the optic axis are imaged with the same characteristics as a point on the axis. In the aberration-corrected case the isoplanatic patch becomes large if the illumination is spread so that it approximates a parallel beam. However, if the application requires working with a convergent beam at very high magnifications (generally $> 10^5$), in order to achieve the short exposure times necessary to record the best high resolution images, then the effects of anisotropic aberrations can be significant and can reduce the size of the isoplanatic patch so that it does not cover the entire image area. If there is any significant tilt of the illumination, that is, if the coma-free condition has not been properly set, then the center of the isoplanatic patch can be outside the imaged area, leading to a continuous change in the character of the image over the entire field of view, as shown in Fig. 3.9. Here, the tilt was misadjusted so that the center of the isoplanatic

Figure 3.9. Image showing misaligned beam tilt, with isoplanatic patch off left edge of image. There is a continuous change in image character (apparently from underfocus through minimum contrast to overfocus) from lower left to upper right on the image.

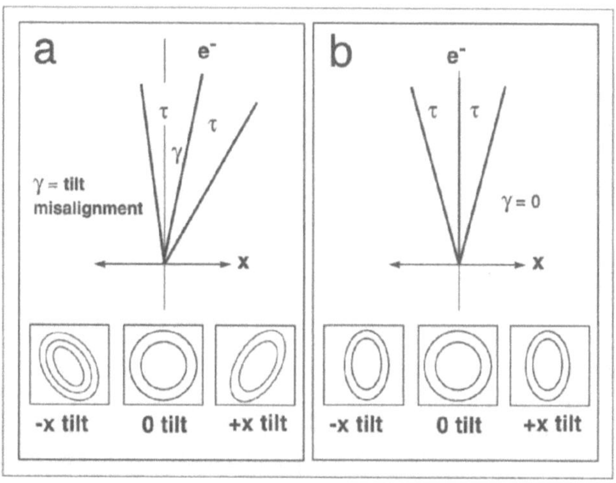

Figure 3.10. a) Schematic showing beam tilt misalignment and appearance of diffractograms in the as-misaligned (and properly stigmated) condition (center) vs. equivalent plus and minus tilts about this condition. b) Properly aligned beam and corresponding diffractograms in zero-tilt (properly stigmated) and plus and minus tilt orientations.

patch fell off the left edge of the image. Thus there is a continuous change in image character from one side of the image to the other, as illustrated by the diffractograms taken from the left, center and right sides of the the image. In practice, setting the beam tilt using the method of high tension centering will usually bring the isoplanatic patch near to the optic axis. However, only a true coma-free adjustment suffices to assure that the isoplanatic patch is correctly centered. Therefore, it is important to first align the illumination for the coma-free condition, and then to assure that the illumination is spread sufficiently to produce an isoplanatic condition over the entire micrograph.

Figure 3.10a illustrates schematically the optical conditions which produce coma in the image. Assume that the illumination enters the OL at some angle γ with respect to the true optic axis of the OL. Also assume that axial astigmatism is corrected as necessary for recording a high resolution image, and that the image is set to a minimum contrast condition. If the beam is then tilted in the x- direction by an amount τ and then in the $-x$- direction by the same angle, the incident beam deviates from the axis of the OL by $\gamma + \tau$ in the x- direction, and $\gamma - \tau$ in the $-x$- direction. Images of an amorphous film taken at a magnification of say 300 kX in both tilt orientations, as well as with no induced tilt, would yield Fourier transforms that would show elliptical patterns for the tilted images, and a circularly symmetric pattern for the no-tilt condition.

When the beam does not coincide with the OL axis, the tilted diffractograms will show marked differences from each other, as shown in Fig. 3.10a. However, if the beam is tilted to coincide with the axis of the OL, and astigmatism is eliminated, then the positive and negative x-tilts would be equal in magnitude, and the elliptical diffractograms would be essentially identical for both tilt directions, as shown in Fig. 3.10b.

To correct for coma, it is important to adjust the pivot point of the tilt so that as the beam is tilted in the plus and minus x- and y- directions the center of the illumination remains centered accurately about the optic axis, particularly at the high magnification (300-500 kX) needed for final coma-free alignment. The capability to do this is standard on modern microscopes, and will not be detailed here.

Before adjusting the pivot point of the beam tilt, it is necessary to do a standard alignment, which, on the HF-2000, includes an adjustment of both the second- and third-order condenser stigmators to form a circular beam with a stigmated caustic. The OL current is adjusted to its optimum set point, and the sample height is adjusted to bring the image to a minimum contrast condition. Finally, the pivot point adjustment is set, the image is stigmated approximately, and the HT centering is performed to set up the starting point for the coma-free tilt alignment.

The coma-free tilt is set maually while observing the image through the binoculars at a magnification of 300-500 kX, with the beam spread to fill the full viewing screen, and the image at the minimum contrast point. The beam is then tilted in the plus and minus x- directions with a frequency of 1 cycle every 2 seconds, usually utilizing the tilt wobble function provided on the microscope. The character of the image is observed as first the x- tilts are alternated (through a tilt angle of about 5 mrad), and the x- tilt control is slowly adjusted until the image character is observed to be identical in both tilts. This process is duplicated using the y- tilt control. With a final trim of the astigmatism, the alignment condition shown in Fig. 3.10b is obtained. It should then be possible to produce a tilt 'tableau,' or Zemlin tableau,[508] as in Fig. 3.11. Fourier transforms of the plus and minus tilt images show similar form and symmetric orientation, confirming that the coma-free condition has been achieved for both the $\pm x$- and $\pm y$- directions. This is sufficient to assure that the variation in image character will be radially symmetric about the center of the image, which in turn assures that the isoplanatic patch is centered about the optical axis. Note that once set, the tilt alignment should not change appreciably over a day's observation, unless the tilt controls are inadvertantly adjusted.

To assure a highly accurate coma-free tilt alignment, it is also useful to check the plus and minus x and y tilt angles to determine that opposing tilts are indeed identical in degree. This can easily be done while observing a focused selected area diffraction pattern, e.g., from sputtered Au on a thin carbon film, through the binoculars. The first diffraction ring (Au (111) reflections) represents a tilt angle of 5.34 mrad. If the diffraction pattern is carefully centered using shift coils in the intermediate lens, it should be easy to observe whether the opposing tilts are precisely identical in both positive and negative directions. The ability to read out the current in the tilt coils

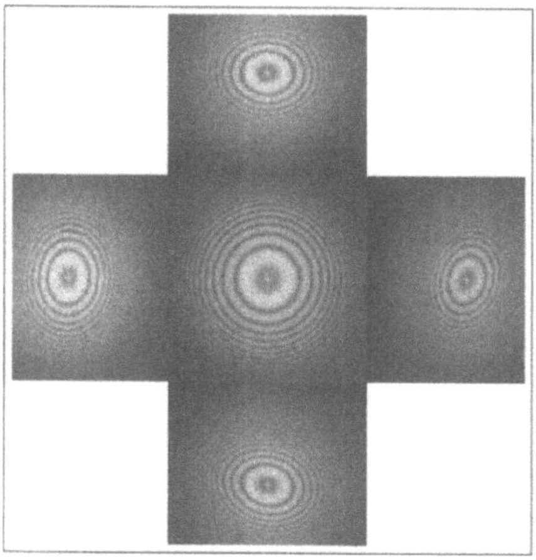

Figure 3.11. Tilt tableau showing similar diffractograms in $\pm x$- and $\pm y$- tilt directions (tilt angle about 5 mrad) after coma-free alignment.

directly would permit a quantitative determination of the accuracy with which the plus and minus tilt coil currents replicate the opposing tilt angles of the beam. On the HF-2000, while the diffraction rings are circular to better than 1 percent, the tilts in both x- and y- directions vary by 7-9%. Thus, the coma-free condition for each axis might not be correctly set if the standard tilt wobbler is used during the alignment procedure. Through computer control using, for example, Gatan's DigitalMicrograph™ scripting,[140] it is possible to accurately adust the x- and y- tilts so that both positive and negative tilt angle in both directions are correctly set. The tilt oscillation is then run using the scripting program rather than using the standard tilt wobbler.

Coma-free tilt alignments can also be assisted by the computer. For example, the software DigitalMicrograph™ offers an automated tilt alignment as part of the Advanced Imaging package. This function requires the same set-up of the image as above, before running the program, so that the iterative procedure will not 'blow up.' It is best to have a significant thin amorphous area available for best results (not always present on many thin foil specimens). A description of a coma-free alignment procedure utilizing a standard TV camera typically found on most modern microscopes has been given by Ishizuka.[214] It would be a useful exercise to test any automated technique by making the diffractogram tableau to assure that the proper results were being obtained before proceeding to do high resolution imaging or holography.

Assuming the coma-free condition is properly set, it remains then to investigate the illumination conditions under which the recorded image shows an isoplanatic character over the entire area.

To observe the isoplanatic patch, an image is recorded at an intermediate magnification with the beam condensed so that it extends just beyond the edges of the recorded area. If a digital camera such as a Gatan 794 is being used. a beam diameter should be set to about 3 cm at an image magnification of 200 kX. If film is used, the beam diameter can be extended to about 12 cm for a film area of 8 cm by 8 cm. Note that it is not important whether the beam is set up with the crossover above vs below the specimen, since the appearance of the isoplanatic patch will be similar for convergent as for divergent illumination, because the change in the convergence angle from above to below crossover is negligible. Note also that microscopes with a double-condenser lens operating mode for normal imaging have the optical geometry that essentially guarantees a parallel beam and therefore an isoplanatic patch that extends over the entire

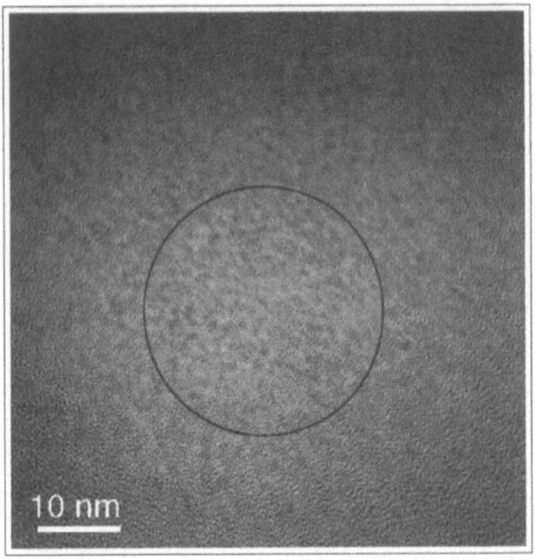

Figure 3.12. Illustration of the appearance of the isoplanatic patch at low magnification. The area of the patch is about 20% of the area of illumination.

illumination area.[161,334]

Figure 3.12 shows a coma-free 1k by 1k pixel digital image taken as above, with an isoplanatic patch roughly 30 nm in diameter in the center of the field, and a radial deviation from isoplanicity away from the edge of the center patch. Diffractograms from the center and the corners illustrate the change in image character related to this deviation. The size of the isoplanatic patch is related to the beam convergence, so it is a straightforward matter to spread the beam until the isoplanatic patch extends beyond the area of the recorded image. As an example, the image of Fig. 3.13a was recorded with the beam spread to a diameter of 15 cm. Because the entire image area now satisfies the isoplanatic condition, diffractograms from the center and corners are all equivalent (Fig. 3.13b). The trade-off is the decrease in illumination intensity and a concomitant increase in exposure time because of the increase in beam diameter. This does not create serious problems in practice, however, for an exposure time of 1-2 sec is typically found to yield images with acceptable counts/pixel (in the range of 500-1500) for high resolution images, when using beam currents in the 15-30 nA range.

5.1. Coma and isoplanatism in electron holograms

When recording high resolution holograms, as will be discussed in detail in Chapter 4, it is usually necessary to adjust the incident illumination so that it is highly astigmatic with a major axis in the direction normal to the axis of the electron biprism. This satisfies the requirement for maximum coherence normal to the interference fringes, giving fringes of highest contrast. A proper coma-free tilt alignment is important so that an astigmatic beam with axial symmetry can be produced. It is also necessary to make sure that the condenser stigmators are adjusted so that the stigmator axes are coincident with the axis of the imaging system. When these conditions are fulfilled, it should be possible to vary the second-order condenser stigmators and observe no sweep in a focussed beam when moving from one side to the other of a focussed, stigmated condition. Most microscopes provide the capability for this alignment, whether by adjustment of external potentiometers, or accessible internal

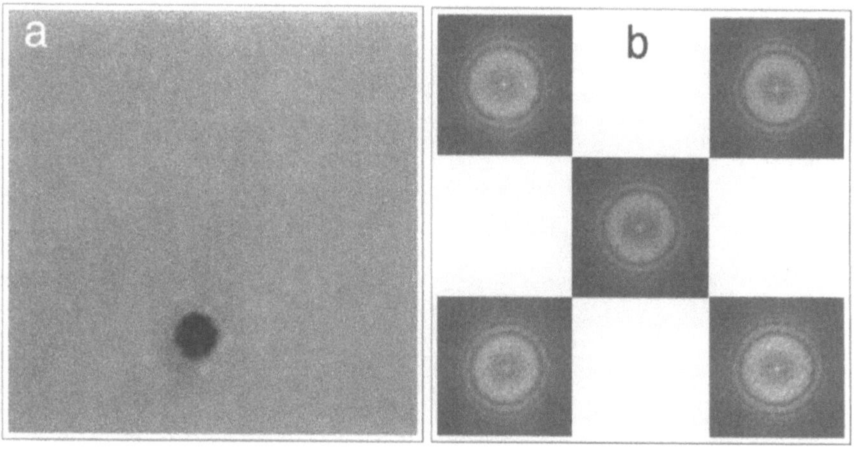

Figure 3.13. a) Image of amorphous tungsten thin film recorded at 300 kX. b) Diffractograms from the center and corners showing isoplanicity over entire image area.

potentiometers as on the HF-2000. If one of the condenser stigmators is then set to a fully clockwise or anti-clockwise position, and the beam is brought to a crossover, a thin straight line of illumination should result. Then, by setting the condenser to focus the opposite astigmatic crossover, the illumination should switch to a thin line normal to the original axis. With proper alignment of the stigmator axes, as described above, these lines should intersect at the centerpoint of each, when observed at low magnification.

To adjust the astigmatic beam before recording a hologram, the condenser aperture should be removed to allow the full illumination caustic to be observed in its astigmatic condition. The astigmatic illumination shows a thin, bright backbone corresponding to the caustic crossover point. This bright line is made symmetrical with respect to the surrounding illumination field, using a combination of adjustments of the second- and third-order condenser stigmators. Figure 3.14 compares a misadjusted beam with a properly adjusted beam. Figure 3.14a shows the beam at low magnification with the condenser aperture removed, to illustrate a mis-aligned condition. Figure 3.14b shows the aligned condition, with the illumination symmetrical normal to the long axis. After the condenser aperture is properly centered, at high magnification with the beam spread just off crossover, the mis-aligned beam shows an asymmetrical profile, as indicated in Fig. 3.14c. This condition is not optimal for recording hologram, as the hologram fringes show contrast changes from one edge of the hologram field to the other. The properly centered beam has a profile as indicated in Fig. 3.14d, and, when spread to the width suitable for recording the hologram, produces uniform fringe contrast over the entire hologram field.

Proper centering of the condenser aperture is difficult to achieve in the astigmatic condition, so it is useful to return to a "round beam" condition for aperture centering. The transition between round and astigmatic beams can be done most effectively by controlling the stigmator settings from the computer, as provided via Gatan DigitalMicrograph™ scripts, for example. After the aperture is centered in the usual manner, the settings are returned to the previously set astigmatic condition, and the beam is ready for hologram recording via the methods outlined in Chapter 4.

The degree of isoplanicity in this astigmatic beam when it is set for reasonable

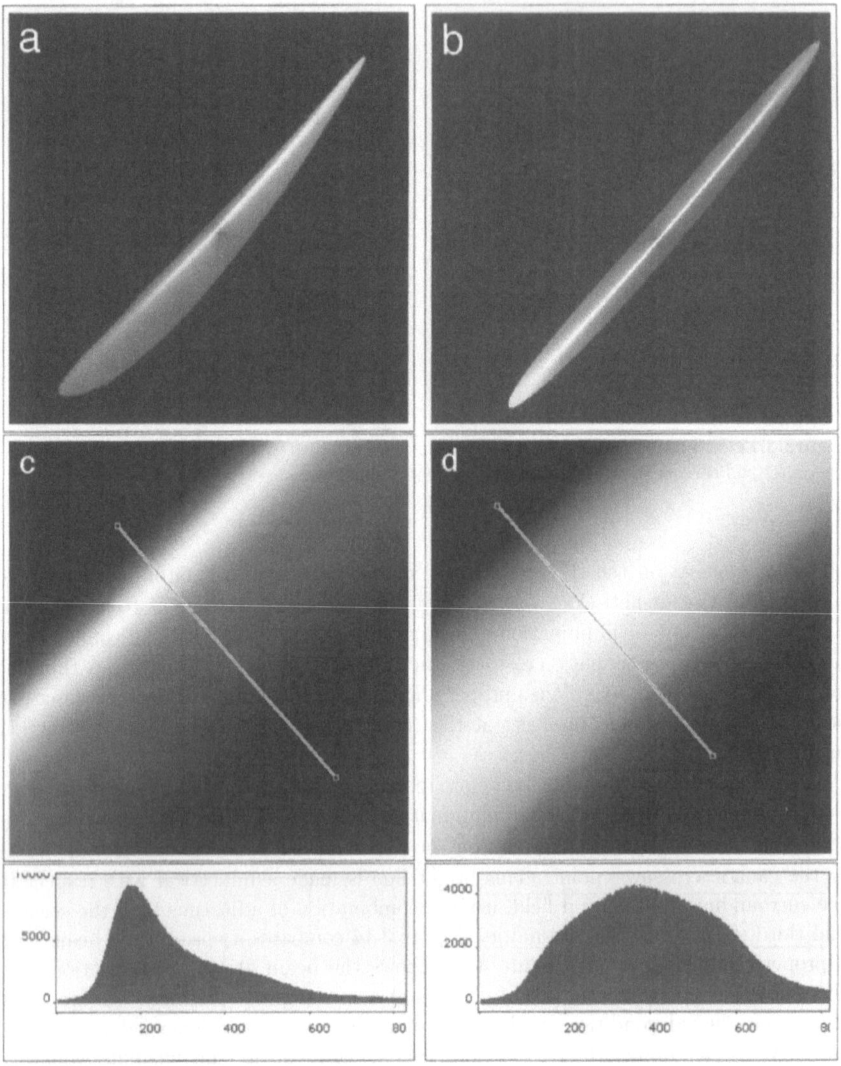

Figure 3.14. a) Astigmatic illumination with no condenser aperture and with 2nd- and 3rd-order stigmators mis-aligned; b) astigmatic illumination with correctly aligned stigmators. The bright line in the caustic is symmetrical with respect to the illumination field; c) nearly focused astigmatic illumination with condenser aperture in and centered, showing assymetry due to incorrect setting of condenser stigmators; d) same condition but with correct stigmation, showing symmetry of illumination in the nearly focused condition.

exposures of high resolution holograms is shown in Fig. 3.15. From the crossover position, the beam is spread to a width of about 10 cm or more at a magnification of 0.7 - 1 MX prior to exposure. The image of Fig. 3.15a was recorded by reducing the magnification to 200 kX so that the width of the astigmatic beam fit within the area of the CCD chip on the camera. Diffractograms from 128 by 128 pixel areas along the long axis direction and normal to this direction, as outlined in the figure, show that the

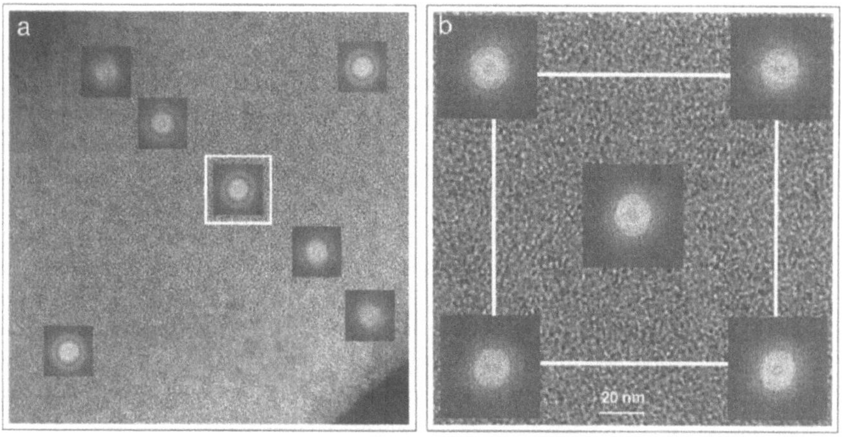

Figure 3.15. Astigmatic illumination showing maintenance of isoplanicity along the long axis direction but strong deviation from isoplanicity in the transverse direction. White square outline shows area that the CCD camera is exposed to at 1000 kX magnification: b) Image at 700 kX under beam conditions of (a), showing area outlined: inset diffractograms show that isoplanicity is maintained fully along long axis, and that there is only a slight deviation from isoplanicity in the transverse direction over the central region of the illumination field.

isoplanicity (i.e., the degree of lateral coherence) is maintained along the long axis of the illumination, as expected, but that the degree of transverse coherence is strongly affected. Medina and Pozzi[300] suggest the measurement of a 'coherence parameter' (E in their annotation), given by the ratio of the area of the isoplanatic patch (also called the coherence patch) and the total area of illumination, allows the minimum width of the illumination to be properly adjusted for recording holograms. The coherence parameter can be measured quite simply in the circularly symmetric beam case, and it is invarient in the astigmatic beam. Therefore, as the beam is made astigmatic using the condenser stigmator controls, the ratio of the width of the isoplanatic region to the width of the illumination stays constant. The coherence parameter measured from Fig. 3.12 for the HF-2000 is about 0.2, so it can be expected that the region of transverse coherence in the astigmatic beam of Fig. 3.15a is about the central 20% of the width of the illumination.

As indicated in the figure, the square outlined in white shows the area of the CCD detector relative to the width of the astigmatic illumination at a magnification of 1000 kX used for high resolution holography. It is outside of this area that significant changes in the diffractogram occur, indicating a change in image character due to non-isoplanicity. These changes occur rapidly, however, so care should be taken to spread the illumination to an appropriate extent off of the fully focused condition, in order to avoid as much as possible significant artifacts introduced by the effects of non-isoplanicity in the hologram. Figure 3.15b shows the image at 700 kX, with the diffractograms inset to show the isoplanicity in both the lateral and transverse directions. Isoplanicity is reasonably maintained over the entire area, indicating that holograms recorded under these condition should give reliable phase image reconstructions.

6. Contrast transfer considerations in the field emission microscope

It is now appropriate to examine the benefits in contrast transfer that can be gained using a field emitter rather than a thermionic emitter, and the other instrumental parameters that also affect the microscope's contrast transfer capabilities. For this purpose, we can compare a series of contrast transfer function plots calculated systematically for a microscope with a thermionic emitter such as the 4000EX, with a similar series calculated for the HF-2000 field emission microscope. Only instrumental effects on the transfer function of the objective lens are considered for these plots; specimen effects are ignored since they have no bearing on the contrast transfer properties of the microscope alone. Image calculations that are used to illustrate the differences in the imaging characteristics will be made assuming the sample satisfies the weak phase object approximation (WPOA), as defined in the previous chapter. As we are using digital cameras only for recording images, and because we use a direct Fourier transform instead of the power spectrum (square of the Fourier transform) the modulus of the CTF will be plotted to show the results of these calculations. These plots can then be compared directly to the modulus of the Fourier transforms of images of thin amorphous films that are used to characterize the contrast transfer of the microscope.

First, it is interesting to examine how the transfer function can actually be measured for a particular microscope. The most common technique is to make images of very thin amorphous films at different levels of defocus under carefully controlled optical conditions. For 100 kV instruments, thin carbon films are suitable, but at higher voltages it is better to use higher atomic number materials because they give higher scattering intensity into the higher spatial frequency components. These films must be extremely thin to approximate, as well as possible, weak phase objects. Amorphous films of germanium 5 nm or less in thickness have been used for characterization of microscopes from 200-400 kV.[144] Amorphous tungsten films, coated with Pt crystallites by evaporation, on the order of 7 nm thick self-supported on holey carbon have also been used (see Ref. 168 and Chapter 9). Recently, we have produced films of amorphous tungsten 0.8-1.5 nm thick,[9] using a thin film ion-sputtered deposition system, which are self-supporting on holey carbon. These films have proved to be excellent samples for approximating weak phase objects with high scattering power to illustrate the contrast transfer capabilities of the field emission microscope, and have been used in several of the examples that follow.

Diffractograms of the amorphous film images are produced either by using a laser optical bench for images that are recorded on film, or by direct computation for images that are recorded digitally. Note that film images can be digitized using an appropriate digitizing camera (a standard flatbed scanner is probably not appropriate, because of questions of linearity) and the diffractograms subsequently computed. The most common technique is simply to record the images directly on a CCD camera and compute the diffractogram (using, e.g., the FFT function available in DigitalMicrograph™), then match the computation to the computed modulus of the CTF. The best correspondence with computed CTFs is obtained by using rotational averaging of the diffractogram, either by directly rotating a negative on the optical bench, or by computing a rotational average from the computed diffractogram (e.g., via DigitalMicrograph™ scripting).

It is instructive to explore first the CTF characteristics of the 4000EX because of the dramatic differences in factors controlling contrast transfer between the illumination system on the thermionic emitter microscope and that of the FE-TEM. Figure 3.16a shows the transfer function computed with temporal coherence effects alone contributing to damping. The horizontal line added to this figure shows approximately the $1/e^2$

or 14% contrast level,[326] which generally corresponds to the level of contrast needed for a feature to be detectable above noise in the image. Temporal coherence effects thus allow an information limit of 0.13nm. Figure 3.16b shows the damping produced by spatial coherence effects for an illumination semi-angle of 0.7 mrad (the smallest aperture which can be used and still allow exposure times short enough to minimize sample drift effects). Figure 3.16c shows the damping effects from both the temporal and spatial coherence envelopes combined. Clearly the spatial envelope limits the ultimate information limit of the 4000EX to about 0.17nm, but the temporal envelope also controls to some extent the contrast transfered in spatial frequencies below about 0.25nm. Figure 3.16d shows the static and rotationally averaged Fourier transforms from an image of a 1 nm thick tungsten film taken at about -50nm underfocus, near the Scherzer value of -49nm. The radial variation of intensity in this diffractogram, shown in Fig. 3.16c, matches well the modulus of the CTF. Thus for the 4000EX, the coherence envelopes damp contrast transfer so that only those diffracted beams at spacings within the Scherzer passband will contribute to the image, even if no objective aperture is used. This allows the image to be directly interpretable more easily, because if spatial frequencies beyond the Scherzer limit were also transmitted, undesirable random reversals of contrast in the image could occur. Fig. 3.16d shows a rotationally averaged diffractogram from an image of a 1 nm thick tungsten film acquired near Scherzer defocus (Δz_{Sch} = -49 nm), at 600 kX in a JEOL 4000EX, with a C_S of 1 mm, a defocus spread of 6.5 nm, and an illumination semi-angle of 1.4 mrad. The radial variation of intensity in this diffractogram matches well the modulus of the CTF shown in Fig. 3.16c which was computed for the above conditions, incorporating both spatial and temporal envelopes.

We can now investigate the behavior of the transfer function for a 200 kV field emission microscope. For the HF-2000, $C_S = 1.2$ mm, defocus spread = 3.5 nm (computed for high voltage and OL stabilities at $1 \cdot 10^{-6}$, and an energy spread of 0.5 eV), and the illumination semi-angle = 0.03 mrad. Figure 3.17a shows the transfer function calculated with just spatial coherence effects. We see that because of the very small illumination angle which is obtained with the FE instrument, the transfer function for only spatial coherence effects extends well beyond the 0.1 nm level. As shown by the CTF of Fig. 3.17b, which was calculated for damping only by temporal coherence, these effects totally control the degree of contrast transfer beyond the Scherzer limit. Thus the transfer function plotted in Fig. 3.17c which was calculating by including both temporal and spatial damping envelopes is not significantly different from the one in Fig. 3.17b with damping due to temporal effects alone. The 14% contrast line intersects the envelope curve at about 0.15 nm, indicating the information limit of the microscope.

Contrast transfer calculations for the FE-TEM show graphically the relative effects of the HT and OL instabilities on the CTF. Earlier we computed that an improvement in stabilitiy of the HT supply from 1 ppm to 0.5 ppm produced only a 3.3% improvement in defocus spread, while a similar improvement in objective lens current stability resulted in a 14.3% improvement in defocus spread for OL improvement. Figure 3.18a shows overlapping transfer functions, computed as before but comparing the effect of reducing the HT ripple from $1 \cdot 10^{-6}$ to $0.5 \cdot 10^{-6}$. This improvement in HT stability extends the information limit for 14% contrast to about 0.14 nm, giving only a slight improvement. However, as shown in Fig. 3.18b, a similar improvement in OL stability produces a significantly greater extension of the information limit. Improving the stabilities of both the OL and the HT in a 200 kV FE instrument could extend the information limit to about 0.12 nm, as shown by Fig. 3.18c.

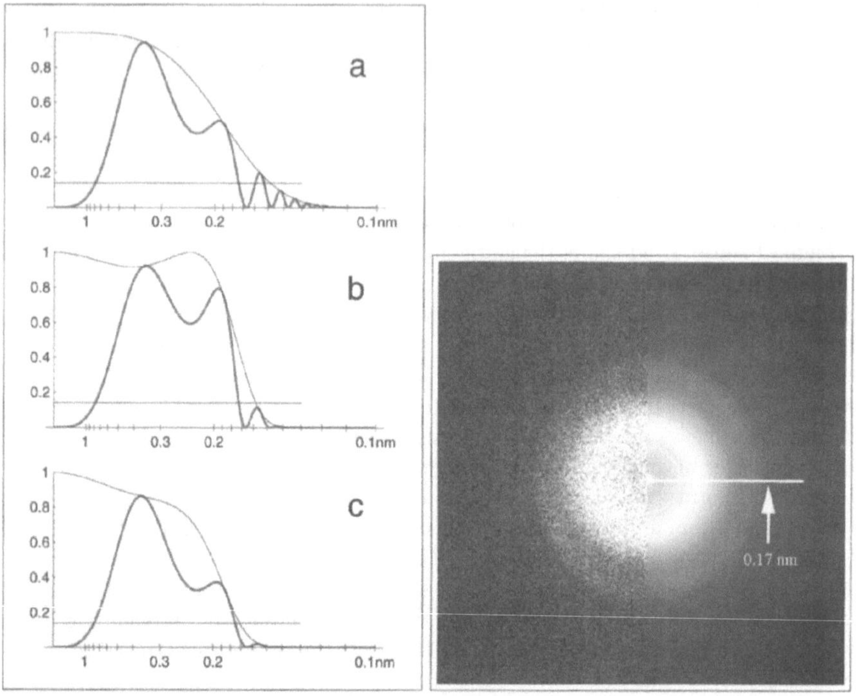

Figure 3.16. CTF plots for the 4000EX TEM at Scherzer defocus. a) CTF with damping from temporal coherence effects only. For this plot, $C_S = 1\,\text{mm}$, HT and OL supply stabilities are $1 \cdot 10^{-6}$, $C_C = 1.7\,\text{mm}$, $\Delta E = 1.5\,\text{eV}$, yielding a defocus spread = 7.8 nm and Δz = -49 nm. b) CTF with spatial coherence effects only, for both 1.4 mrad and 0.7 mrad convergence angles calculated. c) CTF including both temporal and spatial coherence effects, for both convergence angles. The small difference in real contrast transfer indicates that the larger convergence angle can be used effectively for high resolution imaging. At this defocus, contrast transfer to just under 0.17 nm is indicated. d) Static and rotationally averaged diffractograms from image of 1 nm W film taken in the 4000EX near Scherzer focus, at -50 nm, with 0.7 mrad beam convergence. The 0.17 nm point as calibrated using 10 nm gold nanoparticles dispersed on the film is shown.

The fact that the damping envelope of an FE-TEM passes contrast well below the Scherzer limit makes it possible to demonstrate additional 'pass bands' in contrast transfer that cannot be observed with typical thermionic emitters. For example, Fig. 3.19a shows a rotationally averaged diffractogram computed from a 1k by 1k pixel image of a 1 nm thick tungsten film, recorded at 360 steps of the focus control (i.e., 'clicks') below minimum contrast on the HF-2000. The positions of minima in the diffractogram, as calibrated using 10 nm gold particles dispersed on the same film, along with the shape of the passband allow a precise match of the diffractogram with a computed CTF (Fig. 3.19b). In comparison, the CTF for a 200 kV thermionic emitter with the same OL optical parameters and with a nominal 0.6 mrad convergence semi-angle as used for the HF-2000 is shown overlaid in Fig. 3.19b. Since there is significantly greater damping for the thermionic emitter (CTF computed for 1.5 eV energy spread and 0.7 mrad illumination angle), it is clear that the details of the passband and the positions of rings at higher spatial frequencies cannot be determined from images taken in this type of instrument. The match of diffractogram to CTF then allows the

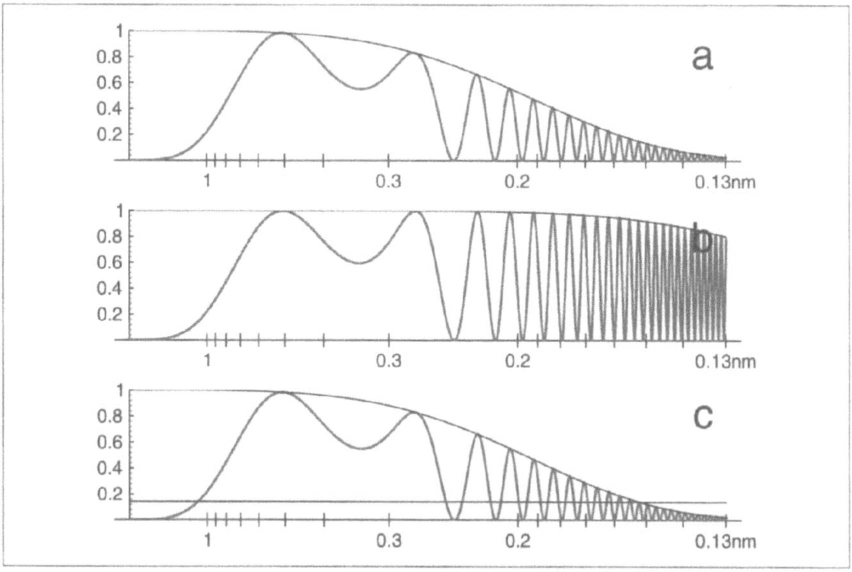

Figure 3.17. CTF plots for the 200 kV cold field emission microscope at Scherzer defocus. a) Temporal coherence effects only. For this plot, $C_S = 1.2$ mm, HT and OL supply stabilities are $1 \cdot 10^{-6}$, $C_C = 1.4$ mm, defocus spread is 3.5 nm and Δz = -65 nm. b) Spatial coherence effects only, for illumination semi-angle of 0.03 mrad. c) Both temporal and spatial coherence effects computed. This shows that, while temporal coherence effects strongly control the behavior of the transfer function for the field emission microscope, the potential information limit is extended to spatial frequencies well beyond the Scherzer limit.

nm/click of the OL fine control to be determined. In the present case, given that the minimum contrast condition[326] is at a focus of $\Delta z_{min} = -0.44\,(C_S\,\lambda)^{1/2}$ at a nominal C_S of 1.2 mm for the HF-2000, then $\Delta z_{min} = -25$ nm. The total defocus amount from the CTF match is -228, so there are 203 nm per 360 clicks or 0.586 nm/click for this microscope.

Digital imaging allows a series of such images and associated diffractograms to be generated, each at a known number of clicks below minimum contrast, which can then be matched accurately to computed CTFs. By carefully adjusting parameters of defocus and even C_S (around the nominal value provided by the manufacturer for the OL pole piece used), matches of CTFs to diffractograms can easily be made with an accuracy of $\pm$ 1.5 nm. The defocus spread as controlled by energy spread in the beam and by instrument instabilities can also be estimated from these diffractograms. If done with care, a plot of clicks below minimum contrast vs defocus can be generated, which should show a straight line, giving confidence in the accuracy with which the diffractogram/CTF pairs were matched. It is important to make a calibration like this, since the nominal value of defocus step size supplied by the manufacturer may be inaccurate. In the present case, the nominal 1 nm step size was determined to be 0.58 nm (per click), a considerable discrepancy.

The availability of a digital imaging system on the microscope allows a similar task to be completed in a very short time. For example, using DigitalMicrograph™ software with the Advanced Imaging package installed, the computer will assess the defocus level of several images and report the defocus amount and the number of nanometers per

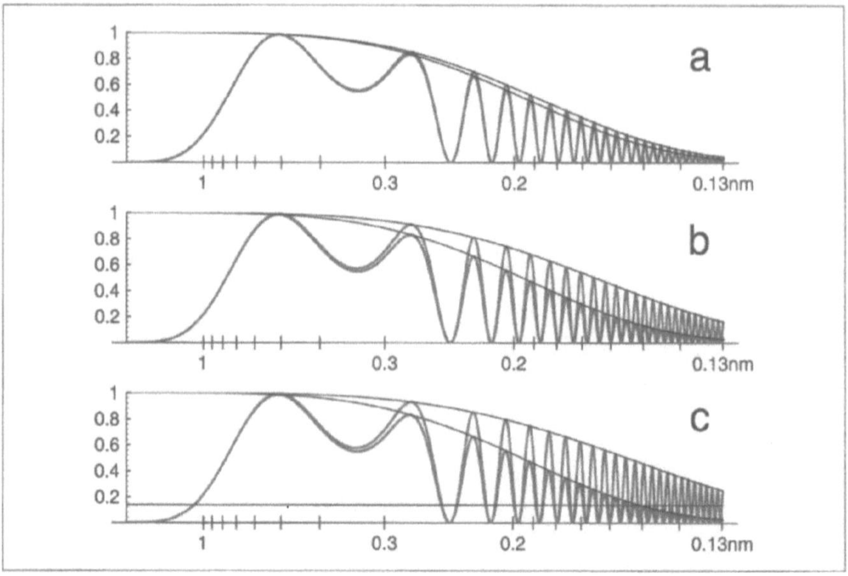

Figure 3.18. CTF plots computed for temporal coherence effects showing effects of HT and OL stabilities on the information limit. a) HT stability at 1ppm vs 0.5 ppm; b) OL stability at 1ppm vs 0.5 ppm; c) Combined effects of improving the stability of both the OL and HT to 0.5 ppm, vs 1 ppm stability of present microscope. An improvement in the information limit from 0.15 nm to 0.12 nm is indicated.

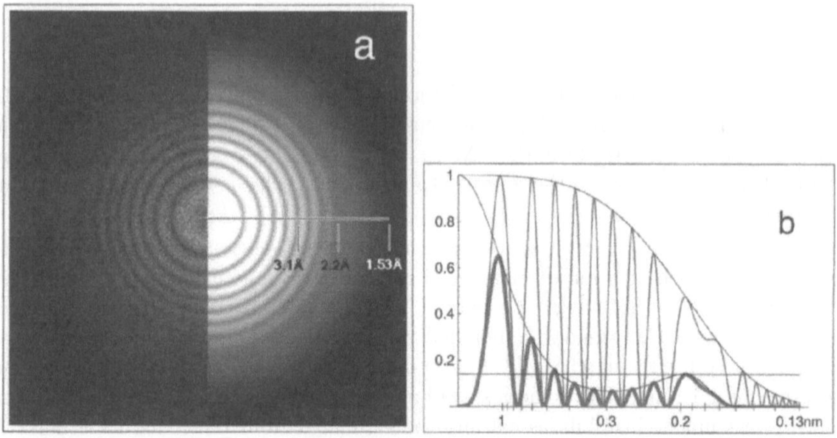

Figure 3.19. a) Diffractogram from an image of thin amorphous tungsten 360 clicks below minimum contrast, showing static computation (left) versus computed rotational average (right) at a large defocus. The positions of several minima calibrated using gold particles on the same film are shown. b) Contrast transfer function plot at Δz = -228 nm, which matches precisely the diffractogram, both in position of minima and in the position and appearance of the passband. The CTF for a thermionic emitter at even this moderate defocus (shown overlaid) does not display details of the passband or subsequent minima.

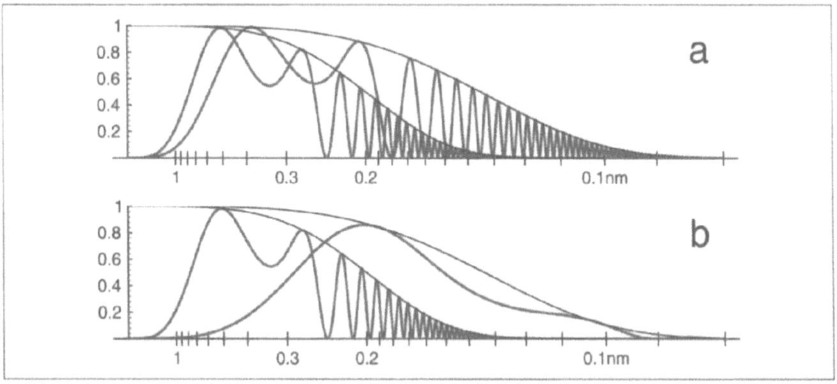

Figure 3.20. CTF plots comparing the anticipated contrast transfer capabilities of a next-generation 300 kV field emission microscope with the present-day contrast transfer of the HF-2000 (from Fig. 3.17c). a) CTF for the 300 kV microscope computed with $C_S = 0.8$ mm, $C_C = 1.4$ mm, Scherzer focus = -52 nm, defocus spread = 1.6 nm, OL and HT stabilities = 0.5 ppm. b) CTF for the 300 kV microscope computed with $C_S = 0.05$ mm as provided by a C_S-corrector, and at the corresponding Scherzer defocus of -12.2 nm. This shows significant contrast transfer at a spatial frequency of 0.1 nm, without accommodating the possibility of correction of 3-fold astigmatism and 5th order C_S, which can improve the information limit even further.

click automatically. This process does not return values for Δ or C_S, however, leaving those parameters to be determined by other techniques.

6.1. The future

A final contrast transfer calculation illustrates the potential for improved imaging with the 'next-generation' (at this writing) field emission microscope. This instrument might operate at 300 kV, have an objective lens with an inherent C_S of 0.8 mm and a C_C of 1.2 mm, and have HT and OL stabilities of $0.5 \cdot 10^{-6}$. With a cold field emitter to give an energy spread of 0.5 eV at an operating emission current of 30 μA (or perhaps a beam monochromator to give even lower energy spread), the defocus spread parameter would be about 15 eV. Using an illumination angle of 0.03 mrad, the CTF for this instrument is shown in Fig. 3.20a, and compared to that of the present-day HF-2000, both calculated for the respective Scherzer focal conditions. The ability to correct C_S to the level of 0.05 mm, as has recently been demonstrated on a prototype 200 kV instrument,[154,155] would yield a transfer function at Scherzer focus shown in Fig. 3.20b. Here we see that there is contrast transfer at 0.1 nm of about 40%, which should provide a capability to not only resolve atomic spacings at this level, but also to determine in a weak phase object the location of different atomic species. This computation does not take into account improvements in contrast transfer due either to correction of 5th-order spherical aberration or to 3-fold astigmatism; these and higher order corrections are also permitted by the C_S corrector. Accommodating these corrections and setting the focus to the optimum focus condition leads to an even better CTF.[369]

The images shown in Figs. 3.21 compare simulations of the appearance of a 6H-SiC crystal 2 nm thick under the contrast transfer conditions shown in Fig. 3.20a and Fig. 3.20b, respectively. It is clear that the Si and C atomic dipoles are unresolvable in the 200 kV case, while they are clearly separated in the 300 kV C_S-corrected case, with the atomic columns of Si easily differentiated from those of C.

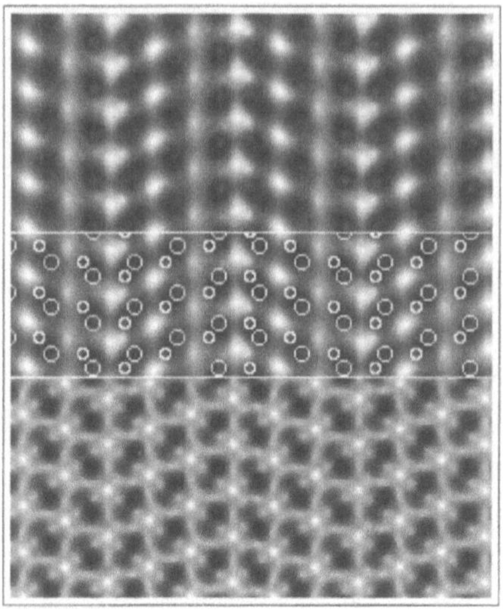

Figure 3.21. Simulated crystal structure images from 6H-SiC in a (110) zone axis orientation.
Top: Image computed at Scherzer focus using the optical parameters of the present HF-2000. Si-C dipoles are not resolved.
Center: Atom overlay; large open circles are Si atom positions.
Bottom: Image computed at Scherzer focus for next-generation 300 kV FE-TEM, with C_S = 0.05 mm. Si-C dipoles are clearly resolved, and atomic species are easy to identify.

While there is promise for electron holography to also provide the ability to correct for spherical aberration to achieve similar imaging performance, as outlined in Chapter 9, it is likely that the utilization of hardware correction techniques will prove to be more straightforward and facile. However, phase images reconstructed from holograms taken in the next-generation instrument described above should provide an enhanced capability for holographic methods such as nanodiffraction from crystal areas as small as a single unit cell[480] and the concomitant capability to determine local sample tilts with precision to facilitate image simulations. Additionally, the width of the field of the hologram can be significantly enhanced if C_S is reduced, as discussed in Chapter 9.

7. Standard criterion for microscopes for holography

The most critical parameter governing the quality of a high-resolution hologram is the contrast of the fringes recorded in the hologram. For higher fringe contrast, better signal-to-noise ratio occurs in the reconstructed phase and amplitude images. As indicated earlier, fringe contrast is governed, to a first approximation, by the brightness parameter of the electron gun which controls the amount of coherent current available for hologram recording. However, gun brightness is usually not the sole criterion for the performance of the microscope for holography, because of the manner in which the coherence can be degraded by environmental factors such as ground loops, stray magnetic fields, instabilities in the gun and lens power supplies, microphonics, microvibration of the biprism filament, temperature instabilities and so on.

In order to assess a particular microscope for its potential to record holograms, we propose a standard test which is specimen-independent.[478] The test involves recording an 'empty' off-axis hologram (i.e., one taken with the specimen removed from the hologram field) under the following microscope conditions:

- Set the biprism voltage to generate fringes with 0.1 nm spacing, as referred to the object plane.

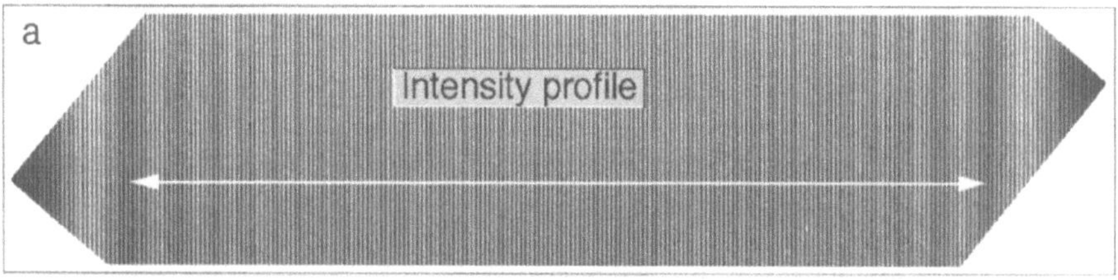

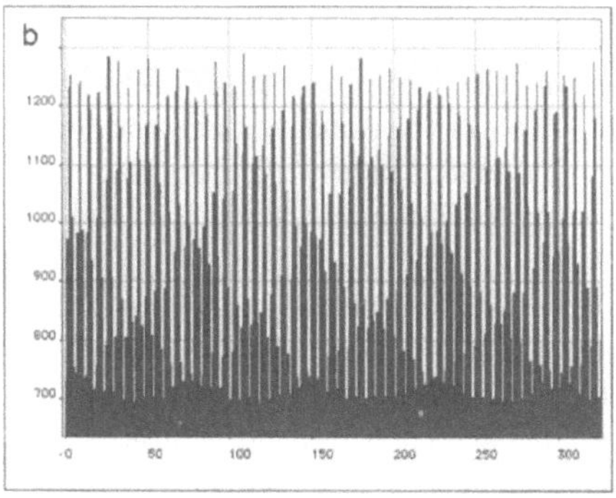

Figure 3.22. a) Empty hologram set up to produce 200 fringes between main minima of Fresnel fringes on each side, at 0.1 nm per fringe nominal spacing. b) Intensity profile of 50 pixel wide line perpendicular to fringe pattern, yielding data for contrast measurement.

- Adjust the imaging lenses to produce 200-300 fringes between the main minima of the Fresnel fringes, as illustrated in Fig. 3.22a.
- Spread the illumination to produce an intensity of $100\,e^-/(\text{A}^2\,\text{s})$. This should assure an isoplanatic patch over the area of 1 inch by 1 inch ($\sim$ 1k by 1k CCD chip) in the (detector) image plane.

With this setup, a number (say 10) empty holograms, all gain- and dark-field-corrected, are recorded using a CCD camera. Then the contrast μ of the fringes for each image is evaluated to obtain the average value $\overline{\mu}$ for the contrast. The number $\overline{\mu}$ gives an estimation of the quality of the microscope for holography. The contrast can be evaluated by using a linescan (integrating over several lines) and the following definition:

$$\mu := \frac{I_{max} - I_{min}}{I_{max} + I_{min}} \tag{3.15}$$

where I_{max} is the maximum in the intensities of the fringes and I_{min} is the minimum in the intensity of the fringes in the central area of the hologram, over which the influence of the Fresnel fringes of the biprism can be neglected (for example, the central 30-60 fringes). Figure 3.22b shows the intensity profile from the center of Fig. 3.22a. Here, I_{max} is 1250 and I_{min} is 700, giving contrast $\mu = 28.2\,\%$. The average from 10 such measurements taken sequentially for this example gave an average fringe contrast of 27.8 %. This is more than sufficient for excellent phase image reconstructions from a hologram taken under these conditions.

In a second method,[476] a Hanning window is first applied to the image, then a Fourier transform is computed and the amplitude of the peaks of the autocorrelation and the sideband is evaluated. The contrast μ is then defined as 2 times the amplitude of the sideband peak divided by the amplitude of the central peak.

Although this standard criterion is not necessarily ideal for the special setup the user may be using, it will provide a sufficient basis with which to compare different microscopes from different vendors, as well as to monitor the performance of a given microscope over some period of time. It is also assumed that the user ensures that the microscope is capable of sufficient magnification to enable a sampling rate of 4 or higher per interference fringe.[472]

8. Conclusion

In this chapter we have attempted to compare the various optical parameters that characterize the field emission microscope with those of the standard thermionic emission microscope, to provide a basis for more fully understanding the operation of the FE-TEM. Special attention has been paid to some specific operational details and calibrations that are necessary and important for doing effective electron holography, in order that the new holographer can produce holograms with greater confidence in their reliability. Because of the unique differences in the illuminating systems between FE and thermionic emission microscopes, concepts of brightness and coherence have been highlighted. The effects of coherence and other parameters on contrast transfer in the FE-TEM have been quantified using CTF plots. Finally, a useful method for monitoring the behavior of a given microscope, or for comparing two different microscopes for doing electron holography, has been described. This background should help the beginning holographer better understand the next chapter, which details specific operational aspects for recording holograms using a variety of holography techniques.

PRACTICAL ELECTRON HOLOGRAPHY

David J. Smith[1] and M.R. McCartney[2]

[1]Center for Solid State Science and Department of Physics and Astronomy, Arizona State University, Tempe, AZ 85287-1504
[2]Center for Solid State Science, Arizona State University, Tempe, AZ 85287-1504

1. Introduction

The underlying basis and motivation for electron holography is that the relative phase change of the electron beam that has passed through the specimen region of interest can be determined by comparison with a coherent reference wave. Thus, unlike the conventional electron microscope image, which represents the intensity of the electron wave without any direct phase information, electron holography enables the phase and amplitude at any point of the coherent wavefront to be obtained. This capability has many useful applications. For example, as will be evident from work described in later chapters, the reconstructed phase image provides the possibility for direct imaging on the nanometer scale of the electric and magnetic potentials within the sample, thereby enabling truly unique information to be obtained.[451] In addition, once the phase image is obtained, an inverse phase plate can be applied so that phase changes imposed by the transfer function of the objective lens, such as spherical aberration and two-fold astigmatism, can be corrected. It then becomes possible to exceed the conventional resolution limit of the electron microscope for bright-field, axial illumination imaging.[261] Finally, it seems appropriate to mention here that the reconstructed amplitude image, which was overlooked by the electron holographic community for many years, can be used to obtain valuable inelastic scattering data.[297] Further details and other applications of the amplitude image can be found in the later chapter by Gajdardziska-Josifovska and Carim, pages 267 ff.

The recent development and availability of reliable field-emission-gun (FEG) electron sources has created an upsurge of interest and activity in electron holography. Instead of being a technique utilized by relatively few specialized laboratories, electron holography has become widespread.[453] Many analytical and high-resolution transmission electron microscopes (TEMs) are nowadays equipped with FEGs: with comparatively minor effort, these microscopes can be reconfigured in several different novel ways to produce electron holograms. The primary purpose of this chapter is to provide a detailed but not-too-technical guide to practical electron holography, especially for those people with access to a FEG instrument but with little or no prior experience of the technique. After describing some of the instrumental requirements, detailed descriptions and explanations are given for the processes of generating and recording electron holograms. Some of the important parameters that affect the quality of the holographic reconstruction process are also considered.

2. Setting up the instrument

There are a multiplicity of conceivable schemes for electron holography involving both the conventional and scanning TEMs, with in-line and off-axis operating modes as well as bright- and dark-field operation.[66] The theoretical bases for many of these modes were outlined in the earlier chapter by Cowley and Spence, and the focus-variation method, which is an in-line form of electron holography, is described later by van Dyck and Op de Beeck. Our focus here is directed primarily towards the practical implementation of off-axis electron holography in the conventional TEM, since it is the most widely used, and perhaps the most easily understood version. We also briefly describe the STEM equivalent of off-axis electron holography as well as providing details of the recently developed, differential form of TEM electron holography which involves insertion of an electrostatic biprism in the condenser aperture plane.[241,299] This latter variant has the considerable attraction that it does not require a vacuum reference wave as part of the recorded field of view. These three modes of operation are illustrated schematically in Fig. 4.1, and described in more detail in the following sections.

It is usually assumed for simplicity that a coherent FEG electron source is available that provides monochromatic incident illumination. In practice, as described in Chapter 3, some loss of coherence is inevitable because of the available brightness of the electron gun. Finite beam divergence (effective source size) and energy spread (temporal coherence) will inevitably have some detrimental effect on the holographic imaging process through loss of contrast of the interference fringes.

This deterioration in quality can be understood by reference to the basic intensity equation (Eq. (2.28)) for off-axis TEM holography which, from Chapter 2, can be written as

$$I_h(\vec{r}) = 1 + |\,o(\vec{r}) \otimes t(\vec{r})\,|^2 + 2\,|\,o(\vec{r}) \otimes t(\vec{r})\,| \cdot \cos(2\pi\,\vec{q}_c\,\vec{r} + \phi(\vec{r})) \tag{4.1}$$

where $o(\vec{r})$ is the object function, $t(\vec{r})$ is the so-called point-spread function of the objective lens (which is the Fourier transform of the lens transfer function referred back to the object plane), $\vec{q}_c$ represents the relative tilt between the incident vacuum reference and specimen plane waves and $\phi(\vec{r})$ represents the phase shifts. This function describes the pattern of cosinusoidal interference fringes with carrier frequency q_c which are modulated in contrast and position by the amplitude $A(\vec{r}) = |o(\vec{r}) \otimes t(\vec{r})|$ and phase $\phi(\vec{r})$. The effect of partial coherence is then easily incorporated

$$I_h(\vec{r}) = 1 + A^2(\vec{r}) + 2\,\mu\,A(\vec{r}) \cdot \cos[2\,\pi\,\vec{q}_c\,\vec{r} + \phi(\vec{r}) + \theta] \tag{4.2}$$

In this expression, μ represents the coherence of the illumination, and is often termed the fringe contrast or visibility, and θ is an additional phase term that is usually independent of position. The problem for the experimentalist is to maintain the fringe visibility as close as possible to its theoretical maximum value of unity. Experimentally, some decrease in μ is inevitable but should it decrease entirely to zero, for whatever reason, then no reconstruction of the complex image wavefunction will be possible.

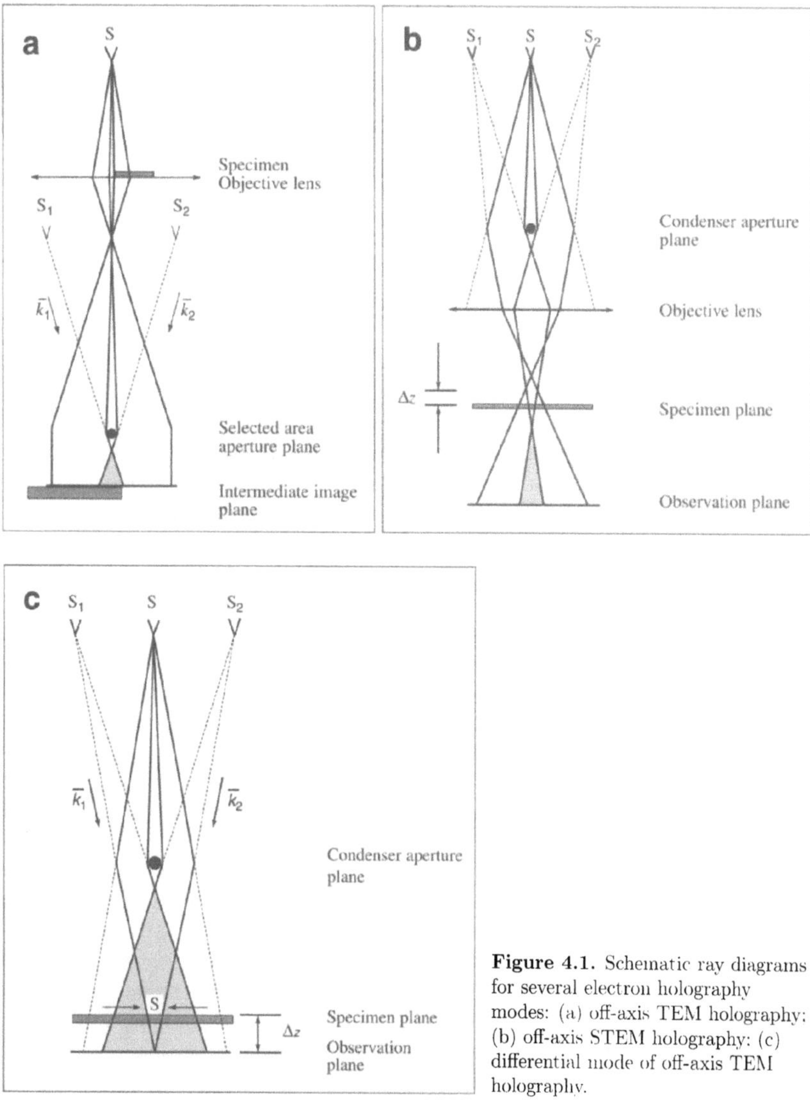

Figure 4.1. Schematic ray diagrams for several electron holography modes: (a) off-axis TEM holography; (b) off-axis STEM holography; (c) differential mode of off-axis TEM holography.

Measurements of μ and experimental factors that may prevent it from being maximized are discussed in later sections.

2.1. Off-axis TEM holography

For off-axis electron holography in the conventional TEM, interference between the specimen wave and the reference vacuum wave is achieved by means of an electrostatic biprism that is located beyond the sample, as illustrated in Fig. 4.1a. The usual location for the biprism is in one of the selected area aperture positions, since it can then be inserted easily and positioned accurately as required. Other locations within the imaging lens system are also possible. However, it should not be overlooked that the biprism can never be located exactly in any intermediate image planes since it then can not have the desired effect of causing overlap between the object and vacuum reference waves. Since most TEMs are normally configured so that the first intermediate image is formed in the selected area aperture plane, some suitable readjustment of the electron optics must be made if the biprism is to be located there. For example, the first intermediate lens could be over-excited and then refocussing of the objective lens will allow the image to be located below the biprism. Reversal of the polarity of the biprism voltage will be required in the event that the first intermediate image is located above the biprism.

In the configuration drawn in Fig. 4.1a, the specimen is situated such that the reference wave arriving at the intermediate image plane from apparent source S_1 has travelled entirely through vacuum. A positive voltage is applied to the biprism, creating a region of overlap between this reference vacuum wave and the specimen wave. As described by Eq. (4.2), the resulting hologram resembles an interference fringe pattern in which sample amplitude information is contained in the relative amplitude of the cosine-like fringes and phase information is contained in their position. This geometry can be considered as analogous to the classic Young's double-slit experiment where one of the two slits is covered by a material having spatially varying refractive index.

2.2. Off-axis STEM holography

The STEM off-axis holography mode is actually a highly defocussed, point projection mode with the scanning coils switched off.[75] When the objective lens is defocussed in the STEM, a shadow image of the object would normally be produced. This defocussed image is actually the in-line hologram proposed by Gabor,[129] which was initially evaluated in limited fashion by Haine and Mulvey well before the advent of the field emission electron source.[157] As illustrated in Fig. 4.1b, the effect of the electrostatic biprism in the condenser aperture plane is to split the stationary illumination before the sample so that it appears as though there are two, mutually coherent, electron sources. The resulting off-axis hologram has an appearance somewhat similar to that of a TEM hologram. However, it is highly distorted as a result of the defocus and spherical aberration of the objective lens, and it is necessary to apply an appropriate correction factor during the restoration process in order that the image features can be interpreted. In this configuration, the image magnification and the separation of the sources relative to the sample are quite flexible, and can be easily adjusted by changing the biprism voltage and/or the objective or post-specimen lens settings.[279]

2.3. Differential phase contrast holography

By placing a biprism in the condenser aperture plane, as drawn in Fig. 4.1c, a TEM configuration can be achieved that is analogous to the off-axis STEM mode.[299] In this case, application of a positive potential to the biprism results in two, closely spaced, overlapping plane waves that create an interference fringe pattern at the specimen level. By defocussing the observation plane by some small distance Δz with respect to the specimen plane, then the two coherent beams produced by the beam splitter, labelled k_1 and k_2, will impinge upon different regions of the specimen. In the special case of a magnetic material, the relative phase shift of the holographic fringes will provide the component of the magnetization parallel to the biprism wire between these two points in the specimen plane. In the case of a uniform domain, a constant field produces a constant phase shift. Contrast in the phase image is thus produced whenever the field changes in either direction or magnitude. This configuration can, in effect, be

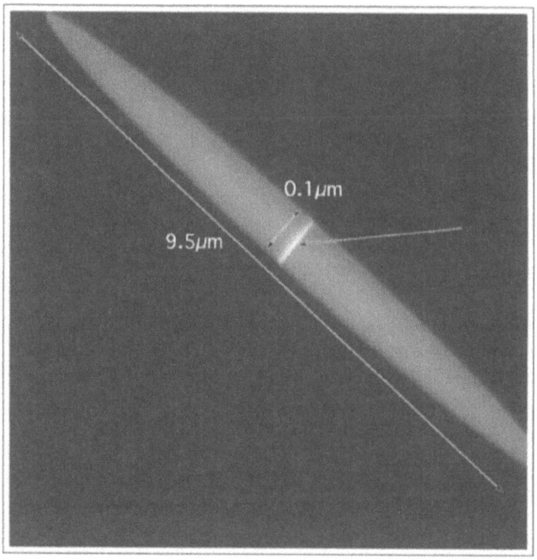

Figure 4.2. Highly elliptical incident illumination produced as a result of deliberate condenser astigmatism showing typical optimum condition used for recording electron holograms. Major axis and minor axes dimensions of 9.5 μm and 0.1 μm, respectively. Position of holographic interference fringes indicated by arrow.

considered as producing differential phase contrast (DPC), although it should be noted that a full two-dimensional DPC map is only available through use of a rotatable biprism or sample holder.

2.4. Condenser astigmatism

As pointed out earlier, perhaps THE most important experimental factor that impacts electron holography is the lateral coherence of the electron beam since this directly determines the holographic fringe contrast.[453] For highest quality electron holograms, it is essential that the coherence is maximized as far as possible by minimizing the beam convergence angle. Usually this requirement would imply demagnification of the electron source as far as practicable within the constraints on available current density per unit solid angle imposed by the finite brightness of the source. Careful consideration of the imaging geometry shows, however, that the use of rotational symmetric illumination is not usually required.[219,453] In fact, it could even be considered as wasteful since excessive numbers of electrons are not contributing usefully to the holographic recording. What is essential for hologram reconstruction is the coherence across the incident illumination between the points that are being exactly overlapped by the biprism.

A significant improvement in the coherence available with any given electron source and condenser lens system can be achieved by employing highly elliptical illumination, i.e., the condenser lens stigmators and focus settings are deliberately adjusted to produce an illumination patch that is very wide in the direction perpendicular to the biprism when the condenser lens is overfocussed, but relatively narrow in the parallel direction. For example, in a typical, medium resolution situation, with the image located below the biprism, the minor axis of the ellipse might be adjusted to be a factor of 2 to 5 times greater than the region of overlap between the reference wave and the object wave, while the major axis could be a factor of 50 to 100 times greater yet again. For the situation shown in Fig. 4.2, before an adjustment was made to fit the entire illumination patch onto the recording medium, the major and minor axis dimensions

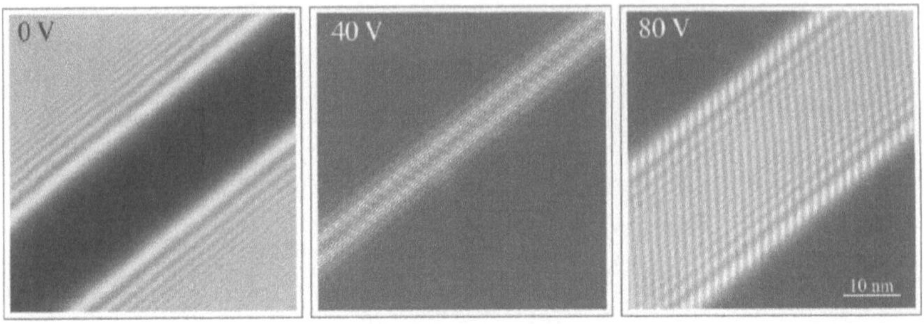

Figure 4.3. Succession of images that show effect of biprism voltage on hologram formation: (a) Zero *V*. Visible fringes result from Fresnel diffraction around edge of biprism; (b) 40 V. Object and reference wave have overlap region of 8 nm; (c) 80 V. Note modulation due to Fresnel diffraction fringes at edges of overlap region which has been increased to 27 nm.

were 9.5 μm and 0.1 μm, respectively, whereas the overlap region of interference fringes (arrowed) was measured to be about 40 nm.

It is important, when using elliptical illumination, to ensure that the major axis is indeed perpendicular to the biprism wire: otherwise, some loss of beam coherence and fringe contrast will result depending on the angular misalignment. This particular geometry can be routinely achieved by suitable adjustments of the condenser stigmator strengths when the condenser lens crossover position has been adjusted so that the major axis of the elliptical illumination patch is instead parallel to the biprism wire. It is convenient in practice to make this adjustment with a relatively low biprism voltage since the holographic fringes will then have higher contrast and larger spacings.

2.5. Biprism adjustment

The electrostatic biprism is integral to almost all forms of electron holography. It usually consists of a very thin wire, preferably less than 0.5 μm in diameter, that is conducting and capable of maintaining potentials as high as 500 V or more for high-resolution applications. Its position should be adjustable, and a capability for rotation is also desirable for many applications. A quartz fiber, coated with a thin, evaporated metal film to make it conducting, is most commonly used,[157] although one microscope manufacturer nowadays provides a biprism made from a platinum wire. Adjustment flexibility, including the possibility for insertion or removal from the beam path, is provided by mounting the biprism across one of the openings of a regular multi-aperture holder strip, which is normally the selected-area holder for the case of off-axis TEM electron holography.[219]

The effect of increasing the voltage on the biprism is illustrated by the series of images in Fig. 4.3. At zero voltage, the shadow of the biprism wire is surrounded by coarse fringes which are due solely to Fresnel diffraction effects at the edges of the wire. At 40 V, the two waves from either side of the wire overlap over a distance of about 8 nm to produce the cosinusoidal interference fringes: these are, however, still strongly modulated by the Fresnel diffraction at the wire. At 80 V, the overlap region of the interference fringes has expanded to about 27 nm, the fringes have decreased in spacing, and the Fresnel diffraction effects only affect the outside edges of the interference pattern. At even higher biprism voltages, the fringe separations will decrease further but the overlap region will continue to expand.

Figure 4.4. Exposures of increasing duration as labelled showing build-up of electron interference fringe pattern.

3. Recording holograms

It is interesting to document the accumulation of intensity in the holographic fringe pattern, as illustrated in Fig. 4.4, which shows a set of recordings on a CCD camera in which the time of exposure has been successively increased by factors of 10. As well as illustrating the statistical nature of the recording process, the quantum mechanical nature of the interference process for individual electron wavepackets is demonstrated. For a very short exposure of 0.001 s, the image appears to consist entirely of random noise but the fringe pattern becomes increasingly apparent with longer exposure times. Since the precision of any holographic phase measurement at any point depends inversely on the square root of the total number of counts, i.e., the signal-to-noise ratio,[87] then it is obviously preferable to expose for longer periods. Instrumental factors such as the stability of beam deflectors or specimen drift will, however, impose an upper limit on the useful exposure time.

3.1. Off-axis TEM holography

Before proceeding with off-axis TEM holography, it is first necessary to properly adjust the microscope for normal operation using the lens settings required for holography. For example, the condenser aperture should be centered, the objective lens astigmatism should be corrected and the incident illumination should be aligned onto the coma-free axis.[508] It is also helpful to locate the specimen feature of interest and carry out the usual tilting and coarse focusing operations. An additional requirement that cannot be avoided in this mode of operation is that the edge of the specimen, or a convenient hole if available, must be somewhere in the field of view in order to provide the reference beam. This restriction is removed for the differential phase contrast TEM mode. For moderate resolution operation, final focusing is not really too critical although it is often convenient to record holograms somewhere between the Scherzer defocus and the Gaussian defocus, which is usually easily recognized. For off-axis holography with atomic resolution, an optimum defocus can be defined that maximizes the resolution of the reconstructed object wave after correction of aberrations.[262] Further details can be found in Chapter 9 (pages 201 ff) on high-resolution electron holography by Rau and Lichte.

After attending to these necessary preliminaries, the electron holography part of the observation process can begin. As a precautionary measure, if the sample is liable either to beam damage and/or contamination, it may be displaced just outside the electron beam. The biprism is inserted into the beam path, the desired biprism voltage is chosen according to the dimensions of the feature of interest (see later), and the elliptical illumination patch is adjusted as described in the preceding section. The magnification should also be chosen to ensure adequate sampling of the interference fringes. These fringes, though faint, should be observable either through binoculars or on the monitor screen when a low-light-level TV or slow-scan CCD camera is available. Where possible, the latter should preferably be used at least for hologram recording since it provides quantitative information in a digital format that is convenient for subsequent signal processing.

As an illustrative example of off-axis TEM holography, Fig. 4.5 shows part of a Si(110) crystal wedge with superimposed holographic interference fringes having a spacing of about 0.75 Å. Note the local changes in amplitude and the sideways shifts of the fringes. The corresponding Fourier transform of this hologram is displayed in Fig. 4.6. In addition to the usual auto-correlation function that is visible in the central region, two additional sidebands that are associated with the cosine term in Eq. (4.1) are also visible. The separation of these sidebands from the central peak varies inversely with the interference (or "carrier") fringe spacing. The next step in the reconstruction process, i.e., retrieval of the complex image wave, involves taking the inverse Fourier transform from a region centered on one of these sidebands. It is then important to ensure that there is actually no overlap of the sideband information with the outer edge of the central peak, unlike this particular example where some overlap occurs. Typically, this requirement means that the carrier fringe spacing in the original hologram should be at least three times smaller than the highest spatial resolution detail from the object that is contained in the image.[261] Finally, it is interesting to note that, since holographic fringes can only be formed by the interference of two coherent electron waves, the sidebands do not contain any appreciable fraction of the inelastically scattered electrons that have emerged from the specimen. Thus, the complex image reconstructed from the sideband is in fact an energy-filtered image.[167,297]

Figure 4.5. High-resolution off-axis TEM hologram showing edge of (110) Si crystal recorded at electron-optical magnification of 700 kX with Hitachi HF-2000 FE-TEM operated at 200 keV, also showing superimposed 0.75 Å interference fringes.

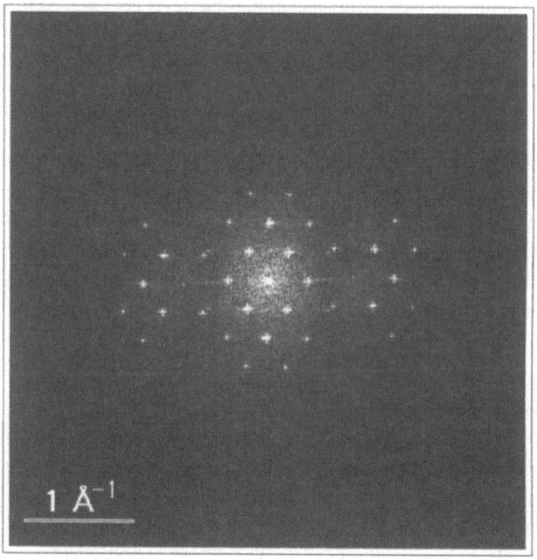

Figure 4.6. Fourier transform of Fig. 4.5. Note presence of two sidebands with separation related inversely to fringe spacing. Inverse transform of sideband produces complex image wavefunction.

3.2. Off-axis STEM holography

Similar setting-up considerations apply for off-axis STEM holography. Beam alignment, specimen tilting, image focusing and stigmating should all be done before the biprism and the incident illumination are readied for holographic operation and recording. A significant difference of the STEM mode is that the interference fringes are formed in the plane of the sample. However, Fourier transformation of the recorded hologram again yields a central auto-correlation peak and two sidebands, either one of which can be used to recover the complex image wavefunction.[279] It still remains to

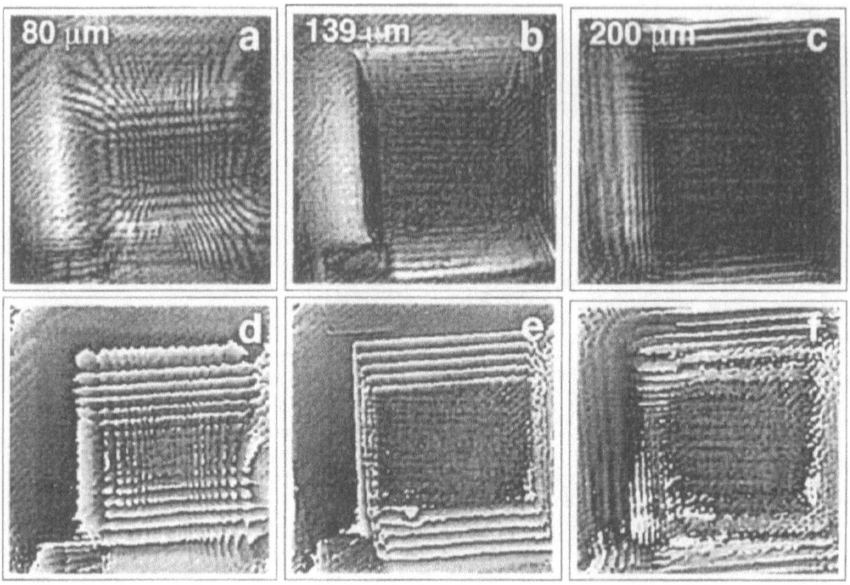

Figure 4.7. Defocus series of phase and amplitude reconstructions from cubic MgO crystal used in determining focus correction factor. Note disappearance of Fresnel fringes in amplitude image (b). Courtesy of J.M. Cowley.[75]

remove the influence of the lens aberrations which otherwise cause considerable distortion, resulting in Fresnel fringes around edges and holes that could obscure image features. In the event that the defocus value is accurately known, then the aberration correction is quick and simple. Otherwise, as illustrated by the example of a cubic MgO crystal shown in Fig. 4.7, it may be necessary to resort to an empirical search in order to determine the correct defocus. The disappearance in an amplitude image of the Fresnel fringe at a crystal edge has been successfully used as a criterion.[75]

3.3. Differential phase contrast mode

It is a severe drawback of the normal off-axis TEM holography mode that the vacuum reference wave must occupy a sizeable fraction of the field of view. In many applications, the feature of interest is not conveniently located near the specimen edge. Implementation of the differential phase contrast mode in the TEM enables this restriction to be overcome.

In DPC imaging, the two effective sources produce slightly displaced but overlapping images, with a separation determined by the amount of objective lens defocus. Since the holograms must be acquired under out-of-focus conditions, they are in effect the superposition of a pair of Fresnel images. The typical appearance of the DPC image is illustrated by the 30 nm thick Co film shown in Fig. 4.8a, with an enlargement in Fig. 4.8b. Outside the region of the interference fringes due to the biprism, the image shows the dominant black and white lines which delineate large magnetic domains, with magnetization ripple visible within the domains. The enlargement clearly shows the doubling of image features which is due to the split incident beam, as well as bending of the holographic fringes as they cross regions of changing magnetic fields.

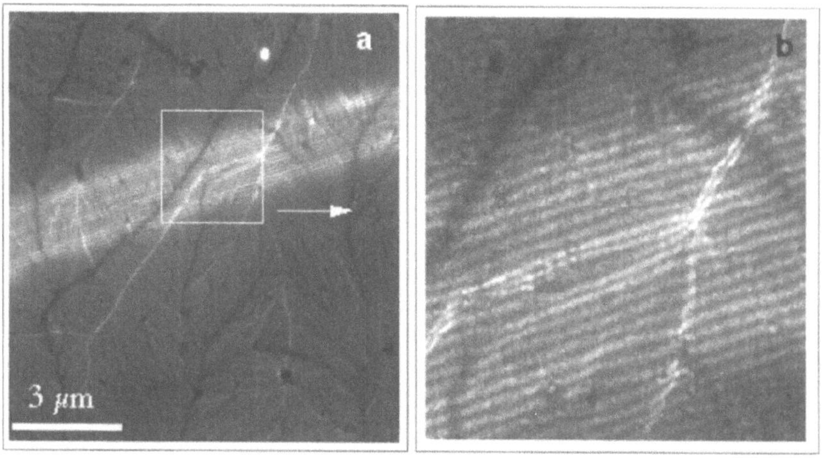

Figure 4.8. (a) Differential mode hologram of thin Co film; (b) enlargement of indicated area showing fringe bending and doubling of image features. From Ref. 299.

4. Important parameters

During the process of holographic imaging, there are several experimental parameters that have a significant influence on the integrity of the subsequent reconstruction process. Factors ranging from the source brightness and coherence to the effective dimensions of the recording medium (pixel size) need to be taken into consideration because of their interrelationship with the dimensions of the specimen feature of interest and/or the resolution level required in the final reconstruction. The requirements are more demanding for high-resolution electron holography.[265]

4.1. Fringe spacings

In general, higher voltages on the biprism mean smaller interference fringe spacings (which are needed for small object dimensions) and larger fields of view (region of fringe overlap) but the fringe visibility will generally be less and the microscope magnification may need to be increased to ensure adequate fringe sampling. Geometrical analysis indicates[258,261,265] that the interference fringe spacing varies inversely with the biprism voltage, with the constant of proportionality depending on several instrumental parameters including the accelerating voltage (electron energy), the focal length of the objective lens and the physical separation between the back focal plane of the objective lens and the biprism. The relationship between fringe spacing and biprism voltage can be calculated if the relevant dimensions are known but it is also straightforward simply to plot the measured fringe spacing as a function of the applied voltage, as shown in Fig. 4.9 for the Philips CM200-FEG in our laboratory. Depending on the geometry of the particular instrument, biprism voltages between about 150 to 240 V are typically required to attain a fringe spacing of 1 Å. To reach a resolution of 1 Å in the restored wavefunction requires fringe spacings on the order of 0.33 Å,[471] implying a biprism voltage of perhaps 500 to 700 V. Special care in fabrication of the biprism is then required to ensure freedom from any electrical breakdown.

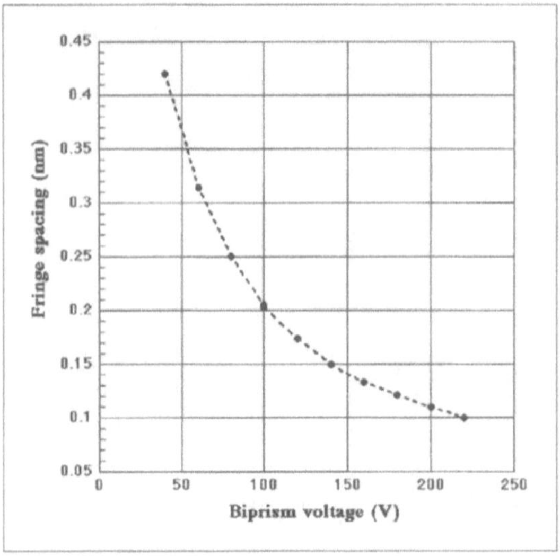

Figure 4.9. Dependence of interference fringe spacing on biprism voltage for Philips CM200-FEG at 200 keV.

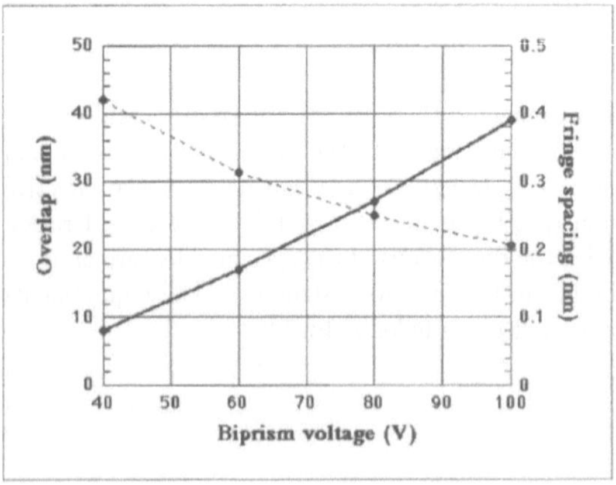

Figure 4.10. Plot of fringe overlap vs. biprism voltage for normal off-axis TEM holography operating mode of Philips CM200-FEG at 200 keV. Also shown (dotted) for comparison is corresponding trace for fringe spacing over same voltage range.

4.2. Fringe overlap

Analysis is again useful in establishing that the region of overlap of the interference fringes is strongly influenced by the biprism voltage, in addition to several geometrical factors which, most importantly, include the distance from the intermediate image plane to the plane of the biprism.[265] Since the biprism voltage may already be predetermined by the fringe spacing required for a particular application, the only available option left to the operator for adjusting the width of the hologram is to change the strength of the subsequent lens after the biprism. The availability of "free lens" control, at least for this lens, is thus an important consideration for anyone contemplating doing off-axis electron holography. If the instrumental parameters are known, then the extent of the overlapping fringes can be determined. However, as shown in Fig. 4.10, it is not experimentally difficult to measure the width of the overlap region as a function of the biprism voltage, and then to plot the result graphically for future reference. In practice, because the coherent current available from the electron source is ultimately limited by the finite source brightness, the biprism voltage cannot be increased indefinitely to expand the field of view. Multiple exposures from adjacent regions might then be possible provided that proper care is taken to ensure mutual registration of the images.

4.3. Fringe visibility

The holographic interference fringe visibility is all-important to the reconstruction process since it directly determines the precision attainable in phase measurements.[87] Since phase determination is usually the primary objective of the electron holographic studies, all possible steps should be taken to maximize the fringe visibility within the limits imposed by the partial coherence of the electron beam. In addition to all the usual instrumental factors such as high-voltage stability, beam energy spread and specimen stage motion, the realizable fringe contrast is especially sensitive to the mechanical stability of the biprism and the electrical stability of the beam deflection supplies, as well as stray ac magnetic fields. We also find that the fringe visibility in our Philips CM200-FEG is quite sensitive to the extraction voltage of the Schottky FEG electron source, presumably because it shifts the position of the gun crossover, and hence changes the effective beam divergence.

There are several alternative ways to determine fringe visibility.[478] The standard method is based on a line trace across the recorded interference fringe pattern in the absence of a sample, using the relation

$$\mu = \frac{I_{max} - I_{min}}{I_{max} + I_{min}} \tag{4.3}$$

However, as illustrated by the example in Fig. 4.11, an averaged profile is preferable when a CCD detector is used because of the likelihood that the interference maxima and minima will lie between pixels. An alternative method is to take the Fourier transform of the hologram and measure the heights of the central peak and the two sidebands using interpolation methods.[85] The fringe visibility is then defined in terms of twice the amplitude of a sideband divided by the amplitude of the central peak.

4.4. Effective pixel size and image magnification

The possibility remains that the detection properties of the recording medium could have a serious detrimental effect on the holographic reconstruction process. The usage of film, for example, is not recommended due to its non-linear features.[475] Many of the issues related to quantitative electron holography are discussed in a later chapter: our concern here is the immediate practical problem that considerable loss of information is liable to occur unless the image magnification is suitably adjusted to accommodate the finite size of each recording element (or pixel). The two solid lines in Fig. 4.12 show plots of interference fringe contrast as a function of biprism voltage at two different electron-optical magnifications at the CCD detector of 490 kX and 1.4 MX. At the biprism voltage of 100 V, the only change made to any of the microscope settings was an increase in the image magnification using lenses located beyond the biprism. Thus, the factor of almost 3 drop in measured fringe contrast can be attributed solely to the modulation transfer function (MTF) of the CCD camera. In this case, the attenuation of high-spatial frequency periodicities is due primarily to the characteristics of the scintillator that precedes the CCD array, as well as the finite pixel size of the detector.[85] Provided that the MTF of the camera has been previously

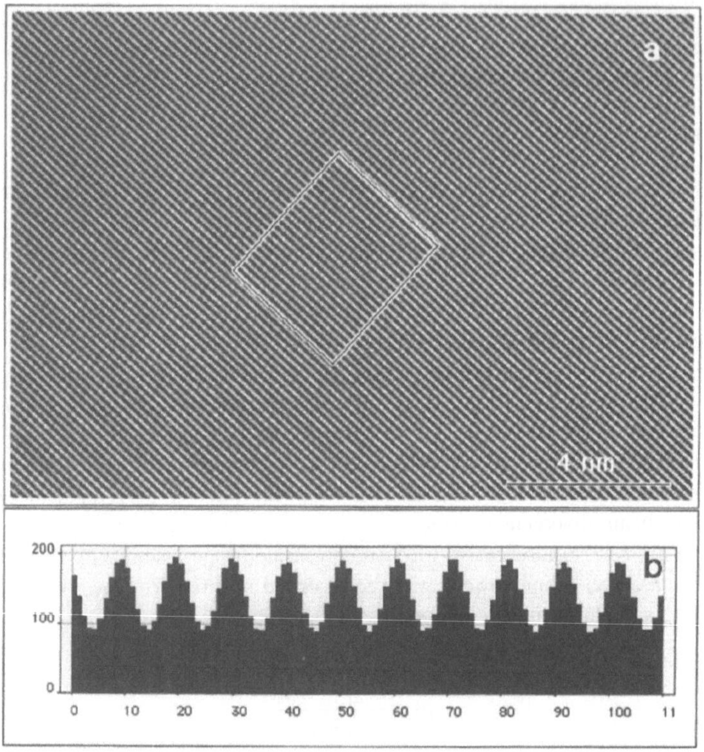

Figure 4.11. (a) Central region of interference fringe pattern recorded in off-axis TEM holography mode with Philips CM200-FEG at 200 keV with biprism voltage set at 120 V. (b) Line trace of fringe intensity profile after pixel averaging from boxed area shown in (a). Measured fringe visibility of 38%.

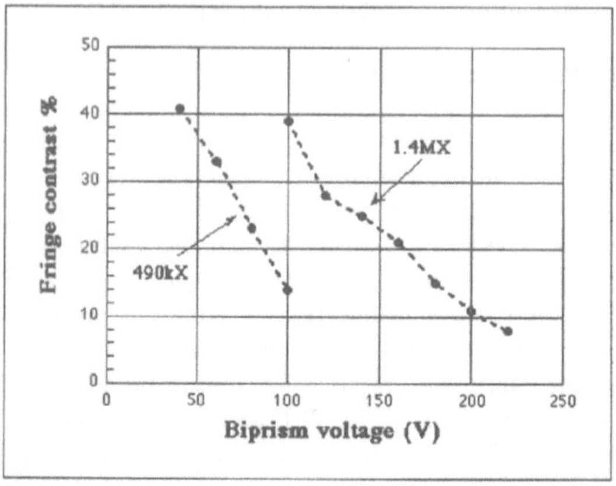

Figure 4.12. Measurements of fringe visibility versus biprism voltage at magnifications of 490 kX and 1.4 MX as shown. Note apparent major loss of fringe visibility at 100 V due to undersampling of interference fringes.

characterized, a restoration filter can, in principle, be applied to compensate for the attenuation. However, in addition to contrast enhancement, the effect of applying the filter during the reconstruction process will be to cause artefactual asymmetries in the

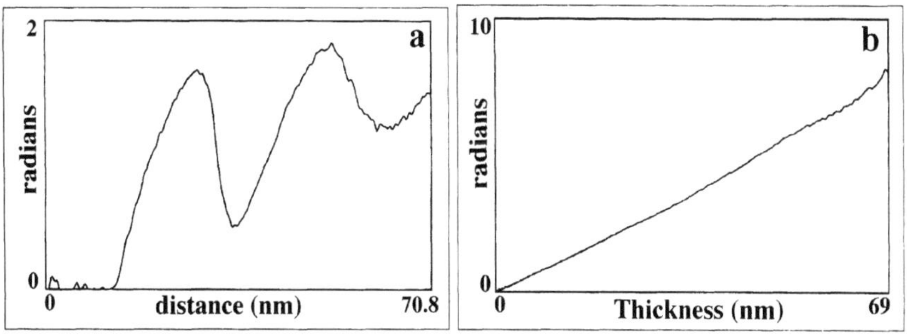

Figure 4.13. Reconstructed phase profiles from 90° GaAs wedge: (a) at [100] zone axis; and (b) tilted by 40.4 mrad away from [100] zone axis along a systematic row of reflections.

information extracted from the sideband. Thus, use of a restoration filter is not usually recommended for applications involving off-axis electron holography.

This attenuation at higher spatial frequencies is a common characteristic of all recording systems irrespective of the incident electron energy. If excessive loss of fringe contrast is to be avoided, especially at higher voltages, then the image magnification must be increased to ensure better sampling of the fringes. A common rule of thumb is that four pixels per hologram fringe is the minimum acceptable sampling rate.[219,254,265] However, quantitative MTF measurements[85] indicate that a factor of about two loss of fringe visibility can then be anticipated unless a suitable restoration filter is applied. For sensitive phase measurements, where the highest possible fringe visibility is needed, our recommendation would be to aim for an image magnification that allows a sampling rate of 6 or even 8 picture elements per fringe period.

5. Further considerations

5.1. Specimen thickness and orientation

In the absence of dynamical diffraction, or any electric or magnetic fields, the phase change $\Delta\varphi$ of the electron wave that has travelled through a specimen of thickness t and mean inner potential V_0, relative to an electron that has travelled through the same distance in vacuum is given by[357]

$$\Delta\varphi = \frac{2\pi e\,(E_0 + E)}{\lambda\,(2E_0 + E)} \cdot V_0\, t = C_E\, V_0\, t \tag{4.4}$$

where e is the electron charge, λ is the electron wavelength. E is the electron kinetic energy, E_0 is the rest energy, and C_E is a constant that depends on the electron energy. Provided that the specimen thickness is known, for example, as a consequence of its particular geometrical shape, then measurement of the phase should enable the mean inner potential to be determined.[254] In practice, this representation proves to be a considerable oversimplification. Dynamical diffraction effects are always likely to make substantial contributions to the phase of the transmitted beam, particularly when the crystal is tilted to a major zone axis orientation. These effects must be compensated before a reliable estimate for the mean inner potential can be obtained.[135]

As an example, Fig. 4.13 compares experimental phase profiles, as a function of distance from the crystal edge, for a GaAs crystal cleaved along two {110}-type planes

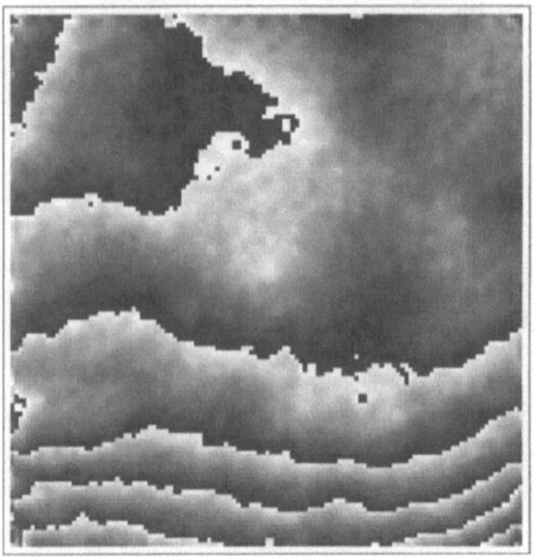

Figure 4.14. Reconstructed phase image from hologram recorded in absence of sample, showing phase variations across the field of view originating from projector lens distortions and fiber-optic shear of CCD camera. Each black-to-white contour corresponds to $\pi/8$ phase step.

to achieve a 90° wedge angle. For Fig. 4.13a, the crystal is oriented very close to a [100] zone axis whereas, in Fig. 4.13b, it is rotated by 40.4 mrad along a systematic row away from this projection so that it is no longer in such a strongly diffracting condition. Large phase jumps, which correspond to extinction contours in the amplitude image, are visible in the former profile whereas a steady, monotonic increase in phase with crystal thickness occurs for the latter, misoriented case. Despite this linear behavior, which is apparently consistent with Eq. (4.4), Bloch wave calculations are still necessary to evaluate the contributions due to residual dynamical diffraction.[135]

In summary, the point here is that these dynamical effects should always be kept in mind when undertaking holographic studies of novel crystalline materials. For example, rapid changes in phase across an interface between two crystals should never be naively interpreted in terms of an equally rapid change in composition unless care is taken to ensure that both crystals are tilted well away from strongly diffracting conditions. Reconstruction of a second hologram recorded after tilting by several degrees about the interface normal is an alternative method of checking for the sensitivity of the particular materials to dynamical effects.

5.2. Reference hologram

Phase information is unavailable in conventional TEM micrographs because only the image intensity is recorded. Thus, geometric distortions due to the imaging lens system, which may have a considerable influence on the phase information at any point, will generally have only a very minor effect on the recorded image. The small local shifts of the interference fringes will, however, affect the apparent phase of the reconstructed wavefront and possibly also the attainable resolution.[269] A rough estimate of the magnitude and importance of the effect can be easily made. Consider a typical example where one period of the cosinusoidal interference fringes, which corresponds to a phase shift of 2π, measures about 100 μm (i.e., about 4 times the size of a single CCD pixel). In order to record reliably a phase shift of $\pi/50$, distortions at the level of the detector should be considerably less than 1 μm. As shown by the example in Fig. 4.14, routine processing of reference holograms, recorded in the complete absence of any sample, typically reveals that phase shifts in excess of $\pi/2$ across a field of view are common.[87] These phase shifts originate from projector lens field distortions as well as fiber-optic shear in the CCD system. Since the distortion phase pattern is purely geometrical in origin and invariant in time, it is a simple matter to subtract these distortions from the reconstructed phase of the image wave when a CCD camera is used for recording a pair of holograms with and without the specimen present. Residual phase errors of $\sim \pi/100$ are then attainable.[87] Even though it has been suggested that the phase errors should be eliminated by recording the electron hologram on high-resolution photographic film at low magnification, it should be apparent from the above that complete phase correction will not be possible without a reference hologram.[450]

5.3. Phase unwrapping

After the sideband has been selected and the inverse Fourier transformation has been completed, the phase image is obtained from the complex image by calculation of the function atan2 (i/r) (standard function of the C language), where i is the imaginary part of the image and r is the real part. Because the principal values of this function are only defined between $-\pi$ and π, phase images where the phase varies by more than 2π will be ambiguous, and thus the phase will need to be unwrapped. Phase wraps will appear as abrupt jumps in phase of less than 2π radians.

Various phase-unwrapping algorithms can be used to locate and unwrap these discontinuities. The most straightforward of these involves searching the image by row or column for adjacent pixels whose phase difference is more than some prespecified threshold value, say 4 or 5, and then adding or subtracting 2π to subsequent pixels. Note, however, that for very steep phase gradients, the phase difference per pixel across the phase wrap may be lower than the threshold. In such cases, simple algorithms may fail and more sophisticated unwrapping algorithms are needed, for example one that keeps track of the derivative of the preceding phases. In an alternative scheme,[87,479] a second phase image may be calculated in which a constant (α) may be added to the phase by multiplying the complex image wave by a factor of the form $(\cos\alpha + i\,\sin\alpha)$. After the phase is calculated, the phase shift a is then subtracted from the phase image. This procedure moves the 2π phase jumps to different positions in the phase image. In this manner, areas of the first phase image which were difficult to unwrap, may be replaced by identical areas of the second phase image (possibly with $\pm\pi$ added) where there are no wraps. This unwrapping procedure is illustrated by the pair of phase images shown in Fig. 4.15, which originated from the edge of an Ni crystal.

5.4. Artefacts

Phase distortions arising from the imaging and detection system, as described in a previous section, can be a serious source of artefactual information, and correction should therefore be done on a routine basis. In addition, there are several other instrumental sources of artefacts that are likely to cause errors in the reconstructed image wave. These have been analyzed in some detail elsewhere because of the potentially serious consequences for high-resolution electron holography unless they are properly accounted for.[269] The following is a short summary intended to alert the newcomer to their existence.

Fresnel diffraction at the biprism filament will inevitably cause phase and amplitude modulation of both the image and reference waves, as for example visible at the outside edges of the fringe pattern in Fig. 4.3c. The disturbances can be avoided by utilizing the undisturbed part of a broader hologram for the recording and reconstruction process.

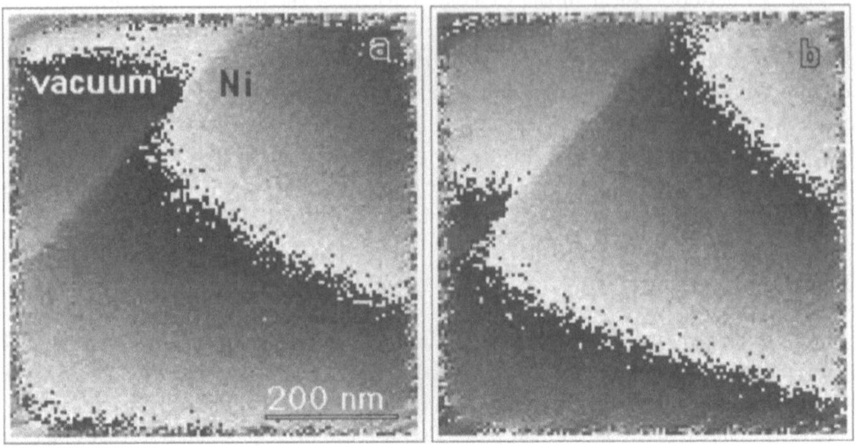

Figure 4.15. Phase unwrapping as applied to edge of Ni crystal: (a) original phase image; (b) phase image everywhere shifted by π.

A further geometrical effect known as vignetting also results from the presence of the biprism. In effect, a thin strip of the hologram close to the shadow of the biprism will contain only single sideband information about certain spatial frequencies because of screening by the filament: without complete phase and amplitude information, the object wave in this region cannot be uniquely determined, and the reconstructed wave should be interpreted with care.

Yet another effect, usually known as windowing, arises during high-resolution electron-wave reconstruction as a direct result of using a finite window to select one of the sidebands. Deconvolution of the wave aberration function of the objective lens then leads to a reconstructed wave that will be considerably different from the true object wavefunction around the outer edges of the original hologram area. Analysis indicates that about three-quarters of the field of view should be correctly reconstructed.[269]

5.5. Problems

In its usual configuration, as depicted in Fig. 4.1a, off-axis TEM electron holography has two major limitations: the somewhat restricted field of view and the need for a vacuum reference wave. As we have seen, the field of view is constrained first by the requirement that the carrier fringe spacing should correspond to about one-third of the maximum image wave resolution, and second by the need for adequate sampling of the interference fringes by the detector. For an order of magnitude estimate, consider that a resolution of 3 Å would require 1 Å fringes, and even with four pixels per hologram fringe, one pixel would represent 0.25 Å. For a 1024 x 1024 pixel array, the total field of view would only be $\sim$ 25 x 25 nm^2, of which perhaps one-third would be given over to the vacuum wave needed for reference purposes. Of course, the available field of view could be increased considerably by using photographic film rather than a CCD detector but at the expense of a non-linear response and loss of the opportunity to provide a reference hologram.

One possible method for enlarging the field of view would be to acquire a series of holograms successively with either the sample or (preferably) the fringe system shifted betweeen exposures. This technique has been adopted successfully, albeit at

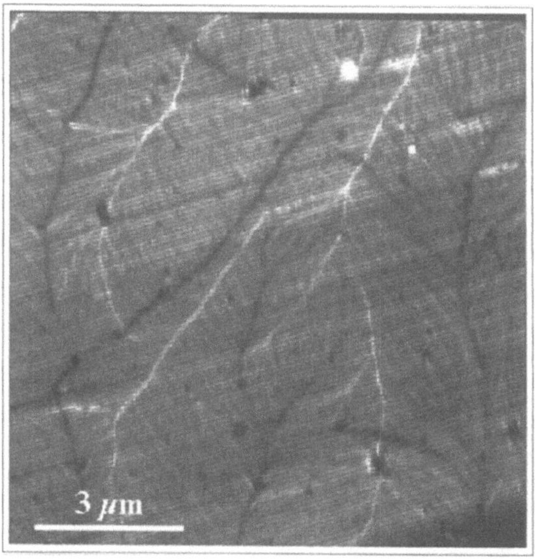

Figure 4.16. Composite hologram of thin Co film produced from a series of eight separate holograms. (From Ref. 299)

low magnification, using the differential phase contrast TEM mode as shown by the composite hologram of a thin Co film in Fig. 4.16. In this particular case, the entire series of 8 holograms was aligned using cross-correlation techniques that were applied to high-contrast image features outside the holographic fringe region. Phase continuity from region to region was also ensured using cross-correlation methods. Although the technique is processing intensive, similar methods involving multiple exposures could very well be used as a means to track phase (and amplitude) changes as the sample edge is progressively further removed from the field of view during a series of multiple exposures.

Another significant feature of the DPC holographic imaging mode, demonstrated nicely by the reconstructed image of the 30 nm Co film shown in Fig. 4.17, is that the requirement for a sample edge to be somewhere in the field of view can be completely removed. In this phase image, the image contrast has been adjusted so that it is proportional to the magnetic field component parallel to the holographic fringes, as indicated by the grey scale bar below the image. Interpretation of the image features indicates, for example, that several vortices are present, such as the one arrowed in the lower right portion of the image. Finally, as an aside, it is worth pointing out that, by using a rotatable biprism or sample holder, this imaging method would allow all in-plane magnetic field components to be mapped out.

Finally, for the benefit of anyone seriously contemplating high-resolution electron holography, it must be emphasized that wave aberrations introduced by the objective lens must be properly taken into account in order that the reconstructed image wave can be related reliably back to the object wave. It cannot be assumed that the phase of the reconstructed image wave is simply transformed to the phase of the object, nor the amplitude of the image to the amplitude of the object. In fact, the transfer characteristics of the objective lens will normally cause mixing of phase and amplitude information at higher spatial frequencies beyond about 0.2 to perhaps $0.3\,\text{Å}^{-1}$ depending on the particular lens and microscope settings. Careful unscrambling is required in order to achieve the true object wavefunction. The usual method is simply to divide the Fourier

Figure 4.17. Differential phase contrast image of 30 nm Co film. Scale bar indicates direction of magnetic field analyzed. Vortex indicated by arrow. (From Ref. 299.)

spectrum of the reconstructed image wave by the lens transfer function,[1,261] although the accuracy with which the aberrations have been predetermined will ultimately limit the resolution of the restored object wave.[263] Further discussion of this issue can be found in the later chapter by Rau and Lichte.

6. Conclusion

We have concentrated our attention primarily on the technique of off-axis TEM electron holography since it is by far the most widespread variant and the one that most newcomers to electron holography are most likely to investigate first. Nevertheless, much of the discussion about instrumental parameters and further considerations is equally applicable to many of the other forms of electron holography and should still be useful. An important message to remember is that electron holography provides convenient access to both the phase and the amplitude of the electron wave after it has passed through an electron-transparent sample. Many novel avenues for fruitful research, only some of which are currently being explored, thereby become accessible to study.

QUANTITATIVE ELECTRON HOLOGRAPHY

D.J. Smith,[1] W.J. de Ruijter,[2] J.K. Weiss[2] and M.R. McCartney[3]

[1]Center for Solid State Science and Department of Physics and Astronomy, Arizona State University, Tempe, AZ 85287-1504
[2]EMiSPEC Systems Inc., 2409 S. Rural Rd., Suite D, Tempe, AZ 85282
[3]Center for Solid State Science, Arizona State University, Tempe, AZ 85287-1504

1. Introduction

Electron holography is necessarily a two-step process. First, the object and reference waves must be coherently superimposed in recording the electron hologram. Second, reconstruction of the hologram is required in order to extract the desired phase and/or amplitude of the electron wave which has passed through the object. Historically, hologram recording has been carried out photographically, and the subsequent reconstruction has been done optically.[261,451] In addition to inevitable time delays, the use of these techniques in electron holography can result in unsuspected artefactual information, thus motivating the utilization of alternative recording and processing methods that offer greater reliability and the possibility of more accurate quantification.

In common with many other experimental techniques within the physical sciences, electron holography benefits greatly from the use of quantitative recording. Rather than qualitative interpretation of reconstructed holograms based mainly on appearance, meaningful information of a quantitative nature can eventually be obtained about important physical characteristics of the specimen such as its mean inner potential or any inherent electric and magnetic fields. The use of off-axis electron holography as a means to achieve atomic-level resolution via aberration correction is also expedited.[329]

The convenient combination of digital recording with computer processing has facilitated novel and worthwhile applications of electron holography to a wide range of materials. It is not the purpose of this chapter to describe these applications in detail, especially since many are considered in later chapters. Our primary objective here is instead to outline the physical basis for quantitative electron holography. Our attention is concentrated first on the detection properties of the slow-scan charge-coupled device (CCD) camera, which include digital output with high linearity over a wide dynamic range. Some of the more important factors that could limit the ultimate reliability and accuracy of hologram recording and reconstruction based on use of the CCD camera are considered. Finally, representative examples of quantitative electron holography are briefly described, primarily to demonstrate the precision and spatial resolution attainable with the technique.

2. Electron detectors

The ideal detector for electron holography should detect every incident electron at the proper place in the recording plane to ensure that the recording process itself does not have a detrimental effect on the holographic reconstruction. This fundamental requirement is not satisfied by all detectors. For example, it is well known that photographic material has a non-linear response to electrons, so that complicated and time-consuming *a posteriori* correction procedures are required before reliable reconstruction results can be obtained.[475] Much greater efficiency can obviously be achieved when the need for such tedious corrections has been eliminated by using a recording medium that has already been optimized so that the contribution of each incident electron to the recorded hologram is easily quantifiable. In this respect, the use of the slow-scan CCD camera permits the recovery of phase and amplitude information with high accuracy and high spatial resolution. Its fixed location with respect to the microscope imaging system is an additional advantage.

Several quantitative criteria are commonly used as a basis for comparison of electron detectors.[193] These include the modulation transfer function (MTF) which compares the relative strengths of the input and output signals as a function of spatial frequency, the detection quantum efficiency (DQE) which measures the efficiency of signal output for each incident electron, and the dynamic range which indicates the maximum number of distinct output signals or "gray levels." Further relevant properties in practice include the linearity (which affects the ease with which intensity levels can be compared), the size of each picture element or "pixel" (which predetermines the minimum practical magnification), and the total array size (which determines the size of the field of view). Additional factors for consideration are geometric distortions and positional stability. Practical concerns include cost, ease of use, and the accessibility of the stored data for subsequent computer processing.

The photographic negative remains the recording medium of choice for most electron microscopists. For typical operating conditions, it has a very large number of picture elements (perhaps 5,000 x 5,000 or more) and a DQE of about 0.6 (meaning that it is a comparatively efficient electron detector). Although the grain size is small, on the order of 1 μm, the effective pixel size is perhaps 10-30 μm since several grains are typically exposed by each incident electron.[96] Despite the very low cost and immense storage capacity, the lack of linearity over a wide range of incident intensities represents a serious disadvantage for quantitative electron microscopy and electron holography. Additional processing is required to compensate for the non-linearities.[225,475] Moreover, digitization of the negative is obviously required before computer processing can be initiated.

The imaging plate (IP), well-known in x-ray radiography, offers a possible route to quantitative recording.[317,513] Careful measurements[215,216,317] have established that the output of the IP provides excellent linearity over a wide dynamic range of about four orders of magnitude, and it has a sensitivity that is voltage-dependent but typically about three times greater than the conventional photographic negative. The MTF depends most strongly on electron dose, rather than electron energy, but drops sharply at very low dose levels.[216] A major drawback of the IP for applications to electron holography relates to the lack of ready accessibility to the stored data. A separate processing system apart from the microscope is required to convert the stored information

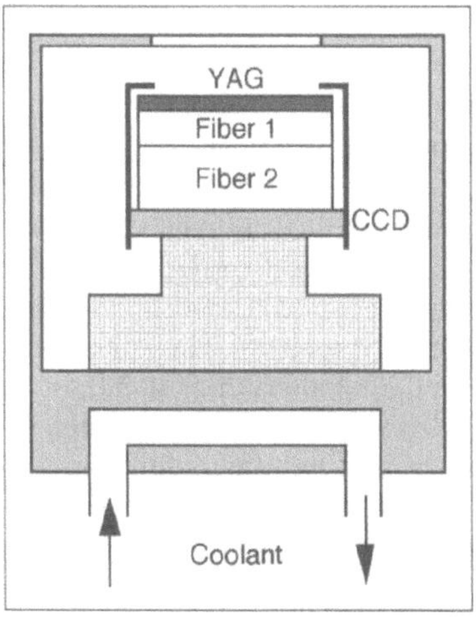

Figure 5.1. Schematic illustration of typical slow-scan CCD camera suitable for electron holography.

into an electronic format suitable for processing and more convenient storage. Another major problem is that the IP can not be conveniently corrected for local gain variations ("gain normalization") resulting in non-linear noise.

The development of the slow-scan CCD camera has revolutionized much of electron microscopy, particularly in the area of quantitative recording for electron holography.[84,85] The sensitivity, wide dynamic range and suitability for digital processing of the CCD camera has long been recognized.[191,193,407] These detection properties of the CCD can be described in quantitative terms,[79,85,209,242] allowing ready comparisons to be made with alternative electron detectors. In normal operation, light originates from single-crystal YAG (yttrium-aluminum-garnet) or phosphor scintillators that are coupled fiber-optically to the CCD detector, as shown in Fig. 5.1. Typical detectors are based on pixel arrays of 512 x 512 or 1024 x 1024, although 2048 x 2048 versions are now available commercially. The principal detection properties of the slow-scan CCD camera have been tabulated for comparison with those of the intensified TV camera and the photographic camera.[85]

3. Characteristics of CCD cameras

The slow-scan CCD camera is rapidly becoming the instrument of choice for all electron microscopists interested in quantitative recording. A particular attraction of the CCD camera is that it provides a digital signal that is readily available for on-line computer processing. Before discussing experimental limits and applications of CCD cameras to quantitative electron holography, it is first useful to consider the detection properties in some detail. Informed choices can then be made about experimental parameters, such as primary magnification, beam current density and biprism voltage, that are appropriate for studies involving electron holography. A comprehensive overview of CCD cameras, including detection parameters, characterization methods and applications has recently been published.[92]

3.1. Readout properties

The basic operation of a CCD camera can be simply understood as follows. The incident photon from the primary electron detector or scintillator creates electron-hole pairs, from which electrons are collected in a square array of potential wells that represent the individual picture elements (or pixels). The recorded image element is then readout, amplified, digitized and stored in computer memory. It is customary to adjust the gain of the analog-to-digital converter so that one digital count corresponds to one primary electron (although a different adjustment may sometimes be more appropriate depending on the particular intended application of the CCD camera). This adjustment is nowadays usually done by the manufacturer of the CCD camera prior to its installation on the electron microscope. Before proceeding to recording features of interest, it is essential to carry out further adjustments known as bias correction (dark current) and gain normalization to compensate for local nonuniformity and variability in either the detector system or the readout electronics. Appropriate software to execute these procedures is usually provided by the manufacturer with the combined CCD camera plus computer package. Figure 5.2 demonstrates this gain normalization process for a Gatan 694 CCD camera attached to a 120 keV Philips CM12 electron microscope.[92] Finally, note that it is customary to cool the CCD camera sensor array to about -40°C in order to minimize the generation of thermally-induced electron-hole pairs which are otherwise a source for readout noise (dark current).

3.2. Modulation transfer function

The modulation transfer function (MTF) of a detector effectively describes its spectral transfer or spatial frequency response. In the case of a CCD camera and electron holography, the rate at which higher spatial frequencies are attenuated, either by the YAG scintillator or the finite pixel size of the camera, will determine how faithfully the digital hologram resembles the electron hologram incident on the detector. The MTF is usually found by analysis of the measured response produced by a known input test signal. These signals can either be stochastic (e.g., white noise) or deterministic (e.g., sinusoidal function or step function). Because the CCD detector is based upon a square array of finite-sized pixels, there is an obvious physical limit to the fineness of the detail which can be discriminated. In fact, it can be easily shown that if d represents the pixel size, then a periodicity in the image with the Nyquist frequency (corresponding to a spacing of $2\,d$) will be attenuated to $\sim 64\,\%$ of its true intensity.[92]

In practice, the MTF of a CCD camera is determined primarily by the YAG scintillator because incident electrons cause light emission from areas that measure perhaps 50 to 100 μm across, which are considerably larger than the usual CCD pixel size of 24 μm. In effect, this attenuation of higher spatial frequencies by the scintillator can be attributed to the finite width of its point spread function (PSF). The shape of the PSF can be determined by taking an intensity profile along a line perpendicular to a sharp electron-opaque edge. As an example, Fig. 5.3a shows a 50-line averaged profile from an image of the beam stop, as recorded with a Gatan 694 slow-scan CCD camera equipped with a YAG scintillator and mounted on a 120 keV Philips CM12 microscope.[316] Each primary electron incident onto the scintillator generates light in a finite volume, eventually giving rise to a roughly Gaussian-shaped PSF with a full-width half-maximum on the order of 2-3 pixels. Some light is also channeled sideways by the scintillator, resulting in a long tail that can contain as much as 50 % of the total signal produced by the electron. The overall MTF due to the PSF of the scintillator combined with the finite pixel size of the CCD array is shown in Fig. 5.3b. The spike

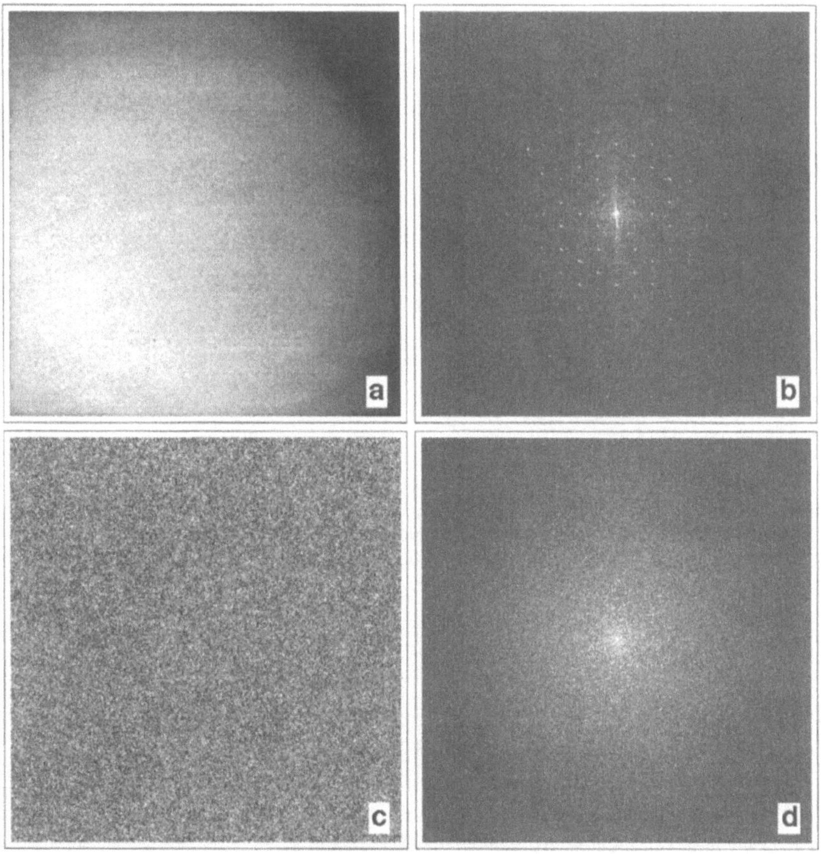

Figure 5.2. Effect of gain normalization: (a) raw image of evenly illuminated YAG scintillator, showing effect of scintillator thickness variations and fixed pattern structure of fiber optics; (b) corresponding Fourier transform; (c) gain normalized image (no sample); and (d) final Fourier transform. Camera: Gatan 694. Microscope: Philips CM12 operated at 120 kV.

at the origin is caused by the long-range tail.

The loss of higher spatial frequency detail clearly has serious consequences for electron holography. For example, in many practical applications, it is common to have only 3 to 4 pixels sampling an entire interference period (corresponding to sampling frequencies in inverse pixel units of $\sim$ 0.25-0.33 pixel^{-1}). According to Fig. 5.3b, the actual output signal could then drop by a factor of 4 or more relative to the input signal. Restoration to the ideal value of unity for all spatial frequencies can be accomplished by Fourier filtering the image with an appropriate restoration filter. An example is the filter MTF_{min}/MTF, as shown in Fig. 5.3c, where MTF_{min} is the value of the MTF at the spatial frequency of 0.5 pixel^{-1} in Fig. 5.3b. Application of this procedure attenuates low frequencies and overemphasizes high-spatial-frequency noise but it proves to be very effective in practice as demonstrated by the corrected profile of the beam stop shown in Fig. 5.3d.

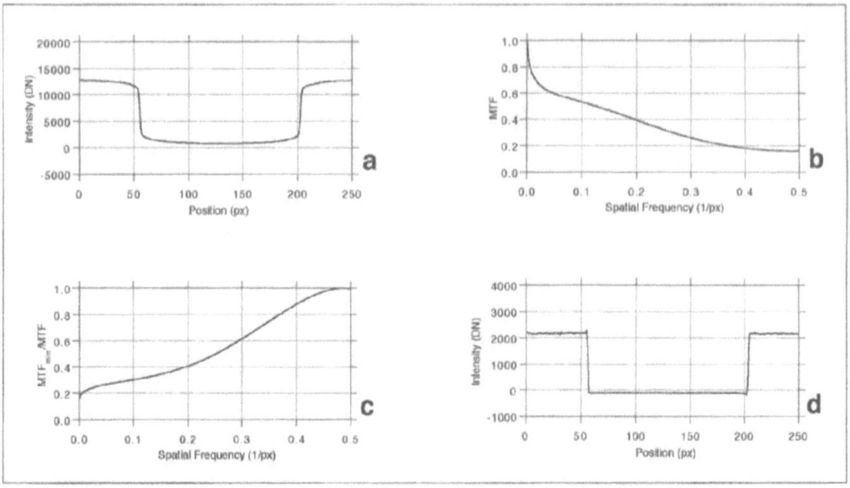

Figure 5.3. Correction of MTF demonstrated with Gatan 694 slow-scan CCD camera attached to a 120 keV Philips CM12 electron microscope: (a) profile of image of electron beam stop; (b) measured MTF; (c) MTF restoration filter; and (d) profile of beam-stop image after MTF correction.

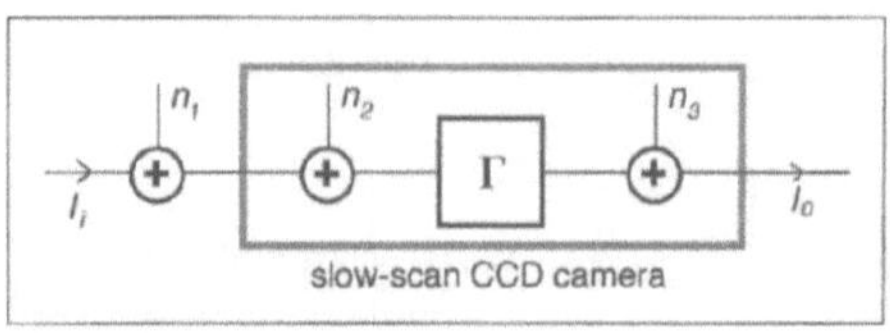

Figure 5.4. Schematic diagram showing noise sources for slow-scan CCD camera: electron shot noise n_1; Fano noise in scintillator n_2; thermal dark current and electronic readout noise n_3. I_i and I_0 represent input and output intensities respectively, and Γ represents MTF of the camera.

3.3. Detection quantum efficiency

The detection quantum efficiency (DQE) of a detector is a quantitative measure of its effectiveness in producing a detectable output response for each incident electron. The ideal DQE value should be unity but its practical value could be considerably less due to the effects of instrumental noise. As illustrated in Fig. 5.4, this noise has three primary sources: shot noise, which is proportional to the square root of the number of electrons per pixel as described by Poisson statistics; so-called Fano noise, which is due to path-length variations of the incident electrons in the scintillator material;[105] and electronic readout noise, which is dose-independent and hence usually dominant at low dose levels. The Fano noise is the primary reason that the DQE value fails to reach unity even at high dose levels.

Traditionally, DQE has been defined in terms of the quotient of squared signal-to-noise ratios (SNR) at the output and input of the camera[192]

$$\mathrm{DQE} = \frac{\mathrm{SNR}_{\mathrm{out}}^2}{\mathrm{SNR}_{\mathrm{in}}^2} \tag{5.1}$$

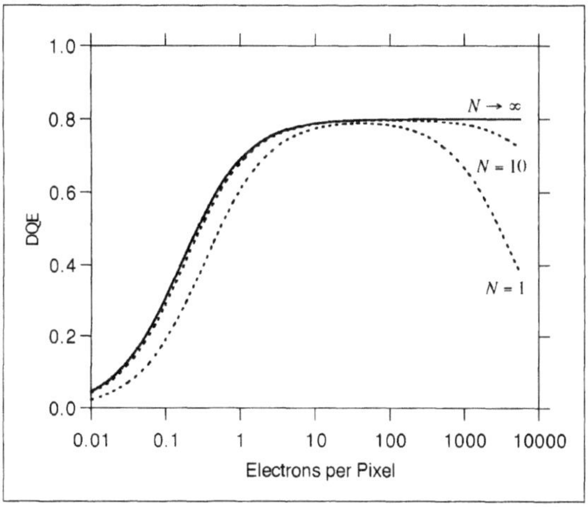

Figure 5.5. Calculated DQE for zero spatial frequency as function of electron dose, according to Eq. (5.3), for $F = 0.25$ and $\sigma_e^2 = 0.2$. Dotted DQE curves account for noise introduced by bias and gain correction. N value indicates number of dark images and gain reference images that have been averaged to reduce noise.

This expression implicitly assumes that there is no lateral spread of photons with respect to the position of the incident primary electron, which ignores the effect of the scintillator PSF, as described in the preceding section. Thus, it appears more appropriate to consider a function $\mathrm{DQE}(\vec{q})$ which is dependent on spatial frequency $\vec{q}$, so that the definition above needs to be slightly modified:

$$\mathrm{DQE}(\vec{q}) = \frac{\mathrm{SNR}^2_{\mathrm{out}}(\vec{q})}{\mathrm{SNR}^2_{\mathrm{in}}(\vec{q})} \tag{5.2}$$

As described elsewhere,[92] this version of the DQE provides an excellent instrument for optimizing CCD camera design for specific applications. In the current context, useful insight into CCD performance for electron holography can still be achieved by a simple evaluation based on Eq. (5.1).

The signal-to-noise ratio of the input signal is given by $\mathrm{SNR}^2_{\mathrm{in}} = N_e$, where N_e is the number of electrons incident on the pixel. Expressing the standard deviation of the instrumental noise in terms of a primary electron equivalent σ_e, and letting F represent the Fano noise, then the DQE for zero spatial frequency is given by the expression

$$\mathrm{DQE} = \frac{1}{1 + F + \sigma_e^2/N_e} \tag{5.3}$$

Figure 5.5 shows a plot of this DQE as a function of electron dose for typical values of the constants in Eq. (5.3).[92] Note that the DQE values range from about 0.6 for detection of single electrons ($N_e = 1$) up to about 0.8 for doses in the approximate range of 10-200 el/px, but with a falloff at higher electron doses because of noise added during bias and gain correction. Addition of multiple dark images and gain reference images can help to restore the DQE value at high doses as shown in Fig. 5.5.

In practice, the measured DQEs of typical CCD cameras follow the general trends predicted by this equation. For example, for electron doses in the range of 100-200 el/px, the detector quantum efficiency for the GATAN 679 CCD camera equipped with a thin YAG scintillator was measured to have the constant value of 0.8.[88] Moreover, in another more recent experimental study[512] it was demonstrated that a considerable drop in DQE from about 0.7 would occur for doses in excess of about 200 el/px unless an anti-reflection coating was applied to the YAG scintillator. A phosphor detector does not have this particular problem of drop in the DQE at high doses.

Overall, these results imply that the DQE of the typical CCD detector should not represent a serious limitation on the outcome of most electron holography studies, certainly not for applications carried out at medium resolution where typical electron counts per pixel are adjusted to be in the 100-500 range. In choosing the optimum experimental conditions (experiment design), the DQE value should, however, be taken into account, as we describe later in Section 4.3.

3.4. Linearity

A highly desirable characteristic of an electron detector intended for quantitative recording is linear dependence between the input signal and the output response. A series of increasing exposures over a wide range with a set of computer-controlled light-emitting diodes was used to evaluate the linearity of a CCD sensor array.[102] It was found that the largest deviation from a straight line, as determined using a least-squares fitting procedure, was negligible in comparison with the shot noise.

These types of measurements account for the linearity of the CCD array and readout electronics but our interest here lies in the overall linearity with respect to the primary electron beam intensity. Thus, the light conversion of the YAG scintillator should also be included in the analysis. Linearity can be established in this situation by monitoring intensity differences between separate areas of the YAG scintillator as a function of incident intensities. In one experiment,[85] a specimen feature partially blocked the incident electrons over part of the scintillator, transmitting about 73 % of the total. The incident illumination intensity was varied over a wide range but the ratio of the measured intensities on two fixed areas remained constant to within less than half of one percent. This variation is within the deviation level expected due to noise. The experimental conditions used in these observations encompassed those typically used for electron holography. It was therefore concluded that the linearity of the CCD camera should be more than adequate for quantitative holographic studies.

3.5. Dynamic range

The final characteristic of the CCD camera to be considered here is the dynamic range. For imaging and holographic applications, electron intensities are seldom larger than about 1000 el/px. A 12-bit analog-to-digital converter is then more than adequate (maximum of 4096 counts). On the other hand, for applications involving electron diffraction patterns, a larger dynamic range is often required. The factors that determine the range are identifiable as follows. The upper end of the intensity scale is limited by the total charge capacity of the individual potential wells (typically $\sim$ 3-5 x 10^5 el/px); the lower end depends on the readout noise (standard deviation is typically $\sim$ 5-20 electrons). Thus, the dynamic range of the slow-scan CCD camera is comfortably in excess of 10^4 which is far in excess of any requirements for electron holography.

4. Factors affecting detection limits

In the preceding section, the detection properties of the slow-scan CCD camera were considered in detail, and it should have become apparent that these features of the device make it an excellent medium for digital recording of electron holograms. In this section, we first discuss additional practical factors that have a strong influence on the quantification of electron holograms recorded with the slow-scan CCD camera. We then consider the reliability and accuracy of phase and amplitude determinations. This discussion is confined to applications of the off-axis technique but similar considerations will apply to other forms of electron holography.

4.1. Geometric distortion and shear

Under most normal operating conditions, geometric distortions of the electron microscope imaging system do not affect the recorded image intensity distribution. In off-axis electron holography, however, local shifts of the interference fringes are used to derive the phase of the electron wave. When high accuracy in the phase measurements is required, corrections must be made for the geometric distortions as well as fiber-optic shear in the CCD recording system. For example, consider the situation where the holographic imaging system has been adjusted so that one interference-fringe period, which corresponds to a phase shift of 2π, is recorded at a spacing of $4\,d$ (where d = pixel size), corresponding to a distance of about 100 μm. Phase precisions approaching $\pi/100$ are obtainable in carefully executed holography experiments with suitable specimens,[87] which implies that distortions on the order of 1 μm at the detector are likely to become significant. Our experience is that distortions of this magnitude or more are always present and must be taken into account before high precision can be obtained (and they will be considerably worse for a CCD camera with a recording area of 2048x2048 pixels). This task can be easily accomplished by recording a reference hologram without any object present, so that the combined distortions of the imaging and recording systems are immediately revealed in the reconstructed phase, which should otherwise be constant across the field of view. In fact, division of the reconstructed distorted complex image by the reconstructed reference complex image enables the effect of the distortions to be removed entirely. This convenient method for distortion removal is a significant advantage that stems directly from the fixed location of the CCD camera with respect to the microscope imaging system: distortion correction is difficult to achieve for holograms recorded on photographic film or with the imaging plate.

As an example, Fig. 5.6a shows the reconstructed phase image from a wedge-shaped crystal of InP before distortion correction, with $\pi/3$ phase contours superimposed. Figure 5.6b shows the corresponding vacuum reference phase image acquired with the sample removed from the field of view but with all other imaging parameters kept fixed. The $\pi/3$ equiphase contour lines delineate the distortions that originate solely from the imaging and detection system (and possibly from some charging on the biprism wire). Finally, the vacuum phase image is subtracted from the original phase image to obtain the distortion-corrected image, as shown in Fig. 5.6c. The substantial flattening in the vacuum phase across the left of the field of view is significant since the reference vacuum phase is essential for determining relative phase changes occurring within the specimen.

4.2. Phase unwrapping and flattening

In the process of reconstructing the phase of the image wave using standard image processing programs, there is always the possibility of ambiguity because the phase can usually only be defined between the limits of $-\pi$ and $+\pi$. Since it is usual for phase variations across the field of view of a hologram to exceed 2π, abrupt phase jumps are inevitable. The problem for the observer is to ascertain whether any visible

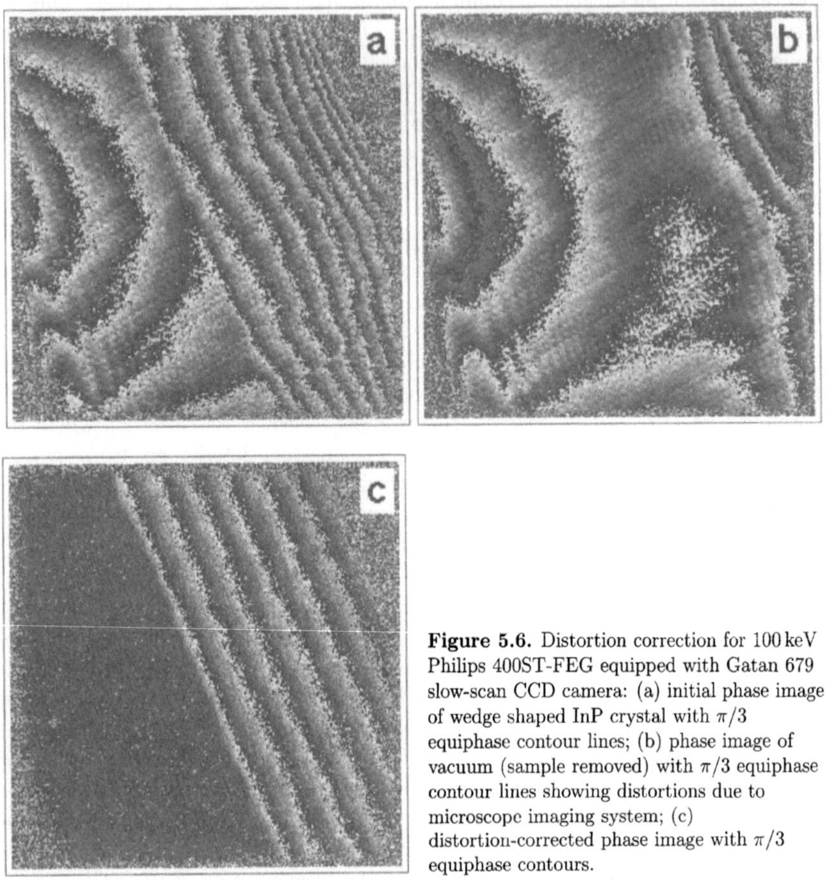

Figure 5.6. Distortion correction for 100 keV Philips 400ST-FEG equipped with Gatan 679 slow-scan CCD camera: (a) initial phase image of wedge shaped InP crystal with $\pi/3$ equiphase contour lines; (b) phase image of vacuum (sample removed) with $\pi/3$ equiphase contour lines showing distortions due to microscope imaging system; (c) distortion-corrected phase image with $\pi/3$ equiphase contours.

phase jumps are occurring as a result of the processing or whether they are genuinely related to characteristics of the specimen. The simplest method of separating the two effects involves shifting the phase of the complex wave across the entire field of view by multiplying by a complex constant. A simple computer routine allows a tableau of phase-shifted image waves to be generated within seconds. As demonstrated for a wedge shaped InP crystal in Fig. 5.7, the 2π phase shifts are moved to different positions in the phase image, and are thus easily diffentiated from those due to the specimen which would obviously remain fixed in position.

Several phase-unwrapping algorithms have been discussed that can detect the discontinuities of 2π, which can then be removed by appropriately adding or subtracting 2π from the areas which were determined to be phase-wrapped. This process is usually accomplished by, e.g., monitoring the difference between two adjacent phase coefficients on a line-by-line basis. When the difference exceeds some pre-specified threshold, then it is established that a discontinuity is present. This particular algorithm works in many cases but it may fail due to low signal-to-noise ratio (due to, for example, a thick sample, or poor interference fringe contrast) which could cause random phase deviations that are greater than the wrapping detection threshold. For a situation, where the holographic field of view is smaller than the field of view of the CCD camera, the phase unwrapping should only be applied to the region of the hologram.

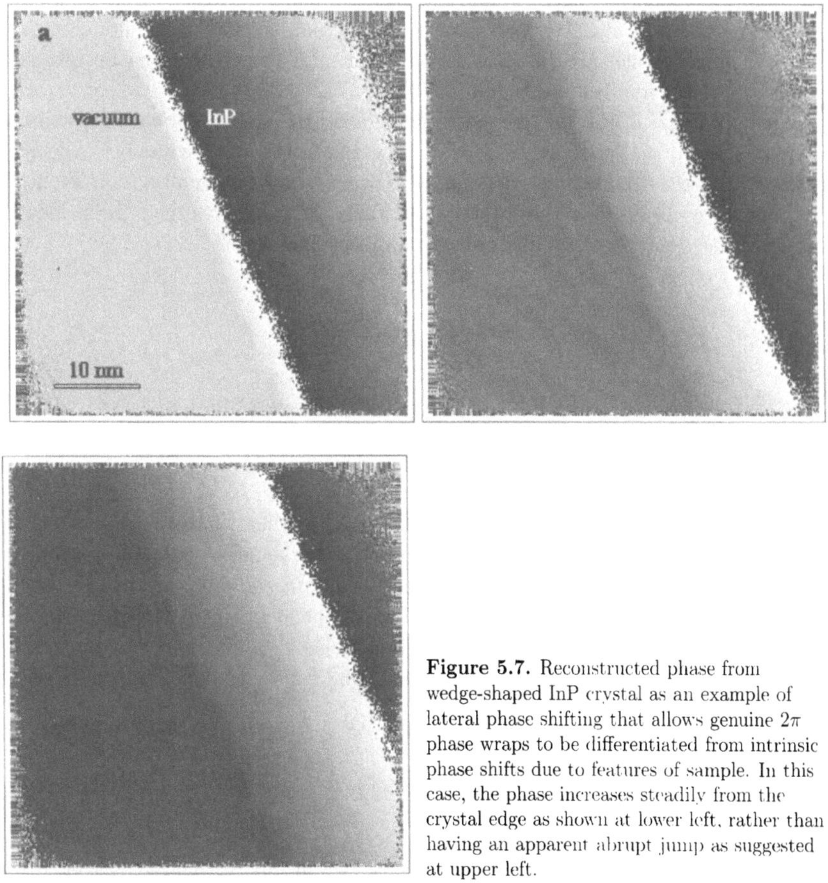

Figure 5.7. Reconstructed phase from wedge-shaped InP crystal as an example of lateral phase shifting that allows genuine 2π phase wraps to be differentiated from intrinsic phase shifts due to features of sample. In this case, the phase increases steadily from the crystal edge as shown at lower left, rather than having an apparent abrupt jump as suggested at upper left.

At this stage of the hologram reconstruction process, one might reasonably expect that all remaining phase changes are solely due to the sample. However, the position of the sideband following Fourier transformation of the original hologram is only located to within one pixel: an additional phase shift occurs if the sideband is not exactly at the center of the extracted region.[476] The reason for this phase shift can be understood in terms of a coordinate translation in reciprocal space by a vector $\vec{q}$ which then corresponds to a multiplication in real space by a plane wave $\psi(r) = \exp(2\pi i \vec{q}\vec{r})$. This additional residual phase is in the form of a tilted plane, with the amount of tilt increasing by 2π over the width of the image for every pixel that the extracted sideband is off-center. Removal of this additional tilted phase plane can be achieved in practice using a least-squares fit to determine the coordinates of a plane of the form $I(x, y) = Ax + By + C$, and thereby flatten the vacuum part of the phase image. This intensity is then subtracted across the entire image so that the remaining (flattened) phase would finally be representative of that resulting from an incident plane wave.

4.3. Precision and experimental design

Off-axis electron holograms are typically recorded at electron doses of 100-500 el/px, so that the recording process is dominated by shot noise. Thus, a statistical parameter estimation should be used to determine the attainable precision of any phase and amplitude measurements. We first summarize the more significant results of such an analysis,[89] and we then discuss how the experimental parameters can be chosen to optimize the measurement precision.[87,254,260,261]

The shot noise in the electron hologram can be considered as Poisson-distributed with a standard deviation given by $\sigma^2 = N_e$ where N_e is the average number of detected electrons per pixel. In the most general case, when no *a priori* information about the spacing and orientation of the interference fringes is available, then the achievable precision for estimation of the phase is given by the expression[87]

$$\mathrm{var}[\varphi] \geq \frac{14}{\mu^2\, N_t} \tag{5.4}$$

where N_t corresponds to the total electron dose in the measured area. Conversely, if *a priori* information about the spacing and orientation of the interference fringes is available, then the factor of 14 in Eq. (5.4) can be replaced by 2,[169,254,260] so that higher precision is possible.

The variance bound achievable in practice is larger than what would be calculated using Eq. (5.4) because of the inevitable decrease in the signal-to-noise ratio associated with detection of the electron hologram by the CCD camera. The camera DQE should ideally be one but, as noted earlier, its value is dependent on spatial frequency and unity is not achieved in practice even at high dose levels. In order to take the effect of the DQE into account, the minimum variance bound Eq. (5.4) should then be rewritten as

$$\mathrm{var}[\varphi] = \frac{14}{\mathrm{DQE}(q_c)\, \mu^2\, N_t} \tag{5.5}$$

where it is presumed that the DQE value to be used in this calculation corresponds to the spatial frequency q_c of the holographic interference fringes.

The amplitude of the interference fringes can also be expressed in terms of the electron dose by $A = \mu\, N_e$ where μ is the fringe visibility.[31] Using statistical analysis,[89] and incorporating the DQE of the CCD camera, then the relative bound for the amplitude A can be written in the form

$$\frac{\mathrm{var}[\vec{A}]}{A^2} = \frac{2}{\mathrm{DQE}(q_c)\, \mu^2\, N_t} \tag{5.6}$$

The bounds for estimation of both phase and amplitude have thus been shown to depend on the visibility (squared) of the interference fringes, as well as the total electron dose in the measurement area. The former dependence reiterates the need to take all possible steps to optimize the fringe visibility. The latter suggests that the precision of the phase measurement could be improved by increasing the dimensions of the measurement area, but this implies a loss of spatial resolution except in certain special circumstances (e.g., one-dimensional line averaging).

The product $\mu^2\, N_t$ can be shown to depend upon a collection of experimental variables, which should preferably be chosen where possible to optimize the measurement precision (experimental design). For example, the total number of electrons N_t incident on an area A_r can be expressed as[78]

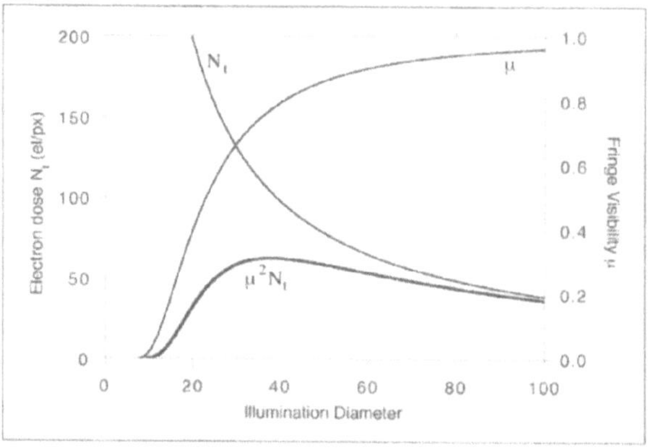

Figure 5.8. Optimization of product $\mu^2 N_t$ with respect to illumination diameter.

$$N_t = (\pi^2 \beta R_S^2 \alpha_0^2 \tau / e) \cdot (4 A_r / \pi d_1 d_2) \tag{5.7}$$

where β is the gun brightness, R_S denotes the projected Gaussian source size at the specimen, α_0 is the convergence half-angle of the illumination. τ is the exposure time, and d_1, d_2 are the major and minor axes, respectively, of the astigmatic illumination. Moreover, the fringe visibility μ depends on the length d_1 of the illuminated region in the direction perpendicular to the interference fringes, as described by[300]

$$\mu = e^{-(2\pi R_S \alpha_0 D/\lambda d_1)^2} \tag{5.8}$$

where D is the distance between the areas at the specimen plane which interfere beyond the biprism, otherwise known as the overlap region. The product $\mu^2 N_t$ can then be maximized with respect to d_1 by suitably combining these two equations, as also illustrated in Fig. 5.8. The result is a specification for d_1 as given by

$$d_1 = 4\pi R_S \alpha_0 D/\lambda \tag{5.9}$$

Further careful consideration leads to the conclusion that the variance for the phase estimate is proportional to U_B^2/τ where U_B is the voltage applied to the biprism. Experimental constraints such as specimen stage stability and biprism drift, and the need to avoid sideband overlap with the central autocorrelation function, in turn impose limits on the range of choices for these variables. After further analysis, taking into account such factors as the size and number of pixels in the CCD array and the position of the biprism relative to the objective lens, recommendations for the optimum magnification and biprism voltage settings appropriate for specific microscopes can finally be made.[87] For example, in the case of a 100 keV Philips 400ST-FEG equipped with a Gatan 679 CCD camera, the optimal standard deviation in the measured phase is calculated to be on the order of $\pi/100$ for a measurement area A_r of 1 nm^2.

An important conclusion of this analysis of phase and amplitude precision is that it is the total measurement area that is of relevance but not its actual shape. Thus, in cases where one-dimensional variations are of interest, averaging along a particular direction will improve the signal variance while preserving spatial resolution in the original direction.

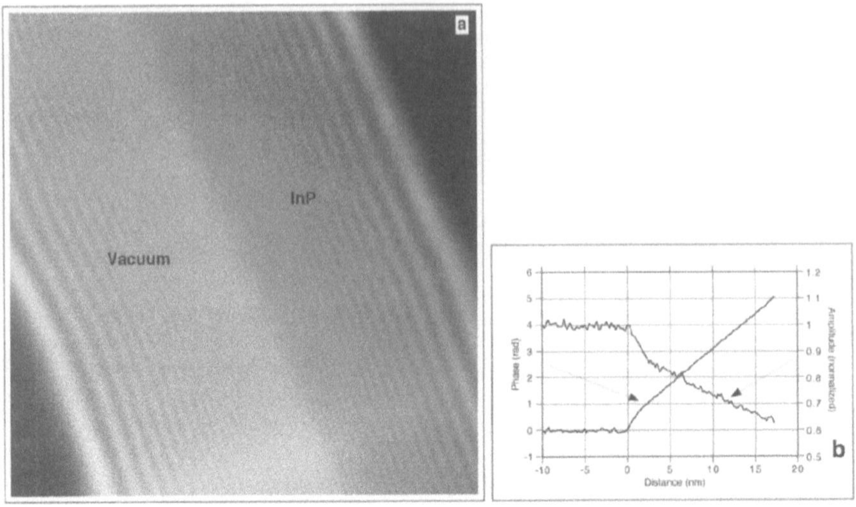

Figure 5.9. (a) Off-axis electron hologram from cleaved wedge of InP. (b) Averaged phase and amplitude thickness profiles extracted from reconstructed image wave.

5. Applications of quantitative electron holography

The slow-scan CCD camera enables off-axis electron holograms to be recorded digitally with high accuracy and reliability. Care must be taken to differentiate the effects of intrinsic electric and magnetic fields, and to avoid complications due to dynamical electron diffraction which can be substantial for crystalline materials in certain projections. Nevertheless, as we now briefly demonstrate, accurate quantification of the phase and amplitude of the electron wavefront via electron holography facilitates the extraction of novel information about a wide range of specimen types.

5.1. Measurement of mean inner potential

In the absence of dynamical diffraction effects, for samples without electric or magnetic fields, the phase change φ of the electron wave passing through the sample relative to the phase in vacuum is related to the mean inner potential V_0 through the expression[357]

$$\varphi = C_E V_0 t \tag{5.10}$$

where C_E is a constant depending on the incident beam energy and t is the local sample thickness projected in the beam direction. For a sample of known geometry, such as the cleaved wedge of crystalline InP shown in Fig. 5.9, this simple relationship provides a direct means for determining the V_0 value provided that the phase has been accurately measured using the methods outlined above. Detailed discussion of mean inner potential measurements using electron holography can be found in the later chapter by Gajdardziska-Josifovska and Carim (pages 267 ff.); our purpose here is just to demonstrate the increased accuracy that accrues from digital recording and processing based on the CCD camera.

The mean inner potential of a solid represents the volume average of the atomic potential, with values typically ranging between -5 and -30 V depending on composition, bonding and structure.[357] Previous methods used for measurement of V_0 have had relative errors in the range of 2.5 to 10 %.[417] However, by using crystal wedges of known angle, and hence thickness, as well as digital recording and processing, measurement accuracies of about 1 % or slightly better have been achieved using off-axis electron holography.[135] Dynamical interactions have to be minimized by deliberately tilting the crystal away from the zone axis, and non-systematic reflections must be avoided by reference to the Kikuchi line pattern. As an example of the technique, Fig. 5.9a shows the off-axis electron hologram from an InP wedge. Based on the slope of the phase profile, and before dynamical corrections, the preliminary estimate of V_0 in this case is 15 V. Table 5.1 lists the V_0 values calculated for crystals of MgO, GaAs and PbS before and after correction for dynamical diffraction using Bloch wave calculations.[135] The measured and dynamically corrected V_0 values for Si were found in the same study to have a significant discrepancy from earlier studies involving reflection high-energy electron diffraction, for reasons that were unclear at the time. In a more recent investigation,[256] weak-beam dark-field imaging was used to provide an independent check of the crystal thickness: analysis of the results implied that, in the case of Si, cleavage did not usually occur along {111} planes as had been presumed in the earlier holography study.

5.2. Interface abruptness

Interface characteristics have a crucial influence on the mechanical properties of multi-grained materials, especially structural features such as interfacial roughness and any compositional gradients. Since off-axis electron holography enables access to the phase of the electron wave, which in turn depends on the (amount of) material present, the technique can be utilized to provide valuable information about interfaces. In the absence of strong dynamical diffraction, with the additional assumption that the specimen thickness does not change appreciably across the interface, then the extracted phase profile from an off-axis electron hologram can be interpreted directly in terms of local variations in the mean inner potential and thus the composition. In practice, the sample thickness will steadily increase away from the specimen edge, although a reasonable thickness profile can often be estimated.[495] For example, with a parabolic thickness correction, the respective Mo-Si and Si-Mo interfacial regions in an Mo/Si multilayer were measured to have widths of 1.2 nm and 0.6 nm respectively, with precisions on the order of 0.2 nm.[495] These values were in close agreement with an earlier study using measurements based on high-resolution electron microscopy.[495] Another interesting result was achieved by processing the off-axis electron hologram of a Si_3N_4/Si_3N_4 grain boundary. As shown in Fig. 5.10, this phase boundary is almost invisible in the reconstructed amplitude image (a) although it is clearly visible in the phase image (b). After averaging across a 40-pixel-wide area to improve the signal-to-noise ratio and

Material	Measured V_0 [V]	Corrected V_0 [V]
MgO	$12.78 \pm 0.6\,\%$	$13.01 \pm 0.6\,\%$
GaAs	$14.13 \pm 1.2\,\%$	$14.53 \pm 1.2\,\%$
PbS	$16.35 \pm 0.7\,\%$	$17.19 \pm 0.7\,\%$

Table 5.1. Values of mean inner potential extracted from electron hologram phase images before and after dynamical correction.

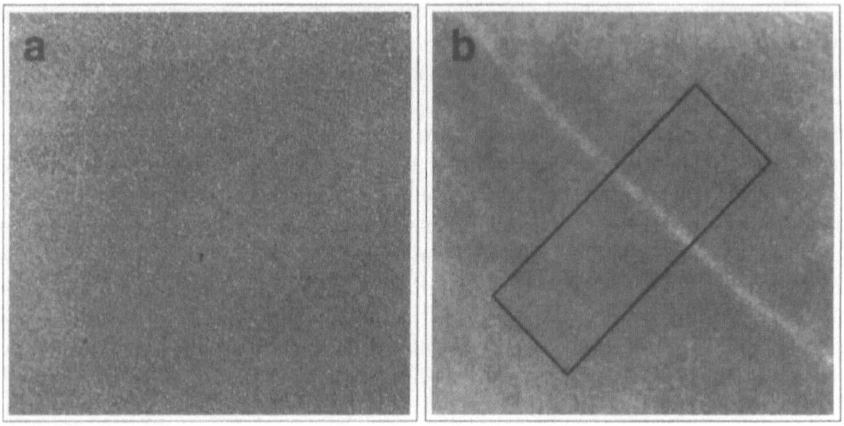

Figure 5.10. Reconstructed amplitude (a) and phase (b) images from an Si_3N_4/Si_3N_4 grain boundary. Amorphous layer has measured width of (1.2 ± 0.2) nm.

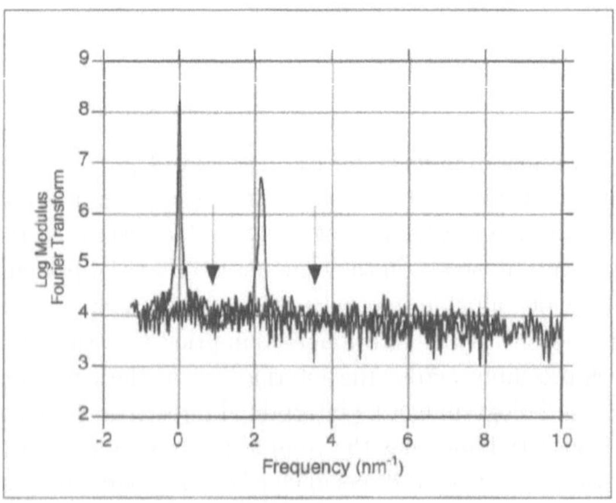

Figure 5.11. Logarithmic plot showing modulus of FT from MgO hologram. Arrows correspond to positions of numerical aperture used for reconstruction.

thus enhance the measurement precision, the amorphous layer between the two grains was measured to have a structural width of $(1.2 \pm 0.2\,\text{nm})$.

In these studies of interfaces using off-axis electron holography, observation was restricted to thin regions close to the edge of the sample, diffraction effects had to be avoided and specific information about elemental composition was unavailable. Nevertheless, these results again demonstrate the high spatial resolution and sensitivity attainable with the technique.

5.3. Determination of inelastic mean-free-path

The angular distribution of elastically and inelastically scattered electrons contained in an off-axis electron hologram becomes spatially separated as a result of the first Fourier transform step of the holographic reconstruction process.[167] Figure 5.11 shows a line scan taken across the central autocorrelation peak and one sideband of the FT from the hologram of an MgO crystal wedge plotted on a logarithmic scale.[297]

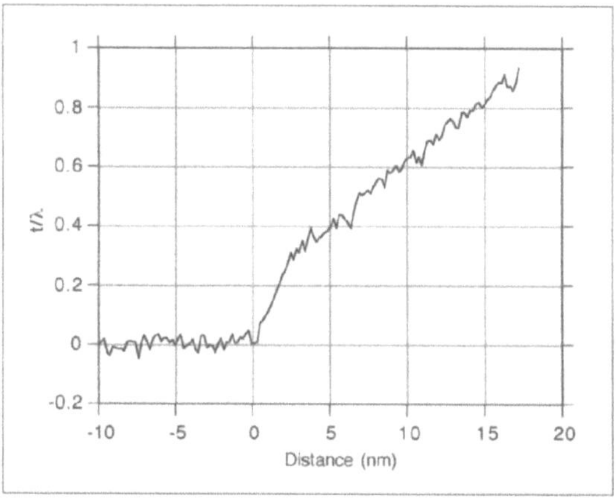

Figure 5.12. Averaged thickness profile of t/λ_i image extracted from off-axis hologram of InP wedge.

At the positions of the numerical aperture used for reconstruction, as arrowed, the amplitude of the sideband has dropped by three orders of magnitude. Since most of the inelastic scattering from the sample is concentrated close to the central peak, the sideband can be regarded as energy-filtered. The corresponding reconstructed amplitude image, which was neglected for many years by the electron-holographic community, has some interesting and entirely novel applications,[136,137,297] which we briefly summarize.

In electron-energy-loss spectroscopy (EELS), the materials scientist is often interested in the scattering parameter t/λ_i where t is the thickness and λ_i is the mean-free-path for inelastic scattering. Analysis of the contributions to the hologram intensity reveals[297] that normalization of the reconstructed image amplitude by reference to the vacuum signal, followed by taking the natural logorithm, gives another image that explicitly displays the spatial distribution t/λ_i. This process is summarized by the following equation

$$t/\lambda_i = -2 \ln \frac{A_0}{A_r} \tag{5.11}$$

where A_0 is the energy-filtered (zero-loss) amplitude and A_r corresponds to the amplitude of the (unscattered) vacuum reference wave. In the event that the thickness is known, for example by reference to a crystal wedge of known cleavage angle, then λ_i can be determined absolutely. The method is illustrated in Fig. 5.12 which shows the averaged thickness profile from the t/λ_i image for the single crystal wedge of InP shown earlier in Fig. 5.9. Despite the overall noisy appearance, which is related to the noise present in the amplitude image, a value of 60 nm can be extracted for the bulk mean-free-path of InP. Analysis of similar wedges of Si and MgO gave values of 92 ± 7 nm and 71 ± 5 nm,[297] respectively, which compare favorably with calculated and experimental values from EELS measurements. Further analysis[137] shows that, by dividing the phase image by the logarithm of the normalized amplitude, the thickness dependence can be eliminated entirely, leading to a composition image that is independent of specimen topography. The method should thus be particularly useful when off-axis electron holography is applied to specimens of unknown topography.

These initial studies were seriously limited by the poor quality of the amplitude images in comparison with the phase images. Analysis at the time indicated that statistical errors resulting from using noisy data represented the major contribution to the total uncertainties ($\sim 5\,\%$) of the experiments. With increased specimen stage and biprism stability, plus enhanced fringe contrast, substantial improvements in the precision should be possible. Meanwhile, it seems worth reiterating that these types of quantitative experiments involving the amplitude image would be totally impractical without the accurate recording and convenience for digital processing afforded by the CCD camera.

Conclusion

Utilization of the slow-scan CCD camera for acquisition of electron holograms has permitted quantitative experiments involving the phase and amplitude of the reconstructed image wave with high precision. In order to obtain precise phase measurements, correction procedures must be implemented to remove residual geometric distortions, and to unwrap and flatten the phase reliably. Enhanced precision of the phase and amplitude measurements can be achieved by careful attention to the selection of key experimental parameters which include the biprism voltage, the image magnification, the exposure time and the use of suitably astigmatic illumination. Averaging techniques may reduce statistical errors but at the possible expense of spatial resolution except for favorable specimen geometries.

THE RECONSTRUCTION OF OFF-AXIS ELECTRON HOLOGRAMS

E. Völkl[1] and M. Lehmann[2]

[1]Oak Ridge National Laboratory, Oak Ridge, TN 37831-6064
[2]Institut für Angewandte Physik und Didaktik,
Technische Universität Dresden, D-01062 Dresden, Germany

1. Introduction

The full digital reconstruction of off-axis TEM holograms involves algorithms (FFTs) which use significant CPU time even on fast computers. For example, on a Macintosh computer (G3 at 307 MHz), the reconstruction time for a 1024 by 1024 pixel image (including the display of the phase and amplitude image) is slightly below 4 seconds. On a DEC 3000 AXP 600 workstation the reconstruction time is about 8 seconds. For the live display of phase images, reconstruction procedures which speed the process by either making assumptions about the object or by using a mixture of optical and digital methods are an important tool for electron holography.

The field of electron holography has been moving over the last few years from a concentration on technique development into an emphasis on applications. Requirements for quantitative results have had significant impact on the evolution of the technique. For example, reference holograms are now considered an essential part of the effort to obtain quantitative data, although extracting quantitative data from holograms remains a complex task.

Additional processing of the reconstructed amplitude and phase images is often considered part of the overall reconstruction procedure. For instance, there are many different ways to highlight phase information. Also, phase images may need phase unwrapping and/or phase amplification to best display their information. Another example for additional processing are images (complex data) which may be further processed to correct for isoplanatic aberrations introduced by the electron microscope. A significant part of this chapter will discuss the aspects of additional image processing.

It is presently understood that the technique of electron holography requires digital recording and processing capabilities to be most effective. If a digital camera is not available, and photographic film cannot be avoided, the nonlinear behavior of film,[2,475] details for double exposures[294] and tips for optical reconstruction (see page 162) should be considered.

2. Basic reconstruction process

We will begin by assuming the object (specimen) in the microscope is illuminated by a plane wave of unit amplitude and with a wave vector parallel to the optical axis. The object modifies the plane wave $\exp(i\vec{k}\vec{r})$ (i.e., the illumination) and creates the so-called object wave $o(\vec{r})$, which is described as

$$o(\vec{r}) = a(\vec{r}) \cdot e^{i\varphi(\vec{r})} \tag{6.1}$$

Both $a(\vec{r})$ and $\varphi(\vec{r})$ are real functions and describe the object amplitude and the object phase, respectively. Note that the vector $\vec{r}$ is a vector in the (x, y) plane only. For reasons of simplicity, magnification factors and image rotations are ignored, so vectors in the image plane and the object plane can be considered equivalent.

A successive set of imaging lenses images the object wave onto successive image planes. In the final image plane, the image function

$$O(\vec{r}) = A(\vec{r}) \cdot e^{i\phi(\vec{r})} \tag{6.2}$$

is defined by its real positive image amplitude $A(\vec{r})$ and real image phase $\phi(\vec{r})$ which can be quite different from the object amplitude and phase due to lens aberrations. For a schematic illustration of the effect of lens aberrations, see Fig. 9.4, page 207.

When recording an off–axis hologram (see pages 93, ff.) the image intensity $I(\vec{r})$ is, in principle, given by Eq. (4.2), page 88

$$I_{hol}(\vec{r}) = 1 + A^2(\vec{r}) + I_{inel}(\vec{r}) + 2\mu\, A(\vec{r}) \cdot \cos\left(2\pi\, \vec{q}_c \vec{r} + \phi(\vec{r}) + \theta\right) \tag{6.3}$$

but with the additional term $I_{inel}(\vec{r})$ which has been added to take the inelastic scattered electrons into account.[167,248] For $A = 1$, and $I_{inel} = \phi = \theta =$const., the term $\vec{q}_c$ describes the spatial frequency of the holographic fringes (sometimes called carrier frequency), while the term μ describes the contrast of the interference fringes.[478] The basic reconstruction process starts with this equation.[113,114]

An example for the intensity distribution described in Eq. (6.3) is displayed in Fig. 6.1, where a small selection of a hologram of gold particles on an amorphous carbon film is shown. The interference fringes run approximately from the top left to the bottom right. The first step in extracting the information about the complex image wave (and from that the image amplitude and phase) from the actual hologram is to perform a Fourier transform on $I_{hol}(\vec{r})$

$$\begin{aligned} \mathbf{FT}\left\{I_{hol}(\vec{r})\right\} = & \\ & \delta(\vec{q}) + \mathbf{FT}\left\{I_{inel} + A^2(\vec{r})\right\} + \\ & \delta(\vec{q} + \vec{q}_c) \otimes \mathbf{FT}\left\{A(\vec{r}) \cdot e^{i\phi(\vec{r})+i\theta}\right\} + \\ & \delta(\vec{q} - \vec{q}_c) \otimes \mathbf{FT}\left\{A(\vec{r}) \cdot e^{-i\phi(\vec{r})-i\theta}\right\} \end{aligned} \tag{6.4}$$

where $\otimes$ denotes convolution and the new image $\mathbf{FT}\{I_{hol}(\vec{r})\}$ is a complex function, the modulus of which (i.e., $|\mathbf{FT}\{I_{hol}(\vec{r})\}|$) is displayed in Fig. 6.2. As the intensity $I_{hol}(\vec{r})$ is real, the modulus of its Fourier transform (see Fig. 6.2) exhibits central symmetry with respect to the origin of the $\vec{q}$ plane.

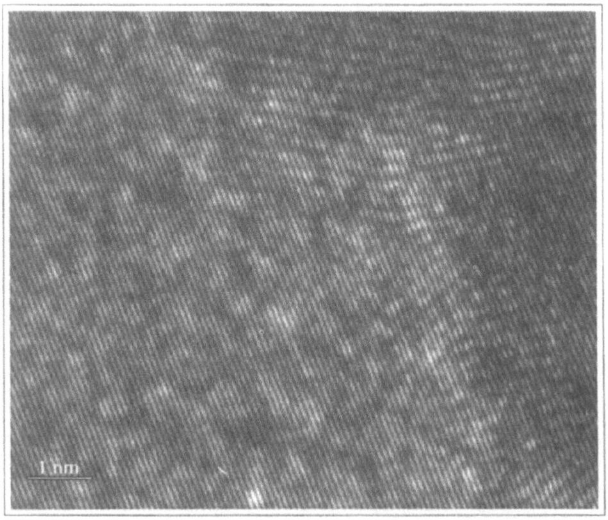

Figure 6.1. Selected area of a hologram of gold particles on amorphous carbon film, showing an example of the general intensity distribution as described in Eq. (6.3).

Both Fig. 6.2 and corresponding Eq. (6.4) contain three distinct elements. The central area in Fig. 6.2 is known as the 'autocorrelation,' and corresponds directly to the second line of Eq. (6.4). The other two elements are called 'sidebands,' which result from the third and fourth lines of Eq. (6.4) and are represented in Fig. 6.2 as the left and right elements. The holographic information, i.e., the information about the amplitude and phase of the image wave are present in those sidebands, which differ only insofar as one is the complex conjugate of the other. Therefore, they contain the exact same information.

At this point it should be noted that each image contains a so-called highest spatial frequency necessary to describe the smallest detail in the image. An important number for recording a hologram is the number of interference fringes per smallest detail.[210,471] In Eq. (6.3), the $A^2 + I_{inel}$ term represents the standard image intensities found in a regular image (not counting the interference fringes). Assuming the highest spatial frequency to be $\vec{q}_{max}$, with an amplitude of A_{max}, its contribution to the image intensity can be written as

$$\begin{aligned} &|1 + A_{max} \cdot \cos(2\pi \, \vec{r} \, \vec{q}_{max})\,|^2 = \\ &1 + 2A_{max} \cdot \cos(2\pi \, \vec{r} \, \vec{q}_{max}) + \underbrace{A^2_{max} \cdot [1 + \cos(4\pi \, \vec{r} \, \vec{q}_{max}]/2} \end{aligned} \tag{6.5}$$

From Eq. (6.5) it becomes clear that, due to the recording process (i.e., the $|\,|^2$ term), twice the maximum spatial frequency (underbraced term) is contained in the image intensity and therefore in its Fourier transform. As the information around the side-band is not affected by the $|\,|^2$ term, the number of interference fringes per smallest detail should be at least three. However, as we deal with real images and significant signal/noise ratios in the recorded holograms (due to the limited brightness of even field emission electron sources) two interferences fringes per smallest detail can be sufficient if the contributions from the artificial higher order spatial frequencies (underbraced term) become buried in noise. This usually happens because of one of two reasons; either the object can be considered a weak object (i.e., $A_{max} \ll 1$) and the underbraced

Figure 6.2. The autocorrelation (center area) and the two sidebands of the hologram in Fig. 6.1. More accurately, this is a subarea of the modulus of the Fourier transform of the hologram in Fig. 6.1 as described in Eq. (6.4).

term is insignificant in itself, or the signal/noise ratio is sufficiently high so as to bury contributions with significant amplitudes $A_{max}(< 1)$. This last situation usually occurs for either low contrast ($\mu \ll 10\%$) of the interference fringes and/or a low electron dose.

The hologram in Fig. 6.1, for example, had an average count rate of 300 per pixel and its Fourier transform, displayed in Fig. 6.2, shows no contributions of $2\,q_{max}$. Therefore, the slightly more than two interference fringes per smallest detail are sufficient to separate the sideband from the autocorrelation.

In order to extract the holographic information from Eq. (6.4), or Fig. 6.2, one of the sidebands is first isolated. This is usually done by creating a new (usually smaller) image with the center of the selected sideband coinciding with the center of the new image (see Fig. 6.3, left). Then, an aperture $B(\vec{q})$ is applied to retain the selected sideband only.

Several different types of apertures are described in the literature for the reconstruction process. The most simplistic aperture, a so-called hard aperture with value 1 inside and zero outside (see page 37, Eq. (2.7)) is used infrequently because of its tendency to cause artifacts in the reconstructed images. So-called soft apertures cause fewer artifacts and can be found in standard software packages (e.g., HolograFree, HoloWorks).[323,479] Typical examples of soft apertures include, for example, the exponential filter[3,380] $B_E(\vec{q})$, or the Butterworth filter[380,482] $B_B(\vec{q})$

$$B_E(\vec{q}) = e^{-(q/q_{max})^8} \quad ; \qquad B_B(\vec{q}) = \frac{1}{1 + c \cdot (q/q_{max})^{2\tau}} \tag{6.6}$$

where c is a constant (usually $= 1$), τ describes the order of the filter and q_{max} denotes the nominal cut-off frequency* of the filter.

*For example, for the Butterworth filter the cut-off frequency q_{max} is defined at $B_B = 0.5$ for $c = 1$.

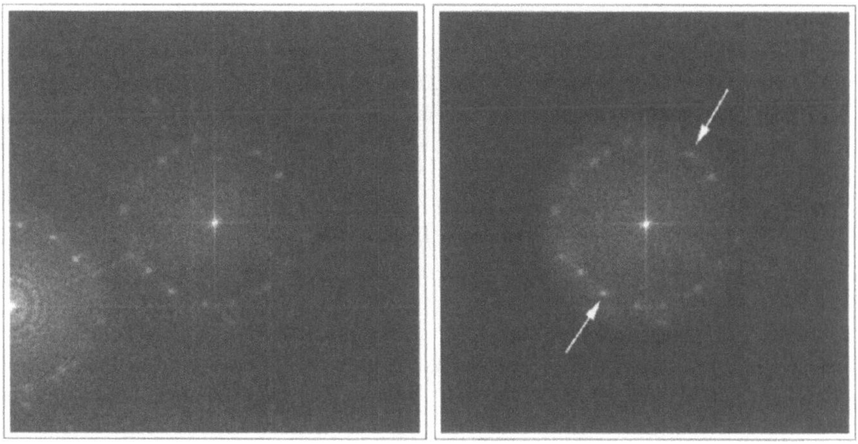

Figure 6.3. Left: The center of the sideband now coincides with the center of the new image. Right: A soft aperture (Butterworth filter, 8th order) has been applied to isolate the sideband. Opposite beams from the Au crystals, as marked with the white arrows, indicate that no major zone axis coincides with the illumination direction.

The effect of a Butterworth filter of order 8, centered on the sideband, is shown in Fig. 6.3, right. If the hologram has been recorded correctly[471] (i.e., for the highest spatial frequency in $A(\vec{r})\exp[i\phi(\vec{r})]$ at least two interference fringes are present), then the aperture truly isolates the information in the sideband from other contributions. In mathematical terms, Eq. (6.4) has been modified to retain only one of the two sidebands, e.g.,

$$\mathbf{FT}\left\{A(\vec{r})\cdot e^{i\phi(\vec{r})}\right\}\cdot B(\vec{q}) \equiv \mathbf{FT}\{O(\vec{r})\}\cdot B(\vec{q}) \tag{6.7}$$

where the convolution with $\delta(\vec{q}+\vec{q}_c)$ is omitted, corresponding to the shift of the origin of the $\vec{q}$ plane from $\vec{q}=0$ to the new origin at $\vec{q}=-\vec{q}_c$. The factor $expi\,\theta$ is usually ignored if θ is constant because it means an additional constant phase shift, and the absolute phase is not observable anyway but only phase terms and phase gradients. Any significant contributions from lines two and four of Eq. (6.4) have been removed by the aperture. It is worthwhile to note that opposite beams in Fig. 6.3, right, exhibit different intensities, i.e., that $\mathbf{FT}\{O(\vec{r})\}$ is not symmetric with respect to the new origin $\vec{q}=-\vec{q}_c$. Mathematically, this indicates that the image in real space is a true complex image. From the materials science point of view this is an indication that the Au crystals were not perfectly aligned along a zone axis.

An inverse Fourier transform is now sufficient to obtain the complex image wave $O(\vec{r})$

$$\begin{aligned}\mathbf{FT}^{-1}\{\mathbf{FT}\{O(\vec{r})\}\cdot B(\vec{q})\} &= O(\vec{r})\otimes\mathbf{FT}^{-1}\{B(\vec{q})\}\\ &\approx O(\vec{r})\otimes\delta(\vec{r}) = A(\vec{r})\cdot e^{i\phi(\vec{r})}\end{aligned} \tag{6.8}$$

At this point, the convolution with the aperture $B(\vec{q})$ usually is ignored, and $O(\vec{r})$ is then displayed in its amplitude $A(\vec{r})$ and phase $\phi(\vec{r})$ components.

3. Minimizing the effects of sampling

So far, the reconstruction process has been discussed based on continuous functions. However, actual holograms are digitized before the computational processing steps take place, and the number of pixels defining a hologram is finite. The hologram as is read from the camera is described by an M by N array

$$I_{hol}(m,n) = \overline{P} \cdot \{1 + A^2(m,n) + I_{inel}(m,n) + 2\mu\, A(m,n) \cdot \cos\left[2\,\pi\,(k_c, l_c) \cdot (m,n) + \phi(m,n) + \theta\right]\} \tag{6.9}$$

where M, N are integer numbers, $m \in \{1, \ldots, M\}$ and $n \in \{1, \ldots, N\}$, $\overline{P}$ is the average pixel value, and corresponds to the average number of electrons per pixel. The pixel value $I_{hol}(m,n)$ is obtained from Eq. (6.3) by integrating over the area of the (m,n)th pixel, i.e.,

$$I_{hol}(m,n) = \overline{P} \cdot \int_{(m-1/2)d}^{(m+1/2)d} \int_{(n-1/2)d}^{(n+1/2)d} I_{hol}(x,y)\ \mathrm{d}x\,\mathrm{d}y \tag{6.10}$$

where the pixel size of the camera was assumed to be square and of size d. To simplify the following discussions, we assume that the recording process does not significantly deteriorate the information in the image intensity from Eq. (6.3) and therefore use the following definition

$$I_{hol}(m,n) := I_{hol}(x,y) \quad \text{for} \quad \begin{cases} x = (m - 1/2)\,d \\ y = (n - 1/2)\,d \end{cases} \tag{6.11}$$

As a consequence of the limited number of pixels in real space, a Fourier transform on $I_{hol}(m,n)$ yields a finite number of pixels in Fourier space. In this case, the center of the sideband will most likely not fall onto a pixel in Fourier space, but somewhere in between. This 'mismatch' of the true center of the sideband with the available discrete pixels in Fourier space can give rise to a number of artifacts in the reconstructed image. The most obvious artifact appears in the reconstructed phase image, where a supposedly flat area, like a specimen-free area, displays a linearly increasing phase value. Other less obvious artifacts include increased signal-to-noise ratios especially in the reconstructed amplitude image.[476]

Also, the use of soft filters, like the Butterworth filter discussed earlier, help reduce artifacts from not perfectly centered sidebands. One possibility to further minimize these artifacts is to apply a so-called Hanning window to the actual hologram in real space before the first Fourier transform. Its purpose is to replace the rectangular window, which is inherent to each image and defines the actual size of the image, by a window with soft edges. The Hanning window of order τ is defined as[476]

$$H^{\tau}(m,n) = H_x^{\tau}(m) \cdot H_y^{\tau}(n) \tag{6.12}$$

where

Figure 6.4. Display of a 3 rd order Hanning window. It is applied (multiplied) to holograms in real space and removed after the reconstruction process. It helps to reduce artifacts in the reconstructed phase and amplitude images by reducing streaking in Fourier space.

$$H_x^T(m) = \frac{N}{2\pi} \cdot \sum_{j=1}^{T} \frac{(-1)^j}{j} \cdot \sin[\pi j (N-1)/N] \cdot [\cos(2\pi j m/N) - 1] \qquad (6.13)$$

$$H_y^T(n) = \frac{N}{2\pi} \cdot \sum_{j=1}^{T} \frac{(-1)^j}{j} \cdot \sin[\pi j (N-1)/N] \cdot [\cos(2\pi j n/N) - 1] \qquad (6.14)$$

An example of a third-order Hanning window is displayed in Fig. 6.4.

Another way to minimize artifacts in the reconstructed images can be achieved by using the Extended Fourier Transform (EFT).[473,474,477,478] While the standard (discrete) Fourier transform is defined as

$$\mathbf{FT}\{I_{hol}(m,n)\} = \frac{1}{N} \cdot \sum_{m,n=1}^{N} I_{hol}(m,n) \cdot e^{2\pi i (km+ln)/N} \qquad (6.15)$$

the EFT is defined as

$$\mathbf{EFT}\{I_{hol}(m,n)\} = \frac{1}{N} \cdot \sum_{m,n=1}^{N} I_{hol}(m,n) \cdot e^{2\pi i [(k+\Delta k)m+(l+\Delta l)n]/N} \qquad (6.16)$$

with the additional parameters Δk and Δl which can be any real numbers. These parameters allow the pixels in Fourier space to be shifted by an arbitrary amount. Therefore, if the true center of the sideband is offset by 0.3 pixels in k- direction, the sideband can be centered perfectly onto a pixel by selecting $\Delta k = 0.3$. This has the advantage that not only is the sideband centered as it should be, but it also allows the apertures used in the reconstruction process to be centered correctly.† It is worthwhile to note also that the **EFT** algorithm is practically as fast as the standard **FT** algorithm.

Eq. (6.16) can be rewritten easily as

$$\frac{1}{N} \cdot \sum_{m,n=1}^{N} \left[I_{hol}(m,n) \cdot e^{2\pi i (\Delta k m + \Delta l n)}\right] \cdot e^{2\pi i (km+ln)/N} = \mathbf{EFT}\{I_{hol}\} \qquad (6.17)$$

†It should be noted that the inverse Fourier transform, which is part of the reconstruction process as discussed earlier, is usually not replaced by an EFT. Note also that the $\Delta k, \Delta l$ values can be replaced with the correct distance from the center of the sideband to the center of the Fourier transform. In this case, the center of the sideband will coincide with the center of the Fourier transform.

from which it is obvious that instead of modifying existing **FT** code, the same effect can be obtained by multiplying the real image $I_{hol}(m,n)$ with a complex wedge $\exp[2\pi\, i\,(\Delta k\, m + \Delta l\, n)]$. The disadvantage, however, is that the **FT** now has to be performed on a complex image, instead of on a real image, which doubles the processing time.

The basic reconstruction process has been outlined above under the aspect of refining the process itself. However, as it turns out, there are additional problems with real electron holograms. These problems are not caused by the digital processing, but originate in the microscope itself or are introduced by the recording process. These problems can outweigh most artifacts discussed so far in the reconstruction procedure.

4. Lens aberrations: Distortions

Distortions are aberrations which depend on the location (x, y) within the image only. This distinguishes them from another type of aberrations, i.e., the isoplanatic aberrations, which depend on spatial frequencies only and which are discussed in detail in Chapter 9, pages 202, ff. Fortunately, aberrations which depend on both the location and the spatial frequencies appear to be of lesser importance in modern electron microscopy and can usually be ignored.

4.1. The cause for distortions

Distortions can originate from many sources. Projection lenses are an imminent source for distortions, as they have to image large areas. CCD cameras cause distortions due to the so-called shear-distortion of their transfer fiber optics[87] and photographic film has a tendency to warp and thus introduce distortions. For most applications, these distortion effects are small enough to be ignored. For the special application of off-axis holography, however, this is not the case. These distortions have a significant impact on the image phase as stored in the hologram. The technique of off-axis holography also faces additional sources for distortions due to an additional optical element, the biprism. Small variations of the thickness of the biprism filament as well as local variations of the biprism voltage due to charging artifacts directly affect the uniformity of the interference fringes.

4.2. The effect of distortions

In most electron microscopes, distortion effects are small enough to be ignored in standard applications. Off-axis electron holography, however, stores information about the image phase in a lateral displacement of the interference fringes, and distortions can cause a similar displacement. In some cases, distortions can be distinguished from actual phase information if the phase information is bandwidth-limited to higher frequencies. As distortions characteristically show up in the lower frequency range, a simple background filtering may be sufficient to remove their contribution.[323] This is in general not such a good idea; there are better ways to compensate for the artifacts from distortions, as discussed below.

To estimate the effect of distortions on the phase image, we presume a sampling frequency (i.e., the number of sampling points per interference fringe.[‡]) of four pixels per one interference fringe, which corresponds on a CCD camera with 25 μm pixel size to a fringe spacing of about 100 μm. A distortion of 25 μm, or the size of 1 pixel,

[‡]For a description of the sampling theorem, see Ref. 81

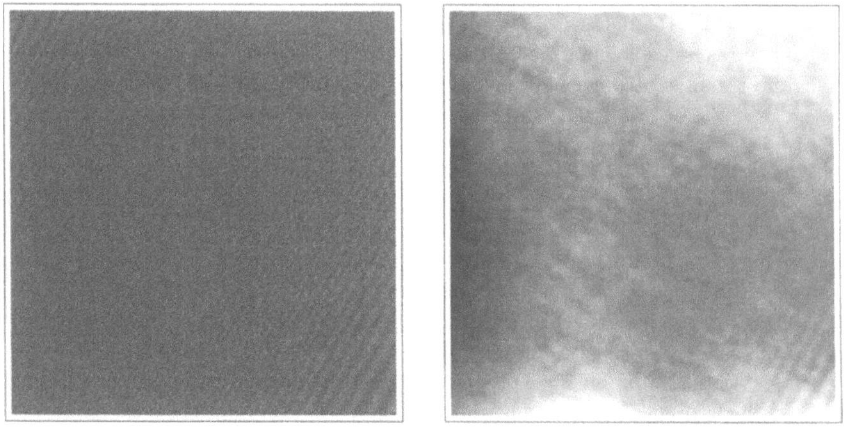

Figure 6.5. Left: A hologram without object. The sampling frequency is 5.5 pixels per fringe (or per 2π). Pixel size is about $25\,\mu$m. Right: The phase image from that hologram shows a total dynamic range of about $2\pi/4$, and corresponds to an overall distortion of less than $34\,\mu$m or less than 1.4 pixels.

offsets the interference fringes by $2\pi/4$. Figure 6.5 shows an empty hologram (left), with a sampling frequency of 5.5, recorded on a Hitachi HF-2000 TEM with a Gatan 694 slow-scan CCD camera, and its reconstructed phase on the right-hand side. The phase dynamic in that image is purely artificial and shows a total phase range of $2\pi/4$, which corresponds to a maximum distortion of about 1.4 pixels throughout the image. This is a reasonable value for most instruments. As an example, for a high resolution image with a lattice spacing of 2 Å and a sampling frequency of 4, the distortion effect in this case would, at maximum, be ~0.7 Å throughout the entire image which can be neglected in almost all cases. Phase resolution requirements better than $2\pi/4$, as standard in holography, require active distortion compensation.[87]

5. The reconstruction process using a reference hologram

In order to include distortion effects into the basic equation for holograms (see Eq. (6.3)), we introduce an additional phase modulation term $\theta(\vec{r}, \vec{q}_c)$.[250] The effect of distortions is a constant at the CCD camera, whereas the width of the interference fringes is a variable. Clearly, any increase in the width of the interference fringes causes a decrease of $\theta(\vec{r}, \vec{q}_c)$, and therefore θ must be a function of the interference fringe width, but $\vec{q}_c$ is a constant for each hologram. Equation (6.3) is therefore modified to

$$I_{hol,dis}(\vec{r}) = 1 + A^2(\vec{r}) + 2\mu\, A(\vec{r}) \cos[2\pi\vec{q}_c\vec{r} + \phi(\vec{r}) + \theta(\vec{r}, \vec{q}_c)] \qquad (6.18)$$

This now includes the variable phase modulation term $\theta(\vec{r}, \vec{q}_c)$ which represents the non-isoplanatic aberrations of the microscope including the image recording device. Subsequent reconstruction of Eq. (6.18) yields the phase image $\phi(\vec{r}) + \theta(\vec{r}, \vec{q}_c)$.

As has been pointed out in earlier papers, the importance of recording a reference hologram along with every hologram taken cannot be over-emphasized.[86,353] The reference hologram is obtained by carefully removing the specimen from the field of view, while not changing any of the optical parameters of the microscope. The interference pattern, now undisturbed by the object, is quickly recorded using the same exposure time, to produce the reference hologram.

If the reference hologram is taken immediately after the hologram, the reference hologram is represented, within limits, by the following equation

$$I_{ref,dis}(\vec{r}) = 1 + 2\mu \cos[2\pi \vec{q}_c \vec{r} + \theta(\vec{r}, \vec{q}_c)] \tag{6.19}$$

where, for simplicity, it has been assumed that the reference hologram has the same fringe contrast μ as the object hologram.

There are two major reasons for recording a reference hologram. First, a reference hologram makes it possible to determine the true center of the sideband in Fourier space, which is essential for producing a properly reconstructed and interpretable phase image. This can be particularly important when dealing with holograms of strong electric and magnetic fields, because such fields have a significant influence on the reference wave and therefore do not exhibit undisturbed interference areas. Second, the compensation of image distortions in the phase image can be successfully performed with a reference hologram.

5.1. Finding the true center of the sideband

Although the interference fringes in a hologram usually form the strongest contribution to a single spatial frequency in Fourier space, the true center of the sideband may not be the pixel with the highest amplitude in the area of the sideband. This is a problem for automated reconstruction procedures as demonstrated in the following figures. Figure 6.6, left, shows a hologram of a standard tip of a magnetic force microscope. The strong magnetic field of the magnetic force microscope (MFM) tip significantly distorts the interference fringes. Both the reference wave and the object wave are modulated strongly by the field. After a Fourier transform, the search for the maximum amplitude in the sideband of the hologram (see marked area in Fig. 6.6, right) results in the selection of this pixel as the center of the sideband. The resulting phase image is displayed in Fig. 6.7, left, and gives the impression that the primary distribution of the magnetic field of this specific MFM tip is off to its right.

Using a reference hologram, however, allows the position of the sideband to be determined within a ± 1 pixel accuracy by selecting the pixel with the highest amplitude from the sideband of the reference hologram. This usually is possible despite the $\theta(\vec{r}, \vec{q}_c)$ term.[§]

Reconstruction of the hologram in Fig. 6.6, left, together with a reference hologram (obtaining the center of the sideband from the reference hologram) yields the phase image as displayed in Fig. 6.7, right. Obviously, in this case the field lines extend in the appropriate direction, although, due to the influence of the magnetic field on the reference wave, a quantitative interpretation without proper simulation is still not possible.

The process for reconstructing a hologram with a reference hologram requires slightly more than twice the computational time, but otherwise follows almost all the steps of the basic reconstruction procedure. After the position of the center of the

[§]Only for unusually large distortions, the $\theta(\vec{r}, \vec{q}_c)$ term may pose a problem, as in this case there is no obvious way to center the aperture (which isolates the sideband) around the true center of the sideband. It is possible that under these circumstances similar artifacts as known from not centering the objective aperture in the microscope arise.

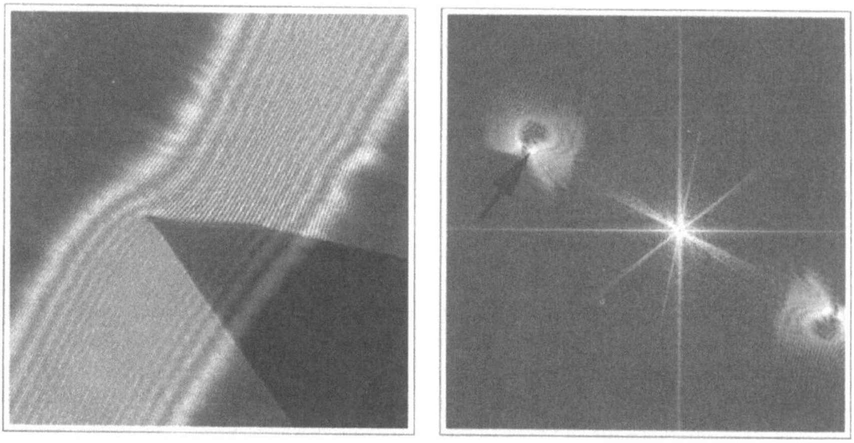

Figure 6.6. Left: The magnetic field of a standard MFM-tip significantly modulates the holographic interference fringes. Right: The pixel with the highest amplitude in the sideband (marked with a black arrow) does not necessarily represent the true center of the sideband. (Hologram courtesy of B. Frost.)

Figure 6.7. Using no reference hologram in the reconstruction process (left) can lead to misinterpretation of the data, whereas the proper reconstruction process provides the expected field distribution around the MFM tip.

sideband is determined, two complex images are computed by applying the identical aperture to both sidebands and performing an inverse Fourier transform. This yields the distorted reference wave

$$O_{ref.dis}(\vec{r}) = \mu\, e^{i\,\theta(\vec{r},\vec{q}_c)} \tag{6.20}$$

and the distorted image wave

$$O_{hol.dis}(\vec{r}) = \mu\, A(\vec{r})\, e^{i\,\phi(\vec{r})+i\,\theta(\vec{r},\vec{q}_c)} \tag{6.21}$$

Both complex waves are now available in amplitude and phase. However, the distortion-induced phase modulation $\theta(\vec{r}, \vec{q_c})$ and an additional phase wedge due to inaccurate centering of the sidebands (not taken into account in phase term of Eqs. (6.20) and (6.21)) are still present in the phase images. This is corrected by complex division of the the image wave by the reference wave, yielding the distortion-corrected image wave[353,479,490]

$$O_{hol}(\vec{r}) = A(\vec{r})\, e^{i\,\phi(\vec{r})} \tag{6.22}$$

The effect of distortion-correction by means of a reference hologram is shown in Chapter 9 (page 217) for an amorphous carbon foil. The reconstructed phase shows a strong long-range modulation due to distortions. After distortion-correction by means of a reference hologram, these artificial modulations of the phase vanish completely resulting in a flat phase in the part of the unscattered wave, which passed the object plane through a hole in the foil, and in fine details of the amorphous specimen.

5.2. High resolution specifics

In high resolution electron holography, the coherent wave aberrations $\chi(\vec{q})$ are corrected by means of a phase plate $\chi(\vec{q})$ in order to unscramble amplitude and phase information and to improve the interpretable resolution beyond the point resolution. This numerically generated phase plate, which models the coherent wave aberrations $\chi(\vec{q})$ of the objective lens, is applied to the complex diffraction pattern of the reconstructed image wave $O_{hol}(\vec{r})$ yielding the object wave $o_{hol}(\vec{r})$. A detailed description of correction of coherent aberrations can be found in Chapter 9.

If the distortion-induced phase modulation $\theta(\vec{r}, \vec{q_c})$ has not been compensated, the correction of coherent wave aberrations of the image wave could only be performed correctly for parts of the field of view since variations of the distortion-induced phase modulation $\theta(\vec{r}, \vec{q_c})$ influences locally both the carrier frequency $\vec{q_c}$ of the interference fringes and the spatial distribution of the object information in reciprocal space. Therefore, the sidebands are smeared out in reciprocal space as is illustrated in Fig. 6.8. As a consequence, the condition for isoplanicity is only locally fulfilled and a correction of coherent wave aberrations would be successful only for parts of the field of view.

Figure 6.9 shows the distortion-induced phase modulation in an experimental example of an image wave of $Ti_2Nb_{10}O_{29}$ in <100> orientation. This hologram was recorded at the Philips CM30FEG ST/Special-Tübingen equipped with a 2k by 2k slow-scan CCD camera from Photometrics. In the amplitude image, the lateral displacement of the imaged object structure cannot be observed since it is smaller than $100\,\mu$m on the CCD chip between two bundles of the fiber optic. However, the phase image shows strong variations in contrast due to distortions of the projector lenses and due to shear-distortion of the transfer fiber optic of the slow-scan CCD camera. Therefore, the information of the diffraction pattern, i.e., the Fourier transform of the complex image wave, is smeared out in reciprocal space.

According to Ref. 269 it can be calculated that the maximum tolerable variation of phase modulation $\theta(\vec{r}, \vec{q_c})$ due to distortions within the diameter of the point-spread-function PSF[262] in an electron hologram of width w_{hol} can be estimated to

$$\left| \frac{\partial\theta(\vec{r}, \vec{q_c})}{\partial\vec{r}} \right|_{max} \leq \frac{2\pi}{3\, w_{hol}} \tag{6.23}$$

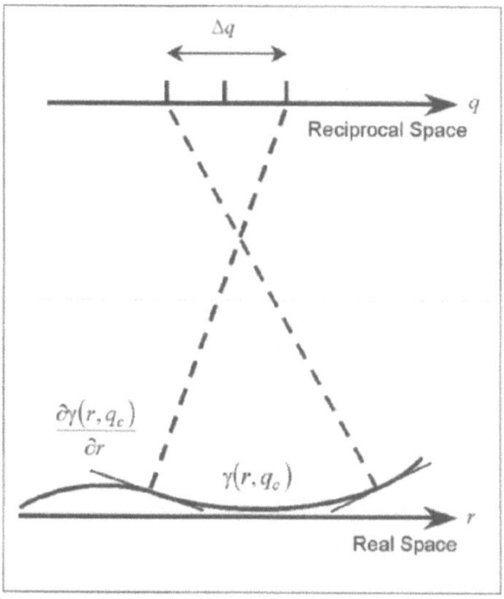

Figure 6.8. Smear out of a reflection in reciprocal space due to distortion-induced phase modulation $\theta(\vec{r}, \vec{q}_c)$.

Figure 6.9. Amplitude (left), Phase (middle), and its diffraction pattern (right) of reconstructed image wave of $Ti_2Nb_{10}O_{29}$ in <100> orientation. The modulation caused by distortions clearly overpowers the phase signal of the object. In reciprocal space, the reflections are smeared out. (Specimen was provided by D. Smith and recorded together with A. Orchowski)

As a consequence, the smearing of the sideband must be corrected better than 1/3 of a pixel and the sideband must be centered within this subpixel accuracy, respectively. Both the centering of the sideband as well as the correction of distortion-induced phase modulation $\theta(\vec{r}, \vec{q}_c)$ are performed by means of a reference hologram which has been described above.

Figure 6.10 shows the image wave and its diffraction pattern of $Ti_2Nb_{10}O_{29}$ in a <100> orientation after compensation of distortion-induced phase modulations by means of a reference hologram. Long-range variations of phase contrast have been completely corrected so that fine object details can now be observed. As a consequence of the distortion-compensation, the blurring of the crystal reflections has vanished.

However, weak phase modulations due to Fresnel scattering at the biprism filament can be found in the lower right corner of the phase image. Practical experience in high resolution holography shows that the amplitude and phase distribution of the Fresnel modulation typically does not fit very well between two successive exposures, even with a slow-scan CCD camera. Additionally, the amplitude of the reference wave is very noisy due to the limited fringe contrast which further degrades the signal-to-noise

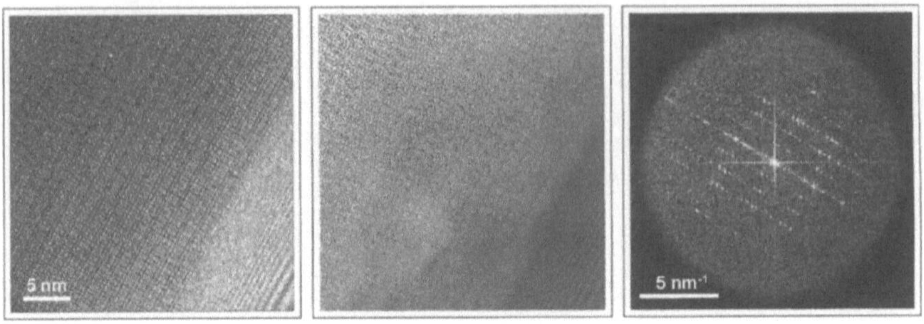

Figure 6.10. Amplitude (left), Phase (middle), and its diffraction pattern (right) of reconstructed image wave of $Ti_2Nb_{10}O_{29}$ in <100> orientation after compensation of distortions by means of a reference hologram. Fine object details can now be observed in the phase image and blurring of crystal reflections has vanished.

ratio of the distortion-compensated image amplitude.[250] Therefore, only the phase of the reference wave should be subtracted from the distorted image phase to yield the distortion-corrected image wave

$$O_{hol}(\vec{r}) = \mu\, A(\vec{r})\, e^{i\,\phi(\vec{r})} \tag{6.24}$$

where only the pixel values of the image amplitude $A(\vec{r})$ are decreased by the constant factor of the fringe contrast μ.

To sum up, two holograms should always be recorded, the hologram with object and its corresponding reference hologram. Then it is possible to correct phase modulations due to distortions and due to inaccurate centering of a sideband by means of the reconstruction steps described above. Afterwards, phase shifts can be quantitatively measured and coherent wave aberrations can be corrected in order to improve the interpretable resolution beyond the point resolution of the microscope.

6. Other reconstruction methods

Looking back into history, Gabor proposed light optical reconstruction as a method for extracting amplitude and phase images from a hologram stored on a photographic plate. This reconstruction method is still used in today's interferometry[159,160,441,485] as it has the advantage that simple, standard laboratory equipment is sufficient for the reconstruction process. However, with the availability of modern slow-scan CCD cameras and inexpensive, fast computers, digital reconstruction of holograms is replacing light optical reconstruction, because data are accessible much quicker than with film-based methods.

However, the full digital reconstruction process, as described earlier, requires a significant processing time even with dedicated computer systems.[353,479] As a consequence, it is presently not possible to observe dynamic processes on a TV-rate basis using a fully digital reconstruction method. An interesting development to compensate for lack of speed is a hybrid-technique which has been developed by J. Chen et al.[46] They record the hologram with a TV-rate digital camera but perform the actual reconstruction on the optical bench and re-record the reconstructed phase images with another TV-rate camera for display on a monitor (see page 143).

Besides the requirement for high speed reconstruction, there are additional reasons to look into reconstruction methods different from the ones mentioned before. For example, a CCD camera with 512 by 512 pixels has a field of view of, say, 256 by 256 nm (at a certain magnification) and therefore a maximum resolution of 1 nm, due to the sampling theorem, which requires at least two pixels per finest detail of the object. When recording a hologram, its highest spatial frequency is given by its interference fringes instead, which must have a spatial frequency of at least twice the highest spatial frequency of the object (see discussion on pages 127, ff). Therefore, the actual field of view of a hologram shrinks to 128 by 128 nm or less, if the spatial resolution of 1 nm is to be preserved. In some cases, non-Fourier methods can maintain a reasonable field of view.

In general, reconstruction methods other than the one described in Section 2 can be divided roughly into two groups: reconstruction methods optimized for high speed reconstruction, and reconstruction methods optimized for an increased field of view. Additional methods not discussed here can be found in Refs. 3,4,163,374.

6.1. The alternative reconstruction method

The alternative reconstruction method proposed by Lenz, Völkl, and Lehmann is based on a statistical method and avoids the Fourier transformations inherent to the standard reconstruction method.[246,248] It therefore has the potential to be significantly faster than the standard reconstruction process. As other techniques can be derived from this algorithm, the alternative reconstruction method is described in somewhat more detail.

A hologram of MgO, recorded with a CCD camera, is displayed in Fig. 6.11. In the vacuum area the interference fringes are parallel to the y- axis and the specimen is significantly defocused to expose the cubic morphology of the MgO. A linescan parallel to the x- axis shows the values of approximately 130 pixels (P_m) along the selected line which intersects 11 interference fringes.

As discussed earlier, at least two interference fringes per smallest detail in the image are necessary to record a hologram correctly. Based on this requirement, the basic assumption for the alternative reconstruction procedure is established: within two to three interference fringes (which correspond to $Z = 20$ to 30 pixels in Fig. 6.11) the image phase and intensity vary sufficiently slowly to be considered constant. Under this basic assumption, Eq. (6.9) simplifies locally in the area of the selected Z pixels to

$$P_{hol}(m) = \overline{P} \cdot (1 + A^2 + I_{inel}) \cdot [1 + \mu' \cos(2\pi k_c m/N + \phi)] \tag{6.25}$$

with the interference fringes parallel to the y- axis and $\mu' = 2\mu A/(1 + A^2 + I_{inel})$. An example of a sub-area of Z pixels for which this approximation holds is marked in light grey in the linescan of Fig. 6.11.

Equation (6.25) describes the theoretically expected output values of the CCD camera, which normally differ from the actually measured pixel values $P_{m,n}$ due to noise as well as the limited validity of the basic assumption of constant phase and intensity values. Thus, by minimizing the expression

$$\sum_{m=m'-Z/2}^{m'+Z/2-1} [P_m - P(m)]^2 = \text{Min!} \tag{6.26}$$

an approximate phase value $\phi(m')$ can be determined, where m' is the $Z/2$th pixel in the sub-area. Fortunately, Eq. (6.26) can be solved analytically to yield

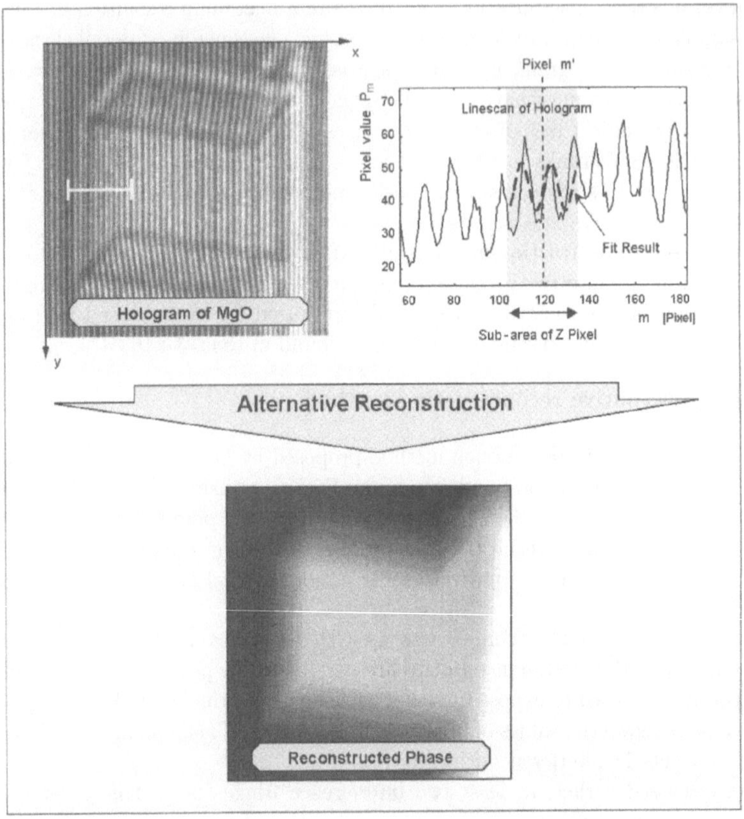

Figure 6.11. Algorithm of the alternative reconstruction method at an example of a Magnesium oxide crystal. The local phase value is determined by fitting a hologram intensity distribution to a 1-dimensional sub-area of Z pixels. Repeating the procedure throughout the hologram yields a phase image without a single Fourier transform.

$$\mu' \exp[i\,\phi(m') + 2\,\pi\,i\,(m'-1)/s_x\,] = X + i\,Y \tag{6.27}$$

which can be, after separating real and imaginary parts, rewritten as

$$\phi(m') = \text{atan2}\left(\frac{X\cdot\cos[2\,\pi\,(m'-1)/s_x\,] - Y\cdot\sin[2\,\pi\,(m'-1)/s_x\,]}{X\cdot\sin[2\,\pi\,(m'-1)/s_x\,] + Y\cdot\cos[2\,\pi\,(m'-1)/s_x\,]}\right) \tag{6.28}$$

where we have made use of the atan2 function,¶ defined in most programming languages, instead of the standard atan function, defined only within $-\pi/2$ to $\pi/2$. Also, the following abbreviations have been used

¶The atan2 function is the single-valued atan function of two independent variables with an output range between $-\pi$ and π.

$$C_1 := \frac{\sin(\pi Z/s_x)}{Z\sin(\pi/s_x)} \quad ; \quad C_2 := \frac{\sin(2\pi Z/s_x)}{Z\sin(2\pi/s_x)}$$

$$X = \frac{C - 2C_1 P}{P(1+C_2) - C\,C_1} \quad ; \quad Y = \frac{S\cdot(1+C_2-2C_1^2)}{(1-C_2)\cdot[P\cdot(1+C_2) - C\,C_1]} \tag{6.29}$$

as well as

$$C := 2\sum_{m=m'-Z/2}^{m'+Z/2-1} P_m \cdot \cos\left[2\pi\,(m - m' + 1/2)/s_x\right]$$

$$S := 2\sum_{m=k'-Z/2}^{m'+Z/2-1} P_m \cdot \sin\left[2\pi\,(m' - m - 1/2)/s_x\right]$$

$$P := \sum_{m=m'-Z/2}^{m'+Z/2-1} P_m \tag{6.30}$$

s_x is the sampling frequency in direction of the sub-area, in this case the x- direction. By moving the sub-area one pixel to the right, the same procedure allows one to determine the phase value $\phi(m'+1)$ and thus, by moving the selected area throughout the hologram, a full set of phase values $\phi(m', n')$ can be obtained without performing a single Fourier transform.

There are two parameters which have to be fed into the algorithm. One is the length of the sub-area Z, which should include at least 1.5 periods of the interference fringes.[246] The other parameter is the sampling frequency s_x, which can be determined once using a one-dimensional Fourier transform of one full line of the hologram.

So far we have assumed that the interference fringes run parallel to the y- axis. However, the alternative reconstruction algorithm works well with interference fringes even at significant angles with the y- axis. The general form of Eq. (6.28) finally is

$$\phi(m', n') = \operatorname{atan2}\left(\frac{X\cdot\cos[2\,\pi\,(\frac{m'-1}{s_x} + \frac{n'-1}{s_y})] - Y\cdot\sin[2\,\pi\,(\frac{m'-1}{s_x} + \frac{n'-1}{s_y})]}{X\cdot\sin[2\,\pi\,(\frac{m'-1}{s_x} + \frac{n'-1}{s_y})] + Y\cdot\cos[2\,\pi\,(\frac{m'-1}{s_x} + \frac{n'-1}{s_y})]}\right) \tag{6.31}$$

Equation (6.27) can also be written in a different form as

$$\phi(m', n') = \operatorname{atan2}(Y/X) - 2\pi\left[(m'-1)/s_x + (n'-1)/s_y\right] \tag{6.32}$$

where the compensation term $(n'-1)/s_y$ for interference fringes not parallel to the y- axis has been included. This is the expression normally used for the computation of $\phi(m', n')$. It should be noted however, that Eq. (6.32) introduces significant phase jumps in the computed phase image. Therefore, an additional phase unwrapping algorithm has to be performed on Eq. (6.32), to obtain the identical results as from Eq. (6.31).[||] Despite the additional phase unwrapping process required for Eq. (6.32)

[||]The phase unwrapping algorithm can be quite simple: at the beginning of each line, the current phase value $\phi(Z/2, n')$ is compared with the phase value $\phi(Z/2, n'-1)$ of the previous line. If the modulus of the difference between both phases exceeds π, the phase of the current line $\phi(Z/2, n')$ is increased or diminished, respectively, by 2π, ensuring that the phase difference between two consecutive lines never exceeds π. Within a line, the current phase value $\phi(m', n')$ is compared with the previous value $\phi(m'-1, n')$ and corrected in the same way.

this route is significantly faster than the direct evaluation of Eq.(6.31). For example, in 1994, on a DEC 3000 AXP 600 workstation with 34 MFLOPS, the processing time for the reconstruction of a phase image was less than 0.2 seconds for a 512 x 512 pixel hologram.[249]

6.2. The hologram-shifting method

The hologram-shifting method, proposed by Ru et al. for real time purposes in electron holography,[375] can be derived directly from the alternative reconstruction method. In this method, a single hologram is recorded digitally and shifted Z times by τ pixels ($\tau \geq 1$ and integer) perpendicular to the interference fringes. Shifting the hologram by more than 1 pixel at a time modifies the effective sampling frequency s_x which should be replaced by s_x/τ. If the hologram can be recorded such that $Z\,\tau/s_x$ is an integer, Eqs. (6.29) simplify to

$$\begin{aligned} C_1 = 0 \quad &; \qquad C_2 = 0 \\ X = \frac{C}{P} \quad &; \qquad Y = \frac{S}{P} \end{aligned} \tag{6.33}$$

For $Z = s_x/\tau$, Eqs. (6.30) can be rewritten as

$$\begin{aligned} C &= 2 \sum_{m=m'-Z/2}^{m'+Z/2-1} P_m \cdot \cos\left[2\pi\,(m' - m\,\tau - 1/2)/Z\right] \\ S &= 2 \sum_{m=m'-Z/2}^{m'+Z/2-1} P_m \cdot \sin\left[2\pi\,(m' - m\,\tau - 1/2)/Z\right] \end{aligned} \tag{6.34}$$

The value for the reconstructed phase at pixel m' is then simplified to

$$\phi_{m'} = \text{atan2}(S/C) - 2\pi\left[(m'-1)\,\tau/s_x + (n'-1)/s_y\right] \tag{6.35}$$

which compares directly to the 'phase shifting formula' as given by Ru et al.,[375] with a slight difference in the origin of the coordinate system and the additional compensation for interference fringes not aligned parallel to the y- axis. In the same paper, Ru et al. pointed out that a significant additional simplification of Eq. (6.35) is obtained for the special case of $Z = s_x/\tau = 4$, which could provide sufficient speed improvement for the computational process to allow reconstruction rates close to TV-rates.

6.3. TV-rate reconstruction performance

Almost all prior described reconstruction methods are presently too slow to simultaneously process holograms and display their phase images at TV-rate. To overcome this problem, Chen et al. have refined the old approach of reconstructing a hologram on the optical bench.[46] They replaced the hologram (usually on photographic film) with a liquid crystal panel, which was fed directly from a TV-rate CCD camera. Their basic set-up is shown in Fig. 6.12.

In this approach, a Mach-Zehnder interferometer supplies two sets of parallel beams which transmit through the LC panel at different angles. A linearly polarized He-Ne laser is used as the light source to assure that the two beams are coherent. The hologram

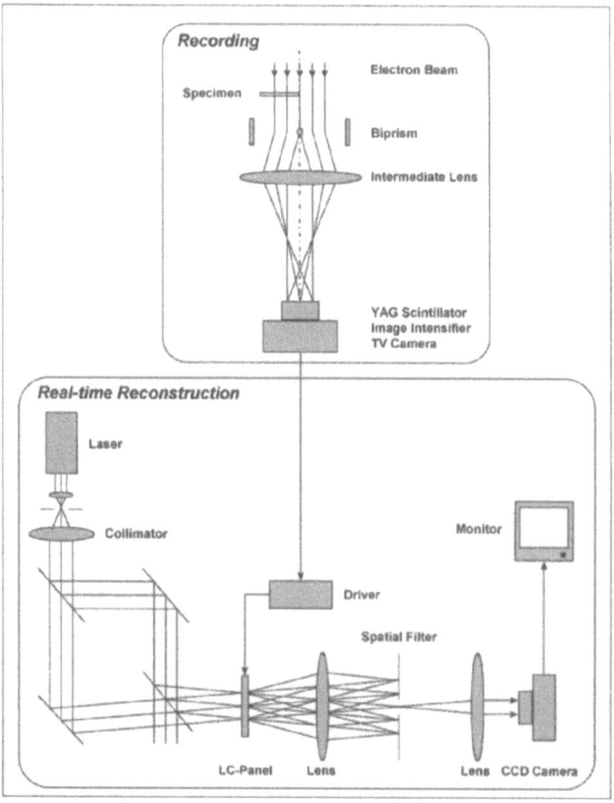

Figure 6.12. Schematic diagram of the real-time electron-holographic interference microscope system using an LC-panel.

itself is recorded at the microscope by means of a TV-rate CCD camera, and the video signal feeds directly into the driver of the LC panel (whose plastic polarizers, which are normally attached to both sides of the LC panel, have been removed). Thus, the device, which is located at the output of the Mach-Zehnder interferometer, works in the phase hologram mode.[47] A subsequent spatial frequency filter (consisting of two lenses and an aperture) is used to transmit only the zero-order beam. The resulting image is the phase image reconstructed from the hologram in the LC-panel and is recorded by means of a CCD camera and is either displayed live on a separate monitor or stored on video tape.

This TV-rate performance was successfully demonstrated during an in-situ observation of a domain wall motion in thin Permalloy film caused by an external magnetic field.[46,47]

7. Methods for lower carrier frequencies

A hologram, to be reconstructed by the conventional Fourier optic technique, must have a spatial frequency of the interference fringes which are higher than the highest spatial frequency of the object under investigation. The spatial frequency of the finest fringes must be two times higher for a weak object and up to three times higher for a strong object in order to separate a sideband, which carries the information of the elastically scattered image wave, from the autocorrelation.[171] However, due to the fi-

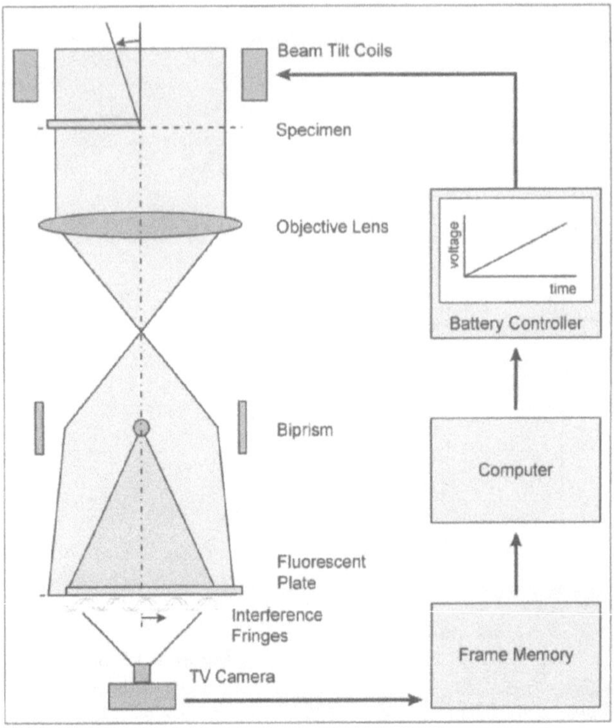

Figure 6.13. The phase-shifting method: Electron optics and instrumentation for shifting the initial phase of a hologram.

nite coherence of the electron source, the fringe contrast decreases when the spatial frequency of the interference fringes is increased. But a high fringe contrast is required to have reasonable sensitivity and precision of the phase reconstruction, which is limited primarily by the quantum noise.

In order to achieve both high resolution and high precision measurements without having fringes at a high spatial frequency, a phase-shifting technique (also called fringe scanning technique) is used which has found wide application in optical interferometry and which has been proposed by Rau and Ru in electron holography.[355,372,377]

The method is based on the use of a number of holograms with properly different initial phases which can be obtained by tilting the incident beam perpendicular to the biprism filament as shown in Fig. 6.13. The specimen must be in-focus in order to avoid an image shift while tilting. The initial phase is determined by measurement of the phase distribution of a sideband-peak in reciprocal space and must be shifted by at least 2π. This, together with the object-independent fringe spacing, determines the number of required holograms. At least three holograms are required for reconstruction of the image wave and the use of up to 300 holograms recorded by means of a TV-rate camera has been reported.[378] Because a sum over many hologram images is included in the calculations of the image phase, the random quantum noise of electrons and the shot noise of detectors are both significantly reduced, allowing high precision in phase measurements.

Several applications of the phase-shifting method have been reported. In low-magnification, the observation of quantized magnetic fluxes, leaking out from a superconducting Pb thin film held at a temperature of 5 K, using 12 hologram images has been demonstrated.[373] The phase-shifting method has also been successfully applied to investigations of the projected mean-inner potential distribution of single-shell carbon nanotubes, which causes phase shifts of roughly $2\pi/100$.[378]

A new approach to reconstruction of phase-shifting holograms has been proposed by Heindl et al., who perform an analytical calculation of the reconstructed image phase using a Kohonen neural network.[178] After carrying out the learning process, the readout process in this neural network reduces the error in the reconstruction of noise-corrupted hologram images since it uses the entire information contained in the input holograms.

8. Display of phase information

Phase information as contained in an off-axis hologram can be obtained only modulo 2π, as the original phase information is contained within the cos-term representing the interference fringes and therefore faces the ambiguity: $\cos(\phi) = \cos(\phi \pm 2\pi)$. This ambiguity then leads to so called 'phase jumps' in the reconstructed phase images as can be seen easily from a simulated phase ramp, whose dynamic range extends from 0 to 2π (see Fig. 6.14a); all values of $\phi \geq 2\pi$ are converted to $\phi - 2\pi$, which then creates the phase jump at 2π.

Initially, the complex image which is computed from a hologram is present in its real and imaginary part. In order to compute the phase information, the atan2 function** is used, which yields the phase image in the range $[-\pi, \pi]$. This image is often adjusted to fit the range $[0, 2\pi]$ by adding π to the image. As a side note, the range $[-\pi, \pi]$ obviously includes the completely equivalent values $-\pi$ and π and it is good practice to convert all π values to $-\pi$, or vice versa.

In general, there are two different types of phase jumps, 'virtual' phase jumps and 'real' phase jumps, both of which are quite inconvenient and should be removed to produce the best phase images.

8.1. Virtual phase jumps

Virtual phase jumps occur in phase images where the 'true' phase dynamic is less than 2π. For example, if the 'true' phase range is from 0.7π to 2.3π, then the displayed dynamic range is from 0π to 0.3π and from 0.7π to 2π. Consequently, phase jumps occur at the 2π level, i.e., all phase values $\phi \geq 2\pi$ are converted to $\phi - 2\pi$.

Virtual phase jumps can easily be removed. The mathematical method is to manipulate the complex image $O = A \cdot \exp(i\,\phi)$ by multiplying O with a phase constant $\exp(i\,\phi_c)$ (e.g., $\phi_c = -0.5\pi$), thus adjusting the true phase range from $[0.7\pi, 2.3\pi]$ to $[0.2\pi, 1.8\pi]$. Extracting the image phase from these modified image data will create no phase jumps.

Requirements for computational speed, however, suggest a different route. For example, by setting an arbitrary (but variable) starting value $\phi_c = 0.5\pi$, all pixels in the phase image with a value $> \phi_c$ are replaced by $\phi - 2\pi$. Then, the compensation value $\tilde{\phi} = 2\pi - \phi_c$ is added to the phase image to restore its original dynamic range from $[0, 2\pi]$. The resulting phase image in our example then has the dynamic range $[0.2\pi, 1.8\pi]$ – with no phase jumps. This operation can be done "on the fly" while interactively changing ϕ_c, even on a standard desktop computer, and provides a simple way to remove virtual phase jumps.

**Standard C-code function; the arctan() itself, is defined only in the range $[-\pi/2, \pi/2]$, whereas the atan2 function is defined over the full range $[\pi, -\pi]$.

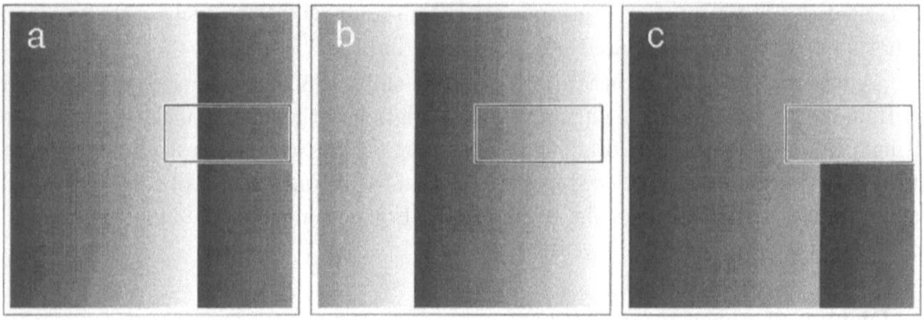

Figure 6.14. a) An artificial phase wedge with a 'true' phase dynamic of 3π displays a phase jump at π within the normal phase range of $[-\pi;\pi]$. b) The same phase wedge modified to display the phase jump at a different location. Composing a new image from a) and b) can yield a phase jump free phase image with an extended dynamic range.

8.2. Real phase jumps

Real phase jumps, i.e., phase jumps in images whose true dynamic range is $> 2\pi$, are much more difficult to deal with, and, although several methods (so-called phase unwrapping methods) offer themselves, none appears to solve all the problems. The most obvious and most widely used method is to search the phase image line by line for phase jumps; that is by setting a certain threshold value and then declaring the difference between two neighboring pixels a phase jump if that threshold is exceeded.[353] Once a phase jump is detected, it is remediated by adding or subtracting 2π to the following pixels, until the next phase jump occurs. Methods like these work well under circumstances where noise is not significant, where phase gradients do not exceed 2π within three pixels, and where the original interference field is equal to or larger than the actual image.

For many holograms (see Fig.6.6) the interference pattern covers only part of the actual image, and part of the specimen is opaque. In this case, an automated phase unwrapping process must be able to distinguish areas within the original interference pattern from areas which are not. As can be seen from the top-left and bottom-right areas in both images of Fig. 6.7, areas that are not part of the original interference pattern, or areas where the sample is opaque, are conspicuous visually for their high frequency noise. These areas pose a significant problem for automated phase unwrapping, although they are easily recognized visually.

Semi-automated phase unwrapping methods[479] take advantage of the ability of the human brain to distinguish real phase jumps from artifacts in critical areas and may well provide the key to fully automated procedures in the future. To demonstrate the semiautomatic phase-unwrapping process, a complex image containing a simulated phase wedge with a true dynamic range of 3π was created. Application of the atan2 function creates the phase image displayed in Fig. 6.14a, where the phase increases linearly (from left to right) from $-\pi$ to π, then resets to $-\pi$ and continues to increase to zero.

From this image, i.e., ϕ_1, another phase image ϕ_2 is created according to

$$\phi_2 = \begin{cases} \phi_1 & \text{for } \phi_1 \geq 0 \\ \phi_1 + \pi & \text{for } \phi_1 < \pi \end{cases} \tag{6.36}$$

which results in a new image (see Fig. 6.14b), where the phase increases from π to 2π, resets to zero and increases to 2π. Both images carry the same information, but display phase jumps in different areas, and exhibit a different range of values. By selectively copying areas from Fig. 6.14b into a copy of Fig. 6.14a, i.e., Fig. 6.14c, the phase jump can be remediated and the full original phase range can be displayed without artifacts. For images with many phase jumps, i.e., a (true) dynamic range of $> 3\pi$, the procedure is simply an extension of the procedure discussed. Once the image phase is free of phase jumps, image displays can be achieved in many different ways.

8.3. Phase amplification

The phase of a complex image can be displayed to show more phase jumps than in the original phase display, i.e., in the $\text{atan2}[O(x, y)]$. Increasing the number of phase jumps is called *phase amplification*. It makes it possible to visualize such features as equi-potential lines or lines of equal thickness more readily. Phase amplification can also help to make an image more quantifiable. By adjusting the amplification factor such that a phase jump occurs, for example, at certain thickness intervals, specimen parameters can be read from the image easily.

Figure 6.15 gives an example of the effect of phase amplification. The object was an aggregate of ZrO_2 particles; the hologram of these particles is displayed in Fig. 6.15a. The image phase as reconstructed[††] from Fig. 6.15a is displayed in Fig. 6.15b and exhibits one major phase jump (the almost diagonal line) as well as additional phase jumps in smaller areas. After the phase unwrapping process, all phase jumps are removed in Fig. 6.15c. The unwrapped phase image has a total dynamic range of $3.1\,\pi$.

Phase amplification is achieved in a manner similar to the phase unwrapping process. A phase amplification by a factor X is produced by replacing all values $\phi \geq 2\pi/X$ with $\phi - 2\pi/X$ until there are no more values $\phi \geq 2\pi/X$. For all negative values (if present), ϕ is replaced by $\phi + 2\pi/X$, until there are no more values < 0.

The phase value $\phi = 0$ remains untouched throughout the phase amplification procedure. This is rather interesting, since any part of the phase image with $\phi = 0$ is somewhat accidental. In general, any offset in a phase image has little physical meaning, as the absolute phase of an electron wave can never be determined: only phase differences can be determined. It is therefore useful to assign the phase value zero to a certain area. One logical choice would be the area that is not influenced by the object (e.g., the vacuum area), and to select an offset for the phase image so that the vacuum area is displayed such that it is closest to $0\,\pi$ without displaying phase jumps.

In Fig. 6.16, a slightly different route was chosen. Here a phase offset was selected such that the edge of the particle would be 'framed' by the phase jump. A phase amplification of 3.53 was then applied to create phase jumps at intervals corresponding to 14.1 nm so that the phase jumps here now corresponds to thickness contours which differ by this amount. This direct interpretation of phase as thickness parameter should, however, be considered with great care, as many parameters (e.g., crystal orientation) can affect the result.[‡‡]

It should be noted that for integer phase amplification, the phase unwrapping process can be omitted. Any existing phase jump at 2π or integer multiples of 2π remains untouched by integer phase amplification. For non-integer phase amplification, e.g., $X = 3.2$, the phase jumps in the phase amplified image appear at $2\pi/3.2$, $4\pi/3.2$,

---

[††]The reference hologram is not shown.

[‡‡]In this example, the theoretical (using the non-binding approximation) mean inner potential $V_0 = 18.1\,\text{eV}$ establishes a phase shift of $0.042\,\pi$ for 1 nm assuming the crystals are not aligned along one of their major crystallographic axes. For more details on this topic see Chapter 12.

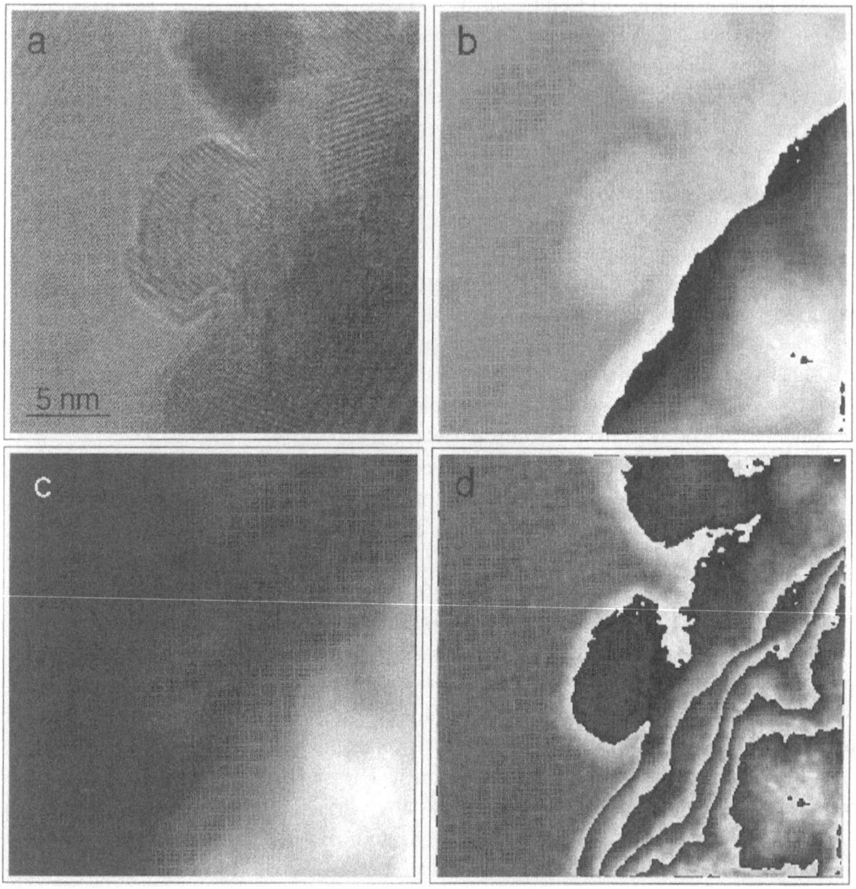

Figure 6.15. Phase amplification process starting from a) the original hologram (reference hologram not shown), b) the reconstructed phase image with a dynamic range of 3.1π, c) the phase unwrapped phase image, and d) the phase amplified (x 4) phase image.

..., but not at 2π anymore. In this case, omitting the phase unwrapping steps is prone to cause artifacts.

In order to point out artifacts that can appear with non-integer phase amplification, two phase-amplified images are displayed in Fig. 6.16. The image on the left was phase-unwrapped before the phase amplification, while the image on the right was amplified without prior phase unwrapping. Two artifacts can be made out quite clearly. The first artifact, the unusual contrast, from black to gray, at some of the phase jumps (see white arrows), is one clear sign of an incorrectly phase-amplified image. The second and most important effect is that the phase jumps appear in the wrong places as indicated by the dark arrows.

8.4. Contour line techniques

Contour line techniques make it possible to visualize equi-potential lines or 'equi-thickness' lines on top of an image without obstructing the view as is the case with the phase-amplification technique. Contour lines are commonly used for topographic maps

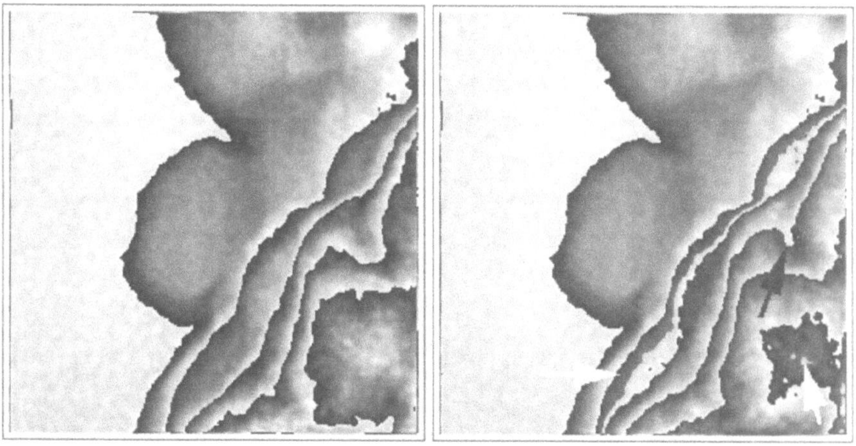

Figure 6.16. Left: Non-integer phase amplification (here x3.53) relies on the preceding phase unwrapping process. Right: Phase jumps with unusual contrast (see white arrows) are a sign that the phase unwrapping process was omitted. Also, phase jumps appear in the wrong places, as can be seen (black arrow) from a comparison with the left phase image.

(e.g., hiking maps) and are also useful for the investigation of phase maps.

Contour lines can be implemented by expanding the phase amplification procedure. Just as for the phase amplification procedure the phase image should be phase unwrapped, unless only integer values (in units of π) for the contour line process are desired. The first step is to create a phase-amplified image with the desired steps as described in the previous section, and then to differentiate this image in both the x- and y- directions, plus some additional refinement steps which will not be discussed here. The result is an image with contour lines only. This contour line image is then overlayed on the phase-unwrapped image.

For images of size 256 by 256 pixels, the procedure is still fast enough to perform the task interactively; the phase difference between two consecutive equiphase lines can be adjusted freely to correspond to a certain convenient step size.

As an example, two phase images are displayed in Fig. 6.17 with contour lines of step size 5 nm (left) and 2 nm (right). In both cases, the zero value of the original phase image was chosen such that the center particle was 'framed.' A comparison of the center ZrO_2 particle with the particle in the top center part conveniently reveals a similar thickness for both areas, which is not immediately obvious from the original hologram.

8.5. 3-D displays

Although three-dimensional displays of two-dimensional phase images do not carry any new information, they can provide a real sense for the structure and can help to bring a large dynamic range into perspective without losing the details. Generating an image of a surface that approximates the appearance of a real physical surface is known generically as 'rendering,' and requires significant computational effort. Therefore, 3-D displays are presently chosen only for final data presentation.[181] Phase unwrapping is a requirement of 3-D displays of phase images.

A surface plot that approximates the appearance of a real, physical surface requires the establishment of the location of at least one light source, as well as the location of

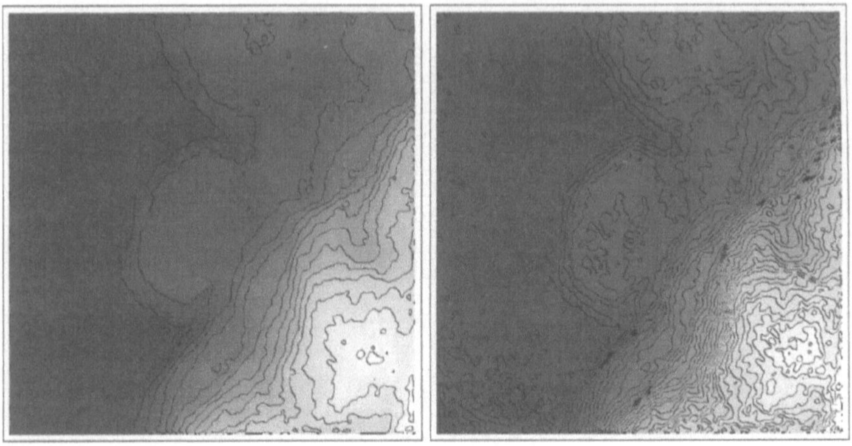

Figure 6.17. Two phase images of hologram in Fig. 6.15a with an overlay of contour lines. Step size between two contour lines is 5 nm, on the left, and 2 nm on the right. The thickness of the center particle can be estimated to around 8-10 nm.

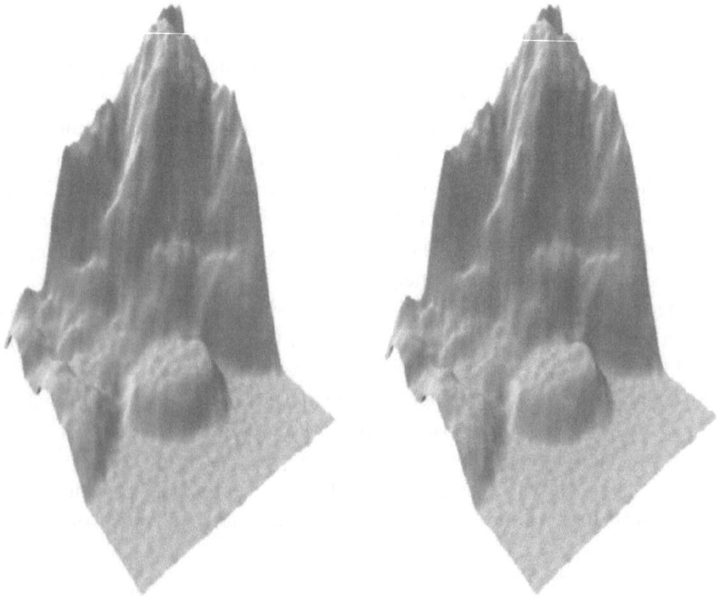

Figure 6.18. Stereo pair as created from the phase unwrapped phase image of the ZrO_2 agglomerate in Fig. 6.15. For some readers, a 'viewer' may be helpful to recognize the 3-dimensional pattern.

the viewer, to render the display. The absolute reflectivity of the surface is an additional parameter that must be adjusted to create the most impressive view. Further, the local roughness of the surface controls the degree of variation in the angle of reflection and thus determines if the surface reflects more specular or more diffuse. An example of this type of surface plot is displayed in Fig. 6.18, left or right.

By changing the coordinates of the viewer slightly, a stereo pair can be created (see Fig. 6.18). For some, this may require an inexpensive 'viewer' that allows one to focus on the separate images while keeping the eyes looking straight ahead. Other readers may have mastered the trick of fusing such printed stereo views without assistance.

Conclusion

The reconstruction procedure for electron holograms is a rather complex task, especially if quantitative data are desired. Image distortions which are "invisible" in standard images are clearly visible through holography and must be compensated. Therefore, digital recording and processing of electron holograms are critical in this area. Despite today's reasonably fast computers (1998), the reconstruction time for electron holograms is still a major drawback. Although combinations of optical Fourier filtering with digital recording, processing, and final display have achieved reconstruction speeds sufficient for TV-rates, they do not include phase unwrapping and correction of image distortions, which are considered an essential part of the reconstruction process. Hopefully, within a few years it will be possible to fully use the "phase-window" electron holography provides to look at specimens in the "phase world" rather than the "intensity world."

ELECTRON HOLOGRAPHY OF ELECTROMAGNETIC FIELDS

J.E. Bonevich,[1] G. Pozzi[2] and A. Tonomura[3]

[1]National Institute of Standards & Technology, Metallurgy Division, Gaithersburg, MD 20899
[2]Department of Physics and Instituto Nazionale per la Fisica della Materia, University of Bologna, viale B. Pichat 6/2, I-40126 Bologna, Italy
[3]Hitachi Advanced Research Laboratory, Hatoyama, Saitama 350-03, Japan

1. Introduction

Electron holography was originally proposed by Gabor[130] as a means to compensate for the lens aberrations and improve the resolution of transmission electron microscopes (TEMs). Within the last decade dramatic improvements in coherent field-emission electron sources,[440] and in TEM instrumentation, have allowed the promise of holography to come to fruition.[451] Improvements in TEMs notwithstanding, it has been the application of holographic techniques and coherent-beam imaging to the investigation of magnetic and electric fields that has sparked the interest of materials scientists. Holography is a unique tool to probe phenomena in the microscopic world in that all the information about the object is contained in the hologram (both the amplitude and phase), whereas conventional microscopy records only the intensity. It is in this sensitivity to the phase that materials scientists find holography's true potential.

Electromagnetic phenomena in a wide range of materials systems have been observed via electron holography. Furthermore, fundamental properties such as the physical meaning of electromagnetic potentials predicted by Aharonov-Bohm[5] (the AB effect, reviewed in Ref. 337) have been conclusively demonstrated for the magnetic case by electron holography.[445] Even very small magnetic fields flowing through superconductors (fluxons), difficult to detect with other techniques, have been successfully observed.

We present here an overview of electron holography as applied to electromagnetic phenomena in materials. First, the theoretical background is outlined as well as the experimental conditions and configuration of the electron microscope. Then, some applications of holography to the electric fields in p-n junctions,[109,110] domain structures in small magnetic particles,[196] the AB effect[445] are presented briefly, whereas the main emphasis is put on the observation of fluxons in superconductors.[25,285,505]

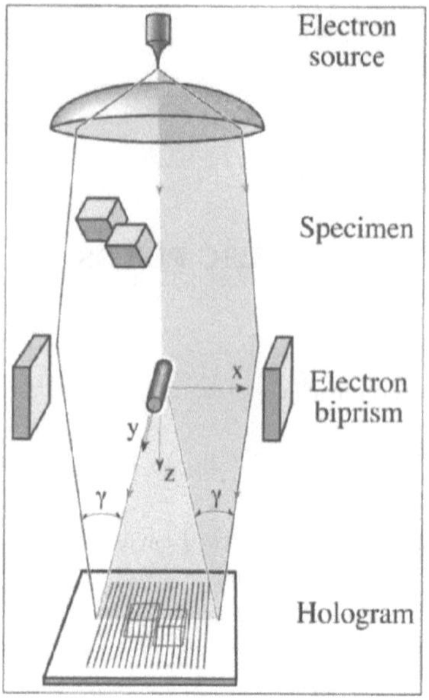

Figure 7.1. A hologram is formed when coherent electrons illuminating a specimen (object wave) are interfered with those passing through the vacuum (reference wave) via the electron biprism. Intensity modulations in the hologram record the amplitude whereas phase information is contained in fringe shifts.

2. Theory

We begin our discussion of electron holography by recalling that a hologram is simply a fringe-modulated image typically in the form of

$$\begin{aligned} I(x,y) = 1 + &|a(x,y)|^2 \\ &+ a(x,y)\, e^{i\,[\varphi(x,y)+2\pi\, q_c\, x]} \\ &+ a(x,y)\, e^{-i\,[\varphi(x,y)+2\pi\, q_c\, x]} \end{aligned} \tag{7.1}$$

where I is the intensity defined in the x- and y-plane, and a and φ are the object amplitude and phase, respectively. The hologram is obtained by coherently interfering the object wavefunction with a vacuum reference wave by means of a Möllenstedt-Düker electron biprism.[311] The carrier frequency q_c of the hologram is defined as $\sin\gamma/\lambda$, where γ is the interference angle produced by the electron biprism and λ is the wavelength of the incident electrons. The biprism is assumed to be aligned along the y-axis and the z-coordinate defines the optic axis. We also assume no contribution from inelastic events and a unity fringe contrast. The amplitude information is stored simply as the intensity variation of the hologram fringes whereas shifts in the fringes reproduce the phase variations, see Fig. 7.1. We also here note the assumption throughout this chapter that the microscope faithfully transfers the object and reference waves to the final intensity in the image plane, justified by the fact that in the spatial frequency range of interest here the contribution of spherical aberration can be safely overlooked.

The spatial variations of an object's electric and/or magnetic fields will manifest themselves mainly in the phase distribution of the hologram.[45,128] We can understand how this information is transferred in the electron beam by noting that the interaction of high energy electrons with electromagnetic fields is described by the Schrödinger equation

$$\left(\frac{1}{2m}\cdot(-ih\nabla + e\vec{A}\,)^2 - eV\right)\psi = E\,\psi \tag{7.2}$$

where m, e, h, ψ, $\vec{A}$, V and E are the electron mass, the absolute value of the electron charge, the Planck constant, the wave function, the vector potential, the scalar potential and the electron energy, respectively. We consider only elastic scattering events and assume an object electron wave function form of

$$\psi(x,y) = a(x,y)\,e^{i\,\varphi(x,y)} \tag{7.3}$$

By applying the WKB approximation (Wentzel, Kramer, Brillouin) to solve Eq. (7.2 the phase of the electron wave can be derived as

$$\varphi(x,y) = \frac{\pi}{\lambda E}\cdot\int_L V(x,y,z)\,\mathrm{d}z - \frac{2\pi e}{h}\cdot\int_L A_Z(x,y,z)\,\mathrm{d}z \tag{7.4}$$

where the integration is carried out along a straight line L parallel to the z-axis corresponding to the classical unperturbed electron trajectory for an illuminating plane wave. Equation (7.4) is remarkable in that the phase is dependent upon the presence of potentials instead of fields. While this may at first seem paradoxical in the magnetic case, it is the basis for the Aharonov-Bohm effect[5] which states, in essence, that between two coherent electron beams a phase shift is produced proportional to the amount of magnetic flux they enclose. We shall return to this effect in our discussion of the magnetic fields of superconductors.

To explore electric fields by holography we assume the vector potential in Eq. (7.4) to be zero. That is,

$$\varphi(x,y) = \frac{\pi}{\lambda E}\cdot\int_L V(x,y,z)\,\mathrm{d}z \tag{7.5}$$

The exact form of the phase shift is dependent upon the form of V. In the case of an object's mean inner potential, V_0, we may assume that $V(x,y,z)$ is constant within the specimen. Integrating over the thickness, t, we obtain a relationship for the phase shift $\Delta\varphi = (\pi/\lambda E)\,V_0\,t$ which, since $E = (U_A/2)(2m_0\,c^2 + e\,U_A)/(m_0\,c^2 + e\,U_A)$, results in

$$\varphi = \frac{2\pi}{\lambda U_A}\cdot\frac{2\,m_0\,c + e\,U_A}{m_0\,c^2 + e\,U_A}\cdot V_0\,t = C\,V_0\,t \tag{7.6}$$

where U_A is the electron accelerating voltage and the constant C has a value $1.039\cdot 10^6\ \mathrm{m}^{-1}\,\mathrm{eV}^{-1}$ for 300 kV electrons. From this relationship we thus have a method to determine from the hologram either the specimen's thickness or its mean inner potential. For instance, the hologram of MgO cubes in Fig. 7.2 demonstrates how the phase reveals topographical information.[162]

Similarly, the electrostatic field generated by a charged dielectric sphere on a carbon film can be modeled as a point charge Q localized at the sphere center in front

Figure 7.2. A hologram of MgO cubes (a) and a 4 times amplified interference micrograph of the phase distribution (b). The equipotential contours in (b) can be directly interpreted as a map of the projected thickness. Hologram courtesy of K. Harada from Ref. 162.

of a conducting plane. If R is the sphere radius, the potential in the half space of the sphere can be calculated by means of the image charge method, and is given by

$$V(x,y,z) = \frac{Q}{4\pi\,\epsilon_0} \cdot \left(\frac{1}{\sqrt{x^2+y^2+(z-R)^2}} - \frac{1}{\sqrt{x^2+y^2+(z+R)^2}} \right) \tag{7.7}$$

where ϵ_0 is the permitivity of free space. In this case the phase shift can be calculated analytically[45] and, at points for which $x^2 + y^2 > R^2$, it yields

$$\varphi(x,y) = \frac{Q}{2\,\lambda\,E\,\epsilon_0} \cdot \operatorname{arcsinh}\left(\frac{R}{\sqrt{x^2+y^2}} \right) \tag{7.8}$$

A similar equation can be derived for the case of a charged dielectric sphere on a copper grid, where the conducting plane is vertical instead of horizontal.[119]

According to Eq. (7.4) the electron phase is also sensitive to magnetic field distributions such as those emanating from small particles or thin film materials. In these cases, we explore the results of Eq. (7.4) when the scalar potential, V, is zero and only the magnetic vector potential contributes to the phase as predicted by Aharonov and Bohm. Let us consider the case of an infinite ideal solenoid lying along the y- axis: the z-component of the vector potential can be written as

$$A_z(x,y,z) = \frac{\Phi_m\,x}{2\pi(x^2+z^2)} \tag{7.9}$$

where Φ_m is the enclosed magnetic flux. By applying Eq. (7.4) to calculate the phase shift undergone by the electrons, it turns out that the transmission function of the solenoid is given by

$$T_{sol}(x, y) = e^{\pi\, e\, \Phi_m\, \mathrm{sign}(x)/h} \tag{7.10}$$

so that an observable phase difference of

$$\Delta\varphi = 2\pi\, e\, \Phi_m / h \tag{7.11}$$

is detectable if an interference experiment is carried out, where the wavefunction passing on one side of the solenoid is overlapped onto that passing on the other. The paradox arises because electrons are not locally influenced by the magnetic field, which is zero outside the solenoid, and they always propagate in field-free regions since the solenoid is made impenetrable to them.

This simple calculation can be applied to a fluxon in a superconductor if we approximate it by a flux tube of negligible radius carrying the flux Φ_m.[304] Therefore, by considering a superconducting specimen of thickness t with surfaces perpendicular to the flux tube, the magnetic vector potential within the specimen is given by that obtained for the infinite solenoid, whereas the field outside the specimen is equal to that produced by a single point magnetic charge of strength Φ_m. For the case of the flux tube perpendicular to the electron beam, lying along y- in a specimen extending between $-t < y < 0$, the phase shift can be shown by elementary calculations to be given by Eq. (7.11) for the phase shift inside the specimen and in the half space $y > 0$

$$\varphi = \frac{\Phi_m}{\Phi_0} \cdot 2 \cdot \arctan \frac{x}{y} \tag{7.12}$$

While this model is not suited to describe a fluxon in a thin film parallel to the electron beam, as the specimen thickness is virtually infinite, its external phase shift can be used to simply interpret the results of dynamic experiments in bulk specimens, e.g., reference.[285]

This flux tube model has also been useful for finding an analytical solution for the phase shift when the specimen is tilted.[305] Here the application of Eq. (7.4) is too cumbersome and the phase shift has been calculated using a more intuitive and direct approach based on the phase difference given by Stoke's theorem as

$$\Delta\varphi = -\frac{2\pi\, e}{h} \int_{S_{12}} B_n \, \mathrm{d}S \tag{7.13}$$

Since B_n is the component of the magnetic flux density normal to the surface S_{12} defined by paths L_1 and L_2, the integral stands for the magnetic flux passing through S_{12}. In the case of a flux tube of negligible radius carrying the quantum flux $\Phi_0 = h/2e$, lying perpendicular to the two surfaces of an ideal superconductor of thickness t, and inclined at an angle α with respect to the optic axis z, as shown schematically in Fig. 7.3, the upper and lower poles (i.e., the exit point and entry points of the flux tube) have projected coordinates $(a, 0)$ and $(-a, 0)$ on the object plane (where $a = t\,\sin(\alpha)/2$), and so the expression for the phase shift is given by

$$\varphi(x, y, a, \alpha) = \frac{1}{2}\left[\arctan\frac{y}{x+a} - \arctan\frac{y}{x-a} + \frac{\pi}{2}\,\mathrm{sign}\frac{y}{x-a} - \frac{\pi}{2}\,\mathrm{sign}\frac{y}{x+a}\right.$$
$$\left. - \arcsin\frac{y\,\sin\alpha}{\sqrt{y^2 + (x+a)^2}} - \arcsin\frac{y\,\sin\alpha}{\sqrt{y^2 + (x-a)^2}}\right] \tag{7.14}$$

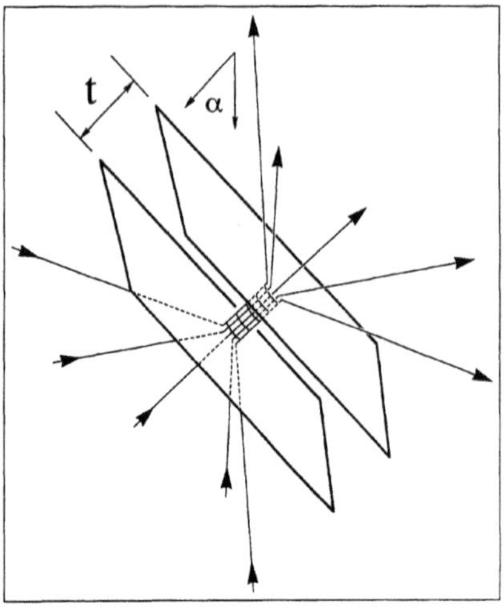

Figure 7.3. Schematic representation of a fluxon in a specimen of thickness t tilted by an angle α to the electron beam.

The true utility of the flux tube model arises from the fact that by overlapping a suitable distribution of flux tubes we can approximately model any flux core structure, in the same spirit as in electrostatics where a given field can be approximated by a suitable arrangement of point charges. We may assume that the model describing the fluxon magnetic core within the specimen is given by the London equation,[204]

$$B = \frac{\Phi_m}{2\pi\,\lambda_L^2} \cdot K_0(r\,\lambda_L) \tag{7.15}$$

where K_0 is the zero-order Bessel function, r the radial distance from the fluxon axis, and λ_L the London penetration depth. Therefore, the previous expression for the phase shift due to the flux tube is convoluted with the two-dimensional projection in the object plane of the London model, in order to find the corresponding phase shift $\varphi_L(x, y, \alpha, t, \lambda_L)$.[26] The results for the phase shift φ_L are shown in Fig. 7.4 for a specimen of thickness 60 nm, tilted at 45° with respect to the beam, and for the ancillary magnetic field for a London fluxon with $\lambda_L = 30$ nm, i.e., $\lambda_L = t/2$.

While the London model describes the external magnetic field of a fluxon, it suffers a divergence in the field distribution at the core of the fluxon ($K_0(0) = \infty$). Clem[48] has proposed a different model for a fluxon, resolving the divergence at the origin, where the magnetic field is given by

$$B = \frac{\Phi_m}{2\pi\,\lambda_L\,\xi_\nu} \cdot \frac{K_0(R/\lambda_L)}{K_1(\xi_\nu/\lambda_L)} \tag{7.16}$$

in which ξ is the coherence length of the fluxon, $R^2 = r^2 + \xi^2$ and K_1 is the first-order Bessel function. The Clem model has a finite value at the fluxon core and converges to the London model far away from the core. Both models provide the same overall fluxon phase shift; however, some differences in the inner-core structure may be discerned in the derivatives of the phase to be discussed later.

It should be noted that whereas the maximum phase difference across the fluxon is π, corresponding to the AB effect due to the flux tube core, for a specimen of finite thickness the phase is broadened and no longer reaches the value of π corresponding

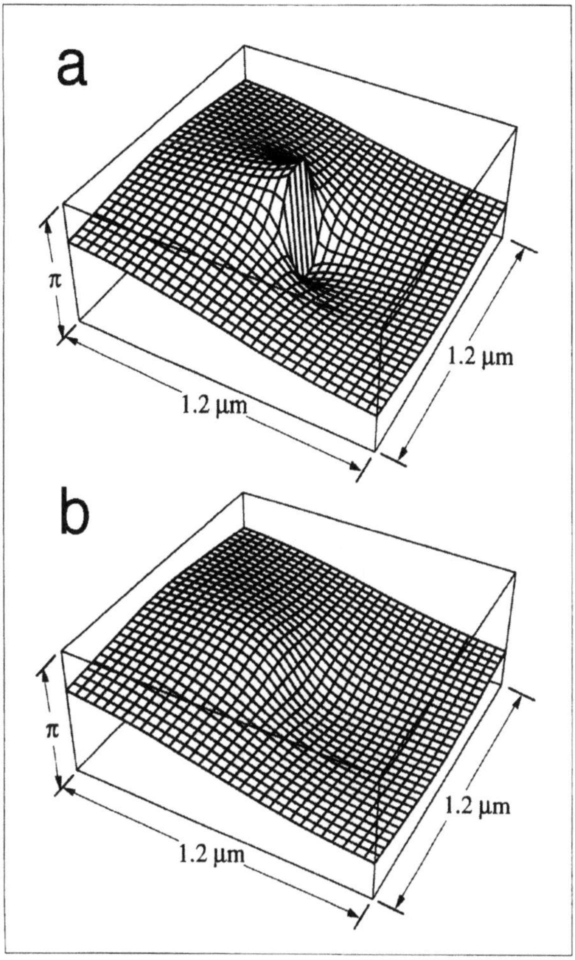

Figure 7.4. (a) Phase shift of a flux tube in a specimen of thickness $t = 60$ nm, $\alpha = 45°$. (b) Corresponding phase shift for a London fluxon with $\lambda_L = 30$ nm.

to the flux quantization. From this unexpected effect it can be ascertained that the broadening due to the finite core effects severely dampens the maximum phase difference and smoothes the slope at the core. It should also be noted that at large distances from the core, the phase difference becomes $\pi/2$ irrespective of the London wavelength.

Geometrical reasoning[25] as well as the analysis of the London phase shift[305] show that this phase difference (which in the more general case of tilt by an angle α is given by α itself) is still present even if the distance between the two poles vanishes, which corresponds to a specimen whose thickness is much lower than λ_L: this is another unexpected effect due to this unusual geometry confirming that the main contribution to the phase shift is due to the external fringing field extending in the two half-spaces above and below the perfectly diamagnetic superconducting film.

3. Holographic electron microscope for superconductivity investigations

The standard electron microscope used for holography consists of a coherent electron source, illuminating optics, an interferometry device, and an imaging system coupled with recording media. For the results discussed herein, the microscope used was a Hitachi 350 kV HF-9000 FE-TEM equipped with two rotatable[222] biprisms to produce a wide range of electron optical conditions for interferometry. Holograms were recorded on photographic film (subsequently digitized) with typical exposure times of 5-10 seconds and the microscope had a TV camera for dynamic observations. The microscope was further modified by addition of a cold stage in the specimen region for dedicated experiments on superconducting materials. Both the optical conditions and the cold stage are discussed in detail below.

In order to observe a single magnetic flux quantum within a superconductor, a penetrating electron beam of high coherence and brightness is required: a field-emission electron source. The cold-tip field-emitter of the HF-9000 has a brightness in excess of $10^9\,\mathrm{A\,cm^{-2}\,sr^{-1}}$ with lateral (spatial) coherence lengths greater than $10\,\mu$m. This high brightness is required to ensure reasonable exposure times for recording holograms even when the illuminating beam convergence angle is in the range ($\sim 10^{-7}$ rads) required to observe fluxons within superconductors. Holographic observation of magnetic fields also fully exploits the free-lens control of the FE-TEM (see Table 7.1 for lens settings). In typical experiments, the objective lens is switched off and the first intermediate (diffraction) lens is used to image the specimen instead.[465] The resulting electron optical magnification of the image can vary from about 500 to 2600 kX. Under these conditions, the first intermediate lens is used to focus on the specimen, and then the first projector lens can be adjusted to obtain the desired magnification. The second intermediate and projector lens excitations are unchanged. When the specimen is observed under this "normal" condition and the upper biprism (located between the first intermediate and objective lenses) is employed, the resulting holograms can have a carrier fringe of 30 nm spacing referred back to the specimen plane over a $4\,\mu$m width.

In contrast, high resolution holograms with electron optical magnifications up to 20 kX can be formed by using the objective lens to focus on the specimen and turning off the first intermediate lens. In this case, the objective lens is only slightly excited owing to the large distance between the specimen and the lens gap (cold stage configuration). Again, the first projector lens is adjusted to obtain the desired magnification while the second projector remains unchanged. By employing the objective lens, we can gain a factor of 3-4 times in the magnification of the specimen relative to the electron biprism. Under these "high resolution" conditions, the upper biprism produces holograms with fringe spacings down to about 7 nm over a $1.5\,\mu$m width. High resolution holography was used to obtain phase information about the vortex diameter, while the lower resolution is best employed for a wide field of view to survey the specimen.

For both the normal and high resolution lens conditions the condenser lens excitations were controlled such that an electron beam of high brightness and coherence necessary for holography was obtained. Typical hologram acquisition (on TEM film) times were around 5-10 seconds. For Lorentz microscopy observations where the co-

C2	OL	I1	I2	P1	P2	Mag
3.00	1.35	0.00	6.00	6.02	8.00	20100 X
3.00	1.35	0.00	6.00	3.01	8.00	2650 X
3.00	0.00	2.15	6.00	8.00	8.00	2650 X
3.00	0.00	2.15	6.00	6.02	8.00	1850 X

Table 7.1. Lens excitations (in amperes) for the HF-9000 FE-TEM. The illumination system (C2) was fixed while the first intermediate and projector lenses (I1 and P1) were adjusted. The objective lens (OL) was switched off for low magnifications whereas higher resolution holograms could be obtained with slight OL excitations.

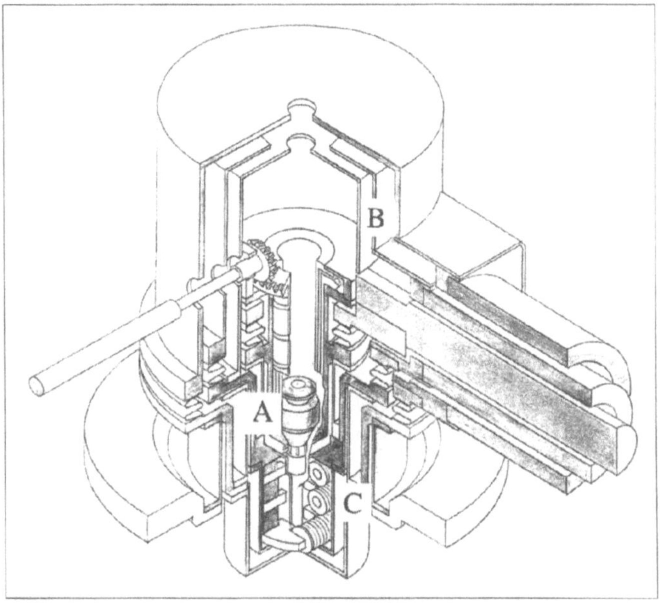

Figure 7.5. Diagram of cold stage. A tilted superconductor is inserted in the specimen holder (A). Cooling shields (B) maintain the specimen down to 4.5 K while magnetic fields (C) up to 200 G may be applied. Stage heaters vary the temperature up to 26 K.

herence requirements were less restrictive, the illumination could be focused to provide the higher current density necessary to obtain real-time imaging of acceptable quality. The obvious tradeoff to increasing the current density is that an overly focused electron beam can cause local heating in the specimen, inducing vortex motion in the thin foil specimen.

The HF-9000 FE-TEM is equipped with a specially constructed liquid helium (LHe) cold stage (see Fig. 7.5) that enables the temperature and magnetic field to be applied to the specimen within the range of 4.5 to 26 K and 0-200 G, respectively. This stage allows superconductors to be observed under both equilibrium and dynamic conditions. The cold stage, connected to an external double-chambered coolant dewar, has three concentric gold-plated heat shields: the outermost shield is maintained by liquid nitrogen (LN_2) at 77 K where the inner two are maintained by LHe at 8 and 4.5 K, respectively. These shields ensure minimal radiative losses, permitting specimens to be continuously observed at 4.5 K for more than 3 hours. The shields are isolated mechanically from the rest of the stage by means of metallic foil supports to minimize transmission of bubbling vibrations to the specimen.

Horizontal magnetic fields are applied to the specimen by means of three coils: the first coil applies the actual field up to 200 G and the subsequent two coils serve to deflect the electron beam back to the optic axis to exactly compensate the first. The DC-powered coils are made of superconducting wire so that there is no discernible temperature rise even at maximum applied magnetic field. Physically, the top-entry cold stage is designed to fit above the upper pole piece of the objective lens. The specimen in the fully lowered position is thus about 30 mm from the lens gap. This distance enables the objective lens to be used for higher resolution experiments and still isolate specimens from the lens' magnetic field. The stage is further designed such that specimen exchange can occur at LN_2 temperatures, enabling fast turnaround between low temperature experiments.

Specimen temperature is controlled by a DC power supply (to eliminate any stray AC field effects) and is measured by a Ge resistance thermometer. One convenient feature of this cold stage is that while the heater coils are designed for specimen temperatures up to 26 K, higher temperatures can be achieved, albeit in a less controlled manner. For example, if LN_2 is used in the inner dewar instead of LHe, in conjunction with pumping, specimen temperatures from 77 K down to about 50 K can be routinely obtained.

The holder used to mount the specimens is also specially designed for use in the LHe cold stage. The components are plated with gold to ensure good thermal contact with the stage while also minimizing any radiative losses. Since the superconductor specimens must be observed at an angle to the electron beam, the holder also has special inserts that hold the specimen firmly at about 45°. Additionally, the holder has provisions to supply an electric current to the specimen for dynamic transport experiments.

3.1. Reconstructions

Electron holograms, recording the phase and amplitude information of the transmitted electron wave, need to be reconstructed in order to recover this information. As the information about the magnetic fields is contained within the phase, it is useful to employ techniques highly developed for light optics to reconstruct holograms by means of a laser bench configured as a Mach-Zehnder interferometer (see Fig. 7.6). Here the laser beam is split into two separate beams that, by proper adjustment of optical mirrors, pass through the hologram at slightly different angles and are recombined within a lens. The fringe modulation of the hologram, acting as a diffraction grating, produces the ± first-order diffraction spots in the back-focal plane of the objective lens. When the + first order spot produced by one laser beam (the reconstructed wavefront) is interfered with the − first order from the other beam (the conjugate wavefront), a 2 times phase-amplified interferogram is obtained. While the separation of the sidebands in reciprocal space is determined by the fringe spacing (being proportional to the biprism voltage), the resolution of the interferogram is determined by the size of the aperture used in the reconstruction. Typically, the apertures used result in interferogram resolutions about three times the fringe spacing of the original holograms.

The phase amplification factor can also be increased beyond 2 times by only using one laser beam to illuminate the hologram. In the back-focal plane, both ±1 order sidebands pass through an aperture, eliminating the auto-correlation (zeroth-order). The sidebands are recombined to form a new hologram with twice the phase sensitivity of the original. By suitable adjustment of the optical components, the 2 times-amplified hologram can be formed with a lateral magnification also twice the original hologram, thus preserving the real-space hologram fringe spacing. After processing, the original hologram may be replaced in the laser bench by this newly formed hologram to repeat the procedure to obtain 4 times phase-amplification. This process may be used to produce high amplifications (e.g., 32 times), but is usually limited by the noise level of the film media. One technique to increase the contrast of the original holograms is to make a positive of the hologram that is subsequently bleached ($K_2Cr_2O_7$ / H_2SO_4) to allow a high transmittance of the laser light through it. Of course, this technique may distort the amplitude information in the hologram, but it also allows diffracted spots

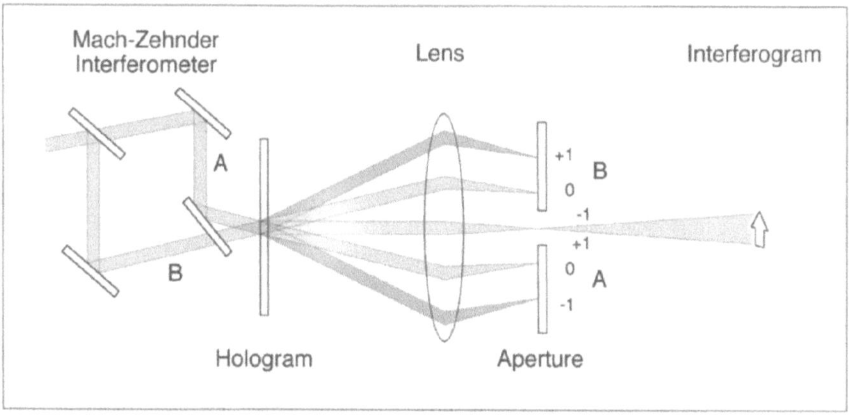

Figure 7.6. Schematic representation of Mach-Zehnder interferometer for phase-amplified interference microscopy.

up to the 3rd order to be observed with sufficient intensity to reconstruct the phase information as well as minimizing any deleterious background intensity.

Digital reconstruction of electron holograms has many advantages over the optical reconstruction techniques. For example, the full amplitude and phase information are retained, and a wide range of digital filtering and processing techniques are available, as well as exploiting the fast "on-line" reconstruction capabilities of modern computers. Recalling the form of the hologram Eq. (7.1), we apply the Fourier transform and obtain

$$\begin{aligned} I(\vec{q}) = {} & \delta(\vec{q}) + \mathbf{FT}\left\{a^2(\vec{r})\right\} \otimes \delta(\vec{q}) + \\ & \delta(\vec{q}+\vec{q}_c) \otimes \mathbf{FT}\left\{a(\vec{r}) \cdot e^{i\varphi(\vec{r})}\right\} + \\ & \delta(\vec{q}-\vec{q}_c) \otimes \mathbf{FT}\left\{a(\vec{r}) \cdot e^{-i\varphi(\vec{r})}\right\} \end{aligned} \tag{7.17}$$

The first term represents the auto-correlation (zeroth order diffraction), the second term the first-order sideband (the electron wave information) and the third term its phase conjugate. One of the sidebands is isolated by means of a digital aperture and the inverse Fourier transform recovers the amplitude and phase information of the original electron wave. We note that the phase image may be amplified by a factor n and the intensity of the phase amplified image is described by

$$I(x, y) = 1 + \cos[n\,\phi(x, y)] \tag{7.18}$$

One difficulty arising in digital processing of holograms is that rather than a continuous fringe modulation, as indicated above, the hologram is a set of discretely sampled intensity points. Furthermore, typical hologram carrier frequencies are such that the peak position (δ-function) of the sideband in reciprocal space does not exactly fall on a pixel, resulting in "leakage" of information into neighboring pixels and a strong streaking effect in the discrete Fourier transform. The effects of this leakage can be minimized by first applying a window function to the hologram. Well-behaved functions such as the Hanning window that strongly damp the intensity near the hologram edge can limit the leakage of information to only the adjacent pixels and also allow the exact position of the sideband to be determined.[90]

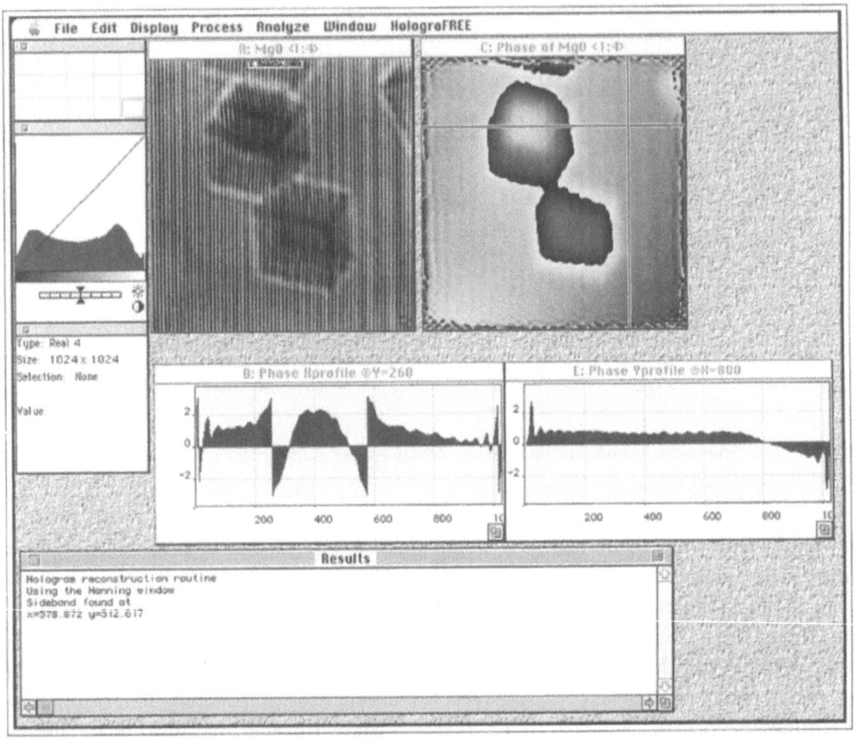

Figure 7.7. Screen picture from a HolograFREE reconstruction of MgO cubes. The software outputs the amplitude (not shown) and phase information from a hologram. Phase profiles at user-specified positions are used to evaluate the tilt of the comparison wave used in the reconstruction to locally correct the phase. The software also allows phase amplification, offsetting, tilting and unwrapping to be performed. The software contains routines to fit the phase data to high-order polynomials for perturbed reference wave correction.

Typically, each holography research group writes its own reconstruction software specifically suited to its particular needs. However, with the increasing use of CCD cameras for image acquisition and powerful PC platforms, it has become feasible for software packages to provide the user with standardized reconstruction routines. One such package, HolograFREE written by the author (JEB), is available from the National Institute of Standards and Technology as free software.[323] This software is designed to run within Gatan's Digital Micrograph™ on Macintosh computers. The routines can reconstruct holograms giving as output amplitude and phase images, remove the contrast transfer function of the objective lens to calculate the exit wavefunction, unwrap phase images, and also tilt the phase to any user-specified angle for localized correction. The routines also fit background phase images to high-order polynomials useful to analyze holograms formed with perturbed reference waves. A typical screen picture from the software is shown in Fig. 7.7 for a reconstruction of MgO cubes.

3.2. Reference waves

Implicit in the reconstruction of electron holograms is the assumption that the reference wave used to form the hologram is a plane wave. However, in some, but

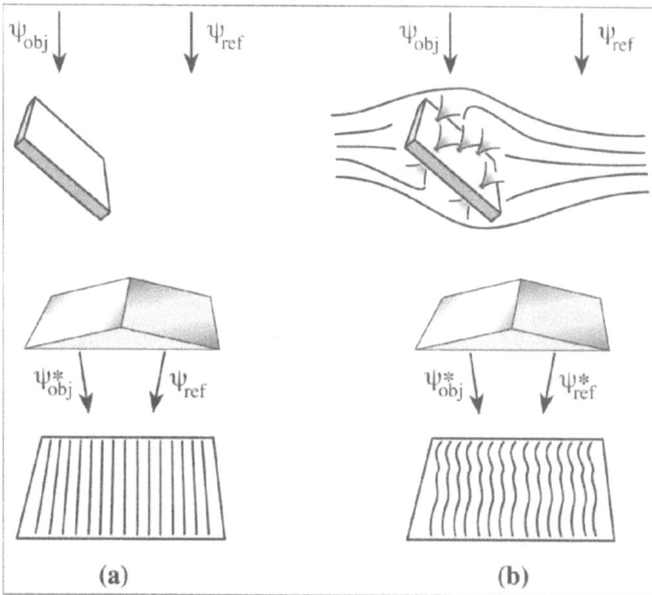

Figure 7.8. Schematic of reference wave perturbation. (a) Normal holography employs an unperturbed planar reference wave, ψ_{ref}, allowing the phase of the object to be determined. (b) In the presence of long-range fields, in the case of superconductors in an applied magnetic field, the initial reference wave is transformed to ψ^*_{ref} complicating the reconstruction of the true phase.

not all, cases the presence of long-range stray fields in the observation of magnetic or electrically charged objects can cause a modulation of the reference wave (see Fig. 7.8). The "hologram" that is formed is, strictly speaking, an interferogram of two perturbed electron waves. The result of this reference wave perturbation is the loss of any objective criteria for obtaining the true phase distribution of the object in the reconstruction. For example, the equiphase contour lines in the phase amplified interference micrograph shown in Fig. 7.2 (MgO cubes) reveal a non-planar phase distribution outside of the particles due to their charging under the electron beam. In the absence of charging, the phase in the vacuum should be perfectly flat and thus there is a dilemma as to which phase distribution in the reconstruction is the correct one. Furthermore, non-uniformities in the reference wave perturbation (e.g., those caused by nearby sources of stray fields or other distortions) may present the situation where the correct phase distribution can be obtained only in localized regions.

The issues involved with the holography of long-range electromagnetic fields were explored by Matteucci et al.[290,293] Their method of alleviating the reference wave dilemma was by double-exposing the hologram. First, the hologram of the object was recorded, then a new hologram with the object removed, but remaining nearby, was recorded. This step ensures that the interfering reference and object waves are recorded with the same tilt angle γ with respect to the biprism. The superposition of the double exposed fringes in the hologram presents the true nature of the phase whose distribution may appear counter-intuitive as can be seen in Fig. 7.9.[291] An important result of the double exposure method is that the long-range nature of electromagnetic fields is

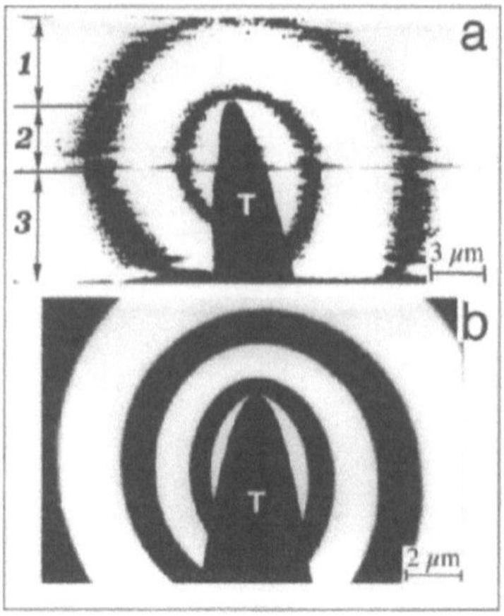

Figure 7.9. (a) Collage of three (1,2,3) double exposure holograms taken from regions adjacent to an electrically-charged microtip showing the trend of the equiphase lines. (b) Computer simulation of the equiphase lines obtained with a perturbed reference wave.

clearly revealed. In fact, whereas typical holograms used to investigate these fields have interference widths around 4 μm, a truly planar (unperturbed) reference wave can in some cases only be obtained with interference widths in excess of 8 μm.[290] Clearly, this stringent requirement is rather difficult to meet in practice due to the limited brightness of even modern field emission guns, instabilities in the field-emission source as well as any mechanical vibrations or stray fields present in the microscope. We should also note that non-planar reference waves can also arise due to non-uniformities in the projector lenses of the microscope. These distortions are evident in electron holography, but their effect can be minimized, as discussed on pages 133 ff., for digitally recorded images.

Digital reconstruction techniques also offer a convenient way of minimizing the problems of a perturbed reference wave. For example, whereas the optical method requires time consuming processing of photographic plates and subsequent reconstruction, a whole series of holograms can be acquired digitally relatively quickly allowing a spatial montage of the field distribution to be obtained. Furthermore, in typical cases the long range "tails" of the field are monotonically decreasing. This fact allows one to fit the perturbed background of the phase data to some elementary functions.[233,371] For example, in the case of a pure magnetic field, the phase satisfies the Laplace function $\nabla^2\Phi_m = 0$, so that harmonic functions that also satisfy this function may be employed to approximate the phase. Likewise, we may use a linear combination of polynomials in fitting the phase data. Subtraction of the fitted background phase image provides a means to compensate the perturbation over large regions of the hologram. However, we should note that although the digital techniques are fast and convenient, we still lack a truly objective criteria for removing a perturbed reference wave. Thus it is necessary to couple the experimental observations with rigorous theoretical modeling to properly understand the distribution of the specimen's electromagnetic fields.

3.3. Applications of holography to electric systems

As elucidated above, electron holography may be applied to a wide range of electromagnetic phenomena in materials. While this work is mainly focused on the observation of magnetic fields in superconductors, we shall endeavor to provide an overview of the results of other groups working with various materials systems.

Reconstructed holograms show in the phase fringes corresponding to projected equipotential contours. In the simplest case of a non-magnetic, non-charging weak-phase object, these contours can be directly associated with the projected thickness contours of the object. Equation (7.6) thus allows a reasonably accurate determination of the mean inner potential, V_0, from a specimen of known thickness (see also Chapter 12 for more details). Thin TEM specimens, cleaved to a well-defined wedge angle, were used to calculate V_0 for materials such as Si, MgO, GaAs and PbS.[135] The measurement of V_0 requires the careful removal of the contribution of dynamical scattering on the phase, achieved primarily by ensuring that the specimen is not in a strong diffracting condition, as well as simulation of the electron scattering. In the case of specimens where the wedge angle is not known *a priori*, Herring has derived a generalized approach to determine the mean inner potential.[182] It should be appreciated that in any determination of V_0, the presence of contact potentials between the material of interest and its support (i.e., Cu grid) will also contribute to the phase and must be taken into consideration; these contact potentials have themselves been investigated.[34] The electron phase sensitivity to changes in V_0 has also been applied to the study of small catalytic particles where the presence of voids within the particles was conclusively demonstrated[6] as well as to resolving the faceted nature of particle surfaces.[38] The inner potential of single-walled tubular fullerenes (buckytubes) have been measured to be around 10 eV.[378] Holography has also been applied to the study of compositional changes in interfaces of heterogeneous interfaces.[195]

The field distributions in electronic materials such as those found in p-n junctions have also been investigated by holography and the depletion layer at the junction could be quantitatively measured.[109,110] The holography results (see Fig. 7.10), sensitive to small variations in the electric field, are dramatically different from those obtained from conventional Lorentz microscopy whose contrast stems mainly from regions of steep field gradients. Interestingly, the built-in potential (0.76 eV) could not be observed and it was necessary to reverse-bias the junction in order to observe effects related to its field. This unexpected result is likely due to the interaction of the electron beam with the specimen. The presence of space charges in the grain boundaries of electroceramic materials such as doped $SrTiO_3$ have been observed. Here, holography is a key technique due to its high spatial resolution and ability to quantify changes in the potential. While strong diffracting conditions and non-uniformities in specimen thickness at grain boundaries as well as any leakage fields are included in the phase shifts, complicating interpretation of the results, the presence of a negatively charged boundary coupled with a positive space charge could be determined.[273,356]

3.4. Applications of holography to magnetic systems

The investigation of magnetic microfields is uniquely well-suited to electron holography as both fields inside and emanating from a specimen may be observed. The reconstructed interference micrographs directly show the projected magnetic lines of force as contour fringes each indicating a constant flux of h/e.[442] These contours are quantitative and their interpretation is straightforward. Furthermore, the contours can be directly related to the microstructure of the specimen, in contrast to the commonly used Lorentz microscopy technique whereby the magnetic phase information is revealed in a highly defocused micrograph. For example, thin ferromagnetic films have been observed and the domain and wall structures revealed.[443] Thin films of magnetic recording media have also been studied revealing bit lengths as small as 70 nm.[331] Even small magnetic particles have been examined where the presence of symmetrical leakage fields in the reconstructed interference micrographs indicate that the particles were

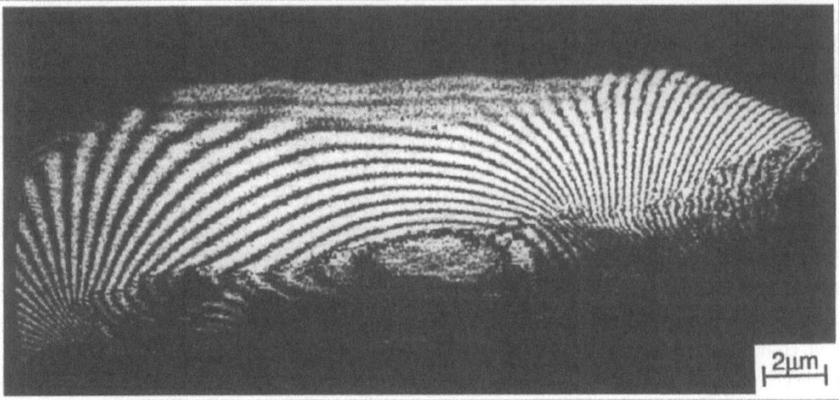

Figure 7.10. Differential contour map between two holograms recorded with the pn junction at 0 and 4 V reverse bias, respectively. Equipotential contour fringes are clearly observed. Courtesy of G. Matteucci.

existing in a single domain state[196] (see Fig. 7.11). Dynamical investigations of domain wall motion in thin films have been achieved by means of real-time interferometry.[195]

Investigations of magnetic specimens also present an additional complication in that not only is the electron phase sensitive to the magnetic field distribution, but it also is affected by the electric potential contribution of the specimen itself. The issue then becomes how to best deconvolve the magnetic and electric phase distributions. We note that the property of time reversal may be applied in these cases. For instance, an initial hologram of magnetic specimen, say a small particle, is recorded. Then a second hologram is acquired with the same specimen inverted in the holder. Recalling from Eq. (7.4) that the phase shift depends on the contributions of scalar and vector potentials, we note that the electric (scalar) potential is unchanged because the specimen has the same projected thickness. However, the magnetic (vector) potential now has the opposite sign and thus its contribution can be resolved in the reconstructed interferograms. This "time reversal" technique was applied to elucidate the magnetization of small Co particles.[448]

One of the most elegant contributions of holography in the measurement of magnetic fields is the confirmation of the Aharonov-Bohm (AB) effect. While in classical electrodynamics the existence of potentials is a convenient tool to describe electromagnetic fields, in quantum mechanics these potentials can not be removed from the Schrödinger equation. The AB effect states that these potentials have a physical significance that may be demonstrated via interferometry experiments; the effect stipulates that coherent electrons passing through field-free regions will nevertheless experience a phase shift if their paths enclose a vector potential. In a series of experiments, small toroidal magnets were completely encapsulated inside a superconductor to eliminate any leakage fields. Electrons that passed through the toroidal hole were interfered with those that passed outside of it, and a phase shift equal to the amount of flux enclosed was revealed.[445] Even though the magnetic field was entirely screened by the superconductor, the AB effect assures that a phase difference is observed as shown in Fig. 7.12. This localized interaction between electrons and vector potentials has significant consequences in the development of gauge theory in quantum electrodynamics.[501]

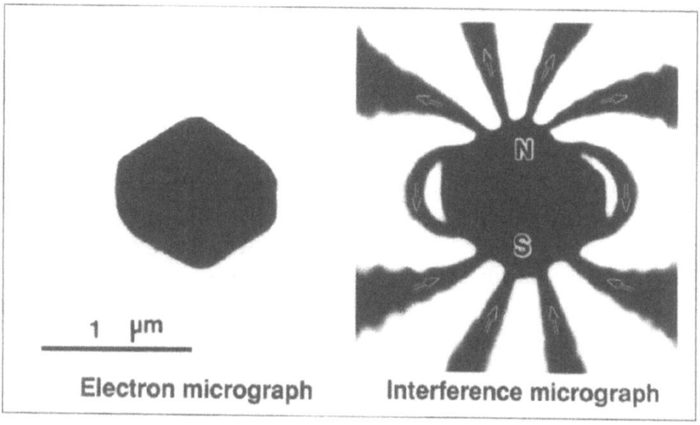

Figure 7.11. Isolated barium ferrite particle observed from the [011] direction (left). The magnetic flux lines are clearly seen emerging from the N-pole and converging to the S-pole (right).

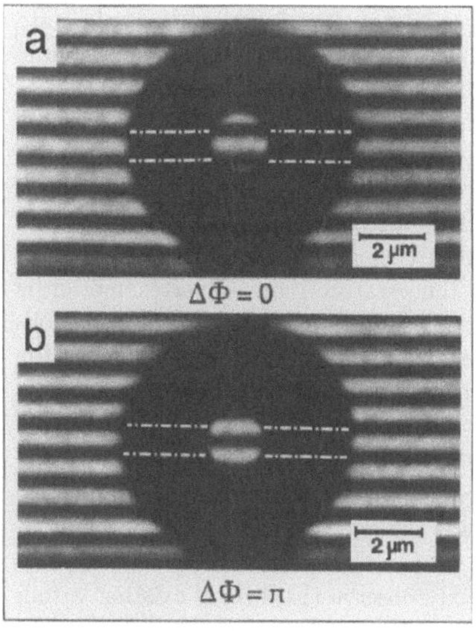

Figure 7.12. Experimental confirmation of the Aharonov-Bohm effect. The phase shifts are quantized according the amount of flux enclosed, either (a) 0 or (b) phase shift of π.

4. Observation of fluxons in superconductors

The AB effect can also be exploited to investigate the magnetic fields emanating from fluxons (or vortices) in thin film superconductors. The behavior of fluxons illuminates many important fundamental aspects of superconductivity, although until recently direct fluxon observation has proved experimentally cumbersome. For example, the traditionally used Bitter technique[94,103] images the static magnetic fields above the surface as a replica and, consequently, is not suitable for dynamic studies. And previous attempts[37,50,148,499,507] to observe fluxons by TEM were stymied by the lack of suitable coherent electron sources with high brightness, as well as the experimental difficulties of stabilizing detectable magnetic fields within the superconductor.

Our success in imaging fluxons directly via electron holographic techniques[25,285,505] has provided a unique perspective of their behavior. The phase information from fluxons are revealed in defocused (Lorentz) electron micrographs.[26] The dynamic behavior of vortices can be observed by this Lorentz microscopy method[42] and their response to changes in temperature, magnetic field and current-driven transport, as well as their interaction with specimen defects such as surface steps, grain boundaries and dislocations can be discerned.[164,165,166] One disadvantage of Lorentz microscopy, however, is that quantitative extraction of data is problematic. For instance, while Lorentz microscopy reveals the location and polarity of a vortex, the degree of flux quantization of the vortex can not be measured. Furthermore, the large lens defocus distorts the image and affects the apparent vortex size. In contrast, electron holography[451] is highly quantitative because both the amplitude and phase information of the entire object are recorded in the hologram allowing precise flux measurement with higher spatial resolution than Lorentz microscopy.

Electron holography has been used to observe magnetic fields emanating from superconducting thin films in a surface profile configuration.[285] From the AB effect, we expect the phase shift caused by a single fluxon ($\Phi_0 = h/2e$) from Eq. (7.11) to be exactly π. Thus, each contour fringe in a 2 times phase-amplified interference micrograph corresponds directly to a quantized fluxon. A thin Pb film was cooled to 4.5 K and the fields from single fluxons could be observed (see Fig. 7.13). In the bulk state (thicknesses $>1\,\mu$m) Pb is a Type-I superconductor and, as such, the fluxons bundle together. The number of fluxons in a bundle is determined directly by counting the contour fringes in the interferogram. If the film is made thin enough ($\sim 0.2\,\mu$m), however, the Pb behaves as a Type-II material whereby each fluxon is isolated and considered to be individual. Since the magnetic field is visualized directly, and not as a replica, the dynamic motion of fluxons is also observable. Fluxons in specimens cooled to 5 K are initially stable, but with increasing temperature thermal excitations cause fluxons to broaden and move about the specimen.[285] Temperature induced vortex motion is random and unpredictable, but when an electrical current is applied to the superconductor, a Lorentz force is created that drives fluxons in a direction perpendicular to the current. Thus the pinning force at a specimen defect can be determined by measuring the current-dependent fluxon motion at the defect site. Dynamical current-driven transport experiments have been achieved by means of real-time interferometry.[505]

Since in the previous experiments the superconductor was viewed in profile, the correlation of fluxons with specimen defects was difficult. Specimen defects play an important role in the mechanism of flux pinning, as well as the development of practical applications, and thus the ability to directly observe the fluxons existing within specimens has many advantages. For these reasons, the specimens need to be viewed in transmission to allow observation of wide areas of the specimen. While we can properly

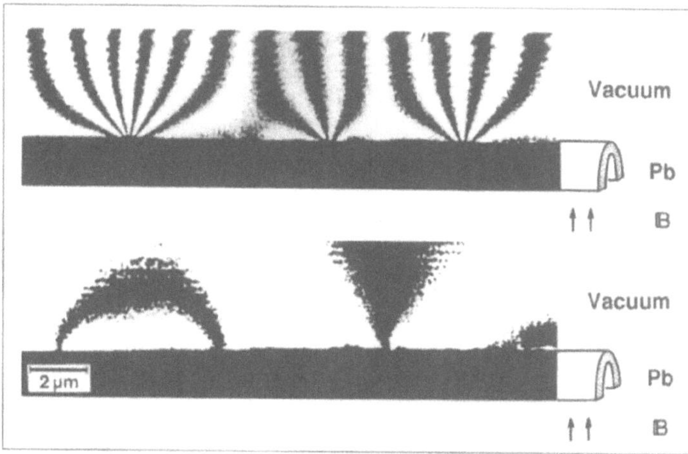

Figure 7.13. Profile observation of fluxons. Each fringe corresponds to a flux of $h/2e$ (single fluxons) in the 2 times phase amplified interferograms. Fluxons bundle together in a 1 μm thick lead film (top). In a film of 0.2 μm thickness single fluxons prevail; a pair of anti-parallel fluxons is also visible (bottom).

understand the phase shift induced by a fluxon from Eq. (7.14). a simpler expression can be empirically derived from Eq. (7.11). Consider a thin, tilted superconducting specimen with a single trapped fluxon.[25] Due to the Meissner effect.[12] the magnetic field spreads out evenly above and below the specimen. Only those components of the magnetic field perpendicular to and enclosed by the electron beam will induce a phase difference

$$\Delta\varphi = 2(\pi - 2\alpha) \cdot \frac{\Phi_0}{h/e} \qquad (7.19)$$

where α is the angle between the specimen surface and the electron beam. Thus, for a usual TEM specimen viewed in transmission ($\alpha = \pi/2$), we determine that the fluxons are undetectable and for this reason the specimen needs to be tilted to the electron beam. For a specimen tilted at 45° ($\pi/4$), we expect a fluxon phase shift of $\pi/2$, a good compromise between the fluxon detectability that also affords wide observable areas of the specimen.

The fluxon transmission experiments[25,387] were conducted with Type-II Nb superconductors. TEM specimens were prepared by annealing the Nb ($T_C = 9.2$ K) that had been rolled to 7 μm thickness at ~2000°C in a vacuum of 10^{-6} Pa. This oxygen-desorbing anneal resulted in a resistance ratio $R_{300\,\mathrm{K}}/R_{10\,\mathrm{K}}$ of ~20 and a foil with grain size of 200-300 μm and a [110] surface texture. Then 2 by 2 mm^2 sections were chemically polished by the window technique (etchant: HNO_3 - 40% HF. 0°C) to final thicknesses of ~100 nm. These electron transparent thin foils were then sandwiched between two Cu grids and mounted in the TEM cold stage. Our experimental procedure was to first observe the fluxons by Lorentz microscopy in a 20 mm defocused image to find suitable regions for holography; the image was then refocused. The biprism was brought into view and aligned roughly parallel with the specimen surface and the interference conditions were established. The experimental configuration is shown schematically in Fig. 7.14. Holograms were usually acquired on film with exposure

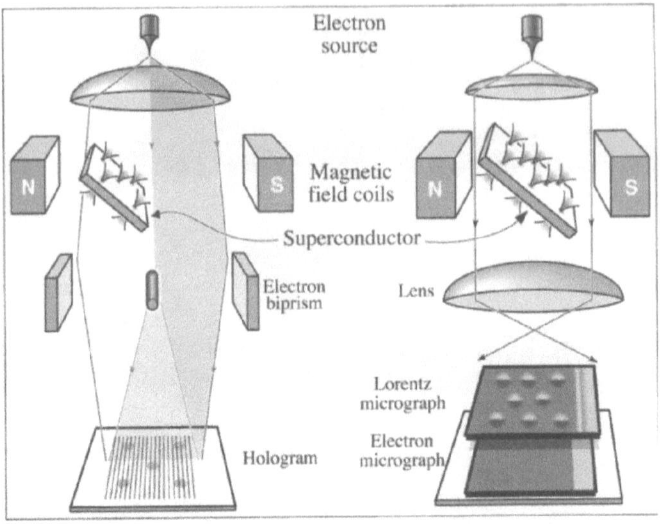

Figure 7.14. Schematic representation of the transmission observation of vortices in tilted superconductors. In holography (left) the object and reference waves are interfered to form a hologram, whereas the phase information is also revealed (right) as a highly defocused Lorentz micrograph.

times of about 5-10 seconds. The holograms were subsequently reconstructed, optically and numerically, to retrieve the fluxon information stored in the electron phase in the form of phase-amplified interferograms.

In the reconstruction process, a planar comparison wave is interfered with the hologram thereby producing an interference map of the phase. The proper interference conditions may be chosen, by tilting this comparison wave, to create a phase map of the fluxons. However, as mentioned above, an implicit assumption in this procedure is that a plane reference wave was used to form the hologram. To satisfy in the electron microscope this condition of an unperturbed reference wave coming from the vacuum region, an interference distance as large as possible between the object and the reference wave should be chosen.[290] Furthermore, the biprism interference fringe spacing with respect to the specimen should be very fine, as this parameter affects the ultimate hologram resolution, about three times the fringe spacing. Given that the interference distance and reciprocal fringe spacing are both proportional to the biprism voltage, by increasing the latter the ideal conditions can be approached. Unfortunately, this is possible only to a limited extent due to the lateral coherence of the electron beam, which diminishes the fringe contrast and hence the hologram quality as the interference distance is increased.

A compromise between the opposing requirements of hologram contrast and resolution, the limitations set by the photographic recording medium, as well as the FE-TEM mechanical and electrical stability, can be achieved by carrying out the observations with the objective lens switched off, and imaging with the intermediate lens, so that the hologram has an overall electron optical magnification of 1800 X, with a carrier fringe spacing referred to the specimen of about 30 nm. These optical conditions comprise relatively low resolution electron holograms as their information limit is about 100 nm.

The resolution can be increased by using the objective lens to focus on the specimen and subsequently forming a hologram in the usual fashion. Since the objective lens magnifies the specimen with respect to the biprism (by a factor of 3-4), the hologram resolution can be improved to better than 30 nm. However, there are drawbacks to the high-resolution approach: By effectively magnifying the specimen with respect to the biprism the interference width of the corresponding hologram will be reduced to ~1.5 μm. This means that the "reference" wave interfered with the object wave can no longer be assumed to be an unperturbed plane wave. In fact, the magnetic field applied to the specimen strongly perturbs the reference wave complicating the reconstruction process. Furthermore, the narrow interference width means that only those vortices near the specimen edge may be examined, increasing the possibility of vortex motion and core broadening due to thin film effects, etc.

Holograms were reconstructed both optically and numerically to extract the phase information about the fluxons. The optical reconstructions, performed by Mach-Zehnder interferometry, were achieved by first making a bleached hologram to improve the contrast. This hologram was then phase-amplified to increase its sensitivity to the phase distribution. Interference micrographs reconstructed from these holograms had wide fields of view revealing many fluxons. However, for precise phase information, the original holograms were digitized and numerically reconstructed with attention focused on individual vortices. While, in our experimental configuration, the optical method allows observation of many vortices, the numerical method is amenable to the application of digital image processing techniques and is better suited for the extraction of phase data as well as the comparison of the experimental data with theoretical simulations.

5. Lorentz observations

The detection of the vortices was first made via Lorentz microscopy, Fig. 7.14(right). A typical example is shown in Fig. 7.15 where the Nb thin foil was cooled to 4.5 K in a field of 100 G. This Lorentz image was defocused by 20 mm to reveal the presence of the vortices. The minimum defocus necessary to image the vortices with enough contrast can be estimated as $\sim 5(\lambda_L^2/\lambda_E)$, where λ_L is the London penetration depth and λ_E the incident electron wavelength, both in nm. Each vortex is composed of adjacent spots of bright and dark contrast with the vortex core being oriented in between them. Bend contours are present due to the slight deformation of the self-supporting thin foil. The vortices have arranged themselves along the applied field direction.

Holograms were then taken of the same specimen region under the in-focus condition. The interference micrograph shown in Fig. 7.16 is 16 times phase amplified so that the phase difference between each dark contour line, given by $\Delta\Phi_m = (2\pi/n)$, n being the phase amplification, is $\pi/8$. The tilt of the planar comparison wave in the reconstruction was chosen in such a way that the contour fringes were running in the same direction as the applied magnetic field with the result that fringe contours become narrowly spaced at the regions corresponding to vortex locations. Comparison with Lorentz images revealed that the spot-like contrast spatially coincides with the regions of finely spaced contours; the contours are widely spaced in between vortices.

A smaller region containing a few vortices was processed via dedicated software. Before digitizing the hologram, it was aligned so that the vortex axis was parallel to the vertical axis. The tilt of the wave in the reconstruction was first chosen in such a way as to have the overall phase as flat as possible over the whole region. Then the vortices in the phase map appear similar to the Lorentz image, namely, adjacent bright and

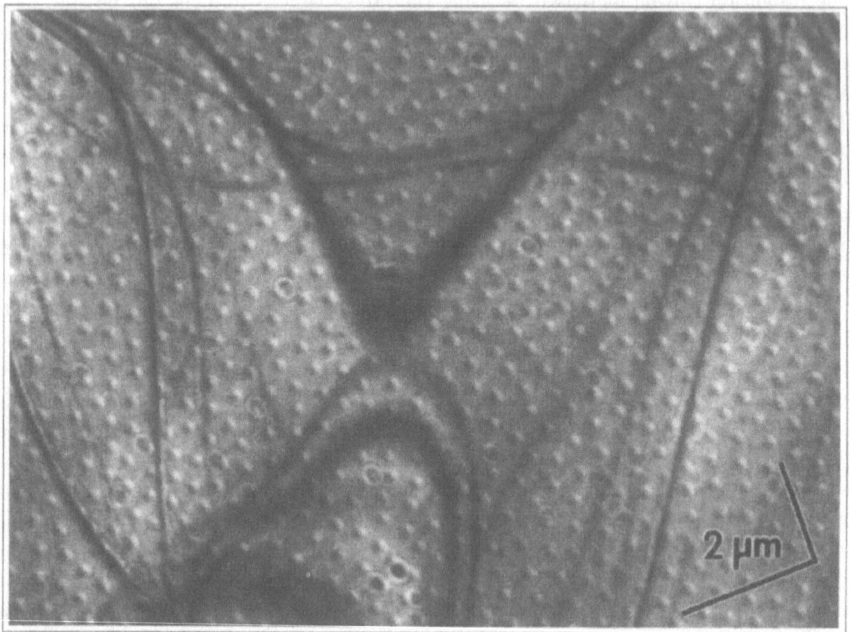

Figure 7.15. Lorentz image of vortices at 100 G and 4.5 K ($\Delta z = 20$ mm). Phase shifts produced by the vortices are manifested as bright/dark spots with the core aligned along the intersection. The dark line bend contours are present.

dark spots (see Fig. 7.17(a)). It should be remarked that while both the Lorentz and holographic phase images look similar, the vortex size and contrast in the former are the result of wave optical effects[26] whereas holography indicates the true vortex size. Then a further tilt was added to the reconstruction, so that the phases on both sides of the vortices were as flat as possible: in this condition the contour map, Fig. 7.17(b), best resembles the expected contour map, Fig. 7.17(c), and the phase difference due to a vortex can be evaluated. The vortices appear as smooth phase steps where the step height indicates the amount of flux enclosed and the step width the diameter of the vortex, in this case measured phase differences were $0.55\,\pi$ and had widths of ~ 150 nm.

A singly quantized fluxon ($h/2e$) in a tilted superconductor should produce an electron phase difference of exactly $\pi/2$. The measured phase difference of $\sim 0.55\pi$ is in good agreement with this value, though it is somewhat larger. The discrepancy can be accounted for by considering that Eq. (7.16) assumes a very thin superconductor, and so the contribution from the vortex core can be neglected. However, our real Nb specimen has a thickness on the order of the penetration depth ($t \approx 2\lambda_L$) and in this case we must consider the vortex core structure. In this situation, the presence of a *finite* length of core will increase the amount of flux enclosed by the electron paths and the phase differences will be slightly greater than $\pi/2$.

5.1. Temperature dependence

It is well known[387] that the London penetration depth, λ_L, of vortices varies with temperature, T, as

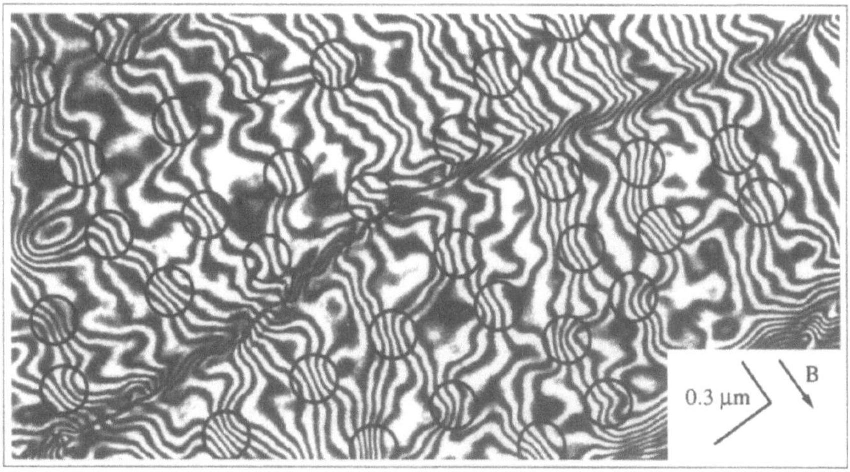

Figure 7.16. 16 times phase amplified contour map of Nb specimen at 4.5 K and 100 G. Contours of projected magnetic lines of force become locally dense at circled regions, i.e., vortex positions.

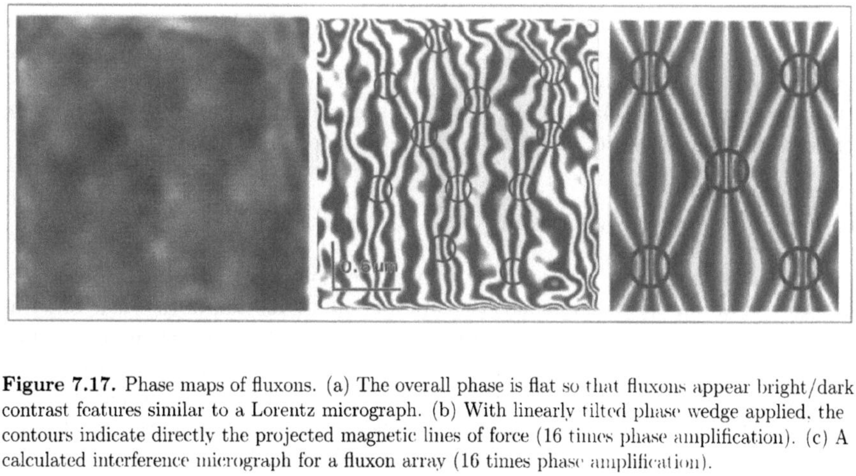

Figure 7.17. Phase maps of fluxons. (a) The overall phase is flat so that fluxons appear bright/dark contrast features similar to a Lorentz micrograph. (b) With linearly tilted phase wedge applied, the contours indicate directly the projected magnetic lines of force (16 times phase amplification). (c) A calculated interference micrograph for a fluxon array (16 times phase amplification).

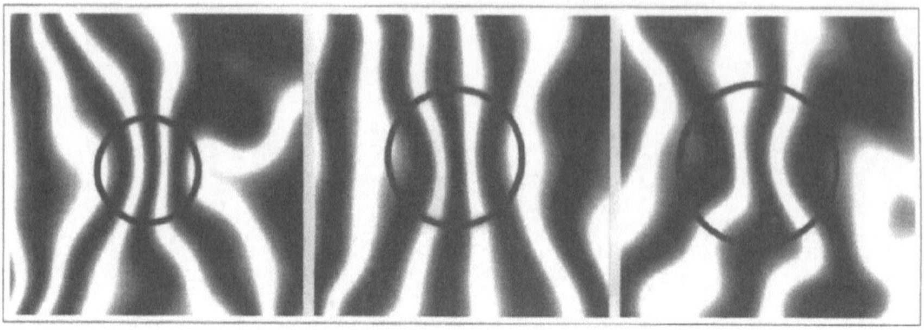

Figure 7.18. Interference micrographs of vortex diameters (12 times phase amplified) broadening with temperature at 4.5, 7 and 8 K (at 20 G) measuring 150, 185 and 230 nm (error: ± 4 nm), respectively.

$$\lambda_L(T) = \lambda_L(0)\left(1 - (T/T_C)^4\right)^{-1/2} \tag{7.20}$$

where $\lambda_L(0)$ for Nb is 31 nm. One expects $\lambda_L(T)$ to vary from 32 nm at 4.5 K to 47 nm at 8 K, an increase of 47%. Theory predicts that the coherence length, ξ, behaves similarly with $\xi(0)$ of 39 nm. Thus the vortex diameter, taken as the width of the phase step,* should also increase with temperature according to Eq. (7.20). To observe this phenomenon, the resolution of electron holography, as described above, is ideally suited.[27] Holograms were taken at 4.5, 7 and 8 K under the condition of constant applied magnetic field (20 G) where the equilibrium spacing between vortices is about 1.1 μm, and they can be considered well isolated.

The linearly tilted interference micrographs (flat phase on either side of vortex) allow the phase differences due to the vortices to be properly evaluated. In this way, not only the broadening of single vortices with temperature, but their actual size at each temperature can be measured (see Fig. 7.18). As before, we have taken the width of the phase step to indicate the vortex diameter. Data compiled from several vortices showed average vortex diameters of 150 nm at 4.5 K, 185 nm at 7 K and 230 nm at 8 K, representing an increase of ~50%. Our values are comparable with those expected from the Clem model: 145, 172 and 212 nm, respectively. Also, while the vortices did broaden with temperature, the total flux enclosed by them was constant.

6. Higher resolution holography

The holograms of the vortices in the previous sections were taken with the objective lens turned off and have a standard resolution of about 100 nm, of the same order as the vortex diameters. To explore the vortex inner-core structure, we have therefore endeavored to examine fluxons under higher resolution conditions by employing the objective as the focusing lens.[28] Despite the information gained from this analysis, it is important to remember that the interference width reduction will cause the reference wave in the hologram to be more strongly perturbed by the long-range magnetic fields

*An alternative way to define the vortex diameter, in terms of the magnetic field distribution $B(r)$, is twice the radius intercepted by a tangential line plotted through the inflection point of $B(r)$. That is, twice the $r-$ intercept of a line of slope $\partial B(r)/\partial r$ evaluated at r^*, where r^* is determined by the condition $\partial^2 B(r^*)/\partial r^2 = 0$.

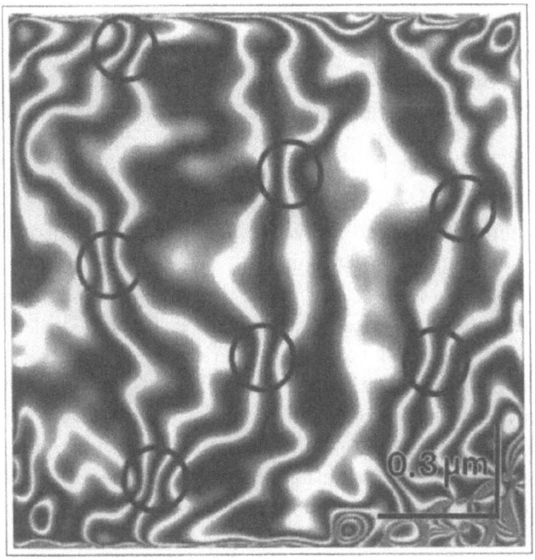

Figure 7.19. Vortices at high resolution (4.5 K and 10 G); 8 times amplified phase was numerically fitted to compensate the perturbed reference wave.

surrounding the specimen. As these fields originate from the specimen itself, it is reasonable that at some distance their spatial variation is somewhat attenuated so that their local influence on the image of a single fluxon can be approximated by a rather smooth function.

Holograms formed under these higher resolution conditions were taken and compared with those at low resolution. The reconstruction process was as follows: a region of the hologram containing vortices was digitized with the vortex axis oriented vertically. The tilt of the planar comparison wave was then adjusted so that the phase across the entire field of view was as flat as possible (because of the perturbation of the reference wave, substantial phase modulation remained). Furthermore, the phase offset was adjusted so that it did not have phase jumps from $-\pi$ to $+\pi$ in the field of view. Removing these discontinuities is essential in order to fit the phase data. Then the phase map was fitted to a 3rd-order polynomial: subtracting the fitted phase from the original data resulted in a corrected phase map. The corrected phase map could then be evaluated for the presence of vortices.

Vortices reconstructed from a higher resolution hologram show several features (see Fig. 7.19). First, as expected, the equiphase contour lines flow smoothly through the vortex and spread out above and below the core. Also the vortex diameter is in rough agreement with the low resolution ones, indicating that although the narrow interference width restricts observable vortices to those near the edge where the specimen is only 30-50 nm ($\sim \lambda_L$) thick, they are similar to those found in thicker regions ($\sim$ 100 nm, or $3\,\lambda_L$). Moreover, the measured phase difference from the hologram is $\sim 0.5\pi$ indicating no change in the degree of flux quantization in the vortices.

6.1. Vortex inner-core structure

An important consideration in the study of vortices is the exact nature of the vortex inner-core, an issue that electron holography is well suited to examine. The models advanced to describe the magnetic field distribution within a vortex, e.g., the London and Clem models, differ in that the distribution in the London model is described by a modified Bessel function depending only upon the penetration depth, whereas the Clem model invokes a ratio of Bessel functions dependent upon both the penetration depth and the coherence length of the vortex.[48] High resolution electron holography offers the possibility of discriminating between these two models through the use of digital reconstruction techniques.[344] For instance, while the phase shifts caused by the vortices are identical for both models, subtle differences between them can be discerned by taking the derivatives of the phase across the vortex core. The London model is expected to have a sharp discontinuity in the 2nd derivative at the core; in contrast, the Clem model reveals a smooth, continuous phase change.

Reconstructed vortex phase distributions were examined: the long-range magnetic field effects in the "reference" wave were numerically fitted and the perturbed background subtracted to reconstruct extended regions of the holograms.[28] Attention was focused upon individual vortices and numerical derivatives of the phase evaluated across the vortex core. The resulting derivatives showed no sharp dicontinuities within the resolution of the phase reconstruction, indicating that the Clem model provides a better physical description of the magnetic field distribution of the vortex. However, the issues of shot noise in the holograms and thin film effects on the broadening of vortices must be considered. For instance, to obtain data smooth enough to perform numerical derivatives on the phase, small apertures must be employed during reconstruction reducing the ultimate resolution of the phase data. Provided that the resolution is better than the London penetration depth, the phase trend across the vortex core is only negligibly affected by the spatial frequency filtering of the reconstruction procedure.[29] Moreover, the effects of specimen thickness can be incorporated into the models of the vortex phase with the result that the London model, using an appropriately broadened penetration depth, can also reproduce the trends of the phase across the vortex core.[29] These results suggest that the computationally simpler London model may be used over the more physically realistic Clem model.

6.2. Fluxon imaging by Lorentz microscopy

While we have mainly focused upon the observation of fluxons via electron holography, Lorentz microscopy is also useful for dynamical studies, for example, under changing temperature[164] or an electrical transport current.[506] Lorentz imaging may also be employed when circumstances do not require the highly quantitative holographic phase data. We therefore present the simulations of fluxons as appearing in a Lorentz image. Beginning with the phase shift for a fluxon (Eq. (7.10)) derived above, we note that the solution of the Schrödinger equation in the space below the specimen is given in the paraxial approximation in the form of a Kirchhoff-Fresnel integral,[145]

$$\psi(X,Y,\Delta z) = \frac{e^{i\beta}}{(\lambda\,\Delta z)^2} \cdot \iint e^{i\left\{\pi\left[(x-X)^2+(y-Y)^2\right]/(\lambda\,\Delta z) + \varphi(x,y,\alpha,t,\lambda_L)\right\}}\, dx\, dy \qquad (7.21)$$

where x and y are the coordinates in the object plane, X and Y the coordinates in the out-of-focus plane at a defocus distance Δz from the object plane, λ the de Broglie wavelength of the incident electrons and β a phase factor of no value in the present case as here only the intensity in the image plane, which is proportional to $|\psi|^2$, is relevant.

Taking λ_L as the scale length, and noting that the London phase shift is a function of x/λ_L, y/λ_L, t/λ_L and α, it turns out that the intensity in the image acquires the following functional form

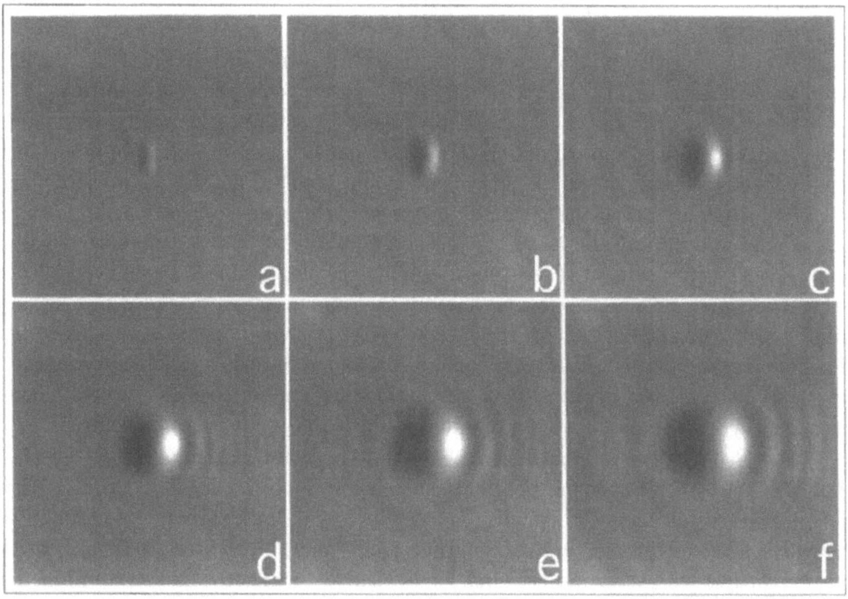

Figure 7.20. Vortices at different defocus distances, described by the London model. $\lambda_L = 30\,\mathrm{nm}$, $t = 5\,\lambda_L$ (150 nm), $\alpha = 45°$. (a) 1 mm, (b) 2.5 mm, (c) 5 mm, (d) 10 mm, (e) 15 mm and (f) 20 mm.

$$I = I\left(\frac{X}{\lambda_L}, \frac{Y}{\lambda_L}, \frac{\lambda \Delta z}{\lambda_L^2}, \frac{t}{\lambda_L}\right) \tag{7.22}$$

Realistic values of the various parameters are used in the calculations; however, the adoption of these dimensionless parameters is convenient because the results obtained are of general applicability, independent of the material, whose superconducting properties are given, in the approximations used, by the single parameter λ_L.

Figure 7.20 shows a focal series calculated for the case of a specimen having a London length of 30 nm (of the same order as that of Nb), a thickness of 150 nm, i.e., $5\,\lambda_L$, at the following values of the defocus: (a) 1 mm; (b) 2.5 mm; (c) 5 mm; (d) 10 mm; (e) 15 mm and (f) 20 mm. The accelerating voltage has been taken equal to 300 kV, corresponding to a de Broglie wavelength of 1.968 pm. The image wavefunction has been calculated by FFT methods, on a square region centered at the fluxon having side 1.2 μm, i.e., $20\,\lambda_L$, with 256 times 256 sampling points. A linear phase term has been subtracted from the phase shift in order to avoid Fresnel diffraction effects from the edge parallel to the fluxon axis, where the phase has a $\pi/2$ jump owing to the truncation. Test calculations done by doubling both size and sampling points show no relevant differences.

A fluxon, being a perfect phase object, obviously shows no contrast in a zero-defocus image. Upon defocus, however, it can be ascertained from the calculated images that the fluxon has the appearance of a tiny globule, with two halves of bright and dark intensity aligned along the fluxon axis, as found experimentally (see Fig. 7.15). Also low contrast fringes surrounding the globules are present. Of interest is the slight elongation of the fluxon contrast in the 1 and 2.5 mm images whose length corresponds to the projected length of the fluxon core within the specimen. This result indicates that the Lorentz images can provide information about the local specimen thickness as well as fluxon position and polarity. Both the contrast and dimension of the globules increase with defocus distance: however, if the patterns are scaled proportionally to $\lambda_L^{1/2}$ it can be ascertained that, apart from the contrast, they look very similar. In the geometric optical approximation the dimensions of the image, being proportional to the Lorentz deflection at the specimen, should increase linearly with the defocus distance, implying that at the defocuses investigated the image contrast and appearance are mainly waveoptical effects due to Fresnel diffraction. The Fresnel effects at large defocus also serve to obscure the elongation of fluxon contrast due to specimen thickness whereas at small defocus the elongation of the spot diminishes with decreasing thickness until it becomes undetectable in the Fresnel image. Calculations have also been done for the cases of specimen thickness equal to λ_L and zero, i.e., negligible with respect to λ_L, and the general appearance of the patterns at larger defocuses does not change significantly with respect to those reported in Fig. 7.20, apart from the contrast.

The Clem model for the fluxon phase shift may also be employed to calculate Lorentz images. By replacing the phase shift in Eq. (7.15), a focal series for a Clem fluxon may be calculated in a similar manner to Fig. 7.20. The results of calculations using the following parameters: λ_L of 30 nm, ξ_v of 40 nm, t of 150 nm, tilted at α of 45° were compared with those from the London model.[345] It was noted that the introduction of a finite coherence length of the same order as the penetration depth has the main effect of lowering the contrast of the fringes surrounding the central globule, leaving the globule itself only slightly affected. The conclusion is that there are no major differences between the two models on the basis of the Lorentz imaging and, therefore, the more computationally simple London model is adequate for most purposes.

7. Conclusions

In summary, the direct observation of electromagnetic field phenomena with high sensitivity and spatial resolution by electron holography has revealed exciting possibilities in physics and materials science. The investigation of electric fields arising from charged objects as well as the precise measurement of mean inner potential and surface topography of various materials is obtained via holography. Magnetic phenomena such as the confirmation of the AB effect, thin film magnetic domain structures, and nanometer-scale magnetic particles may be studied. Even long-range electric and magnetic fields may be observed though the proper treatment of the perturbed reference wave. Fundamental issues such as the phase distributions of single vortices in superconductors were investigated by both high and low resolution holography and the results were found to agree with theoretical simulations. Likewise, the broadening of single vortex diameters with temperature was measured and agreed with the expected behavior. Basic materials science questions, e.g., the dynamic interactions of vortices with specimen defects and the nature of pinning sites as well as the study of high temperature superconductivity, can all be investigated by electron holography.

Acknowledgments

The authors gratefully acknowledge the contributions of Dr. T. Matsuda, Dr. K. Harada, H. Kasai and T. Yoshida of Hitachi and Dr. A. Fukuhara, Dr. U. Kawabe and Dr. T. Onogi of Hitachi and Dr. Q. Ru of the Tonomura Electron Wavefront Project (ERATO) for their stimulating discussions and Dr. J. Endo, T. Furutsu, Dr. M. Igarashi, Dr. H. Kajiyama, Dr. S. Kondo, S. Kubota, S. Matsunami. N. Moriya and Dr. N. Osakabe of Hitachi for technical assistance.

ON RECORDING, PROCESSING AND INTERPRETATION OF LOW MAGNIFICATION ELECTRON HOLOGRAMS

B.G. Frost[1] and G. Matteucci[2]

[1]EM Facility, University of Tennessee, Knoxville, TN 37996-0810
[2]Department of Physics and Instituto Nazionale per la Fisica della Materia, University of Bologna, viale B. Pichat 6/2. I-40126 Bologna, Italy

1. Introduction

In 1948 D. Gabor introduced electron holography, a new microscopic principle to record amplitude and phase of an electron wave.[129] In one of his following publications he discussed the possibility of improving the point resolution of electron micrographs by applying holography to correct for the spherical aberration of the objective lens.[130,131] Phase images obtained from electron holograms contain easily accessible information usually lost by conventional electron microscopy. Therefore, many problems in materials science were already solved by interpreting the reconstructed electron phase[6,25,292,296,314,356] before Orchowski et al.[329] succeeded in 1995, in improving the point resolution by holography.

At low magnification, however, extracting correct information from phase images is a different challenge. The aim of this chapter is to review some possible misinterpretations of phase images by magnetic and electric fields. We discuss figures and formulas only on a basic level and refer to earlier publications for more sophisticated and more comprehensive evaluations.

2. General considerations on the electron phase

2.1. Phase shift by the magnetic potential

The phase shift φ_m of an electron wave compared to a reference wave amounts to:[5]

$$\varphi_m = -\frac{e}{\hbar}\oint \vec{A}\,\mathrm{d}\vec{s} = -\frac{e}{\hbar}\int \vec{B}\,\mathrm{d}\vec{F} = -\frac{e}{\hbar}\,\Phi_m \tag{8.1}$$

where e is the electron charge, $\hbar$ is Planck's constant, $\vec{A}$ is the magnetic vector potential, $\vec{B}$ is the magnetic induction, Φ_m is the magnetic flux, and F is the area bordered by the trajectories of the interfering electrons.

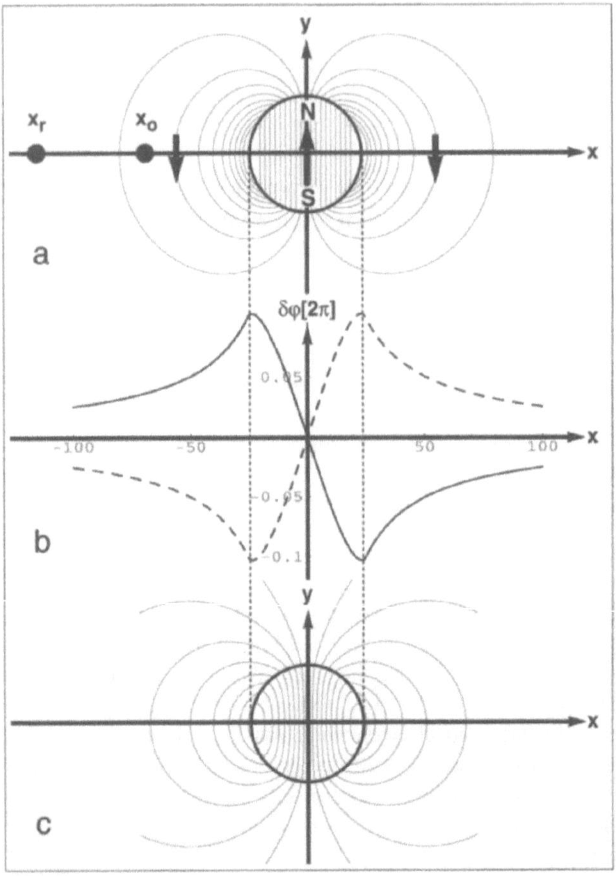

Figure 8.1. Sphere uniformly magnetized in y-direction. a: Cross section with magnetic flux lines. b: Phase shift along x-axis. c: Contour lines of phase image.

Let us schematically draw the phase shift along the x-axis of a cartesian coordinate system where a sphere uniformly magnetized along the y-axis serves as sample. The cross section through the center of such a sphere surrounded by its magnetic flux lines is displayed in Fig. 8.1a. Assume that the electrons travel parallel to the z-axis through points on the x-axis. Let the reference beam be at a very great distance x_r to the left of the sphere such that the magnetic flux on the left side of x_r is negligible. The object beam travels through a point x_o to the right of x_r. As we move the object beam further to the right more and more flux lines between the two beams are enclosed increasing the phase shift according to Eq. (8.1). Moving the object beam further to the right such that the electrons pass through the sphere reverses the trend of the phase shift because the sign of the flux inside the sphere is reversed compared to the flux outside. At the center of the sphere the flux outside the sample is completely counterbalanced by the flux inside, resulting in a phase shift of zero. For reasons of symmetry, the phase on the right side of the center of the sphere is point symmetrical to the phase on the left. This behavior is seen by the solid line in Fig. 8.1b. The dashed line in this figure corresponds to an arrangement where the reference beam is far to the right of the sphere and the object beam is on the left side of the reference beam. We see that the arrangement with the reference beam on the opposite side of the sample results in

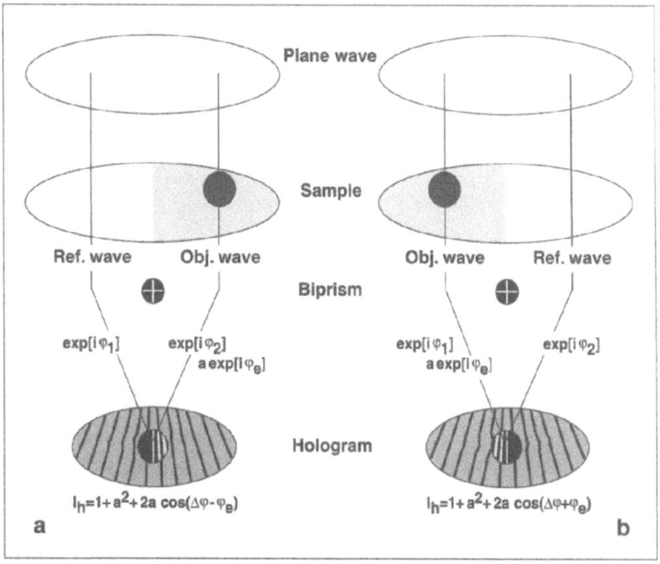

Figure 8.2. Schematic ray path for off-axis holography. Left: Reference wave on left side of sample. Right: Reference wave on right side of sample.

a reversed sign of the phase. The phase plots in this figure were evaluated (see Fig. 8.8) assuming a sphere with 50 nm diameter and a magnetic induction of 1 T inside the sample. The contour lines of the wavefront modulated by the vector potential of this sphere are displayed schematically in Fig. 8.1c.

2.2. Phase shift by the electric potential

The phase difference between a reference wave and an object wave modulated by the electric potential V_e is given by[5]

$$\varphi_e(x,y) = \frac{e}{\hbar} \oint V_e(x,y,z)\,\mathrm{d}t \tag{8.2}$$

where V_e is the electric potential of the sample, and the integration is performed along any closed circuit in time. The object wave below the biprism can be written as $\exp(i\,\varphi_2)\cdot a\,\exp(i\,\varphi_e)$ (see Fig. 8.2a), where a and φ_e are amplitude and phase modulation by the object, respectively. Superposing this wave to the reference wave $\exp(i\,\varphi_1)$ results in the following intensity I_h of the hologram

$$\begin{aligned} I_h &= \left(e^{i\varphi_1} + a\,e^{i(\varphi_2+\varphi_e)}\right)\cdot\left(e^{-i\varphi_1} + a\,e^{-i(\varphi_2+\varphi_e)}\right) \\ &= 1 + a^2 + 2a\,\cos(\Delta\varphi - \varphi_e) \end{aligned} \tag{8.3}$$

Moving the reference wave to the opposite side of the sample (Fig. 8.2b) results in an intensity now given by

$$I_h = 1 + a^2 + 2a\,\cos(\Delta\varphi + \varphi_e) \tag{8.4}$$

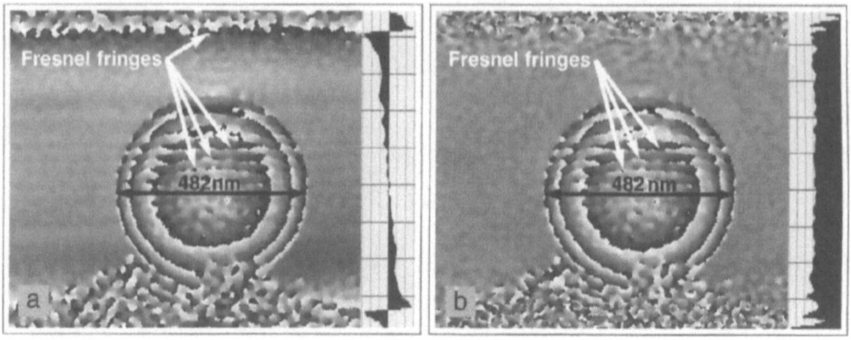

Figure 8.3. Phase image of 482 nm latex sphere. a: Reconstructed without reference hologram. b: Reconstructed with reference hologram.

Comparing Eq. (8.3) to Eq. (8.4) shows that this modification reverses the sign of the object phase, as in the magnetic case. This means that moving the sample to opposite side of the biprism fiber does not offer the possibility of separating 'electric' from 'magnetic' phase shifts.

2.3. Phase shift by the Fresnel fringes of the biprism fiber

The diffraction of a highly coherent electron beam at the edges of the biprism fiber causes Fresnel fringes in an off-axis image plane electron hologram seen mainly at the border lines of the interference region (Fig. 8.3a) and inside the sample (Fig. 8.3a,b). A sample considerably decreasing the amplitude of the object wave also decreases the amplitude of the Fresnel fringes contained in this wave, while the amplitude of the Fresnel fringes contained in the reference wave is not affected by the sample. Therefore, we find in the hologram, inside the sample, Fresnel fringes of both low contrast by the object wave and high contrast by the reference wave. If the Fresnel fringe contrast inside the sample is about the same as the contrast of the interference fringes (or even higher), then the Fresnel fringes strongly falsify the object wave and strongly modify the reconstructed wave. This is seen by the phase distribution of the 482 nm latex sphere in Fig. 8.3 which shows the Fresnel fringes as noisy straight lines superimposed on concentric circles which represent thickness contour lines. Only the simple geometry of this sample allows us to see these as separate and different effects. Samples with more complex structure can mix up both parts in such a way that it becomes nearly impossible to get reliable phase information of the sample.

We can correct for the Fresnel fringes in the empty part of a hologram by acquiring a second hologram without a sample, called a reference hologram. Dividing the complex waves reconstructed from these two holograms yields a phase image without Fresnel fringes outside the sample, as seen comparing Fig. 8.3a to Fig. 8.3b. This procedure does not correct for the Fresnel fringes inside the sample (also seen by comparing Fig. 8.3a to Fig. 8.3b) because the behavior of the Fresnel fringes in the object wave is different from the Fresnel fringes in the empty object wave of the reference hologram.

Other features of reconstructing the object wave by applying a reference hologram are to find the true center of the sideband (discussed later on) and to eliminate distortions in the final image (for more details on reference holograms see pages 133 ff.). Distortions are wide area phase modulations caused by such factors as the imaging lenses and by irregularities of the biprism fiber.

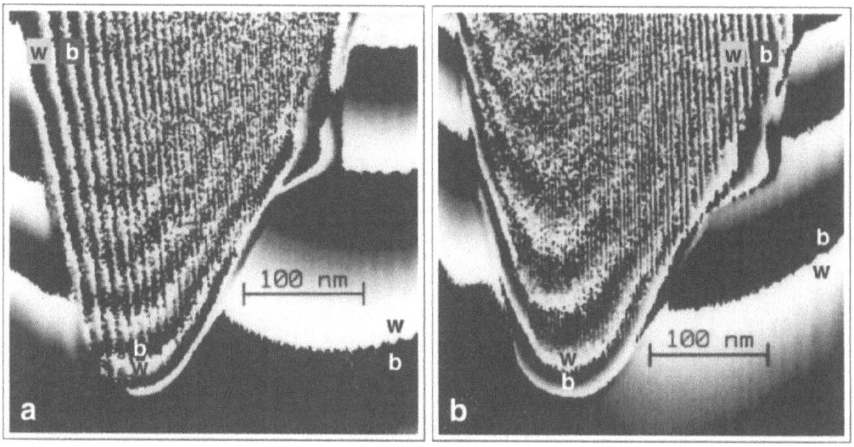

Figure 8.4. Magnetized nickel tip shows phase by electric potential, magnetic potential, and by Fresnel fringes. a: Reference wave on right side of sample. b: Reference wave on left side of sample.

As discussed in Fig. 8.1 and Eqs. (8.3), (8.4), electric and magnetic contributions to the phase image cannot be distinguished by simply moving the reference wave to the opposite side of the biprism. As an example, Fig. 8.4 shows two phase images of a bulk magnetic nickel probe (poled in an external field oriented along the tip axis). For Fig. 8.4a, the hologram was recorded with the reference wave on the right hand side of the sample. For Fig. 8.4b, the hologram was recorded with the reference wave on the left hand side of the sample.

In both cases, the phase is comprised of the 'electric phase,' the 'magnetic phase,' and the 'Fresnel phase.' Outside the tip the phase lines are caused by the magnetic field. Inside the tip two different contributions to the phase are revealed. 1) The phase lines nearly parallel to the probe axis are due to the Fresnel fringes of the biprism fiber. 2) The phase lines parallel to the probe profile are due to thickness variations.

By comparing Figs. 8.4a and b, it is clear that both the electric and magnetic phase contributions have simply changed their sign (see reversal of black/white contrast). The Fresnel fringes, however, change significantly, as expected.

3. Centering the sideband

The reconstruction of the image wave from a hologram is usually performed by means of Fourier optics. Evaluating the Fourier transform of the hologram represented by Eq. (8.4) results in

$$FT(I) = \delta(q) + FT(a^2) + \mathrm{FT}(a\, e^{i\varphi_o}) \otimes \delta(q_c + q) + \mathrm{FT}(a\, e^{-i\varphi_o}) \otimes \delta(q_c - q) \qquad (8.5)$$

where φ_o ($\varphi_o = \varphi_m + \varphi_e$) is the object phase and $q_c = 1/s$ is the carrier frequency (s is the spacing of the interference fringes). The first, second, third, and fourth part of Eq. (8.5) show zero order diffraction, autocorrelation, and the two sidebands representing the Fourier transform of the complex image wave, respectively. Selecting the brightest point of one of the sidebands (e.g., $\delta(q_c - q)$) as center of an area surrounding

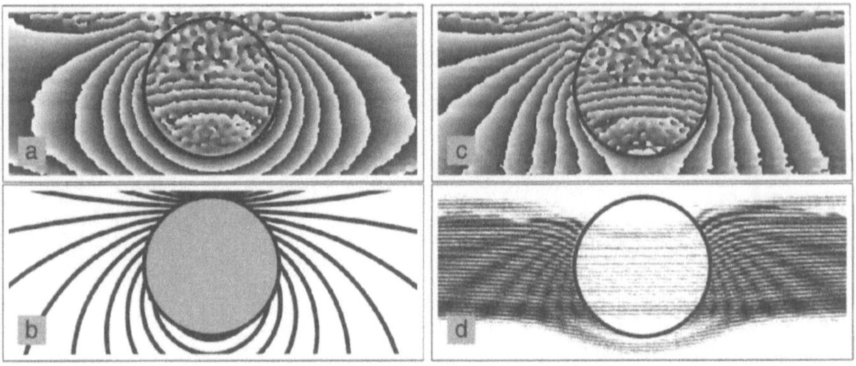

Figure 8.5. Phase of electrically charged 2.68 μm latex sphere. a: Reconstructed from hologram (no reference hologram). b: Simulated equiphase lines. c: Reconstructed from hologram (with reference hologram). d: Double exposure image.

this point and performing an inverse Fourier transform of only this area results in the object wave in its amplitude and phase.

In general, selecting the brightest point of one sideband as its center means that the reconstructed phase image corresponds to the wave front of the image wave in the microscope projected onto the image plane, i.e., the wave viewed directly from top. This assumption, mostly justified, can lead to misinterpretations of the reconstructed image wave when the interference fringes are modulated by the sample such that another bright point, even brighter than the original center, is created. Then, even software for fully automated reconstruction of the image wave (e.g., Holoworks[479]) may select this 'wrong' point. This is seen in Fig. 8.5a by the phase image of an electrically charged latex sphere of a diameter of 2.68 μm. Comparing this image to the phase in Fig. 8.5b, which simulates viewing the wave from top, shows strong differences. However, selecting between different points in the neighborhood of the brightest point of the sideband can yield a phase image similar to the simulation. The most often applied technique to find the best point for reconstruction is to acquire a reference hologram. The brightest point of the sideband of this reference hologram gives the desired center as its brightest spot. Thus selecting the corresponding point in the diffraction pattern of the object hologram yields the reconstructed wave viewed from top (Fig. 8.5c).

Another method to avoid 'incorrect' centering of the sideband, the double exposure technique, is discussed by Matteucci et al.[289] With the double exposure technique, equiphase lines become visible even without the reconstruction process simply by adding the reference hologram to the hologram of the object (Fig. 8.5d).

As discussed above, centering an area to a point close to the true center of the sideband tilts the wave front. The projection of the tilted wave front onto the image plane corresponds to a phase viewed not from top but from an angle. This is seen in the tableau of Fig. 8.6 which shows phase images of a latex sphere with 482 nm diameter. In each image of the tableau only that part of the wave front is tilted which is modulated by the sphere. These images reveal the three dimensional character of the phase reconstructed from a hologram. Two dimensions are given by the plane onto which the phase of the wave (the third dimension) is projected. We note that by this technique not the sample but the wave front is viewed from different angles.

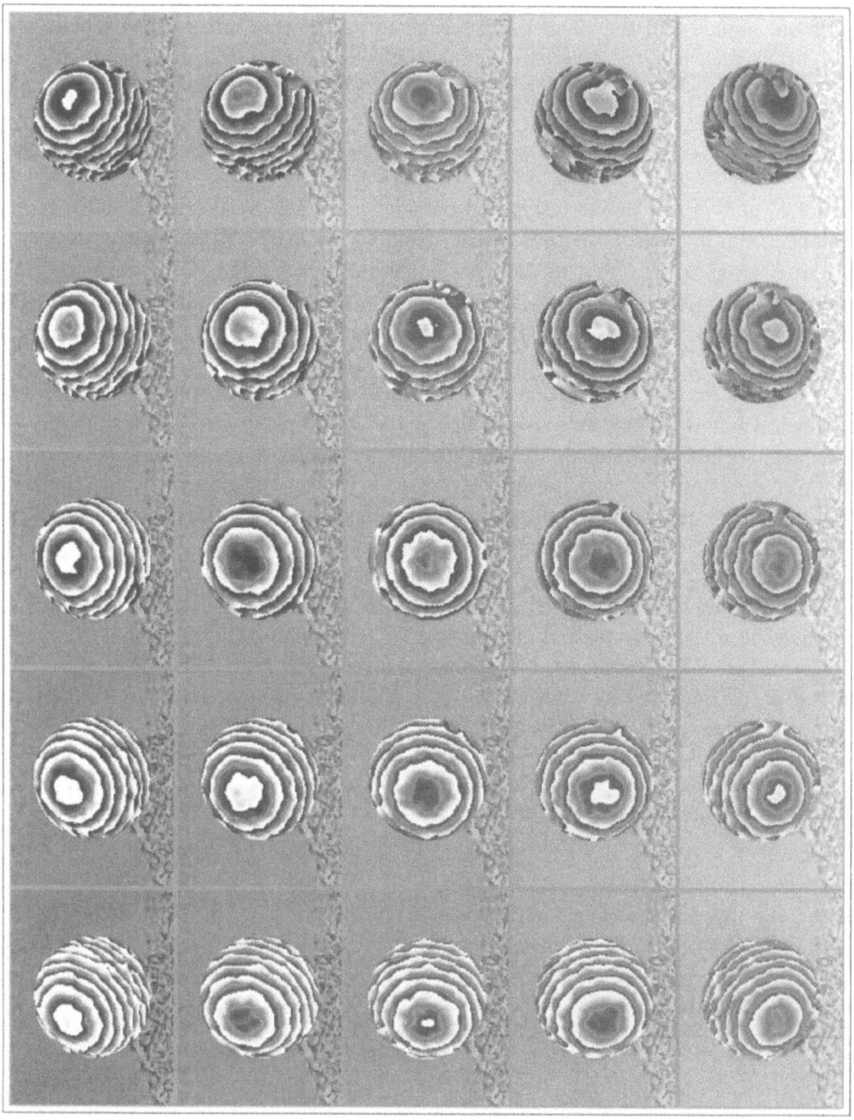

Figure 8.6. Tableau shows phase images of a 482 nm latex sphere reconstructed from one hologram viewed from different angles.

4. Quantification of magnetic and electric fields

Before we are able to simulate phase shifts by magnetic potentials to the extent necessary for a rough quantitative interpretation of the phase image we have to know the magnetic induction $\vec{B}$ inside and outside the sample. In general, the simulation of magnetic fields involves highly complex mathematics. To keep the required equations as simple as possible we will show only the principles involved in simulating a magnetic phase image of an uniformly magnetized sphere.[428] Let the radius of the sphere be r_s and the magnetization $\vec{M}$ parallel to the y-axis of a cartesian coordinate system. In addition, assume that the magnetization leaps to zero on the surface of the sphere. For this arrangement the magnetic scalar potential ψ_m at a point $P(x, y, z)$ results in

$$\psi_m(P_i) = \frac{1}{3} M\, y \qquad \psi_m(P_o) = \frac{1}{3} M\, y\, r_s^3 (r^2 + y^2)^{-3/2} \tag{8.6}$$

where P_i and P_o are points inside and outside the sphere, respectively, and r is the distance from a point P to the axis of symmetry (y-axis). Applying the magnetic potential given by Eq. (8.6) to the basic equation $\vec{B} = \mu_o(\vec{M} - \nabla\psi_m)$ results in the magnetic induction. For points inside the sphere ψ_m depends linearly on y and is independent of r. Thus $\vec{B}$ inside the sample is constant.

We now evaluate (see Fig. 8.7) phase shifts assuming a magnetic field with rotational symmetry along an axis parallel to the image plane, and the trajectories of the imaging electron beams as straight lines perpendicular to the image plane (a plane electron wave and the magnetic field too weak to deflect the electrons). Let the area F bordered by the two interfering beams (the object wave and the reference wave) be perpendicular to the axis of symmetry. Corresponding to Eq. (8.1) only the magnetic flux through this area contributes to the phase shift. If the axis of symmetry is the y-axis then only the y-component of the magnetic induction $\vec{B}$ contributes to the flux through the area F. For reasons of symmetry, $B_y(r, y)$ is constant in the area $\mathrm{d}F = l{\cdot}\mathrm{d}r$ drawn in Fig. 8.7. The length l of the arc is given by $l = 2r \arccos(x/r)$. The flux $\Phi_m = B_y{\cdot}\mathrm{d}F$ enclosed by a reference beam at a great distance from the sample and an object beam at (x_o, y_o) is given by[121]

$$\Phi_m(x_o, y_o) = 2 \int_{x_o}^{\infty} B_y(r, y_o) \cdot r \cdot \arccos\left(\frac{x_o}{r}\right) \,\mathrm{d}r \tag{8.7}$$

where the flux on the right side of the reference wave (Fig. 8.7) is assumed to be negligible. We obtain the resulting phase shift by applying Eq. (8.7) to Eq. (8.1). If the flux on the right side of the reference beam is not negligible we have to subtract from Eq. (8.7) the correction

$$\Phi_c(x_r, y_r) = 2 \int_{x_r}^{\infty} B_y(r, y_r) \cdot r \cdot \arccos\left(\frac{x_r}{r}\right) \,\mathrm{d}r \tag{8.8}$$

where x_r and y_r are the coordinates of the reference wave. The distance between the object wave and the reference wave is defined by the interference width Δ.

Since the phase by a magnetized sample is also modulated by its mean inner potential we also have to evaluate this contribution to obtain a useful simulation. In the case of a sphere, the integration of Eq. (8.2) is straightforward and not performed here. Instead, we show as a result the line scan along the x-axis of a phase image modulated by electric and magnetic potentials. This line scan, shown in Fig. 8.8, was calculated assuming an accelerating voltage of 200 kV and a sphere with a diameter of 50 nm, a mean inner potential of 20 V (about the potential of Ni or Co), and an uniform magnetic induction of 1 Tesla (T) parallel to the y-axis. We see that in this case the

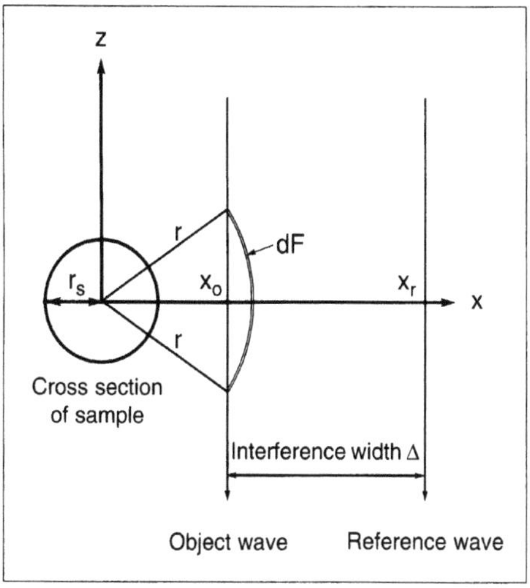

Figure 8.7. Schematic drawing to evaluate phase shift by magnetic fields with rotational symmetry.

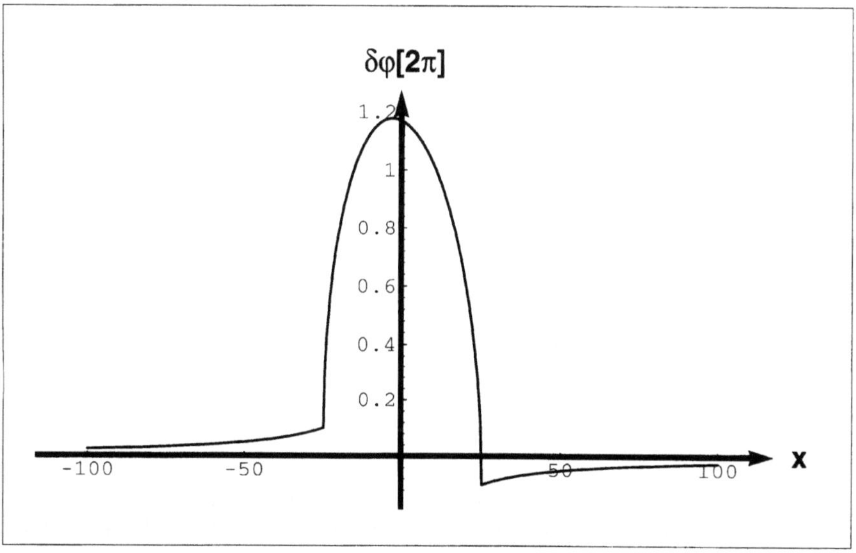

Figure 8.8. Simulated line scan through the center of a sphere shows a phase shift by the inner potential ten times higher than the phase shift by the magnetic flux. (Diameter: 50 nm, inner potential: 20 V, magnetic induction inside the sphere: 1 T, accelerating voltage: 200 kV).

phase shift by the mean inner potential of the sample is much greater compared to the phase shift by the magnetic potential. At lower accelerating voltages the electric phase shift is even higher whereas the magnetic one is unchanged. The phase shift by the magnetic field, though a strong field, is very small, because the flux through a small area is small. Whenever we interpret phase images of small magnetized particles we should never neglect the phase caused by the mean inner potential when comparing simulations to measurements.

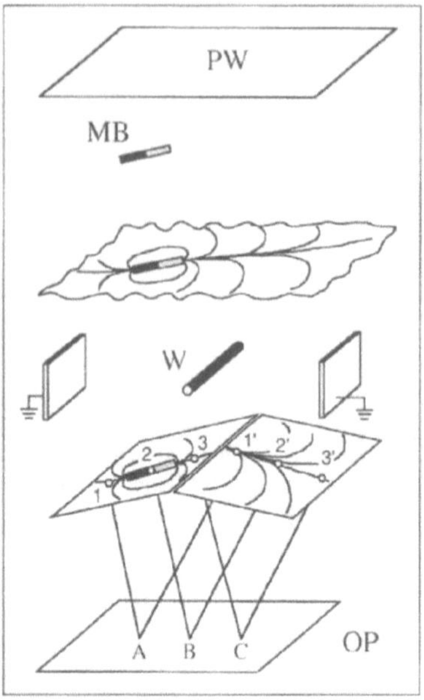

Figure 8.9. Schematic ray path formation of a hologram with an affected reference wave.

4.1. Modulation of the reference wave

Particular attention must be paid when investigating specimens that generate long range electric and/or magnetic fields extending in the vacuum space around the object and perturbing the phase of the reference wave (for a review see Refs. 45,290,294,295). In fact, the hologram stores the phase difference between the object and the affected reference waves. In order to extract reliable phase information, a model of the field source is extremely helpful.[45,119,290,294,295] In the following a simple example is reported to show how a modulated reference wave may affect the phase distribution of the hologram.

Assume a plane electron wave PW illuminates a small magnetic bar MB as shown in Fig. 8.9. The magnetic field extends far away from the bar such that both the object wave and the reference wave are modulated by the field. The biased biprism wire W splits the incoming wave into two parts which overlap in the observation plane OP. Consider the points $1, 2, 3$ of the object wave and points $1', 2', 3'$ of the reference wave. Suppose that the points 1 and 3 are symmetric with respect to the center of the bar. The ray tracing shows that the couple of points 1 and 1', 2 and 2', 3 and 3' interfere in A, B, C, respectively. Since the beam passing through 1' is much more affected by the leakage field than the beams through 2' or 3', the recorded phase shift at A will be different from that at C. Fig. 8.10 shows the simulation of the electron phase assuming the hologram was recorded with a reference wave unaffected by the magnetic field (Fig. 8.10a) and modulated by the field (Fig. 8.10b). The phase lines in Fig. 8.10a are symmetric with respect to the center of the bar, whereas in case of a modulated reference wave the symmetry has changed (Fig. 8.10b). A possible method for numerically correcting experimental phase images for this influence is suggested by Kou and Chen.[233]

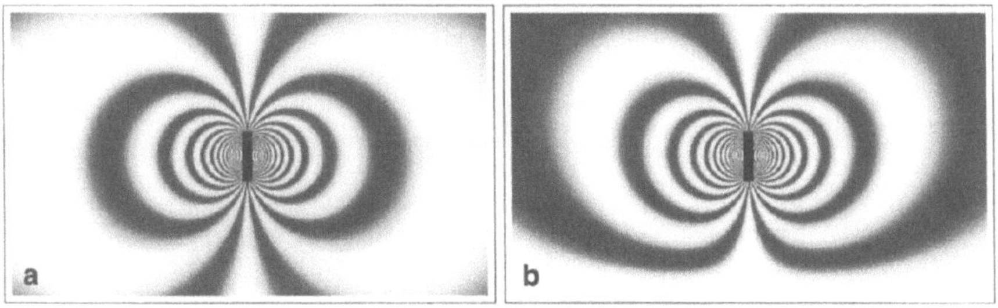

Figure 8.10. Simulation of the phase lines around a small magnetic bar using (a) a not modulated reference wave (b) an affected reference wave.

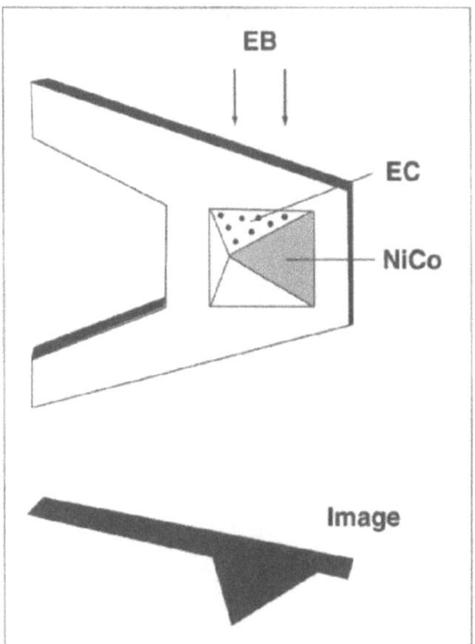

Figure 8.11. Sketch of a pyramidal microprobe on a cantilever.

5. Distinguishing magnetic and electric fields

In a phase image of a magnetic sample the structure by the electric potential is often superimposed to the structure by the magnetic potential. Sometimes it is desirable to separate both effects, e.g., to investigate the fine structure of magnetic domain walls.

An excellent method to separate electric from magnetic phase shifts was suggested by Wohlleben[500] and experimentally performed by Tonomura[448] applying holography. We briefly review the principle of this method and separate electric from magnetic leakage fields arising around magnetic force microscopy sensor tips. Commercially available silicon nitride pyramidal tips are obliquely evaporated with NiCo to form a film which predominantly coats one side of the apex (Fig. 8.11). The tip is then poled in a strong external magnetic field directed perpendicularly to the coated side. During the electron hologram recording step the magnetic tip becomes electrically charged (EC) under the radiation of the electron beam (EB) generating an electric leakage field.

Therefore, the electron wave traveling in the region near the tip apex is phase

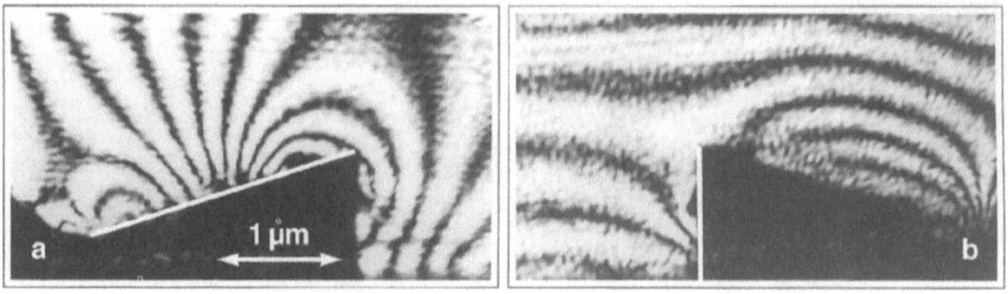

Figure 8.12. Map of equiphase fringes for a: $\varphi_{o1} = \varphi_e + \varphi_m$ and b: $\varphi_{o2} = \varphi_e - \varphi_m$.

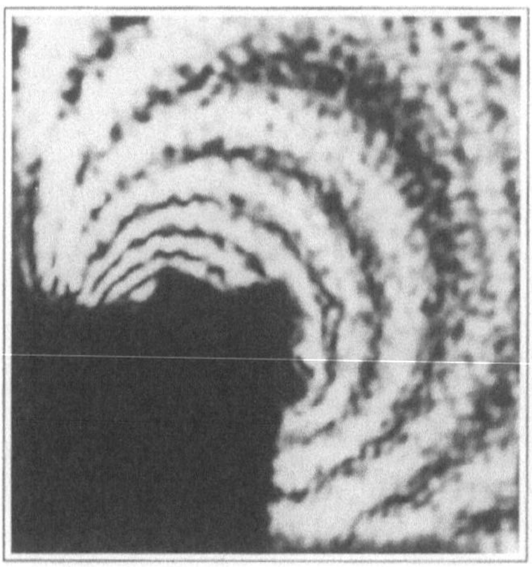

Figure 8.13. Map of projected magnetic lines of force around the tip apex.

Figure 8.14. Simulated contour map of Fig. 8.13.

modulated by the combined effects of both electric and magnetic leakage fields. In the presence of these fields the total phase shift is given by $\varphi_o = 1/\hbar \int (m\vec{v} - e\vec{A}) \cdot \mathrm{d}\vec{s}$. Reversing the direction of the incident electrons (experimentally done by flipping the sample upside down) changes the sign of the velocity $\vec{v}$ and the trajectory $\vec{s}$, but does not reverse the magnetic vector potential $\vec{A}$. The electric phase shift $\varphi_e = (1/\hbar) \int m\vec{v}\, \mathrm{d}\vec{s}$ is therefore invariant, while the magnetic phase shift $\varphi_m = (-1/\hbar) \int e\vec{A} \cdot \mathrm{d}\vec{s}$ changes sign.

In order to experimentally separate electric from magnetic phase shifts we record two holograms. Let the phase shift stored in the first one be given by $\varphi_{o1} = \varphi_e + \varphi_m$. After flipping the sample upside down the phase stored in the second hologram becomes $\varphi_{o2} = \varphi_e - \varphi_m$. In general: $\varphi_{o1} \neq \varphi_{o2}$. The sum or the subtraction of the two phase images φ_{o1} and φ_{o2} can be made by digital or by optical methods. It turns out that both $\varphi_{o1} + \varphi_{o2} = 2\varphi_e$ and $\varphi_{o1} - \varphi_{o2} = 2\varphi_m$ can be obtained separately. The phase difference maps displayed in Fig. 8.12a,b and Fig. 8.13 were obtained using a Mach-Zehnder interferometer and show $\varphi_{o1}, \varphi_{o2}$, and $2\varphi_m$.

In these maps, the dark shadow is the projected image of the pyramidal probe. The cross view of the thin magnetic film is marked by a white line. The difference between Fig. 8.12a and Fig. 8.12b confirms the presence of a significant electric contribution to the total phase shift. Fig. 8.13 shows the two-time phase amplified map of magnetic lines of force alone around the tip apex. Figure 8.14 is a contour map computed

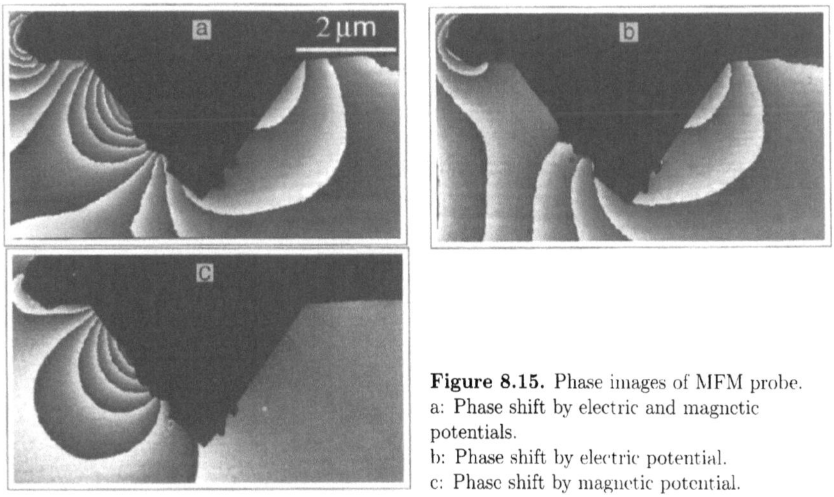

Figure 8.15. Phase images of MFM probe.
a: Phase shift by electric and magnetic potentials.
b: Phase shift by electric potential.
c: Phase shift by magnetic potential.

by simulating the thin magnetic film by a triangular distribution of magnetic dipoles which self-organize themselves in presence of a high poling magnetic field perpendicular to the coated face of the tip.[120,275,276] These results show how the separation of the magnetic from the electric phase shifting effects, generated by the leakage fields, can be successfully achieved.

A further approach to separating phase shifts by magnetic leakage fields of a tip from phases by the electric leakage fields is illustrated in Fig. 8.15, which shows another silicon nitride probe (for MFM) coated on the left side with a ferromagnetic NiCo-layer. The phase image in Fig. 8.15a shows contour lines by the magnetic field due to the thin ferromagnetic layer and by the electric field due to charging of the probe by the electron beam. This image was reconstructed from a hologram acquired in the standard low magnification mode of operation for electron holography (Buhl[35]) with the objective lens off. The magnetic field of a weakly excited objective lens causes the magnetic dipoles of soft magnetic materials to align parallel to the magnetic field of the lens and therefore nearly parallel to the electron beam. In this case the magnetic flux between the two interfering waves is nearly zero. Thus the phase distribution seen in Fig. 8.15b reconstructed from a hologram acquired with the objective lens slightly excited is mainly due to the electric leakage field. Coating the sample with a thin gold layer prevents electric charging by the electron beam. In this case the equiphase lines surrounding the probe are due to the magnetic potential of the thin NiCo-layer (Fig. 8.15c).

Another possible way to separate the electric from the magnetic phase is to acquire two holograms of the same specimen but at different accelerating voltages U_{a1} and U_{a2}. Since the magnetic part of the phase distribution φ_m is independent of the acceleration voltage (Eq. (8.1)) one obtains, on subtraction of the reconstructed phases, an image whose phase information depends on the electric component only. We show this by the following example.

The phase distribution in the first hologram is

$$\varphi_{o1}(x,y) = \varphi_m(x,y) + \frac{1}{\hbar}\sqrt{\frac{em}{2U_{a1}}} \int\limits_{z_1(x,y)}^{z_2(x,y)} V_e(x,y,z')\,\mathrm{d}z' \tag{8.9}$$

The phase distribution in the second hologram is

$$\varphi_{o2}(x,y) = \varphi_m(x,y) + \frac{1}{\hbar}\sqrt{\frac{em}{2U_{a2}}} \int\limits_{z_1(x,y)}^{z_2(x,y)} V_e(x,y,z')\,\mathrm{d}z' \tag{8.10}$$

It easily follows that

$$\begin{aligned}\varphi_{e1} &= \frac{1}{\hbar}\sqrt{\frac{em}{2U_{a1}}} \int\limits_{z_1(x,y)}^{z_2(x,y)} V_e(x,y,z')\,\mathrm{d}z' \\ &= (\varphi_{o1}(x,y) - \varphi_{o2}(x,y)) \cdot \frac{\sqrt{U_{a2}}}{\sqrt{U_{a1}} - \sqrt{U_{a2}}}\end{aligned} \tag{8.11}$$

The disadvantages of this procedure are obvious. Slower electrons are absorbed by thicker parts of the specimen. In addition, at different accelerating voltages the imaging conditions of the microscope are different.

6. Dislocations in a hologram

A strong change of the gradient of electric and/or magnetic potentials of a sample in combination with a poor phase resolution (poor fringe contrast) of the hologram and/or a poor spatial resolution (large fringe spacing) sometimes causes artificial branching of the interference fringes. In the following we simulate this effect using a capacitor as model and we discuss the influence on the reconstructed phase images.

6.1. Phase by capacitor

The phase by a capacitor consisting of two parallel and infinitely thin wires of length $2l$ with uniform but opposite charge distributions on them can be modeled by evaluating first the electric potential of two point charges, then the phase by these two points, and finally the phase by the capacitor.

The potential of an electric dipole, whose point charges Q are placed in the (x,y) plane at a distance $2d$ from each other on a line parallel to the x-axis at a distance y_o from the x-axis, amounts at any point to:

$$V_e = \frac{Q}{4\pi\varepsilon_o} \cdot \frac{\sqrt{(x+d)^2+(y+y_o)^2+z^2} - \sqrt{(x-d)^2+(y+y_o)^2+z^2}}{\sqrt{(x+d)^2+(y+y_o)^2+z^2} \cdot \sqrt{(x-d)^2+(y+y_o)^2+z^2}} \tag{8.12}$$

The phase by these point charges evaluated by Eq. (8.2) can be written as:

$$\varphi_e = \frac{e}{\hbar}\frac{Q}{8\pi\varepsilon_o}\sqrt{\frac{2m}{eU_A}} \cdot \ln\left(\frac{(x+d)^2+(y+y_o)^2}{(x-d)^2+(y+y_o)^2}\right) \tag{8.13}$$

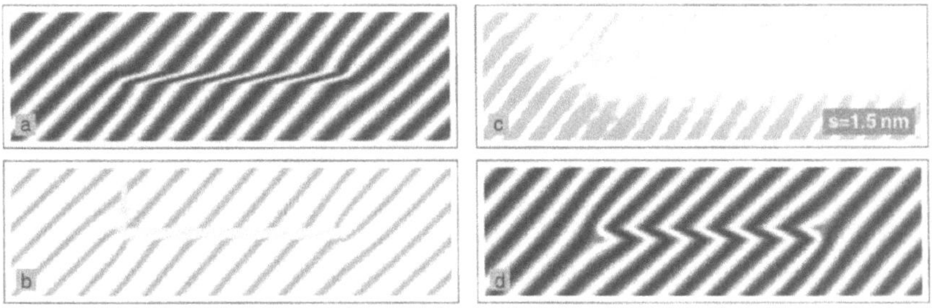

Figure 8.16. Hologram of a capacitor. a: Simulation of high phase resolution. b: Simulation of low phase resolution causes artificial dislocation. c: Hologram of a CrO_2-needle showing an artificial dislocation. d: Simulation of high phase resolution, reference wave on the opposite side of the sample.

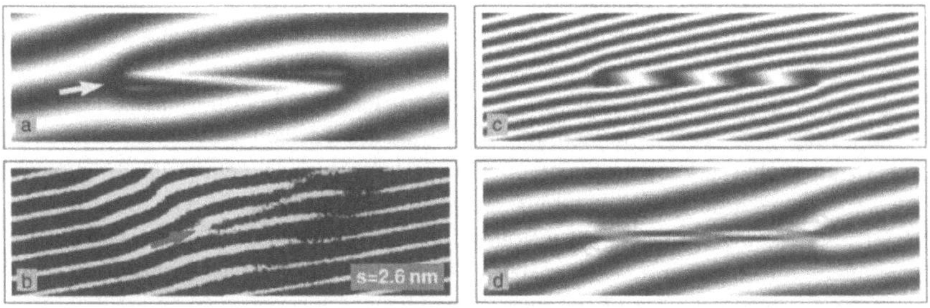

Figure 8.17. Holograms of a capacitor. a: Simulation of an artificial dislocation caused by low spatial resolution. b: Hologram of a nickel tip with artificial dislocation. c: Simulation of high spatial resolution. d: Simulation of low spatial resolution, reference wave on the opposite side of the sample.

when the electron beams are assumed to be parallel to the z-axis.

In order to obtain the phase by the capacitor we have to integrate Eq. (8.13) with respect to y_o with limits from $-l$ to l (the lengths of the two lines amounts to $2l$). This integral follows as

$$\frac{1}{2}\int \ln\left(\frac{(x+d)^2+(y+y_o)^2}{(x-d)^2+(y+y_o)^2}\right)\,\mathrm{d}y_o = \tag{8.14}$$

$$(x-d)\arctan\frac{y-y_o}{d-x}+(x+d)\arctan\frac{y+y_o}{d+x}+\frac{1}{2}(y+y_o)\ln\frac{(x+d)^2+(y+y_o)^2}{(x-d)^2+(y+y_o)^2}$$

The simulation of holograms with different phase resolution, different point resolution, and different positions of the reference wave, applying the phase distribution by a capacitor with $l=1, d=0.05$, and $Q=60\,\mathrm{e}$ ($\mathrm{e}=1.602\cdot 10^{-19}\mathrm{C}$), and assuming an acceleration voltage of $U_A=200\,\mathrm{kV}$, reveals some of the circumstances causing dislocations in holograms.

6.2. Dislocations by poor resolution

The excellent fringe contrast of the simulated hologram in Fig. 8.16a makes it possible to see clearly the behavior of the interference fringes. In this simulation, the spacing of the fringes inside the capacitor is much smaller than outside. In general, the contrast of narrow fringes depends much more strongly on instabilities of the microscope than the contrast of wider fringes. In addition, the visibility of the interference pattern decreases when the spacing of the (narrow) fringes approaches the resolution limit of the electron detector. Therefore, a slight decrease in the fringe contrast outside the capacitor can result in a strong decrease of the visibility of the interference pattern inside. As a consequence, artificial dislocations can be caused in the hologram (branching of the interference fringes) as seen in Fig. 8.16b. This behavior is also seen in an experimental result (Fig. 8.16c) at the border line between vacuum and a CrO_2–needle. Since this result is in good agreement to the simulation of a charged capacitor we assume that the line to the right of the dislocation is electrically charged. The poor fringe contrast along this line causes artifacts in the amplitude image. Moving the reference wave to the opposite side of the sample modifies the interference pattern such that the spacing inside the sample is about the same as outside (Fig. 8.16d). Since the fringe spacing is larger than in the previous example, the artificial fringe branching will not be observed in this case even for poor fringe contrast.

A poor spatial resolution of a hologram can also result in artificial dislocations in the interference pattern. This is illustrated by the simulated hologram of a capacitor in Fig. 8.17a. Here, the impression of the dislocation would additionally be enhanced by the strong decrease of the visibility of the narrow horizontal fringes at the center of the sample thus resulting in a poor phase resolution. The experimental image in Fig. 8.17b shows a hologram of a bulk nickel tip with good phase resolution but poor spatial resolution. The branching of one of the interference fringes is very similar to the simulated dislocation by the capacitor. By increasing the spatial resolution seen in Fig. 8.17c the true behavior of the interference fringes is revealed. The artificial fringe branching would not be observed even if the phase resolution was poor. Moving the reference wave to the opposite side of the sample also shows the correct behavior of the fringes (Fig. 8.17d). However, in this case a poor phase resolution again results in a dislocation of the fringes.

6.3. Reconstructed wave containing dislocation

The phase distribution reconstructed from the hologram of Fig. 8.16c and phase amplified by a factor of two is displayed in Fig. 8.18, left, where a dislocation can be recognized. Following a closed loop (e.g., the white circle in this figure) around this dislocation a continuous increase (decrease) of the phase shift is displayed. While it is possible to 'measure' a phase shift of 2π along one full loop, the total phase shift could be any integer multiple of 2π. In all cases, from the directly reconstructed (and phase amplified) image it is apparent that one starts with the same value as one ends up with. This behavior is similar to an imaginary world depicted in *Ascending and Descending* by M.C. Escher (Fig. 8.18, right). The surface of the reconstructed wave in the real world is similar to the helicoidal form shown in Fig. 8.19a. However, comparing this phase to the original phase of the capacitor (Fig. 8.19b) shows strong discrepancies.

Dislocations in experimental holograms are always connected to a sample. In order to investigate Fourier transforms, phase images, and amplitude images of such holograms without the disturbing influence of a sample we simulated holograms consisting only of dislocations (Fig. 8.20). The numerically performed Fourier transforms of these holograms also show dislocations in the sideband in agreement to the Fourier transform obtained by diffraction of a laser beam.[16] The bright area at bottom of each Fourier transform of this figure shows the autocorrelation. The phase image of Fig. 8.20a which contains only one dislocation is similar to the experimental phase image in Fig. 8.18a, as expected. The amplitude images show dislocations as dark spots.

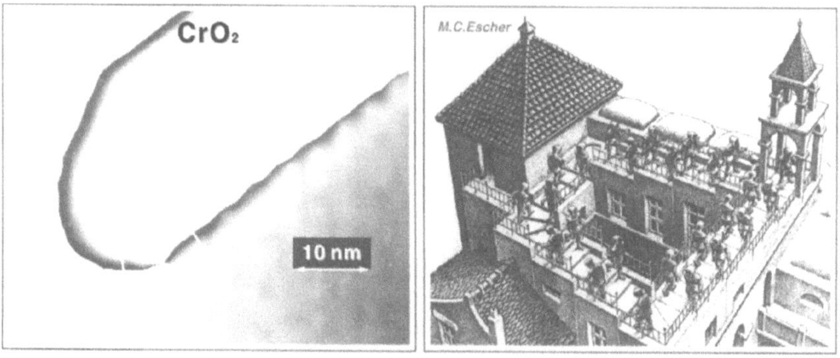

Figure 8.18. left: Phase image of CrO_2-needle containing a dislocation. right: M.C. Escher, "Detail of 'Ascending and Descending'" ©1998 Cordon Art - Baarn - Holland. All rights reserved. Both figures show a striking similarity.

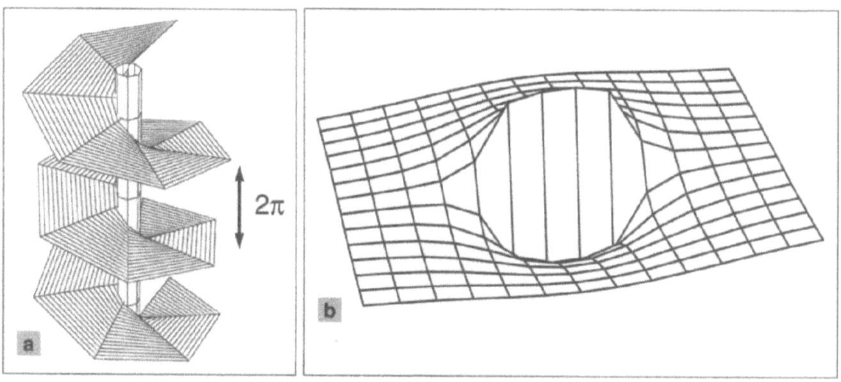

Figure 8.19. Phase images. a: Simulated phase of dislocation. b: Simulated phase of capacitor.

Acknowledgments

This chapter has partly been supported and sponsored by Oak Ridge National Laboratory, managed for the DOE by Lockheed Martin Energy Research, Inc. under DE-AC05-84OR21400 and partly by funds from MURST coordinated by Consorzio INFM and CNR-GNSM. Dr. E. Luncdei is gratefully acknowledged for his help with two of the computer simulations and the optical processing of electron holograms of magnetic tips. Some figures and parts of the text are taken from *Journal of Microscopy*, Vol. 187, Pt2, (1997) 85-95, courtesy Royal Microscopical Society.

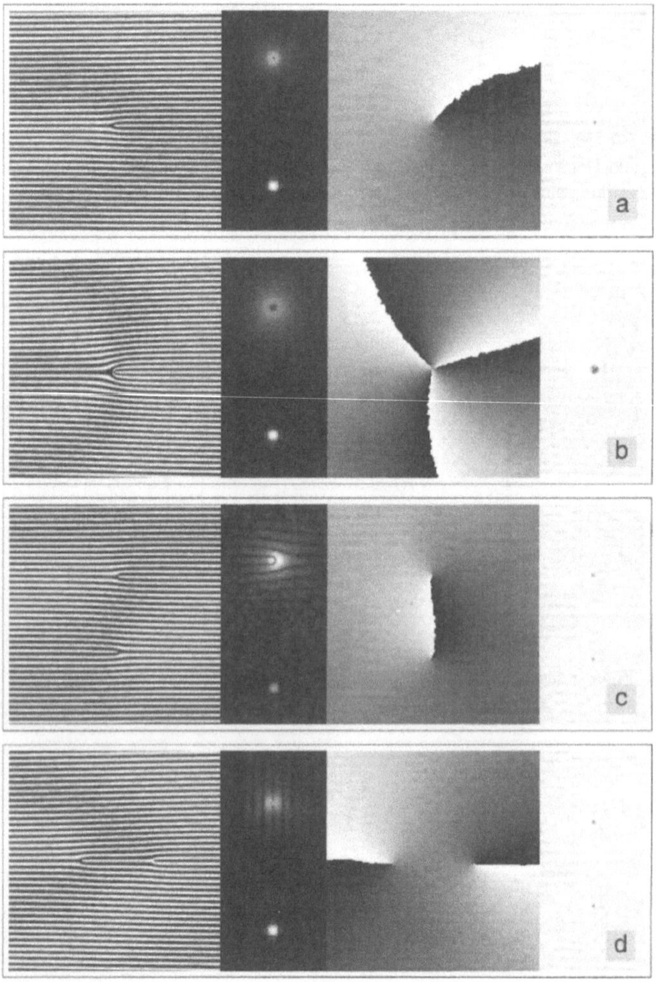

Figure 8.20. Simulated dislocations. From left to right: Hologram, Fourier transform, phase image, amplitude image.

HIGH RESOLUTION OFF-AXIS ELECTRON HOLOGRAPHY

W.D. Rau[1] and H. Lichte[2]

[1]Institute for Semiconductor Physics, D-15230 Frankfurt (Oder), Germany
[2]Institut für Angewandte Physik, Technische Universität Dresden, D-01062 Dresden, Germany

1. Introduction

In contrast to the light optical case, in conventional electron microscopy the lateral resolution is not governed by the diffraction limit, i.e., by the wavelength of the electrons, but instead is limited by the spherical aberration of the objective lens. As already shown by Scherzer[391] in 1936, this error cannot be avoided as long as common rotationally symmetric lens designs and static electromagnetic fields are used to focus the electrons. For nowadays typical intermediate high voltage instruments with accelerating voltages of 200-400 keV, corresponding to relativistic wavelengths of 2.5-1.6 pm, typically point resolutions of 0.24-0.16 nm are achieved.[326] These values are about two orders of magnitude above the physical limit given by the diffraction limit. Various efforts to correct the spherical aberration[313,392,397] by using non-rotationally symmetric lens designs have, unfortunately, not been very successful up to now. However, recent results in this field seem to show promising results.[158,369]

On the other hand, already in 1948 Dennis Gabor proposed a "new microscopic principle" to overcome the resolution limit of electron microscopy (about 0.5 nm at that time).[129] His proposed "electron interference microscope" would register amplitude and phase of the electron image wave in a "two step process, by electronic analysis, followed by optical synthesis". In the second step a wave optical reconstruction procedure[130,133] could be applied, that "allows an a posteriori compensation of the errors by assuming an imperfect recording that has been made with coherent illumination". A basis for this approach – called electron holography – was set with the invention of the electron biprism beamsplitter by Möllenstedt and Düker in 1956.[312] It allowed the application of the off-axis technique[251] in electron holography. Based on the proposals of Weingärtner et al.[492] the first thorough investigations of image plane off-axis electron holography were performed by Wahl[483,486] and others.[159,446] The advent of highly coherent field emission guns[76,466] was of considerable importance to help electron holography enter the high resolution domain.[258,259,441] Replacing the reconstruction on the optical bench by numerical procedures proved to greatly enhance the flexibility in adapting the reconstruction parameters to the experimental situation. First demonstrations of correction of the objective lens aberrations[124,211,261] followed quickly. Recently, off-axis electron holography has been shown to be capable of retrieving interpretable information beyond the conventional Scherzer resolution limit.[168,212,329] Thereby the resolution limit in high resolution transmission electron microscopy (HRTEM) can be improved from about 0.2 nm point resolution down to 0.1 nm, which is the information limit reachable with modern medium voltage microscopes.

The following chapter is organized as follows: In the next part, we will give a short introduction to the contrast transfer theory in HRTEM and point out the most important aberrations. The third section deals with the basic experimental set-up of high resolution off-axis holography. We will then describe the principle of aberration correction in part four and show how the experimental parameters have to be optimized for the off-axis technique in section five. Finally, in the last part we will present some experimental examples and finish with a short conclusion of the whole chapter.

2. Wave optical imaging in HRTEM

In conventional high resolution electron microscopy (HRTEM), a thin specimen is illuminated with a plane electron wave $e^{2\pi i \vec{k}\vec{r}}$. For uniform direction k and magnitude $|\vec{k}| = 1/\lambda$ the illumination is considered to be coherent. In the more realistic case of a distribution of electrons with varying direction and energy the effects of partially coherent illumination have to be incorporated. For the imaging process, only the electrons that have been elastically scattered at the Coulomb potential of the atomic nuclei are taken into account. After passing through the specimen, the wave is modulated in amplitude a and phase φ, resulting in the so-called object wave

$$o(\vec{r}) = a(\vec{r}) \cdot e^{i\varphi(\vec{r})} \tag{9.1}$$

Details of the complex interaction process described by dynamical scattering theory can be found elsewhere.[357,408] Using an objective and several projector lenses, the object wave is then imaged into the highly magnified image wave (Fig. 9.1)

$$b(\vec{r}) = o(r) \otimes \mathrm{PSF}(\vec{r}) \tag{9.2}$$

The aberrations of the imaging system are incorporated as a convolution of the object wave with the Point Spread Function $\mathrm{PSF}(\vec{r})$ characteristic to the microscope. Specifically, $\mathrm{PSF}(\vec{r})$ is the response of the imaging system to a delta function $\delta(\vec{r})$ and represents the error disc that corresponds to the image of a point-like object. This description is restricted to the so called isoplanatic approximation that implies the independence of the imaging process from the position $\vec{r}$ in the image. The magnification of the objective lens is not included in our notation, meaning that the coordinate $\vec{r}$ always is related to the object.

Following the laws of wave optics, the imaging process can be described in a two step Fourier formalism: In the back focal plane of the objective lens the spectrum of the object wave is obtained as Fourier transform of the incoming wave according to

$$O(\vec{q}) = \mathbf{FT}\{o(\vec{r})\} = \int o(\vec{r}) \cdot e^{2\pi i \vec{q}\vec{r}} \, d\vec{r} \tag{9.3}$$

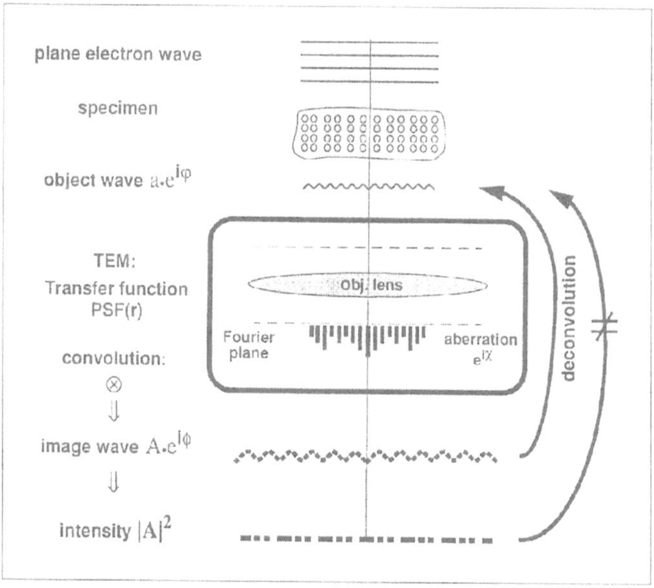

Figure 9.1. Image formation in HRTEM. The influence of the aberrations of the objective lens are described as a multiplication of the object wave spectrum with the phase plate $\exp(i\chi(\vec{q}))$ in the back focal plane. Since in conventional TEM only the intensity of the image wave is recorded, a wave optical back-propagation to the level of the object wave is not possible.

The spatial frequency coordinate $\vec{q}$ is related to the scattering angle θ through $\theta = q\,\lambda$, where the origin of Fourier space is given by the illumination direction. The influence of the aberrations of the objective lens can now be described as a multiplication of the spectrum of the object wave with a Wave Transfer Function $\mathrm{WTF}(\vec{q})$, leading to the aberrated wave spectrum

$$O'(\vec{q}) = O(\vec{q}) \cdot \mathrm{WTF}(\vec{q}) = O(\vec{q}) \cdot e^{i\chi(\vec{q})} \cdot E_S(\vec{q}) \cdot E_C(\vec{q}) \tag{9.4}$$

with $\mathrm{WTF}(\vec{q})$ being the Fourier transform of $\mathrm{PSF}(\vec{r})$. Different spatial frequency components of the object wave spectrum are shifted differently in phase due to the wave aberration $W(\vec{q}) = \lambda/2\pi \cdot \chi(\vec{q})$. In addition, even if highly coherent field emission guns are used, the high frequency information is damped by the envelope functions $\mathrm{E}_S(\vec{q})$ and $\mathrm{E}_C(\vec{q})$, respectively, which stem from the limited spatial and temporal coherence of the illumination. We will address the description of this incoherent attenuation later in this section. In a second step, the objective lens brings all the frequency components to interference again, thereby forming the image wave. This is described by the inverse Fourier transform of the distorted object wave spectrum given by

$$b(\vec{r}) = A(\vec{r}) \cdot e^{i\phi(\vec{r})} = \mathbf{FT}^{-1}\{O'(\vec{q})\} = \int O(\vec{q}) \cdot \mathrm{WTF}(\vec{q}) \cdot e^{-2\pi i \vec{q}\vec{r}}\, \mathrm{d}\vec{q} \tag{9.5}$$

Since $\chi(\vec{q})$ is a strongly varying function of frequency $\vec{q}$, object and image waves usually do not agree. Conventionally, only the intensity $I = |A(\vec{r})|^2$ of the image wave is recorded in HRTEM. Since then the phase of the wave is lost, an unambiguous deconvolution from the transfer function of the objective lens is no longer possible. Therefore, before going for correction of aberrations, the first goal has to be the application of a wave optical approach (like holography) to store and retrieve the amplitude A as well as the phase ϕ of the electron image wave.

The influence of isoplanatic aberrations on the image wave, which can be described mathematically by a phase plate $e^{i\,\chi(\vec{q})}$ in Fourier space, can be illustrated by considering the imaging of a spherical wave emerging from a point in the object plane. Ideally, at the end of the imaging process, this spherical wave should converge again into one single point in the image plane. The aberration of the lens, however, leads to a deviation of the wave front from the ideal spherical shape and therefore introduces a blurring of the corresponding image. The wave aberration $W(\vec{q})$ is simply the geometrical deviation from the ideal shape, and $\chi(\vec{q})$ represents the corresponding phase shift. In expanding the wave aberration into a power series of q, for reasons of symmetry, only even powers in q have to be considered for the common rotational symmetric lens designs. The restriction to small scattering angles (as usual in electron microscopy) leads to the common representation of χ as a series up to the 4th order in q

$$\chi(\vec{q}) = \frac{2\pi}{\lambda} \cdot \left(\frac{1}{4}\, C_S\, \lambda^4\, q^4 + \frac{1}{2}\, \Delta z\, \lambda^2\, q^2\right) \tag{9.6}$$

C_S denotes the coefficient of spherical aberration characteristic of the objective lens. The coefficient of the quadratic term in q is the defocus Δz; it represents the free parameter that can be used to change the transfer function of the microscope. Underfocussing of the lens results in a negative value for the defocus Δz in this nomenclature.

In the case of deviations from rotational symmetry, the description additionally has to include terms that depend on the direction in Fourier space, i.e., the series has now to be calculated using the complex coordinates $q \cdot e^{i\,\alpha}$ which includes the azimuth coordinate α in the back focal plane. The phase shift of the aberrated wave front can then be described as[213,240,385,508]

$$\begin{aligned}\chi(\vec{q}) = \frac{2\pi}{\lambda} \Bigg(&\frac{1}{4}\, C_S^4\, \lambda^4\, q^4 + \frac{1}{2}\, \Delta z\, \lambda^2\, q^2 + \frac{1}{2}\, C_A\, \lambda^2\, q^2 \cdot \cos[2(\alpha - \alpha_A)] \\ &+ \frac{1}{3}\, C_3\, \lambda^3\, q^4 \cdot \cos[3(\alpha - \alpha_3)] + B_C\, q^3 \cdot \cos(\alpha - \alpha_C)\Bigg)\end{aligned} \tag{9.7}$$

The first additional quadratic term in q denotes a twofold astigmatism that can be understood as a defocus dependent on azimuth; $2C_A$ is the distance between the two focal lines, while α_A is the azimuthal direction of the astigmatism. Using the objective lens stigmators, the two line foci can be equalized by adjusting the respective granularity contrast of an amorphous area. The two terms anti-symmetric in q represent axial coma ($\sim e^{i\,\alpha}$) with coefficient B_C and azimuth α_C, and threefold astigmatism ($\sim e^{3\,i\,\alpha}$) denoted by C_3 and α_3. Axial coma is present when the illumination direction does not coincide with the optic axis of the electron optical set-up. The coma-free axis is usually found by symmetrically tilting the illumination in perpendicular directions, thus determining the direction with maximum symmetry with respect to the apparent granularity of an amorphous region. While earlier workers reported the influence of three fold astigmatism to be negligible in the coma free set-up,[508] more recently a considerable influence of this aberration for resolutions below 0.2 nm has been reported.[213,240]

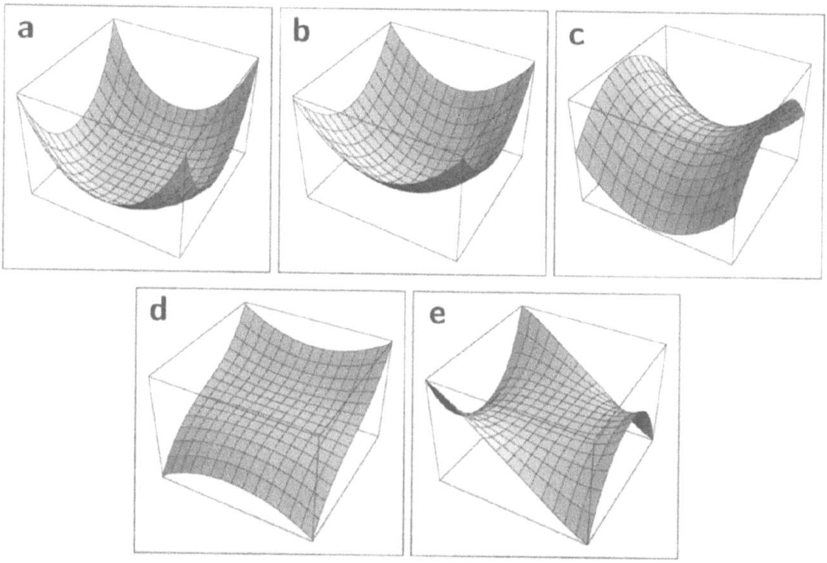

Figure 9.2. Characteristics of the most important aberrations in TEM: a) spherical aberration, b) defocus, c) astigmatism, d) axial coma, e) threefold astigmatism.

A comprehensive description can be found in Ref. 385. The influence of the different components to the wave aberration is shown in Fig. 9.2.

The partial coherence of the illumination leads to damping of high frequency information and therefore sets the general limit to the available information. The energy spread ΔE of real electron guns, together with instabilities ΔU_A of the accelerating voltage U_A and ΔI of the objective lens current I, results in the envelope of temporal coherence given by[357]

$$E_C(\vec{q}) = \exp\left\{ - \left(\frac{\pi\,\lambda\,q^2\,C_C\,f_r}{4\sqrt{\ln(2)}} \right)^2 \cdot \left[\left(\frac{\Delta E}{E} \right)^2 + \left(\frac{\Delta U_A}{U_A} \right)^2 + \left(\frac{2\,\Delta I}{I} \right)^2 \right] \right\} \qquad (9.8)$$

C_C is the coefficient of chromatic aberration and represents a property of a given objective lens; $f_r = (1 + E/E_0)/(1 + 0.5E/E_0)$ is a relativistic correction with the electron rest energy E_0. If an illumination aperture with angle Θ_C is used, additional damping with the envelope of spatial coherency occurs which is to first order given by[111,357]

$$E_S(\vec{q}) = \exp\left(\frac{-\pi^2}{\ln 2} \cdot (C_S\,\lambda^2\,q^3 + \Delta z\,q)^2 \cdot \Theta_C^2 \right) \qquad (9.9)$$

This stems from the fact that the specimen is illuminated with a distribution of wave vectors $\vec{k}$ filling the illumination aperture. The resulting image contrast is obtained by averaging over all contributions. Note, however, that the shape of this envelope is dependent on defocus, and can therefore be changed and optimized by the microscopist.

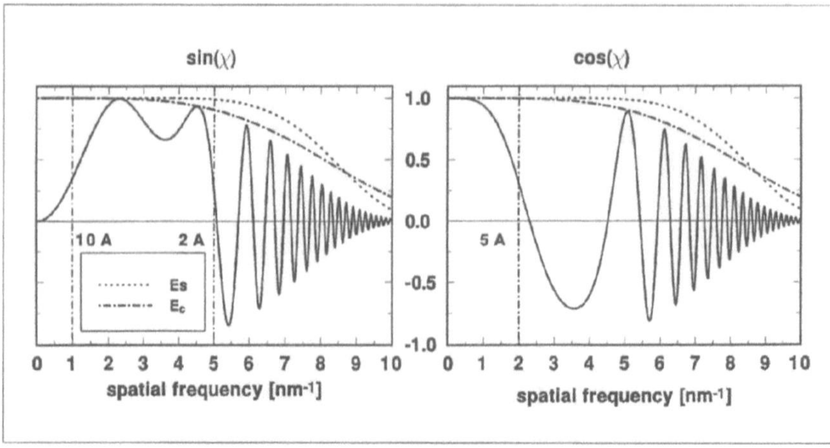

Figure 9.3. Phase Contrast Transfer Function (PCTF) $\sin\chi$ and Amplitude Contrast Transfer Function (ACTF) $\cos\chi$ of the Philips CM30ST FEG microscope at Scherzer defocus (-60 nm). The rapid oscillations of the contrast transfer in the high frequency domain make the interpretation of the corresponding image details difficult. The point resolution is defined as the first zero crossing of the $\sin\chi$ function while the information limit is determined by the damping of information with the envelopes of spatial and temporal coherence $E_s \cdot E_c$, respectively.

The resulting transfer characteristics in Fourier space with respect to both, the real ($\cos(\chi)$) and imaginary ($\sin(\chi)$) parts are shown for the commonly used "Scherzer focus" condition in Fig. 9.3. The parameters used correspond to a typical medium voltage high resolution microscope; in this case the data of a Philips CM30 FEG TEM with $C_S = 1.2\,\text{mm}$, $U_A = 300\,\text{kV}$, $C_C = 1.4\,\text{mm}$, $\Delta E = 1\,\text{eV}$ and $\Theta_C = 0.1\,\text{mrad}$ at Scherzer defocus of $\Delta z_S = -60\,\text{nm}$ were assumed.

As an example to illustrate the contrast transfer from the object wave to the image wave let us consider a weak scattering object that allows us to linearize the object wave function

$$o(\vec{r}) = a(\vec{r}) \cdot e^{i\varphi(\vec{r})} = 1 - t(\vec{r}) + i\varphi(\vec{r}) \tag{9.10}$$

Keep in mind that for most samples of realistic thickness, this assumption will fail and that we use this model only for demonstration of the basic principles. Without loosing generality, let us consider a structure offering only one spatial frequency q_0 in transparency $(1 - t)$ and phase φ according to

$$\begin{aligned} t(\vec{r}) &= t_0 \cdot \cos(2\pi\,\vec{q}_0\,\vec{r} + \epsilon_t) \qquad \text{for } t_0 \ll 1 \\ \varphi(\vec{r}) &= \varphi_0 \cdot \cos(2\pi\,\vec{q}_0\,\vec{r} + \epsilon_p) \qquad \text{for } \varphi_0 \ll 2\pi \end{aligned} \tag{9.11}$$

Later on, we will describe any structure by Fourier synthesis of these spatial components. Following the above description, the aberrations of the objective lens lead to the image wave

$$\begin{aligned} b(\vec{q}) = 1 &- \frac{t_0}{2}\left(e^{2\pi i\vec{q}_0\vec{r} + i\chi(\vec{q}_0) + i\epsilon_t} + e^{-2\pi i\vec{q}_0\vec{r} + i\chi(-\vec{q}_0) - i\epsilon_t}\right) \\ &+ \frac{i\varphi_0}{2}\left(e^{2\pi i\vec{q}_0\vec{r} + i\chi(\vec{q}_0) + i\epsilon_p} + e^{-2\pi i\vec{q}_0\vec{r} + i\chi(-\vec{q}_0) - i\epsilon_p}\right) \end{aligned} \tag{9.12}$$

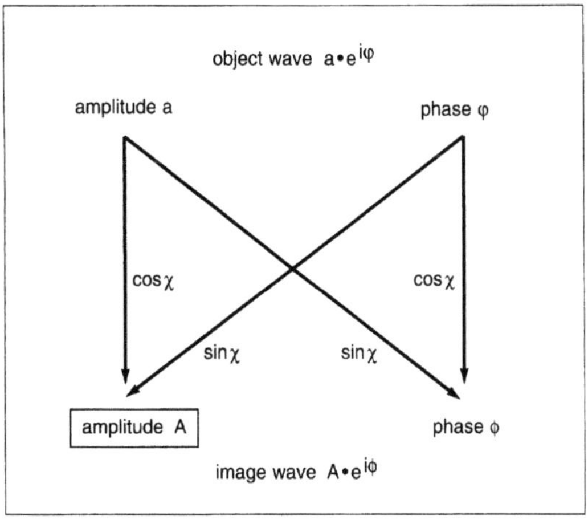

Figure 9.4. For a weakly scattering object, the symmetric aberrations lead to a complex exchange of amplitude and phase information following the $\sin\chi$, while the direct transfer is governed by the $\cos\chi$ function.

If we separate χ into a symmetric part χ_s and an anti-symmetric part χ_a with respect to $\vec{q}$ according to

$$\begin{aligned}\chi_s(\vec{q}) &= \frac{1}{2}\cdot[\chi(\vec{q})+\chi(-\vec{q})]\\ \chi_a(\vec{q}) &= \frac{1}{2}\cdot[\chi(\vec{q})-\chi(-\vec{q})]\end{aligned}\tag{9.13}$$

the image wave can be written as

$$\begin{aligned}b(\vec{r}) &= 1 - t_0\cdot e^{i\chi_s(\vec{q}_0)}\cdot\cos(2\pi\,\vec{q}_0\,\vec{r}+\chi_a(\vec{q}_0)+\epsilon_t)\\ &\quad + i\varphi_0\cdot e^{i\chi_s(\vec{q}_0)}\cdot\cos(2\pi\,\vec{q}_0\,\vec{r}+\chi_a(\vec{q}_0)+\epsilon_p)\\ &= 1-[\ t_0\cdot\cos\chi_s(\vec{q}_0)\cdot\cos(2\pi\,\vec{q}_0\,\vec{r}+\chi_a(\vec{q}_0)+\epsilon_t)\\ &\qquad +\varphi_0\cdot\sin\chi_s(\vec{q}_0)\ \cdot\cos(2\pi\,\vec{q}_0\,\vec{r}+\chi_a(\vec{q}_0)+\epsilon_t)\,]\\ &\quad + i\,[\ \varphi_0\cdot\cos\chi_s(\vec{q}_0)\cdot\cos(2\pi\,\vec{q}_0\,\vec{r}+\chi_a(\vec{q}_0)+\epsilon_p)\\ &\qquad -t_0\cdot\sin\chi_s(\vec{q}_0)\ \cdot\cos(2\pi\,\vec{q}_0\,\vec{r}+\chi_a(\vec{q}_0)+\epsilon_p)\,]\end{aligned}\tag{9.14}$$

A comparison with the linearized object wave (Eq. 9.10) reveals that symmetric aberrations (spherical aberration, defocus and astigmatism) lead to an exchange of phase and amplitude between object and image wave. Fig. 9.4 contains an illustration of the character of this transfer. The direct transfer of information is guided by the real part of the aberrations, whereas the imaginary part leads to a crosstalk between amplitudes and phases of object and image wave, respectively. Since in conventional microscopy only $|A|^2$ is registered, $\cos(\chi_s)$ represents the so-called Amplitude Contrast Transfer Function (ACTF) while $\sin(\chi_s)$ is addressed as Phase Contrast Transfer Function (PCTF). In addition, anti-symmetric aberrations like threefold astigmatism and axial coma give rise to a phase shift $\chi_a(\vec{q}_0)$ anti-symmetrical in $\vec{q}_0$.

Considering only the most commonly regarded symmetric aberrations, the implications for imaging in HRTEM become clear: to ideally image the amplitude component of the wave, a value of $\chi = 0$ is needed; whereas, for the registration of the phase the "Zernike" condition $\chi = \pi/2$ is the obvious choice. Generally, this can not be achieved over the whole range of spatial frequencies which real specimens incorporate. Ideal

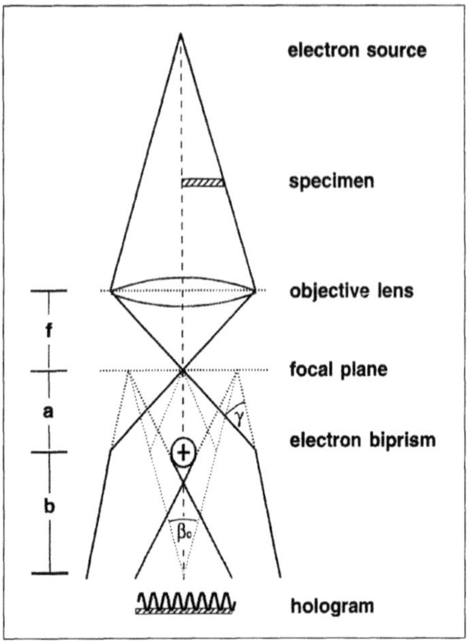

Figure 9.5. Schematic path of rays for the acquisition of an off-axis electron hologram. By applying a positive voltage to the biprism filament, object and reference wave interfere in the intermediate image plane.

contrast transfer for both, amplitude and phase, is only possible in limited pass bands. The definition of Scherzer defocus of $\Delta z_S = -\sqrt{3\,\lambda\,C_S/2}$ aims at counterbalancing spherical aberration and defocus in such a way that a broad range of spatial frequencies (the Scherzer band) is evenly transferred with the optimum phase (Fig. 9.3). Since $\sin(\chi)$ is inevitably close to zero for the low spatial frequencies, large area phase contrast is not possible. Additionally, the heavy oscillations that lead to contrast reversals in the high frequency domain make a direct interpretation of the fine image details difficult. The definition of the point resolution limit as the first zero crossing of the $\sin(\chi_s)$ function reflects this shortcoming. Furthermore, there is always an intermediate spatial frequency regime with a mixture of amplitude and phase contrast. It is therefore necessary to compare the micrographs with image simulations for unambiguous interpretation. Keeping this in mind, HRTEM has been successfully applied and established as an important tool for microscopic characterization in material science, but from the point of view of the optical performance it is still far from being ideal.

This had been realized by Gabor already in 1948.[129] He proposed to use an interferometric approach to record both amplitude and phase of the image wave. During the reconstruction step, a deconvolution of the aberrations could then be performed.[130] Finally, amplitude and phase of the object wave can be investigated in a direct way, ranging from low frequencies up to the information limit given by the damping due to partial coherence.

3. Off-axis electron holography: Principle

To create an off-axis electron hologram, a Möllenstedt-type biprism[312] (a thin conducting wire with a thickness preferably less than 1 μm) is inserted between the focal plane of the objective lens and the first intermediate lens of the microscope (Fig. 9.5). The sample is located in the object plane in such a way that half of the field of view stays free, thereby dividing the electron wave into an image wave and an undisturbed

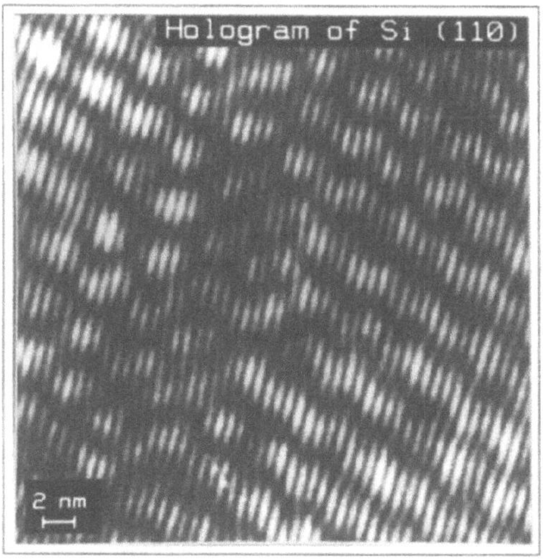

Figure 9.6. Enlarged region of a hologram of Si imaged in [110] projection. The phase of the image wave is captured as a bending of the hologram fringes (spacing 0.07 nm) while the amplitude leads to a contrast modulation of the interference pattern.

reference wave. By applying a voltage U_f to the biprism filament, the two waves are brought to interference. The influence of the biprism can be considered as a coherent splitting of the illumination into two virtual sources that are tilted by the angle $\beta_c/2 = q_c\,\lambda/2$ with respect to the optical axis, with only one of them illuminating the specimen. The hologram is then produced in the image plane as an interference pattern between the tilted object and reference wave,

$$b(\vec{r}) = A(\vec{r}) \cdot e^{i\,\phi(\vec{r})} \cdot e^{2\pi i \vec{q}_c \vec{r}/2} \qquad \text{and} \qquad e^{-2\pi i \vec{q}_c \vec{r}/2} \tag{9.15}$$

respectively. The resulting cosinusoidal intensity distribution with the carrier frequency q_c is given by

$$I(\vec{r}) = 1 + A^2(\vec{r}) + 2A(\vec{r}) \cdot \cos[2\pi\, \vec{q}_c \vec{r} + \phi(\vec{r})] \tag{9.16}$$

Since only elastically scattered electrons will coherently contribute to the resulting interference pattern, the information contained in the hologram fringes can be considered to be "zero-loss" filtered.[167] As an example, Fig. 9.6 shows an enlarged part of a hologram of Si in [110] orientation, taken with a Philips EM420 ST FEG microscope. The amplitude of the image wave is recorded as a contrast modulation, while the phase leads to a local bending of the hologram fringes, which act as a carrier wave for the image wave. The separation of the information on the carrier wave from the simultaneously recorded conventional high resolution image $|A(r)|^2$ is carried out in Fourier space, where the spectrum of the hologram is given as

$$\begin{aligned} I(\vec{q}) \;=\; & \delta_{\vec{q}} + \delta_{\vec{q}} \otimes \mathbf{FT}\left\{A^2(\vec{r})\right\} && \text{AC} \\ & + \delta_{(\vec{q}+\vec{q}_c)} \otimes \mathbf{FT}\left\{A(\vec{r}) \cdot e^{i\phi(\vec{r})}\right\} && \text{SB1} \\ & + \delta_{(\vec{q}-\vec{q}_c)} \otimes \mathbf{FT}\left\{A(\vec{r}) \cdot e^{-i\phi(\vec{r})}\right\} && \text{SB2} \end{aligned} \tag{9.17}$$

The spectrum consists of a center or autocorrelation part that represents the diffractogram of the conventional image and two sidebands separated by the carrier

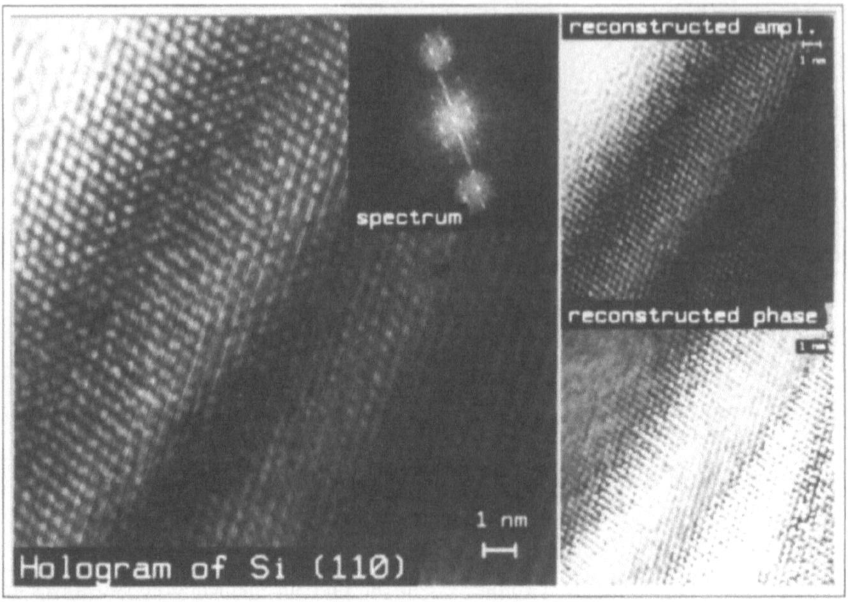

Figure 9.7. After reconstruction of the image wave by selecting the wave spectrum found in the sideband of the hologram spectrum (inset), amplitude and phase are available separately. To allow the separation in Fourier space, the fringe spacing has to be sufficiently small.

frequency q_c. Each of these sidebands contains the full complex spectrum of the image wave (Fig. 9.7, inset). Either one of them can now be separated by an appropriate aperture and centered at the origin of Fourier space. The inverse Fourier transform then yields the complete image wave separated in amplitude and phase (Fig. 9.7). The reconstruction can be performed either numerically from a digitized hologram, or in an analog way on the light-optical bench. In the later case, after illuminating the hologram with coherent light, the image wave spectrum can be selected with a physical aperture in the back focal plane of the lens used. To allow the separation in Fourier space, the carrier frequency has to be 2-3 times higher than the maximum frequency contained in the recorded wave. Specifically, this means that every detail to be resolved after reconstruction has to be sampled with 2-3 hologram fringes. For high resolution applications down to 0.1 nm, the stability and stray field requirements for the TEM rise accordingly, since fringes with spacings below 0.03 nm have to be recorded at sufficient contrast.

The recorded interference pattern is characterized by the fringe spacing σ_{hol} and the width w_{hol} of the hologram. Since object and reference wave interfere at the angle β_c (Fig. 9.5), the fringe spacing can be calculated from simple geometrical considerations as

$$\sigma_{hol} = \frac{\lambda}{\beta_c} = \frac{\lambda\,(a+b)}{2\gamma\,a} \tag{9.18}$$

As indicated in Fig. 9.5, $\gamma = \gamma_0\,U_f$ is the deflection angle of the biprism at the filament voltage U_f with the specific deflection angle of the biprism[252]

$$\gamma_0 = \frac{f_r\, \pi}{2\, U_A\, \ln(r_{bp}/r_f)} \tag{9.19}$$

at accelerating voltage U_A and the previously introduced relativistic correction f_r; r_f is the radius of the filament and r_{bp} its distance from the surrounding ground electrode. The width of the hologram is then given by[259]

$$w_{hol} = 2\, \gamma_0\, U_f\, b - 2\, r_f \cdot \frac{b+a}{a} \tag{9.20}$$

The second term represents the width of the shadow of the biprism filament as projected into the intermediate image plane. To relate these distances to the level of the object, we have to divide by the magnification $M = (a+b)/f$ of the objective lens with focal length f resulting in

$$\sigma_{obj} = \frac{\lambda f}{2\, \gamma_0\, U_f\, a} \qquad \text{and} \qquad w_{obj} = \frac{2\, \gamma_0\, U_f\, b\, f}{a+b} - \frac{2\, r_f\, f}{a} \tag{9.21}$$

for the fringe spacing σ_{obj} and the hologram width w_{obj}. At high magnifications, the focal length of the objective lens is more or less constant so that the fringe spacing, mainly independent from the position of the intermediate image plane used, depends only on the filament voltage U_f. Therefore, for a given microscope, a biprism constant S_{bp} can be defined with $\sigma_{obj} = S_{bp}/U_f$. Since the fringe spacing is predetermined by the resolution desired after reconstruction, the field of view can only be changed by an appropriate excitation of the first intermediate lens, thereby changing the distance b between biprism and intermediate image plane.

In any case, the maximum width of a hologram is limited by the area that can be illuminated coherently. Due to the partial coherence of the illumination, holograms can only be recorded with limited contrast $\mu = \mu_s\, \mu_c$ introduced by the spatial (μ_s) and temporal coherence (μ_c) of the electron optical set-up* resulting in

$$I(\vec{r}) = 1 + A^2(\vec{r}) + 2\, \mu\, A(\vec{r}) \cdot \cos[2\, \pi\, \vec{q}_c\, \vec{r} + \phi(\vec{r})] \tag{9.22}$$

for the intensity distribution of the hologram. The fringe contrast μ represents the dominating factor for the precision at which the image wave can be retrieved. Specifically, the phase detection limit $\Delta\phi$ of the reconstruction is given by

$$\Delta\phi = \frac{c}{\mu\sqrt{N_e}} \tag{9.23}$$

Depending on the situation considered, c is a constant between $\sqrt{2}$ and $\sqrt{14}$,[87,260] and N_e the number of electrons per resolved pixel assuming an ideal (i.e., not noise adding) detector. The influence of spatial coherence on the fringe contrast is explained by the finite size of the electron source, since every point in the source leads to a spatially shifted interference pattern. Let us assume an electron source with normalized Gaussian current density distribution

*Additional reduction of fringe contrast is introduced by instrumental instabilities, inelastical scattering from the specimen and the contrast transfer function of the detector.[252,260,352]

$$i(\vec{r}) = \frac{1}{\pi\rho^2} \cdot e^{-r^2/\rho^2} \tag{9.24}$$

that illuminates the hologram plane at a distance R, where R is large compared to ρ; $\vec{r}$ is the coordinate in the plane of the source and $\rho^2 = \langle r^2 \rangle$ describes the source size. Following the van Cittert-Zernike theorem,[30] the degree of coherence between two points separated by $\vec{d}_{12}$ in the hologram plane which have an angular distance $\alpha = d_{12}/R$ as seen from the source is given by the Fourier transform of $i(\vec{r})$, as

$$\mu_s(k\alpha) = \iint i(\vec{r}) \cdot e^{2\pi i k \vec{\alpha} \vec{r}} \, d\vec{r} = e^{-(\pi k \rho \alpha)^2} \tag{9.25}$$

Consequently, the degree of spatial coherence μ_s is only dependent on the angular distance α of two interfering points in the hologram plane and is the same in the whole interference area. Since α increases with the width of the hologram, the fringe contrast is decreasing. Simply reducing ρ by increasing the magnification of the apparent electron source does not give any advantage, since the current in the image plane will decrease resulting in an undesirable increase in exposure time. The total "coherent" current is given by the axial gun brightness $B_0 = I/(A\,\Omega)$ that is defined as the current emitted from the gun area A into the solid angle Ω which is an optical invariant. Taking $\pi\rho^2$ as source area and $\pi\alpha^2$ as solid angle, this current is given as:[266]

$$I = B_0 \cdot (\pi\,\rho\,\alpha)^2 = -B_0 \cdot \frac{\ln\mu}{k^2} \tag{9.26}$$

The application of highly coherent field emission guns is therefore mandatory for electron holography to avoid prolonged exposure times which increase the demands on the stability of the microscope. Note that the above formula can be used for a rough estimate of the axial gun brightness: While monitoring the fringe contrast in a hologram of given angular width α, the condensor lens is increasingly focused (thereby increasing the effective source size ρ) until the fringe contrast vanishes below visibility (e.g., $\mu \le 0.05 \Rightarrow \ln\mu = -3$). B_0 can then be estimated from the current within the hologram area using the calibrated exposure meter of the microscope.[266]

A similar argument holds for the contrast reduction due to temporal coherence introduced by the finite energy distribution of the electrons. The degree of temporal coherence μ_c is then derived as the Fourier transform of the energy distribution function. It is therefore dependent on the position coordinate in the hologram and determines the maximum number of fringes possible. As experimentally shown by Schmid,[395] the use of field emission guns makes it possible to produce electron interferences up to the 160000 th order. Since usually only about 1000 fringes are needed in high resolution off-axis holography, $\mu_c = 1$ can be obtained in most experimental situations.

Some care has to be taken for the interpretation of border areas in off-axis holograms. The biprism filament represents a foreign body in the path of the electron wave that is neither located in an image nor in a diffraction plane. Therefore, vignetting and Fresnel diffraction from the biprism have to be taken into account.

The effect of vignetting is illustrated in Fig. 9.8: Depending on the distance from the optic axis, higher spatial frequencies from one side of the Fourier spectrum in the back focal plane do not contribute to the formation of the image wave. All the frequency components higher than q_{max} are either filtered out by the filament itself or,

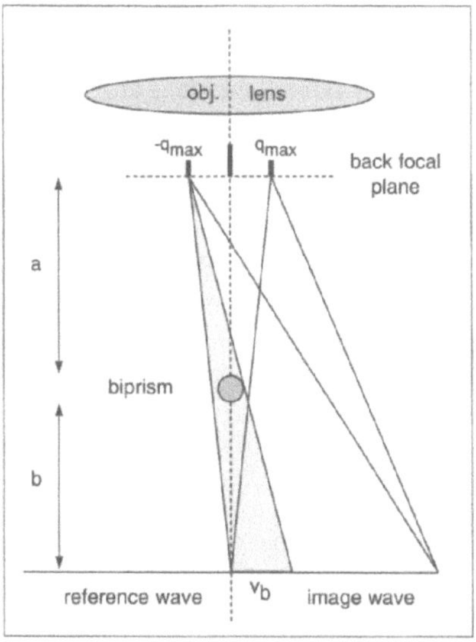

Figure 9.8. The effect of vignetting from the biprism filament. Single sideband imaging occurs in the border area of width ν_b. The respective area therefore has to be discarded for aberration correction.

when applying a biprism voltage, are shifted away from the hologram since they pass at the wrong side of the biprism. This means that in a stripe of width ν_b at the hologram edge only single-sideband imaging occurs. As shown in Ref. 269, the unambiguous retrieval of the object wave function is therefore no longer possible for a specimen that cannot be described by the weak phase object approximation. The width of this area is given by[269]

$$\nu_{b,img} = \frac{b\, q_{max}\, f}{a\, k} \tag{9.27}$$

where q_{max} is the maximum spatial frequency to be reconstructed. Relating this to the object plane yields

$$\nu_{b,obj} = \frac{b\, q_{max}\, f^2}{a(a+b)k} \tag{9.28}$$

The effective width of the hologram $\hat{w}_{obj} = w_{obj} - \nu_{b,obj}$ can then be calculated as

$$\hat{w}_{obj} = w_{obj}\left(1 - \frac{q_{max}}{q_c}\right) - 2\, r_f \cdot \frac{f\, q_{max}}{a\, q_c} \tag{9.29}$$

The first term shows that, for the usual requirement of $q_c = 3\, q_{max}$, only 2/3 of the hologram area on the far side of the biprism contains the full information on the image wave. This region is additionally narrowed by the shadow of the filament represented by the second term. Since the biprism is located at a distance b above the image plane, Fresnel diffraction at the edges of the filament shows up.[357]. This distortion produces artificial structures in amplitude and phase at the edges parallel to the biprism wire. A detailed analysis reveals[269] that at both sides of the hologram a stripe of width

$$p_{fp} = \frac{1}{\pi\,\Delta\phi} \cdot \sqrt{\frac{f^2\,b}{4\,k\,a(a+b)}} \qquad \text{and} \qquad p_{fa} = \frac{1}{\pi\,\Delta A/A} \cdot \sqrt{\frac{f^2\,b}{4\,k\,a(a+b)}} \tag{9.30}$$

has to be discarded in the reconstructed field of view if a precision of $\Delta\phi$ in the phase and a relative accuracy of $\Delta A/A$ in the amplitude has to be maintained.

Finally, one has to consider the fact that the information on every point in the object plane is smeared out to an error disc that corresponds to $\mathrm{PSF}(r)$ in the image plane. Since only part of this information is recorded in the border area of the hologram, the deconvolution of the corresponding image wave details is not possible in this region.[265]

4. Correction of aberrations

Since electron holography allows the retrieval of the complete information on both amplitude and phase of the image wave, the influence of the wave aberrations can be removed by inverting the imaging process of the objective lens described in Section 2. This is preferably performed numerically in the computer, since the reconstruction parameters can be changed and adapted to the experimental situation in a highly flexible way. The deconvolution of the transfer function $\mathrm{PSF}(\vec{r})$ of the electron microscope is performed in Fourier space by division of the aberrated wave spectrum with $\mathrm{WTF}(\vec{q})$, which has to be modelled with appropriate precision. Thereafter, the complex exchange between amplitude and phase of object and image wave introduced by the microscope is reversed, hence resolution is established up to the information limit. The precision of this correction is determined, in the axially symmetric case, by the knowledge of the experimental conditions such as the exact magnification, the coefficient of spherical aberration C_S and the defocus Δz. A highly accurate determination of these parameters is therefore of utmost importance for the success of the correction. Up to now, the most successful measurement techniques for the aberrations coefficients require a weak phase object whose wave function is mainly transferred with $\sin(\chi)$. The two main approaches to determining the coefficients of the wave aberration rely either on the analysis of tilt induced image displacements[243] or on diffractogram analysis in Fourier space,[235] and have been extensively investigated so far.[51,213,232,240,385,508] The most accurate method relies on diffractogram analysis of images from thin amorphous layers that represent weak phase objects (WPO) and, due to their random structure, offer a wide continuous range of spatial frequencies. The diffractogram of such an image directly yields the zero crossings of $\sin(\chi)$ that can be used for measuring C_S and Δz. In the case of a coma free set-up, and neglecting the influence of threefold astigmatism, the zero crossings are found at

$$n_i = \frac{1}{2}\,C_S\,\lambda^3\,q_i^4 - \Delta z\,\lambda\,q_i^2 \tag{9.31}$$

where i describes the index of the transfer gap and the n_i are integers. In the original proposal,[235] this equation is divided by q_i^2 and after plotting n^2/q^2 versus q^2, C_S and Δz are determined by linear regression. The magnification has to be calibrated carefully, e.g., by measuring the spacings of reflections of crystalline structures which can be performed with very high accuracy.[91] Various refinements of the regression technique have been developed[51,168] to avoid the different weighting of each zero that occurs in the original 'Krivanek method.' If twofold astigmatism is present, this procedure has to be applied to different angular "sectors" in Fourier space to determine both magnitude and azimuth of the astigmatism.[239]

Since the diffractogram intensity of a WPO is transferred with $\sin^2(\chi)$, a symmetric function in q, axial coma and threefold astigmatism that are anti-symmetric in $\vec{q}$ cannot be unambiguously determined from one single image, but can only be assessed by taking a tilt series of one image area. Fig. 9.9 shows an example of such a so called "Zemlin tableau," a series of diffractograms from images taken by symmetrically tilting the beam around the (assumed) coma free axis. The corresponding shift of the origin in Fourier space then leads to an additional, apparent astigmatism, and a thorough analysis of the corresponding diffractograms can be used to reveal all the aberration coefficients needed.[213,240,385,508]

For a demonstration of the holographic capability to correct aberrations, Fig. 9.10 shows a hologram of an amorphous carbon foil taken with a Philips EM420 ST and a slow scan CCD camera.[352] The fringe spacing amounts to 0.1 nm. In the autocorrelation part of the hologram spectrum (inset), a strong defocus combined with astigmatism is visible. In addition, the reconstructed phase of the image wave carries an obvious distortion, indicated by the strong long-range phase modulation that is visible as phase jumps in the representation between $-\pi$ and π. The distortion can be described as a slowly varying, additional phase term $\theta(\vec{r})$ in the reconstructed wave (see also Section 5, page 133ff.)

$$b(\vec{r}) = A(\vec{r}) \cdot e^{i\,[\phi(\vec{r})+\theta(\vec{r})]} \tag{9.32}$$

that covers the phase information on the carbon foil. It is introduced by an imperfect flatness of the reference wave, stemming from inhomogenities in charge and thickness of the biprism wire, as well as from distortions of the projector lenses or any parasitic local charging in the path of the electrons (e.g., at the apertures used). The distortion can be extracted from the reconstructed phase of a reference hologram taken under the same EM conditions after removing the specimen from the hologram area (Fig. 9.11). A correction is then possible by a complex division of the images in real space,[87,353] thus revealing the distortion free phase of the image wave. An additional contribution to the distortion is introduced by the numerical precision at which the exact (subpixel) position of the sideband frequency in Fourier space can be determined. It leads to a systematic tilt of the reconstructed wave.[249,478]

In the next step, the astigmatism visible in the spectrum of the amplitude and the phase image in Fig. 9.11 can be eliminated. The upper right part of Fig. 9.12 shows the $\sin^2$ of the applied phase plate χ_{ast} by which the object wave spectrum is divided in Fourier space. Thereafter, the spectra of amplitude and phase are perfectly centrosymmetric, a feature that is used as the criterion for the success of the correction. Note that the complementarity of the spectra of amplitude and phase indicates that no information on the object wave is lost. In the final step, spherical aberration and defocus are corrected with the phase plate shown in Fig. 9.13. After correction, almost all the information is transferred into the phase of the object wave, as expected for a WPO. This shows up in both the images and spectra of the corrected amplitude and phase, respectively (Fig. 9.13). The apparent large area phase contrast reveals the non-uniform thickness of the carbon foil. In addition, a thorough wave optical analysis is now possible, e.g., the calculation of a whole focal series of this wave reconstructed from a single exposure is now possible in the computer and can be analyzed in both amplitude and phase.[124,267,470]

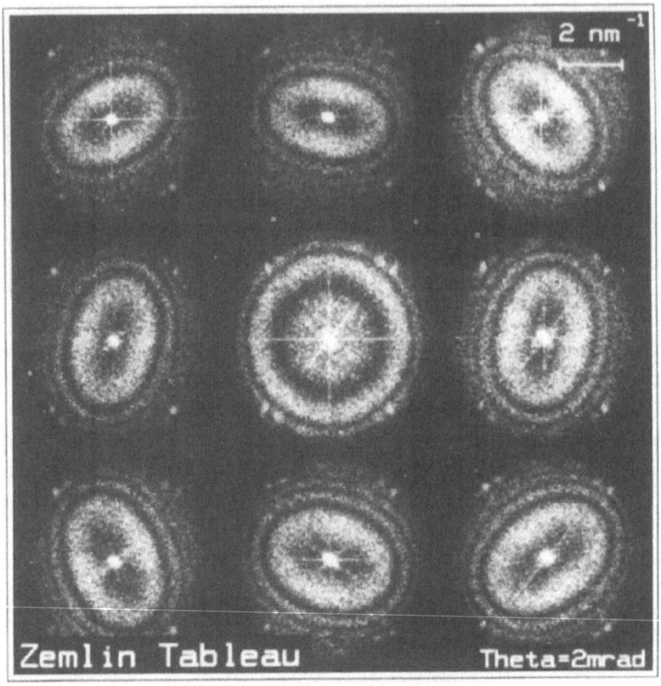

Figure 9.9. A Zemlin tableau taken to confirm the coma free alignment of the microscopical set-up (tilt angle 2 mrad). A careful evaluation of the zero crossings in the diffractograms can reveal the coefficients for spherical aberration, defocus, two- and threefold astigmatism and axial coma.

Figure 9.10. A hologram of amorphous carbon foil with a fringe spacing of 0.1 nm. The hologram spectrum reveals strong astigmatism and high underfocus. The reconstructed phase of the image wave incorporates a strong distortion.

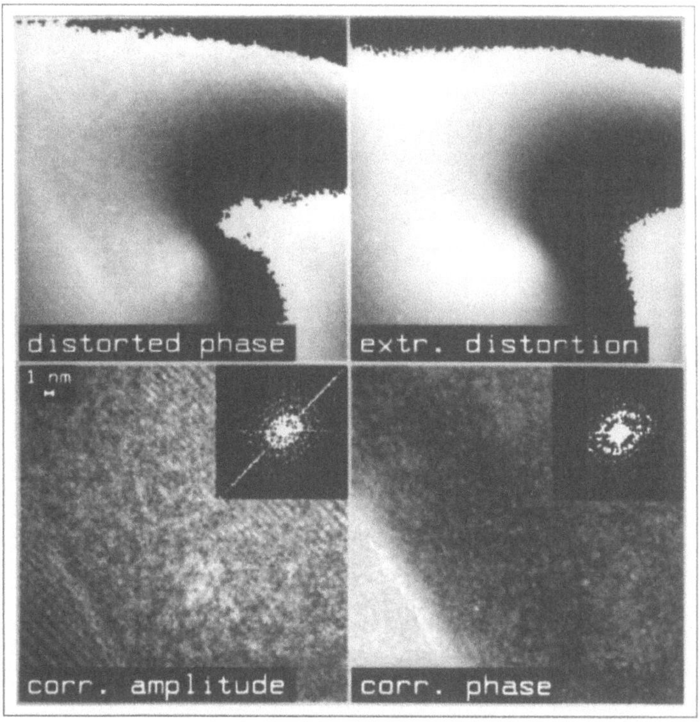

Figure 9.11. Correction of the distortion in the reconstructed phase. The distortion is extracted from an empty reference hologram (upper right). The success of the correction clearly shows in the undistorted phase (lower right)

For a thin amorphous specimen, like the one shown above, a systematic approach is possible to determine and refine the parameters used for correction of spherical aberration and defocus. After careful calibration of the magnification, the free parameters C_S and Δz can be varied starting from a reasonably close guess. The maximum contrast reduction in the amplitude image is then used as the criterion for the optimum parameter set.[125,354] Note, however, that the parameters found by this procedure represent a data set that properly matches the total transfer function present when taking the hologram. The true values for C_S and Δz might be different from an independent calibration since, e.g., an error in magnification calibration can be compensated by altering the other free parameters. The global maximum that defines the optimum parameters for correction can then in principle be found in an automatic way[123] taking due care to ensure that local extrema are avoided.

In the case of crystalline specimens that do not represent weak phase objects, up to now no automated approach for the determination of the proper parameters for correction exists. However, a comparison of the "corrected" wave with image simulations of specimen areas with known structure can reveal the wanted parameters. Since amplitude and phase of the image wave are separately available, the data set used for comparison is considerably enhanced.[329,330] Taking advantage of today's highly increased computational capacity, a computer can be used to emulate focusing, astigmatism and adjusting of beam tilt, which allows a visual on-line comparison with image

Figure 9.12. Correction of twofold astigmatism. The square of the applied phase plate is shown in the upper right part. After correction, the asymmetry in the spectra of amplitude and phase is removed (insets).

simulations.[247] Still, these tedious trial and error procedures incorporate many free parameters, such as specimen thickness and mistilt, as unknowns; in addition, quantitative comparison schemes are needed to replace the "visual" inspection.

Therefore, a reliable criterion for the optimization of the imaging parameters is highly desirable. In the case of crystalline specimen with thicknesses that still allow a description by the channelling approximation (see, e.g., Ref. 357), an interesting approach has been proposed recently.[153] In this case, the regions between the atomic columns are described as pure (not weak) phase objects and again the minimization of amplitude contrast in these areas is used to determine the proper parameters. An experimental verification of this correction scheme has not been reported yet and remains a challenging goal.

5. Optimizing the parameters for high resolution electron holography

Since in electron holography the effects of the oscillations of the transfer functions of the objective lens, $\sin\chi$ and $\cos\chi$, can be removed, the commonly used Scherzer (or pass band) condition is no longer necessary for recording the holograms. Instead, a different choice of focus is preferable that maximizes the transfer of information into the recorded image wave. Specifically, the minimization of the high frequency damping due to partial coherence and the optimization of the sampling of the respective transfer

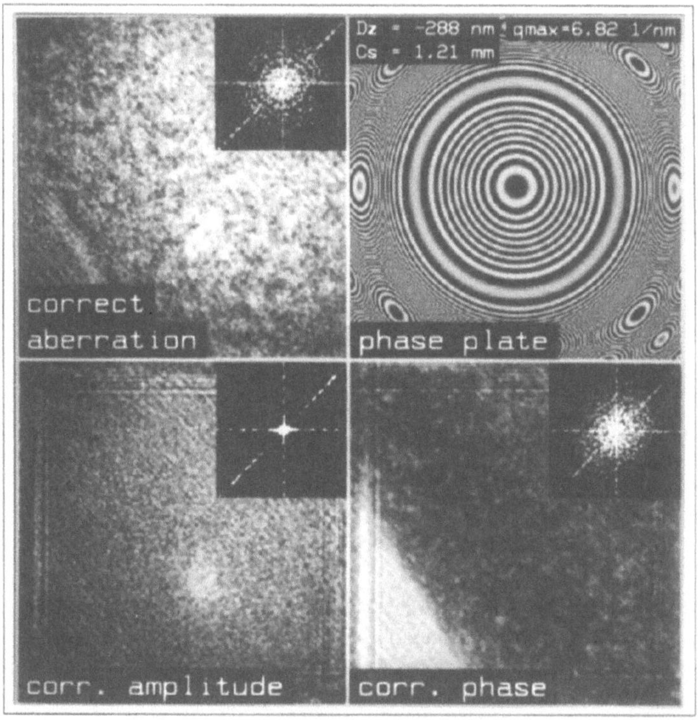

Figure 9.13. Correction of spherical aberration and defocus. Due to the correction the contrast in the amplitude image is minimized, as indicated in the spectra of the amplitude and the phase image.

functions are of major concern. Then, amplitude and phase of the object wave can later be displayed in an optimum way after aberration correction. For a desirable resolution q_{max} well inside the information limit of the TEM, the hologram is preferably recorded at the optimum defocus for holography:[262]

$$\Delta z = -0.75\, C_S \cdot \frac{q_{max}^2}{k^2} \tag{9.33}$$

Fig. 9.14 illustrates the advantages of the optimum focus with respect to the Scherzer condition at the example of the PCTF for the Philips CM30 FEG microscope. The focus is chosen in such a way that $\nabla\chi(\vec{q})$, the quantity defining the slope of the envelope of spatial coherence, is minimized within the region of spatial frequencies of interest. By allowing some attenuation for intermediate spatial frequencies, the contrast transfer is considerably enhanced for the high frequency domain up to 0.1 nm resolution. An additional achievement is the improved sampling of the WTF. Severe restrictions arise from the fact that, due to the cut-out of the sideband in Fourier space during reconstruction, the reconstructed wave is numerically represented with a reduced number of pixels compared to the input sampling capacity of the detector used. Typically, only

$$N_{rec}^2 = \frac{N_{pix}^2}{36} \tag{9.34}$$

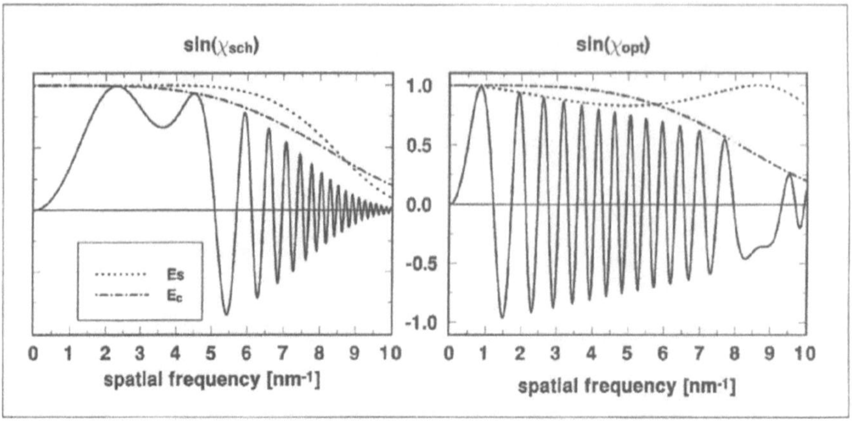

Figure 9.14. Comparison of the PCTF $\sin\chi$ in Scherzer condition ($\Delta z = -60\,\mathrm{nm}$, left) and the optimum focus for a correction up to 0.1 nm ($\Delta z = -348\,\mathrm{nm}$, right) for the Philips CM30ST FEG microscope. Parameters used: $C_S = 1.2\,\mathrm{mm}$, $U_A = 300\,\mathrm{keV}$, $C_C = 1.4\,\mathrm{mm}$, $\Delta E = 1\,\mathrm{eV}$ and $\theta_c = 0.1\,\mathrm{mrad}$.

pixels are available to reconstruct the image wave if the hologram has been digitized with N_{pix}^2 points. Therefore, sampling problems for the rapid oscillations of the transfer functions $\sin\chi$ and $\cos\chi$ in the high frequency domain arise. Due to the minimization of the slope of χ this restriction is considerably reduced in the case of the optimum focus. Fig. 9.15 demonstrates this advantage in the example of a PCTF sampled at Scherzer focus and optimum focus, with 256^2 and 128^2 pixels respectively. The phase plates shown correspond to a correction of spherical aberration and defocus up to $5\,\mathrm{nm}^{-1}$ for a Philips EM420 ST electron microscope ($U_A = 100\,\mathrm{keV}$, $C_S = 1.2\,\mathrm{mm}$). The Moiré structures in the high frequency domain for the Scherzer condition (upper part) reveal that a correction is only possible up to about $4\,\mathrm{nm}^{-1}$ while for the optimum focus (lower part) proper sampling up to the desired resolution can be achieved. Following[265], the improvement in resolution q_{opt} with respect to the value for the Scherzer focus q_{sch} is given by

$$\frac{q_{opt}}{q_{sch}} = 0.3\,N_{pix}^{1/4} \tag{9.35}$$

assuming $N_{pix}/36$ reconstructed pixels.

The use of optimized image detectors is crucial for recording high resolution electron holograms. The most important considerations are a linear sensitivity to electron exposure, the number of pixels available, the signal noise characteristic that is described by the detection quantum efficiency (DQE), and the spatial resolution in terms of the modulation transfer function (MTF) describing the contrast transfer of a specific spatial frequency into the image (see also Section 3).

Despite the unsurpassable pixel number of the conventionally used photographic plates, their non-linear response to electron radiation can cause severe problems in electron holography. Due to the non-linear transfer characteristics, higher order Fourier components of the carrier frequency can be introduced into the spectrum of the recorded hologram. The resulting distortion of the image wave spectrum leads to artifacts under correction of aberrations. A careful correction of these non-linearities is therefore nec-

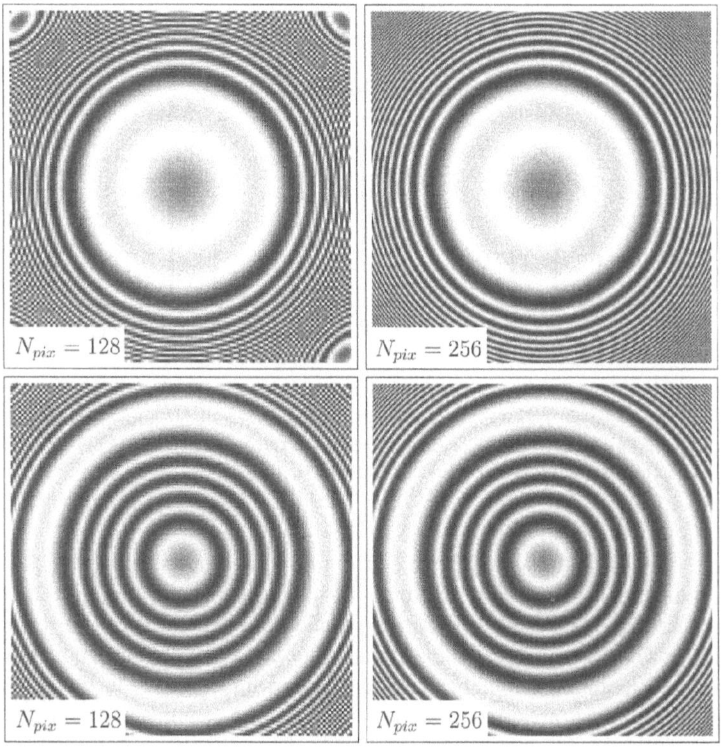

Figure 9.15. With an insufficient number of reconstructed pixels sampling errors in modelling the WTF of the microscope arise. This is indicated by the Moiré structures visible in the $\sin\chi$ function at Scherzer defocus for a Philips EM420 ST (upper part, $\Delta z = -80\,\text{nm}$, $U_A = 100\,\text{keV}$, $C_S = 1.2\,\text{mm}$). At optimum focus (-308 nm) proper sampling is achieved for a correction up to 0.2 nm (lower part).

essary before the holograms can be processed numerically.[475] Electronic image recording with modern Slow Scan CCD camera systems has the advantage of strictly linear response to electron exposure. Since camera systems with up to 2048x2048 pixels are available already, the complete replacement of the photographic plate might be achieved in near future.[79,80,89,193,238,407] With a CCD camera, images can be evaluated and processed on-line at the electron microscope, thereby avoiding time consuming work of photographic development and image digitization. In addition, it allows an immediate feedback to the operator or even digital control of the electron microscope.[237]

The DQE is defined as the relation between signal noise ratio SNR at the output and the input stage of the image detector. Since the electron distribution in the input signal follows a Poisson distribution with the mean number of electrons $\bar{N}$ and the standard distribution $\sqrt{\bar{N}}$, the DQE is given as

$$\text{DQE} = \left(\frac{\text{SNR}_0}{\text{SNR}_i}\right)^2 = \frac{\text{SNR}_0^2}{\bar{N}} \tag{9.36}$$

A detector with low DQE, therefore, needs a higher electron dose in order to achieve a good SNR in the recorded image.[193] This hampers the recording of electron holograms

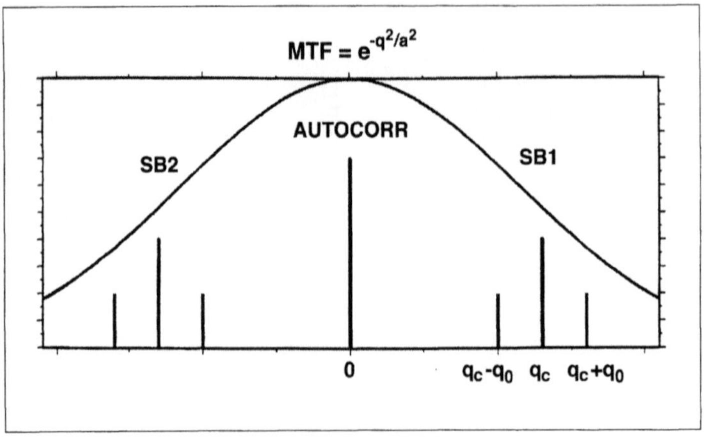

Figure 9.16. The contrast transfer function (MTF) of the detector system used can lead to damping of the information in the holographic sidebands. An asymmetric characteristic within the sideband region can introduce artifacts in the reconstruction.

with high contrast and good SNR, since the total coherent current is limited by the gun brightness. A high spatial resolution of the image detector is of utmost importance for recording high resolution holograms. To allow the separation of the holographic sidebands in Fourier space, each hologram fringe has to be sampled preferably with 4-5 pixels to maximize the number of reconstructed image points in the case of $q_c \geq 2\,q_{max}$. Fig. 9.16 illustrates the sampling scheme for the example of a typical contrast transfer function (MTF) of a Slow Scan CCD detector system for an intermediate (200-400 keV) voltage microscope. The sideband spectrum is damped by the contrast transfer $\mathrm{MTF}(\vec{q} - \vec{q_c})$ of the camera system according to

$$O'(\vec{q}) = O(\vec{q}) \cdot \mathrm{MTF}(\vec{q} - \vec{q_c}) \tag{9.37}$$

The resulting reduction of the fringe contrast of the hologram therefore reduces the precision with which the amplitude and phase of the image wave can be retrieved. In addition, the asymmetrical damping of the sideband spectrum can lead to a mixing of amplitude and phase in the reconstruction. Describing the MTF by a Gaussian $\exp(-q^2/a^2)$, the reconstructed image wave is given as

$$b'(\vec{r}) = b(\vec{r}) \otimes \mathbf{FT}^{-1}\left\{e^{(\vec{q}-\vec{q_c})^2/a^2}\right\} \tag{9.38}$$

Assuming again a phase object containing only one spatial frequency $\vec{q_0}$ according to

$$o(\vec{r}) = 1 + i\,\varphi_0 \cos(2\pi\,\vec{q_0}\,\vec{r}) \tag{9.39}$$

the reconstructed wave is given as

$$b'(\vec{r}) = \pi^2 a^2 \left[e^{-q_c^2/a^2} + \frac{i\varphi_0}{2}\cos(2\,\pi\,\vec{q_0}\,\vec{r}) \cdot \left(e^{-(\vec{q_c}+\vec{q_0})^2/a^2} + e^{-(\vec{q_c}-\vec{q_0})^2/a^2}\right) \right.$$
$$\left. - \frac{\varphi_0}{2}\cos(2\,\pi\,\vec{q_0}\,\vec{r}) \cdot \left(e^{-(\vec{q_c}+\vec{q_0})^2/a^2} - e^{-(\vec{q_c}-\vec{q_0})^2/a^2}\right) \right] \tag{9.40}$$

In addition to the damping with $\exp(-(q_c/a)^2)$, phase information is transferred to the reconstructed amplitude due to the difference in contrast transfer for the sideband reflections at $q_c \pm q_0$. The same argument holds for the amplitude of the image wave. In the case of a steep slope of the MTF within the sideband region, a MTF-"restoration" procedure,[316] i.e., a division of the hologram spectrum with the MTF of the detector system, can be applied to minimize the exchange of information. Note, however, that this will not restore the original SNR and that the information that is lost due to the damping envelope can not be retrieved.

In high resolution applications hologram fringes as narrow as 0.03 nm have to be recorded with sufficient contrast. Therefore, the geometrical arrangement of the biprism in the electron optical set-up of the microscope has to be optimized for maximum robustness against mechanical and electronic instabilities. A high magnification factor of the objective lens with respect to the intermediate image plane used to record the holograms seems to be favorable.[471] By placing the biprism even below the first intermediate lens and thus increasing the effective magnification, fringe spacings as narrow as 9.3 pm have been reported.[223] However, one also has to take other considerations into account. The width of the hologram has to be matched to the problem under investigation. Since the area that is free from artifacts decreases with increasing magnification, an indefinite increase in magnification is not desirable. It can be shown that an optimum position of the biprism can be found that optimizes all of the above concerns. We can describe any imaging system, independent of the number of lenses used, by the imaging distance L between the back focal plane of the objective lens and the intermediate image plane (e.g., $L = (a + b)$ in Fig. 9.5) with a corresponding magnification $M = L/f$. Following Ref. 270, an optimum value for the magnification is given by

$$M_{opt} = \frac{-2\, r_f\, \sigma_{obj} + \sqrt{4\, r_f^2\, \sigma_{obj}^2 + \lambda\, L\, \sigma_{obj}\, w_{hol}}}{\sigma_{obj}\, w_{hol}} \tag{9.41}$$

At this magnification, the deflection angle γ of the biprism is minimized. Therefore, the filament voltage needed can be reduced, and possible problems with voltage stability of the biprism that incorporates structures on the sub-mm scale can be avoided. Additionally, the border regions in the hologram that are distorted by vignetting, Fresnel diffraction at the filament, and the shadow of the biprism are minimized, resulting in an increased effective width $\hat{w}_{obj}$ of the hologram that can be used for aberration correction. Moreover, the sensitivity against mechanical instabilities of the biprism is considerably reduced due to the higher magnification at the level of the filament position.

Finally, for high resolution electron holography, the total magnification of the electron microscope has to be considerably increased with respect to the typical values needed in conventional HRTEM. Since for a 0.1 nm resolution, hologram fringes with a spacing of 0.03 nm have to be sampled with 4 pixels on a Slow Scan CCD camera (typical pixel size 20-24 μm), a total magnification of $4 \cdot 10^6$ has to be available at the level of the detector. In addition, the possibility of free control of the excitation of the intermediate lenses is needed to be able to adjust the electron optical set-up for optimum field of view, fringe spacing, and magnification.

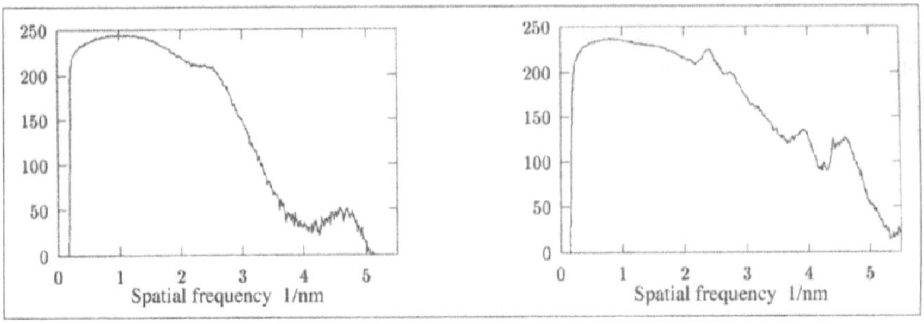

Figure 9.17. Radial intensity distribution in electron diffraction patterns of amorphous carbon (left) and tungsten (right). Tungsten delivers a much higher scattering signal in the spatial frequency region above $3\,\mathrm{nm}^{-1}$.

6. Practical examples

As pointed out in Section 4, amorphous specimens allow a straightforward scheme for correcting aberrations by either calibrating the transfer function from the transfer gaps in the diffractogram, or by minimizing the amplitude contrast in the reconstructed wave. As an example of how electron holography can push the resolution limit of the electron microscope to the information limit, Harscher et al.[168] recorded high resolution holograms of amorphous tungsten foil supported on a carbon grid with holes. The microscope used was a Philips EM420 ST with a point resolution (Scherzer limit) of 0.31 nm. The information limit of the microscope is 0.17 nm, as judged from the ring structures appearing in diffractograms from very thin areas. Amorphous tungsten was used, because it delivers a much stronger signal for scattering angles above $3.5\,\mathrm{nm}^{-1}$ as compared to, e.g., carbon as shown in Fig. 9.17. Figure 9.18 contains a diffractogram of a thin area imaged at a defocus of -291 nm which is reasonably close to the optimum focus of -308 nm for correction up to 0.2 nm. To enhance the signal around the frequency regime at $5\,\mathrm{nm}^{-1}$, as well as to enable a proper calibration of the magnification, small platinum crystallites have been evaporated on top of the tungsten foil. Due care has to be taken to image a thin area with a maximum thickness of about 5-8 nm for taking the holograms. In thicker areas, different transfer functions apply to different specimen slices perpendicular to the optical axis, which, after averaging over the corresponding defocus values, leads to blurring of the rings in the diffractogram in the high frequency regime.[168,266] The position of these rings is much more sensitive to small variations in C_S and Δz than that of the transfer gaps in the lower frequency regime. From the diffractogram, defocus and spherical aberration have been determined as 1.28 and -291 nm respectively. The authors report an accuracy of 0.01 mm and 1 nm in these values, thus leading to a residual maximum difference of $2\pi/7$ between the aberration function used for correction and the actual value when taking the hologram. The right part of Fig. 9.18 shows the frequency spectrum of amplitude and phase after aberration correction. The main contribution in the amplitude spectrum stems from the lattice planes of the platinum crystallites. The phase spectrum, in contrast, contains information from large area features down to 0.2 nm. This shows that by aberration correction, the resolution can be enhanced almost up to the measured information limit of the microscope.

For the demonstration of aberration correction for crystalline samples, the characteristic "dumbbell" shaped contrast of crystalline Si imaged in [110] projection (spacing

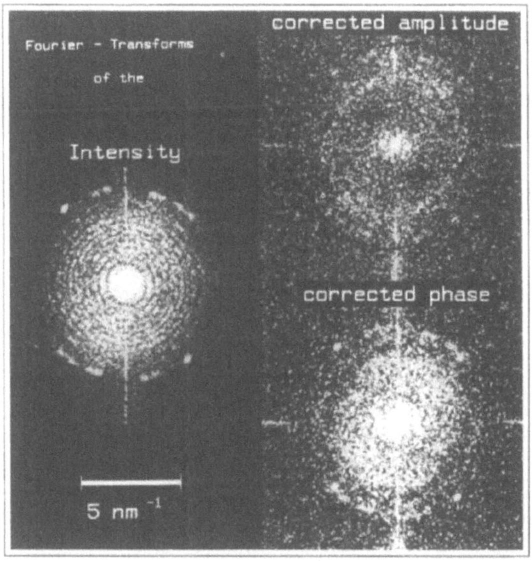

Figure 9.18. Diffractogram of amorphous tungsten foil at a defocus of -291 nm (left part). In the right part, spectra of the corrected amplitude and phase are shown. The phase signal contains information up to 0.2 nm without transfer gaps.

0.136 nm) is commonly used as a test object in HRTEM. The arrangement of the atomic columns together with an image simulated with the EMS program (see, e.g., Ref. 427), where all beams up to a spacing of 0.13 nm have been allowed to pass the objective aperture, is shown in Fig. 9.19. Because of the influence of spherical aberration and defocus, the dumbbell structure is not resolved. Moreover, the white dot contrast which can be easily mistaken as the position of the atomic columns is shifted away from the dumbbells. Fig. 9.20 (left part) shows a hologram taken by Orchowski et al.[329] using a Philips CM30 FEG microscope specially adapted for the needs of high resolution holography.[266] The hologram with a fringe spacing of 0.05 nm has been taken at an underfocus of about -120 nm. Despite the deviation from the optimum focus condition of $\Delta z = -190$ nm for a correction up to 0.136 nm, the phase plate later used for aberration correction is properly sampled up to the desired frequency. The point resolution of the microscope amounts to 0.198 nm while the information limit has been determined as 0.1 nm.[266] The sidebands in the hologram spectrum (Fig. 9.21) include the {400} beams, which are needed to contribute to the imaging process if the dumbbell structure is to be properly resolved.[143]

After correction of aberrations, the information previously spread out in the intensity distribution of the hologram is concentrated at the atomic columns, as shown in the corrected amplitude and phase of the electron wave in Fig. 9.21. The phase image contains the white dumbbell contrast that is located at the position of the atomic columns. In the corrected amplitude image dark "8-shaped" features with faint white dots at the position of the Si atoms appear. A simulated thickness defocus tableau for $C_S = 0$ and a residual error in defocus of -10 nm to 10 nm can be used to confirm the proper correction (Fig. 9.22). Comparing the corrected wave with the simulation reveals a satisfying agreement, since already a small deviation of about 5 nm for the proper focus drastically changes the image contrast of both amplitude and phase. The small asymmetry that remains in the corrected wave may be due to residual crystal or beam tilt.

The ability of electron holography to reveal such small misalignments is of spe-

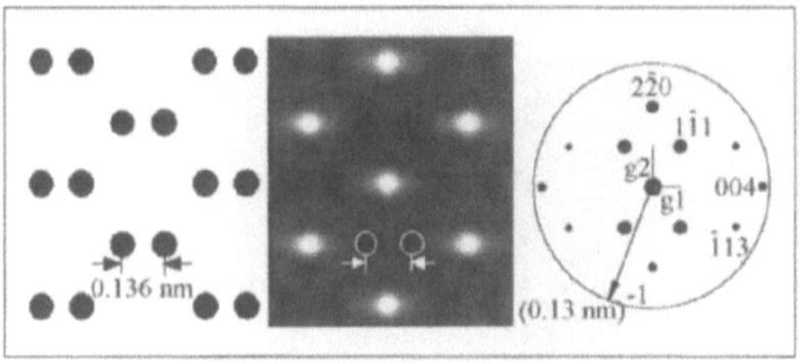

Figure 9.19. Left: [110] projection of Si showing its characteristic dumbbell arrangement. Middle: Si as simulated for a Philips CM30ST FEG at Scherzer focus and an aperture of $0.13\,\mathrm{nm}^{-1}$ which includes the {400} beams responsible for the dumbbell structure (right).

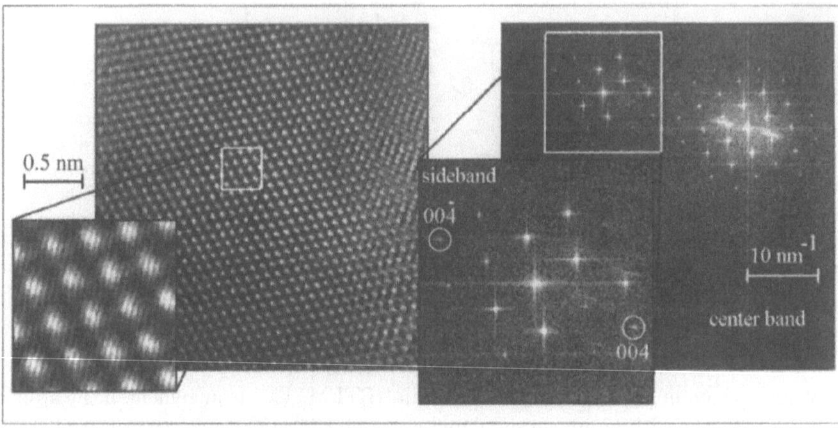

Figure 9.20. Hologram of Si in [110] projection with a fringe spacing of 0.05 nm (left). The sideband used for reconstruction (enlargement) in the spectrum of the hologram (right part) contains the 400 reflections corresponding to lateral information of 0.136 nm.

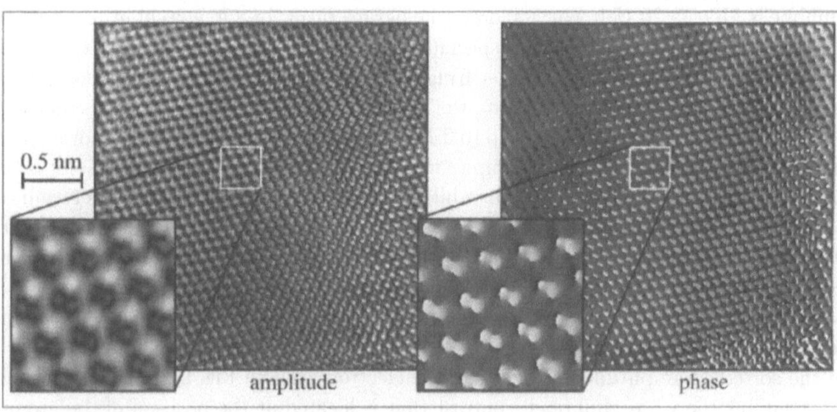

Figure 9.21. Aberration corrected amplitude and phase of the electron object wave reconstructed from Fig. 9.20 (Parameters used: $C_S = 1.23\,\mathrm{mm}$, $\Delta z = -115\,\mathrm{nm}$, $C_a = -3\,\mathrm{nm}$, $\alpha_a = 120^o$). For a better visualization some noise reduction (Fourier filtering of the nonperiodic background) in the wave spectrum has been applied.

cial value for the investigation of imperfect crystals. An example has been given by Orchowski and Lichte who investigated $\Sigma = 13$ tilt grain boundaries in Au by means of high resolution electron holography.[330] Fig. 9.23 shows a hologram of such a grain

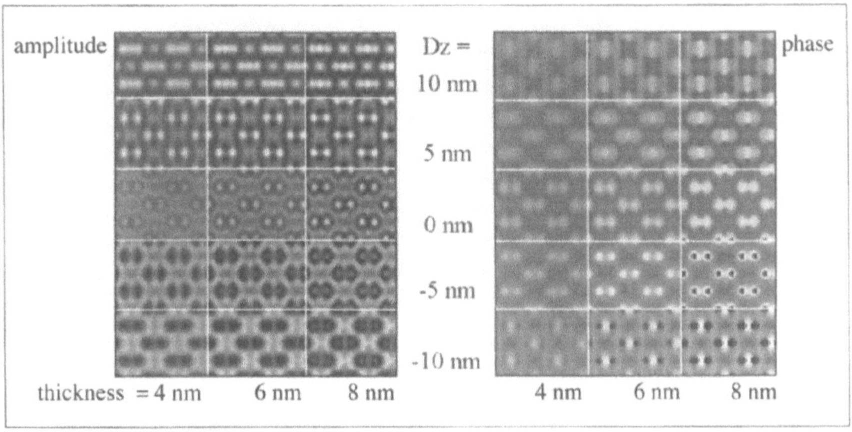

Figure 9.22. Simulated thickness defocus tableau of Si [110] for a correction to $C_S = 0\,\text{mm}$. The comparison with the reconstructed object wave from Fig. 9.22 reveals a satisfying agreement for $\Delta z = 0$.

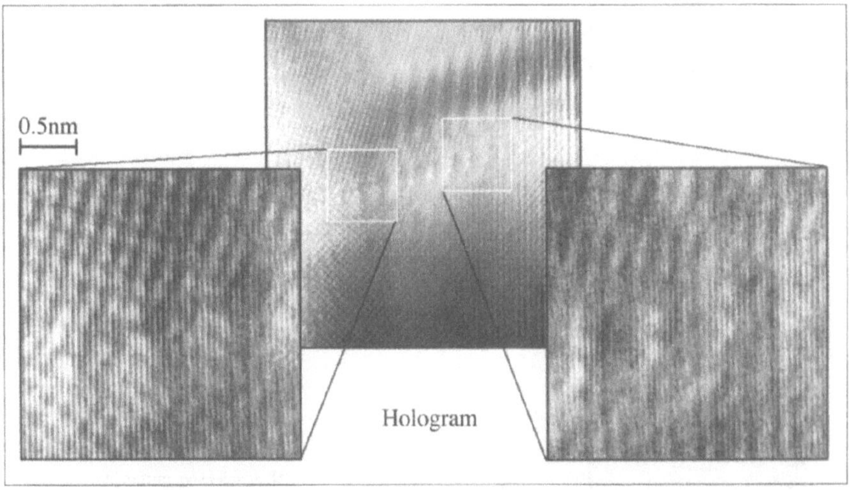

Figure 9.23. Hologram of a $\Sigma = 13$ [100] tilt grain boundary in Au taken with a Philips CM30ST FEG microscope. The grain boundary runs horizontally across the field of view. The two enlarged regions of a thinner (left) and a thicker (right) specimen area reveal the basic periodical core structure of the boundary.

boundary in [100] projection, together with two enlarged regions of a thinner (left) and a thicker part of the specimen. The analysis of the sidebands in the Fourier spectrum of the hologram (Fig. 9.24) reveals an improper alignment of the sample to the zone axis: nearly all of the {200} reflections but only a few of the {220} reflections are present. In addition, opposite reflections show different intensities which indicates some mis-tilt.

Note that, due to Friedel's law, these effects cannot be observed in the autocorrelation (or diffractogram) part of the hologram spectrum. It is a common experience in high resolution holography that even after careful alignment of the sample to the zone axis, residual mis-tilt is observed in the image wave spectrum. Holography therefore

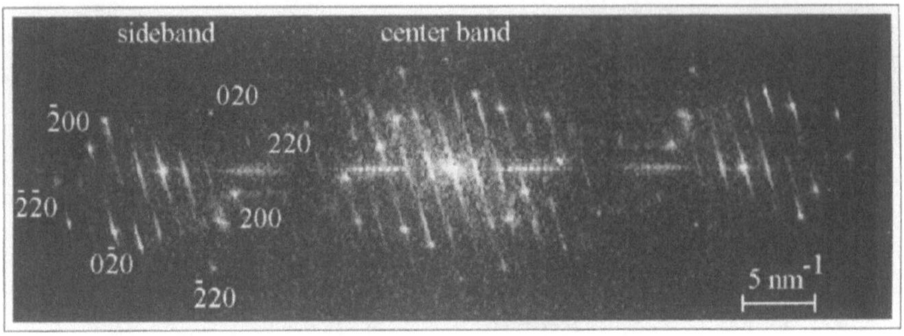

Figure 9.24. Fourier spectrum of the grain boundary hologram. The image wave spectrum in the sidebands reveals an asymmetric excitation of the reflections, indicating a mis-tilt of the crystal orientation.

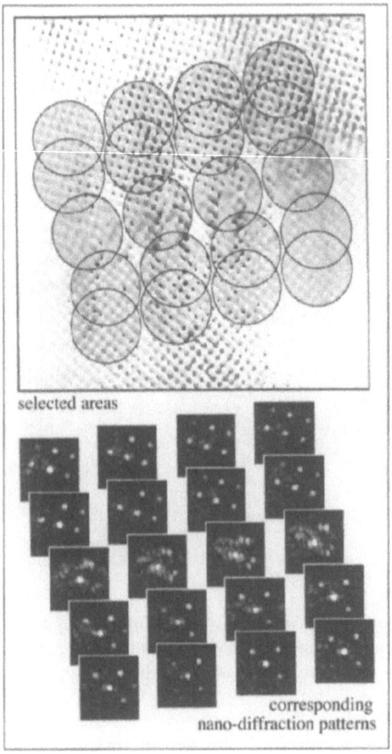

Figure 9.25. Nano-diffraction pattern from 20 different areas with a selected numerical aperture of 1.6 nm in diameter. A general mis-tilt is revealed in the strongly asymmetrical excitation from all areas. The changes from one region to another indicate local bending within each grain. Additionally, some twist between the grains is visible.

can serve as a valuable tool for investigating such mis-tilt effects. Moreover, since the complete image wave is available, nanodiffraction studies of the local specimen tilt becomes possible. In Fig. 9.25, the reconstructed phase image of the grain boundary has been analyzed by placing a numerical aperture of 1.6 nm in diameter in 20 different regions of the specimen. The evaluation of the diffraction pattern from these small areas reveals the local specimen tilt. Despite the fact that there is not yet a theoretical way to translate these findings into measured values for the tilt in the case of dynamical scattering,[357,408] a comparison with image simulations can lead to valuable insight with respect to, e.g., local bending structures of the specimen.

7. Conclusion

Due to the ability to retrieve the complete information on the electron image wave, high resolution electron holography can reveal valuable information in addition to the data available in high resolution microscopy. The advantages are numerous.

In conventional microscopy, the phase of the complex electron image wave is lost. Therefore, the ability to retrieve both amplitude and phase of the electron wave separately considerably enhances the dataset that can later be used for characterization of sample properties. Moreover, since only information on elastically scattered electrons is reconstructed from the hologram, the retrieved images can be considered as "zero-loss" filtered. In a second step, by carefully evaluating the transfer characteristics of the electron microscope, all the information up to the limit which is mainly given by the degree of coherence of the electron optical set-up can be revealed in a directly interpretable way by aberration correction. Since the complete information on the complex wave is retrieved, a thorough wave optical analysis of the retrieved image wave becomes possible. Techniques like nano-diffraction on the sub-nm scale or the production of a focal series from one single exposure are available and can reveal additional valuable insight in the microscopic properties. The availability of large area phase contrast is an additional advantage.

The evaluation of the image wave is preferably performed numerically in the computer. By using Slow Scan CCD image detectors for the registration of the holograms, the reconstruction and investigation of the electron wave can be performed on-line at the microscope. A direct feedback either to the operator or even by using digital control of the electron microscope is possible and can be used to optimize the experimental conditions. For amorphous specimen areas, the minimization of amplitude contrast is used as criterion for the success of aberration correction. In the case of crystalline specimens, no simple approach for the determination of the true correction parameters is available up to now. Quantitative comparison of the corrected wave with image simulations of known specimen details can be used to optimize the correction but includes many free parameters which makes the procedure rather tedious and ambiguous. New approaches for the optimization of the wave optical back propagation to the level of the object wave have to be investigated, if electron holography is meant to be established as a standard tool in high resolution structure characterization. As shown recently, the influence of third order astigmatism, residual axial coma, and mis-tilt will have to be incorporated into the correction scheme, if the final goal of a resolution down to 0.1 nm is to be achieved.

OFF-AXIS STEM HOLOGRAPHY

M.A. Gribelyuk[1] and J. Sum[2]

[1]IBM East Fishkill, Analytical Services Group, Hopewell Junction, N. 12533
[2] Gatan GmbH, Ingolstädter Strasse 40, D-80807 München, Germany

1. Introduction

An extensive review of different modes of electron holography, with the emphasis on their interrelationship, has been presented in Chapter 2 of this volume. It is clear from that discussion that among all methods of STEM holography the major progress has been made in the off-axis modes. Here we review theoretical and experimental results obtained with the off-axis methods and discuss their challenges.

The chapter is organized in the following way. First the formation of an off-axis hologram in the STEM instrument is discussed along with suggested reconstruction schemes. The realization of the off-axis holography requires consideration of a number of experimental factors. These are briefly discussed in the next section with the aim to reveal their interrelationship. A separate section is devoted to one of them, i.e., the accuracy of the phase measurements, due to its particular importance. Ideally, one would prefer to obtain results of reconstruction on-line, i.e., during experimental observation, with the highest spatial resolution. However, the currently available reconstruction schemes for achieving high resolution involve extensive computation, making at present an on-line observation impossible. In Section 5 we discuss the use of special detectors which enable on-line observation with some limited spatial resolution. Recently the far-out-of-focus version of off-axis STEM holography has offered much promise due to its success in the study of magnetic microstructure. Several examples of this work are given in Section 6. Finally, Sections 7 and 8 deal with the theoretical analysis of the off-axis STEM holograms for high resolution applications. In particular, Section 8 discusses determination of imaging conditions from the local area of observation, which is very important for successful correction of aberrations.

2. Hologram formation and reconstruction schemes

The full exit plane wave of an arbitrary strong scattering object can be restored if an electrostatic biprism[312] is placed in the illumination system of a microscope close to the electron source so that two coherent focused probes are produced at the specimen level separated by some distance a (Fig. 10.1). Propagation of one of them through the

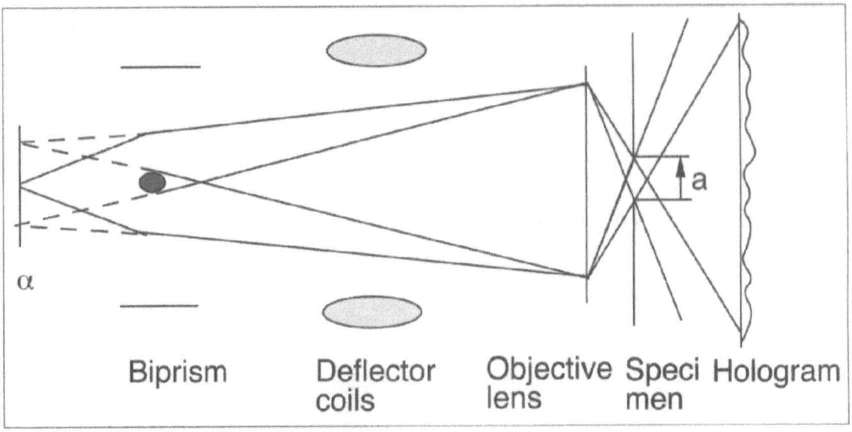

Figure 10.1. Scheme illustrating the formation of an off-axis hologram by use of a biprism in a STEM instrument.

object results in elastic scattering by the atoms with a distribution of electrons at the exit plane $\psi(\vec{r})$. If the second wave $t_e(\vec{r})$ passes through vacuum one finds in the exit plane of the object

$$\psi(\vec{r}) = \psi_e(\vec{r}) + t_e(\vec{r} - \vec{a}) \tag{10.1}$$

Here $t_e(\vec{r})$ is the reference wave at the exit plane of the object. The reference wave $t(\vec{r})$ at the entrance plane of the object is described as the point spread function of the microscope

$$t(\vec{r}) = \mathbf{FT}^{-1}\{T(\vec{q})\} \tag{10.2}$$

Therefore, the reference wave $t_e(\vec{r})$ at the exit plane of the object is formed by propagation of the wave $t(\vec{r})$ by the object thickness t

$$t_e(\vec{r}) = \mathbf{FT}^{-1}\{T_e(\vec{q})\} \qquad \text{with} \qquad T_e(\vec{q}) = T(\vec{q})\, e^{i\pi\lambda q^2} \tag{10.3}$$

The exponential term in the last equation describes the Fresnel propagation of the (monochromatic) wave of wavelength λ by a distance t. The hologram $I(\vec{q})$ is recorded in the plane P located at a distance $R \gg a$ behind the specimen. In this case, the Fraunhofer diffraction condition is satisfied,[71] i.e., the wave $\Psi(\vec{q})$ observed in the plane P can be described as the Fourier transform of the exit plane wave function $\psi(\vec{r})$. Therefore, the hologram intensity $I(\vec{q})$ is

$$\begin{aligned} I(\vec{q}) &= |\Psi(\vec{q})|^2 \\ &= |\Psi_e(\vec{q})|^2 + |T_e(\vec{q})|^2 + \Psi_e(\vec{q}) \cdot T_e^*(\vec{q}) \cdot e^{-2\pi i \vec{q}\vec{a}} + \Psi_e^*(\vec{q}) \cdot T_e(\vec{q}) \cdot e^{2\pi i \vec{q}\vec{a}} \end{aligned} \tag{10.4}$$

Meanwhile, $|T_e(\vec{q})| = K = \text{const.}$ for all reciprocal vectors $|\vec{q}| \leq q_0$, the objective aperture radius, if no chromatic aberration is taken into account. Its value is determined from the requirement that the total incident beam intensity is unity, i.e.,

$$\sum_{|q|\leq q_0} |T_e(\vec{q})|^2 = 1 \tag{10.5}$$

Three reconstruction schemes have been suggested for this geometry. In mode 1, the reconstruction is performed completely in reciprocal space.[255] The maps of the phase and amplitude of holographic fringes are evaluated as the two probes scan the object. The phase change of the probe $\Psi(\vec{q})$ passing through the object causes lateral shifting of fringes. In the case of a thin object (i.e., where the POA holds) the phase change in the holographic fringes is proportional to that of the object potential. The phase of holographic fringes is determined from comparison of a hologram intensity with a fixed optical grating, i.e., mask detector, placed in the detector plane. The amplitude of holographic fringes is found by intensity averaging in the detector plane over the fringe period in the direction perpendicular to the fringes. Some recently obtained results with this and similar approaches will be discussed in Section 5.

The full object wave can be determined if the reconstruction is performed in direct space (mode 2).[63] The hologram intensity $I(\vec{q})$ is multiplied by the product $T_e(\vec{q})\exp(2\pi i\vec{q}\vec{a})/K^2$ and then inverse Fourier transformed. Three conjugate images located at $r = 0$, a and $2a$ are found in direct space

$$\begin{aligned} H(\vec{r}) &= \mathbf{FT^{-1}}\left\{I(\vec{q})\cdot\frac{T_e(\vec{q})\cdot e^{2\pi i\vec{q}\vec{a}}}{K^2}\right\} \\ &= \psi_e(\vec{r}) + t(\vec{r}-\vec{a})\otimes\mathbf{FT^{-1}}\left\{1+\frac{|\Psi_e(\vec{q})|^2}{K^2}\right\} + t_2(\vec{r}-2\vec{a})\otimes\frac{\psi_e^*(-\vec{r})}{K^2} \end{aligned} \tag{10.6}$$

Here $t_2(\vec{r}) = \mathbf{FT^{-1}}\{T_e^2(\vec{q})\}$. Under optimum imaging conditions the object wave $\psi_e(\vec{r})$ can be restored with high spatial resolution. The reconstruction is performed at each probe position while the two probes scan the object. The details regarding evaluation of imaging conditions which allow for separation of conjugate images will be discussed in Section 7. It should be noted that the success in achieving high spatial resolution critically depends on aberration correction. The approaches suggested for determination of imaging conditions directly from the area of observation will be discussed in Section 8.

The third reconstruction scheme (mode 3, see also Ref. 75) considers the scattering by the object in the projection approximation with the object transmission function $o(\vec{r})$ and is applicable to holograms taken under far-out-of-focus conditions. The hologram is taken at one position and no scanning is involved. The total wave is described as

$$\psi(\vec{r}) = o(\vec{r})\cdot t_e(\vec{r}) + t_e(\vec{r}-\vec{a}) \tag{10.7}$$

It can be shown (see Ref. 75) that for large defocus values and in the small-angle approximation the two last terms in Eq. (10.4) take the form

$$[o(\vec{r})\otimes t_e(\vec{r})]\cdot e^{2\pi i\vec{q}\vec{a}} + [o(\vec{r})\otimes t_e(\vec{r})]^*\cdot e^{-2\pi i\vec{q}\vec{a}} \tag{10.8}$$

The Fourier transform of $I(\vec{q})$ produces the sidebands in the reciprocal space of the form $\tau(\vec{q})\cdot t_e(\vec{q})$ separated by $\vec{v} = \pm\vec{a}$ from the origin, which is similar to the off-axis TEM holography case. This sideband is then separated, multiplied by $t_e^*(\vec{q})$ to correct for aberrations and inverse Fourier transformed to produce the ultimate distribution of the object transmission function $o(\vec{r})$. Since it is difficult to determine defocus value from a hologram, its optimum value was used in the experiment. The criterion suggested in Ref. (75) requires that the wave amplitude is minimum in the vacuum region adjacent to a sample edge. The application of this mode to the study of magnetic materials with the spatial resolution $d = 1\,\mathrm{nm}$ will be discussed in Section 6.

3. Experimental considerations

The following is a brief and perhaps incomplete list of parameters which should be taken into account when setting up a holographic experiment. The choice of acquisition time of a hologram at each position of the two probes presents a trade-off between two factors. On one hand, it should be long enough to maintain the required accuracy of wave reconstruction. The parameters controlling the minimum detectable signal in holographic measurements are discussed in the next section. On the other hand, it should be short enough to control sample drift, biprism stability, and radiation damage effects. This choice is particularly difficult for modes involving scanning of the two probes. Depending on the mode, the time at the probe position has been varied from 10^{-4} sec to about 2 sec.

The computer requirements again depend on whether the scanning or standing modes are used. In scanning modes 1 and 2 the processing performed during acquisition should be short enough so that its time does not limit the total acquisition time. This requirement is obviously less stringent for the mode 3. The ability to store large amounts of data, 1 GB and more, is important in case of a mode 2, and this has hindered its experimental utilization.

The field of view of a hologram depends on the voltage applied to the biprism, the distance from the object to the observation plane, and the detector size. The holographic fringe spacing depends on the biprism voltage and should be adjusted to the pixel size of the detector. At least three pixels per fringe spacing are necessary for digitization. To our knowledge there are at present only two systems set up for off-axis STEM holography. The system at the University of Tübingen (Germany) uses a photodiode linear array detector with the pixel size $50\,\mu\mathrm{m}$ x 2 mm and containing 512 pixels.[431] The system at the Arizona State University employs the 512 x 512 pixel Photometrix CH-250 CCD chip which offers almost perfect linearity and a wide dynamic range.[493]

4. Accuracy of phase measurement

The electron hologram of an object can be seen as a set of fringes with the position-dependent amplitude and phase. These vary as the beam passes the areas of an object with different scattering power. In general, the relationship between the phase of a hologram and the object wave is complex and depends on the way the hologram is formed. Below we discuss the accuracy of the phase measurement of holographic fringes. It serves as the starting point in evaluating the accuracy of the object wave reconstruction, which should be performed for each of the holographic modes separately.

The phase variance of the holographic fringes depends on the fringe contrast μ and the number of (coherent) electrons N per pixel of the hologram region[260]

$$\mathrm{var}(\phi) = \frac{2}{N\,\mu^2} \tag{10.9}$$

On the other hand, the fringe contrast is determined by the lateral coherence of the electron source. The effect of temporal coherence can be ignored for FEG instruments. Consider an ideally incoherent Gaussian source with the radius r_0 and two points in the object plane; the source subtends the angle α at the object. If the waves scattered by these two points are now brought to interference by a biprism the fringe contrast μ is

$$\mu = V(\alpha/\lambda) = e^{-(\pi\, r_0\, \alpha\, /\lambda)^2} \tag{10.10}$$

Here V is the coherence function for the two points; α is the angle subtended by the source at the object. The total (coherent) current I across the illuminated pixel area $S = \pi(\alpha/2)^2$, with the coherence function $V(\alpha/\lambda)$ and the source brightness β is

$$I = B\,\pi^2\, r_0^2 \cdot \frac{\alpha^2}{4} = -\beta\,\lambda^2 \cdot \frac{\ln\mu}{4} \tag{10.11}$$

We note that decreasing the source size r_0 or the pixel size (defined through α) has an opposite effect on the fringe contrast μ and current: the current decreases while the fringe contrast increases. Since the total number of electrons per pixel $N = I\,t\,e$ (t, exposure time) the phase variance takes the final form

$$\mathrm{var}(\phi) = \frac{-4\,e}{\beta\,\lambda^2\,\mu^2\,t\,\ln\mu} \tag{10.12}$$

The minimum detectable phase difference corresponds to the minimum of the phase variance $\mathrm{var}(\phi)$ and is set by

$$\mathrm{var}(\phi)_{opt} = \frac{8\,e^2}{\beta\,\lambda^2\,t} \tag{10.13}$$

The minimum Eq. (10.13) applies to the case of an ideal detector. The following two factors are important for the real detector. First, the noise of its electronics and saturation effects limit the number of effectively processed electrons. The decrease of the signal-to-noise ratio at the detector exit with respect to that at its entrance is described by the detector quantum efficiency coefficient, DQE

$$\mathrm{DQE} = \frac{\mathrm{SNR}_0^2}{\mathrm{SNR}_i^2} \tag{10.14}$$

Second, the detector sensitivity falls for the high spatial frequency signals. This effect is described by the frequency dependent modulation transfer function MTF and leads to decrease in detectable fringe contrast μ

$$\mu_{\mathrm{det}} = \mu \cdot \mathrm{MTF}(q) \tag{10.15}$$

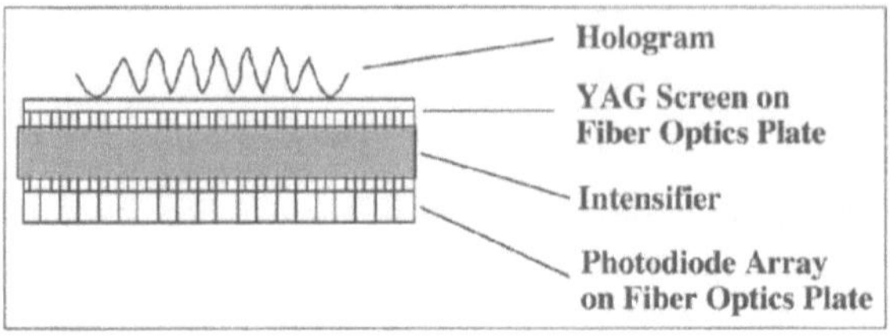

Figure 10.2. Scheme of a photodiode array detector.

Taking into account the detector properties the standard deviation of the phase is given by[260]

$$\sigma_\phi = \frac{\sqrt{2}}{\mu \cdot \mathrm{MTF}(q) \cdot \sqrt{N \cdot \mathrm{DQE}}} \tag{10.16}$$

with N, again, being the number of coherent electrons. The experimentally adjustable parameters are the fringe spacing and the acquisition time. The spacing is controlled by the voltage applied to the biprism and is inversely proportional to it. The detection of finer fringe spacing in turn requires smaller detector pixel size for adequate sampling of holographic fringes. The expressions Eqs. (10.9)-(10.16) provide the framework for accurately evaluating each experimental setup.

5. Application of special detectors

In the method originally discussed by Leuthner,[255] a reference grating, consisting of opaque and transparent stripes, is placed in the entrance window plane of the photomultiplier. The orientation and spacing of stripes matched those of the holographic fringes. At each probe position the hologram intensity is integrated over the detector plane with and without the grating inserted. The comparison of these two allows, in principle, both amplitude and phase of holographic fringes to be determined. However, this has not been realized experimentally due to signal processing limitations.

Recently this work has been extended with development of a new detector system.[431] The photodiode linear array detector consisting of 512 pixels has been used to record the hologram (Fig. 10.2). The initial hologram intensity was first converted into a light signal and then amplified. Each detector pixel is 2.5 mm long and 50 μm wide. The detector was positioned so that the longest pixel side was parallel to holographic fringes, resulting in improvement of the signal-to-noise ratio. The detector MTF does not affect the minimum detectable signal, as about 20 pixels correspond to one fringe spacing. The phase resolution $\Delta_\phi = 2\pi/17$ ($\sigma_\phi > 0.059$, see Eq. (10.10)) has been achieved for the typical case of the number of detected electrons per pixel $N = 2000$, acquisition time at each probe position $t = 4$ msec, and the measured DQE= 0.8.[431]

The hologram intensity at the probe position R can be described as

$$I(\vec{q}, \vec{R}) = 1 + a^2(\vec{R}) + 2\,a(\vec{R}) \cos\left[2\,\pi\,\vec{R}_c\vec{q} - \varphi(\vec{R})\right] \tag{10.17}$$

Here $a(\vec{R})$ and $\varphi(\vec{R})$ are the amplitude and phase of the aberrated object wave Ψ_e. The fringe frequency R_c is measured from a hologram when two probes pass through the

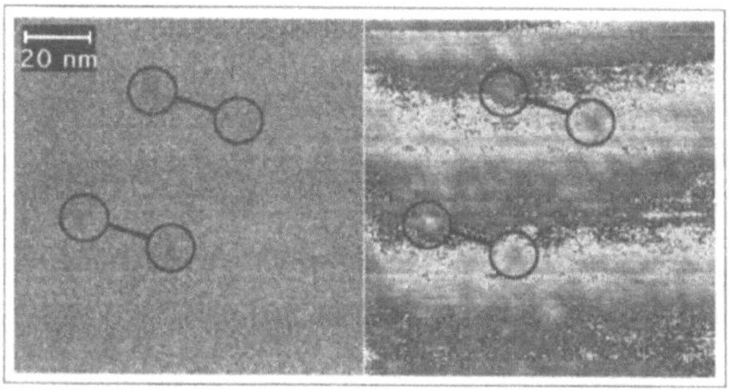

Figure 10.3. Amplitude (left) and phase (right) maps of ferritine molecules (encircled) imaged with photodiode detector. The conjugate images are linked by a solid line. The molecules are supported by a thin carbon film.

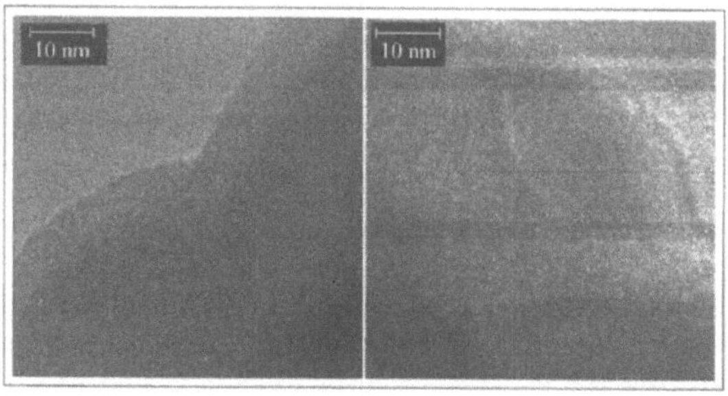

Figure 10.4. Amplitude (left) and phase (right) maps of a hologram of an asbestos microcrystal showing (002) lattice fringes.

vacuum. Its value is then stored in a computer. At each probe position the amplitude and phase are found by Fourier transformation of a hologram carried out for the fringe frequency R_c. The fact that only one Fourier component, rather than the full Fourier transform, is needed speeds up reconstruction. Both amplitude and phase maps can be displayed on-line while the two beams are scanning the object.

Fig. 10.3 presents one application of this mode. The left and right images are the amplitude and phase maps of an unstained ferritin molecule placed on the carbon film, respectively.[431] Since the molecule scatters very weakly it is hardly visible in the amplitude map. This represents a recognized problem in analysis of biological objects by electron microscopy. However, the phase map produces a stronger contrast. This example shows that the method can offer some advantages with respect to conventional TEM for observation of small biological molecules.

The second example shows that the same detector system used in Ref. 255 allows for resolution improvement if modifications are made of the microscope and its electronics (Fig. 10.4). In Ref. 255, $d = 30\,\text{Å}$ was considered as the resolution limit. The left image is the bright field image of asbestos microcrystalls showing (002) lattice fringes with the spacing $d = 9.03\,\text{Å}$.[431] The right image corresponds to the signal after it has passed the grating

$$S_\varphi = \int I(\vec{q}, \vec{R}\,) \cdot D_{im}(\vec{q}\,)\, \mathrm{d}^2\vec{q} \tag{10.18}$$

Since the detector function has the form $D_{im}(\vec{q}\,) = 1 + \sin(2\pi R_c q)$, this image contains the information on both the amplitude and phase. The lattice fringes are visible in both images. The noise present in the right image is largely due to the short acquisition time at each probe position ($t = 61\,\mu$sec).

It is possible to display the amplitude $a(\vec{R}\,)$ and phase $\varphi(\vec{R}\,)$ simultaneously if the new detector design is used (Fig. 10.5). Here the detector plane with the detector $D(\vec{q}\,)$

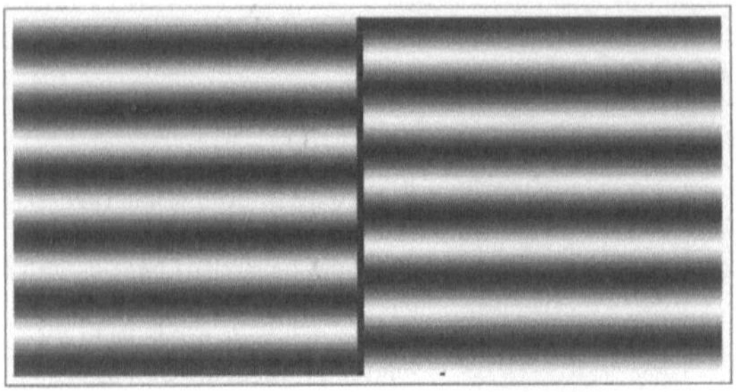

Figure 10.5. Proposal for an improved mask detector design enabling simultaneous reconstruction of amplitude and phase of a hologram.

is divided in two parts which are then shifted by $\pi/4$ with respect to each other. This leaves the left part unchanged while the detector function for the right part becomes $D_{re}(\vec{q}) = 1 + \cos(2\pi \vec{R}_c \vec{q})$. The light passing through each part is processed by a separate photodiode detector. Since in this case both signals, S_φ and S_a

$$S_a = \int I(\vec{q}, \vec{R}) \cdot D_{re}(\vec{q})\, \mathrm{d}^2\vec{q} \tag{10.19}$$

are available, some simple processing allows for simultaneous display of amplitude and phase. The masks can be created on a transparent LCD display with the fringe spacing controlled by the software. In this case the detector could be easily adjusted to any experimentally observed holographic fringe spacing. At present this proposal awaits its experimental realization.

The following comments are concerned with the reconstruction procedure. First, no attempt is made to separate conjugate images. Since the probe separation is smaller than the field of view, one observes double images of the object. Application to large objects requires either a modification of the reconstruction scheme or an increase of the probe separation (and decrease of the pixel size in the detector plane for proper sampling); otherwise, the method is restricted to small objects. Second, the correction for aberrations is not realized although it is in principle possible. Without correction the interpretation becomes difficult if the high resolution work is attempted. Third, for the sake of processing speed, only one Fourier component of the object wave is restored at each probe position. This limits the available data and can be seen as a disadvantage for high resolution work.

We conclude that in its present state this approach is useful for fast data visualization in the medium resolution work. Further development is required if it is going to be applied for high resolution applications.

6. Far-out-of-focus off-axis STEM holography: Application to magnetic materials

The continuing expansion in industrial application of magnetic materials drives the development of novel systems with controlled properties. At present their market ranges from bulky navigation compasses to recording media in microelectronics. Since the properties are controlled by the atomic scale microstructure there is a need for development of new characterization techniques which can quantify magnetic microstructure information at high spatial resolution.

The recently-developed, greatly-defocused, off-axis STEM holography presents a very promising method for these applications.[277] Its principles have been described

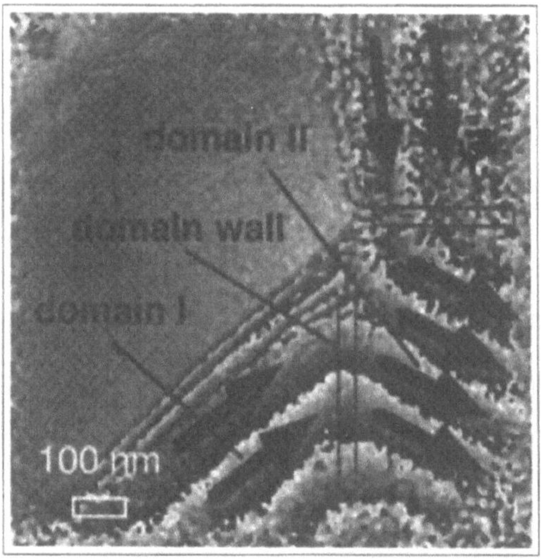

Figure 10.6. Phase image of a 25 nm thick Co film with the magnetic microstructure indicated. The hologram was taken in the absolute mode.

in detail in Chapter 2 of this volume. Here we review some results obtained by this method.

The thin magnetic films are usually polycrystalline and anisotropic with the grain size in the range from 2 nm to 10 nm. Their magnetic structure corresponds to the energy minimum from exchange, stray fields, and anisotropy, and is further complicated due to interaction between structure defects and domain walls. In Ref. 277, nominally 25 nm thick Co films were examined. The hologram was digitally acquired using a 512x512 pixel CCD camera attached to a HB-5 STEM[493] and reconstructed in a way similar to that used in TEM off-axis holography. Fig. 10.6 presents a phase image of adjacent domains of the film. Line-scans along maximum gradients both in domains 1 and 2 have revealed that the phase change is proportional to the distance from the edge of a sample. This is consistent with the constant thickness and magnetization within the domain. The slope of the line-scan provides an absolute measure of in-plane magnetization and was found to be the same in the two domains within experimental error. The analysis of the slope at the wall core has proved less than the saturation value. This suggests that the wall had segments with out-of-plane magnetization or wall vortices. The thickness-averaged magnetization could not be determined as the film thickness was not known accurately. The rotation of the magnetization across the wall and the wall thickness could be measured directly from the phase image and were found consistent with the earlier published results.

Multilayer structures composed of alternating magnetic and non-magnetic metals have been used for a study of antiferromagnetic exchange coupling effects between adjacent layers. This effect is responsible for the observation of giant magnetoresistance and was found dependent on growth conditions.[118] In Ref. 279. the magnetic superlattice Pd(20 nm) / $[\text{Co}(1\,\text{nm})\ \text{Pd}(1.1\,\text{nm})]_{10}$ was used to reveal the magnetic microstructure. Figure 10.7 shows a phase image of two adjacent domains separated by the domain wall which is imaged as the bright line. The domain walls run perpendicular to the film edge; magnetization is perpendicular to the edge of the sample and rotates by 180° when crossing the domain wall. Magnetization was measured directly from the phase map inside each domain from a slope of a line-scan taken parallel to the sample edge.

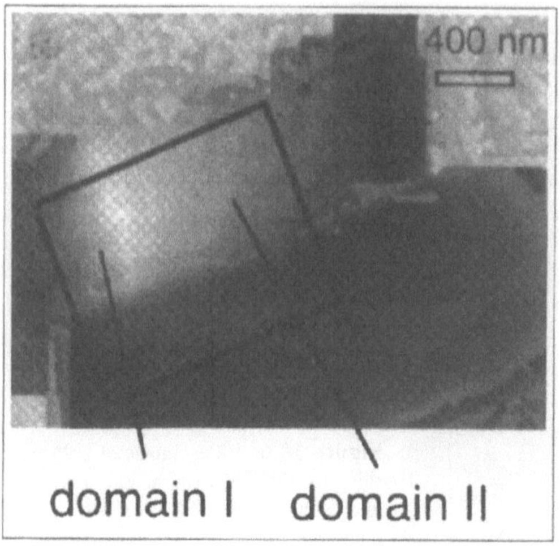

Figure 10.7. Electron holography of Pd(20 nm)/[Co1 nm)Pd(1.1 nm)]$_{10}$ multilayer structure: partially unwrapped phase image of domain structure near the edge of a sample.

The authors showed that inside domain 1 all of the Co layers were ferromagnetically aligned within a multilayer stack. In domain 2, the measured magnetization was 40% of that in domain 1. This suggests that in domain 2 not all Co layers were magnetized in the same direction. In this particular case seven of ten layers were magnetized in one direction and three in the opposite one.

These results have been obtained in the absolute mode, i.e., when one of the beams passes through vacuum while the other passes through the object. The differential mode has also been developed.[277] It assumes that both beams pass through the object and the hologram is formed at a distance $R \gg a$, the beam separation, behind the object. The measured phase difference as

$$\Delta\varphi \propto \iint \vec{B}\ \mathrm{d}\vec{S} \cong B_n\, a\, t \tag{10.20}$$

Here B_n is the component of the magnetic field normal to the plane defined by the wave vectors of the two beams; t is the film thickness. In contrast to the absolute mode the phase difference is constant if two beams pass through a uniformly magnetized domain of a constant thickness. The achievable resolution is limited by the beam separation at the specimen level. At low spatial resolution, the separation between two dual images in the recorded hologram should be less than approximately three holographic fringe spacings. The differential mode allows for a straightforward interpretation of some simple magnetic distributions at medium resolution ($d = 200$ Å) and is particularly useful for the analysis of domain wall structures. The hologram in this case can be taken at an arbitrary distance from the hole in the sample as no reference wave passing through vacuum is required.

These examples show uniqueness of the method which provides both quantification of magnetic microstructure and spatial resolution on the nanometer scale, a combination which is not offered by other characterization techniques. It certainly awaits a wide range of applications. From the point of view of basic research, it can provide insight into the microstructural nature of ferromagnetic interactions. It can also play an important role in industry in development of novel technologies.

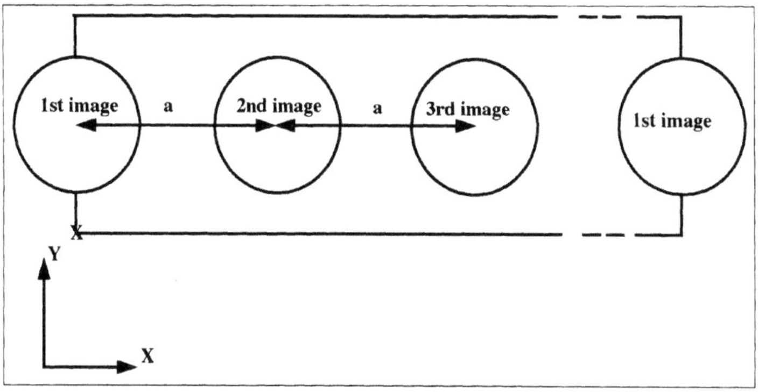

Figure 10.8. Position of conjugate images in the real space within an artificial unit cell of dimensions a_x and b_y. The parameters of the unit cell are determined by the sampling intervals used for digitization of a hologram $I(\vec{q})$. The beam position coincides with the origin of a unit cell and is marked as "X".

7. Optimum conditions for high resolution imaging

The computer analysis involved in the reconstruction scheme proposed in Ref. 63 appeared in Refs. 151 and 231. Reference 151 was primarily concerned with evaluation of the highest achievable resolution in the restored wave function, the minimum beam separation between the two beams a_{min}, and determination of the size of a region of a sample S which can be restored at each probe position. The optimum choice of the beam separation presents a trade-off between the interference effects of conjugate images and the number of sampling points required for processing of a hologram. The greater is the separation of the two beams, the smaller are the interference effects (Fig. 10.8) and the larger is the number of sampling points. Given the beam separation, the contribution of interference effects depends on the shape of the incident probe function $t(\vec{r})$, i.e., on defocus and the objective aperture size. On the other hand, the fringe periodicity $f = 1/a_{min}$ of a hologram decreases as the beam separation increases. Given a highest spatial frequency $\vec{q}_0$ up to which a wave function should be restored, a minimum number N_x of sampling points of a hologram in the direction of beam separation, with

$$N_x = 2\,(3\,a_{min}\,q_0) + 1 \tag{10.21}$$

is required to prevent interference of conjugate images from adjacent unit cells. Here a fringe is represented by 3 sampling points. The interference between conjugate images is caused by the imposed artificial periodicity of $H(\vec{r})$ due to sampling of a hologram $I(\vec{q})$. The optimum beam separation is one for which the interference effects are sufficiently suppressed so that the restoration error of the wave function $R(\vec{r})$,

$$R(\vec{r}) = \left| \frac{\psi_{rest}(\vec{r})}{\psi_{mod}(\vec{r})} - 1 \right| \tag{10.22}$$

is kept below a certain limit. Here $\psi_{rest}(\vec{r})$ and $\psi_{mod}(\vec{r})$ are the restored and the model wave functions, respectively.

The size of a region S_y in the direction perpendicular to the line joining the two beams and the sampling interval of a hologram $1/b_y$ depend only on the shape of the

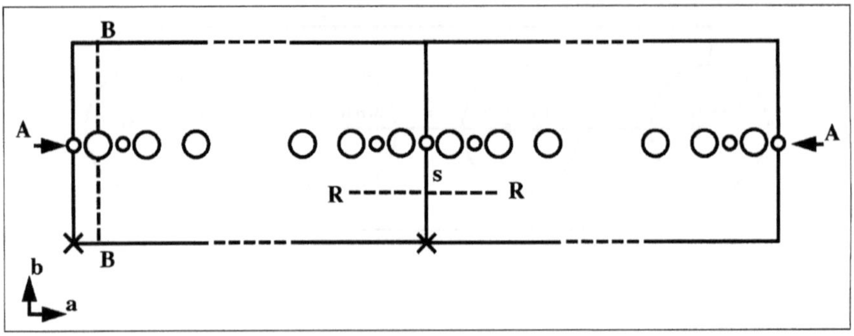

Figure 10.9. Model viewed along [001] direction with lattice parameters of an orthorhombic unit cell $a_x = 250\,Å$, $b_y = 5\,Å$and $c_z = 2\,Å$. Small and large open circles represent oxygen and uranium atoms, respectively. The distance between neighboring columns $r = 2\,Å$. The R-R section is parallel to the row of atomic columns and is located a distance s apart; the section B-B is considered across uranium column position. The A-A section is drawn across all atomic column positions. The incident beam position is indicated as "X". Two unit cells in the a axis direction are shown.

object point spread function $t(\vec{r})$; they have been found from inspection of calculated profiles $t(\vec{r})$ for optimum imaging conditions. The size S_x and the sampling interval $1/a_x$ depend both on the interference of conjugate images within the artificial unit cell and the interference of adjacent unit cells. Given the required resolution of an exit plane wave function they have been evaluated from the analysis of restoration errors $R(\vec{r})$ provided that an optimum defocus and beam separation were used.

The structure model used in simulations is shown in Fig. 10.9. A series of model wave functions has been prepared by the multi-slice method as a function of the objective aperture size q_0 and defocus for a $t = 20\,Å$ thick crystal. The analysis has been carried out for the 100 kV HB-5 ($C_S = 0.8\,mm$) and the prospective 300 kV STEM instrument ($C_S = 0.72$ mm). The model holograms have been calculated from the model exit plane wave functions for defocus range $\Delta z = -200\,Å$ to -1000 Å and $|q| < q_0$ using Eq. (10.4). The restored wave function has also been obtained for all spatial frequencies $|q| < q_0$ from reconstruction Eq. (10.6) and compared with the model.

The results are summarized in Tables. 10.1 and 10.2. The restoration error $R_{thr} = 0.2$ has been chosen as a threshold. In terms of structure, this corresponds to a "loss" of a less than one oxygen atom provided that the POA holds. For a given aperture size q_0, defocus and beam separation were varied until the minimum a_{min} was found which maintained the restoration error $R < R_{thr}$. The corresponding defocus range is considered as the optimum one. As the resolution of a hologram $d = 1/q_0$ gets higher,

d [Å]	Δz [Å]	a [Å]	S_x [Å]	S_y [Å]	$1/a_x$ [Å^{-1}]	$1/b_y$ [Å^{-1}]
3	-200 to 400	≥ 40	8	20	≤ 0.0083	0.01
1.6 to 2.0	-500 to 700	≥ 60	8	≈ 10	≤ 0.0055	0.01 to 0.0083

Table 10.1. Optimum defocus Δz, beam separation a, size of the reconstructed region S_x x S_y at each probe position and the sampling intervals of the hologram $1/a_x$ and $1/b_y$ as a function of resolution of the reconstructed wave function d for 100 kV STEM holography.

the oscillating part of the transfer function $T(q)$ contributes more to the hologram. In the direct space the point spread function $t(\vec{r})$ becomes delocalized with pronounced subsidiary maxima which makes separation of conjugate images difficult. As the result the minimum beam separation increases with the resolution (see Tables. 10.1 and 10.2). Since the separation of the two beams cannot be made arbitrarily high due to practical considerations, oscillations of $T(q)$ imply a limit on the highest resolution of the restored wave function achievable with this reconstruction procedure. Given aberrations of the microscope, its point-to-point resolution increases with the accelerating voltage. Therefore one needs less beam separation with a 300 kV STEM than with a 100 kV HB-5, provided that the wave function is restored with the same resolution (compare Tables. 10.1, 10.2, $d = 2$ Å). The highest resolution is $d = 1.6$ Å for the former and $d = 1.2$ Å for the latter. The examples of restoration are shown in Fig. 10.10 where the amplitude of the point spread function $t(\vec{r})$ and the restoration error in the direction of conjugate images separation (section A-A in Fig. 10.9) are presented. It should be noted that the restoration error is asymmetrical with respect to the beam position. This can be expected since the first conjugate image (Fig. 10.8) used in reconstruction is surrounded by different conjugate images from the right (image 2) and left (image 3) sides.

It follows from sampling interval values (Tables. 10.1, 10.2) that the minimum size of sampling points required for acquisition of a hologram on HB-5 is $N = 128$ x 128 for $d = 3$ Å, and 512 x 256 for $d = 2$ Å; the appropriate figures for 300 kV STEM are $N = 256$ x 256 for $d = 2$ Å, and $N = 512$ x 256 for $d = 1.2$ Å. In Tables. 10.1, 10.2 the size of a region S_x x S_y within which a wave function can be restored at each probe position is symmetrical with respect to probe position. For a resolution of $d = 1.2$ Å scanning steps $L_x = 4$ Å and $L_y = 15$ Å of the probe provide the wave function for the 100 Å x 100 Å large region after 200 reconstructions, with the total computer time $t < 4\,h$ (VAX 11/750); the number of reconstructions decreases for the lower resolution. It should be noted that the beam position is arbitrary with respect to the location of atoms; the method is much less sensitive to the specimen drift in the directions lying in the projection plane in comparison with previously discussed methods. Indeed, the distance between neighboring probe positions should not necessarily be kept the same during acquisition of holograms. It should be only small enough to allow for the overlap of the wave functions reconstructed from the neighboring probe positions.

We conclude that high spatial resolution can be achieved with this form of off-axis STEM holography. The main problems in its experimental realization are concerned with control of sample and biprism instability during scanning and the requirement to handle large amount of data.

d [Å]	Δz [Å]	a [Å]	S_x [Å]	S_y [Å]	$1/a_x$ [Å^{-1}]	$1/b_y$ [Å^{-1}]
3	-200 to 400	≥ 40	8	20	≤ 0.0083	0.0133 to 0.01
1.6 to 2.0	-500 to 700	≥ 60	8	≈ 10	≤ 0.0055	0.01 to 0.0083

Table 10.2. Optimum defocus Δz, beam separation a, size of the reconstructed region S_x x S_y at each probe position and the sampling intervals of the hologram $1/a_x$ and $1/b_y$ as a function of resolution of the reconstructed wave function d for 300 kV STEM holography.

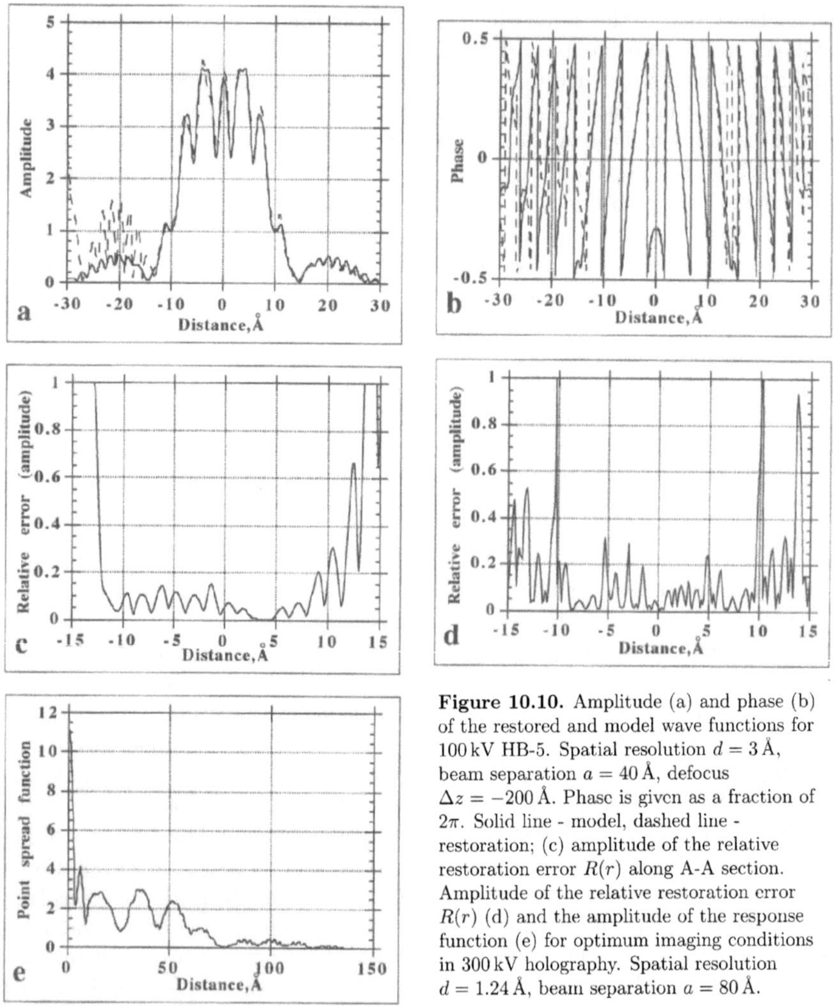

Figure 10.10. Amplitude (a) and phase (b) of the restored and model wave functions for 100 kV HB-5. Spatial resolution $d = 3$ Å, beam separation $a = 40$ Å, defocus $\Delta z = -200$ Å. Phase is given as a fraction of 2π. Solid line - model, dashed line - restoration; (c) amplitude of the relative restoration error $R(r)$ along A-A section. Amplitude of the relative restoration error $R(r)$ (d) and the amplitude of the response function (e) for optimum imaging conditions in 300 kV holography. Spatial resolution $d = 1.24$ Å, beam separation $a = 80$ Å.

8. Correction of lens aberrations

Any reconstruction scheme applied for high resolution work requires an accurate knowledge of experimental imaging conditions, i.e., the defocus and spherical aberration coefficient of the objective lens. Since electron holography is seen as a part of a strategy for solving unknown structures, one ideally needs a method for determination of imaging conditions which would require no preliminary knowledge about the structure, assume some contributions of dynamical scattering, and provide the necessary data from the local region of interest in the hologram. The latter becomes evident since defocus might change across the specimen. The development of a technique which would satisfy these requirements and provide very accurate data (see Refs. 151 and 263 for accuracy requirements) is not trivial, and presents nowadays a challenge to further progress in electron holography.

It is assumed in the method suggested in Ref. 152 that the object of interest is an unknown crystal structure, or some form of defect in a crystal, viewed in a low-index zone orientation so that, for at least some part of the field of view, the appropriate structure projection can be considered as consisting of well separated columns of atoms. This implies that the wave function near some column is not affected much by the neighboring columns. If the dynamical scattering effects are not strong one should be able to find a region R near the column where the exit plane wave function can be approximated by

$$\begin{aligned} \psi_e(\vec{r}) &= [o(\vec{r}) \cdot t(\vec{r})] \otimes P_t(\vec{r}) \\ P_t(\vec{r}) &= \frac{i}{\lambda t} e^{-i\pi r^2/(\lambda t)} \\ o(\vec{r}) &= e^{-i\sigma V(\vec{r}) t} \end{aligned} \qquad (10.23)$$

Here $t(\vec{r})$ and $o(\vec{r})$ are the point spread function of the incident beam and the transmission function of the object, respectively.

The object is then considered as the single slice with the potential $V(\vec{r})$ projected onto the entrance surface. The propagation of the scattered wave $t(\vec{r}) \cdot o(\vec{r})$ from the entrance plane to the exit plane is described as the convolution with the propagation function $P_t(\vec{r})$. For a single column of atoms we expect the dynamical scattering effects to be first revealed at the position of the atoms as the crystal thickness increases; at the same time these effects are less essential in the region located far from the columns in the projection plane. Therefore the approximation Eq. (10.23) should be valid at some distance from the column where the effect of the neighboring columns is still small. As illustration we compare in Fig. 10.11 the profiles of approximated and model wave functions along the line B-B in the structure projection (Fig. 10.9). The approximated wave functions were derived from Eq. (10.23) and the model ones were calculated by the multi-slice method. We note that the discrepancy is high at the atomic sites whereas the two functions coincide starting at a certain distance from the atom.

Now we consider how imaging conditions can be derived from an off-axis STEM hologram. The hologram $I(\vec{q})$ is multiplied by the product $T_{est}(q)\exp(2\pi i \vec{q}\vec{a}_{est})/K^2$ and inverse Fourier-transformed:

$$H_1(\vec{r}) = \mathbf{FT}^{-1}\left\{\frac{I(\vec{q}) \cdot T_{est}(\vec{q})}{K^2} \cdot e^{2\pi i \vec{q}\vec{a}_{est}}\right\} \qquad (10.24)$$

Here a_{est} is the estimate of the beam separation and $T_{est}(\vec{q})$ is the transfer function for the estimates of defocus and spherical aberration coefficient $C_{S\,est}$. We note that the knowledge of crystal thickness is not required in this reconstruction. Indeed, if approximation Eq. (10.23) is applied to the object plane wave function the appropriate terms in $\Psi_e(\vec{q})$ and $T_e^*(q)$, depending on the crystal structure, cancel themselves in the desired image (see Eq. (10.4)) and the conjugate image is given in the direct space as:

$$\psi_1(\vec{r}) = t(\vec{r}) \cdot o(\vec{r}) \otimes t(\vec{r} + \vec{a}_{est} - \vec{a}, \Delta z_{est} - \Delta z, C_{S\,est} - C_S) \qquad (10.25)$$

If the estimated values of defocus Δz_{est} and spherical aberration coefficient $C_{S\,est}$ approach the true ones, the amplitude $|\psi_1(\vec{r})|$ converges to $|t(\vec{r})|$. Therefore, the imaging conditions can be found by minimization of $J(\vec{R})$:

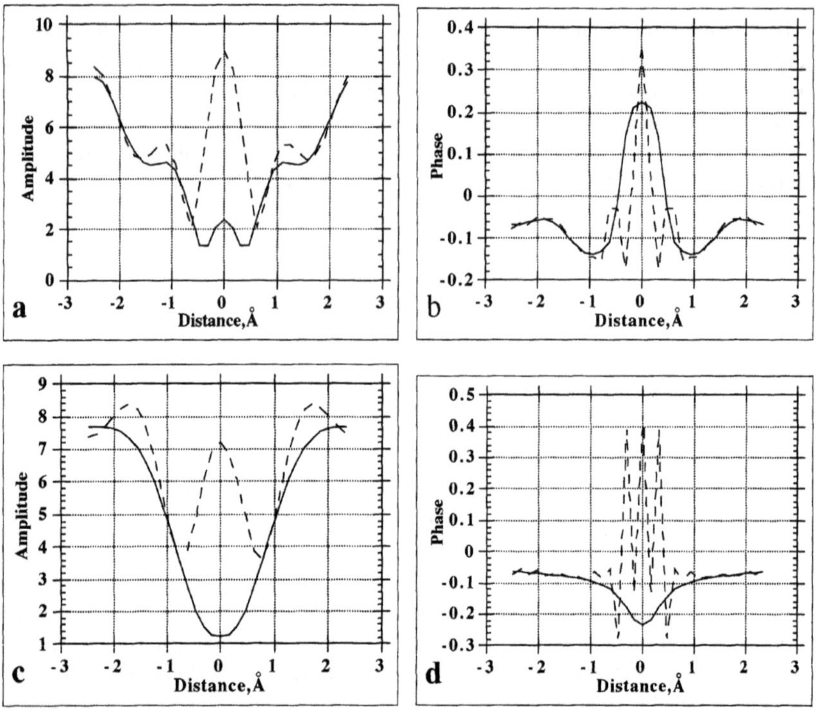

Figure 10.11. Profiles of the model (solid line) and approximated (dashed line) wave functions at the exit plane of a crystal along the B-B section (see Fig.8). (a,b)- spatial resolution $d = 1.2$ Å, crystal thickness $t = 20$ Å; (c,d)- spatial resolution $d = 1.6$ Å, crystal thickness $t = 40$ Å. The phase is presented in units of 2π.

$$J(\vec{R}) = \sum_{\vec{R}} \left| |H_1(\vec{r})| - | o_{est}(\vec{r}) + o_{2est}(\vec{r} - 2\vec{a}_{est}) + 2o_{est}(\vec{r} - \vec{a}_{est}) | \right|^2 \qquad (10.26)$$

Here $J(\vec{R})$ is averaged over the points of a line R-R (Fig. 10.8) considered at some distance s from the row of atoms. In the derivation of Eq. (10.26) an exit plane wave function of the form $\psi_e(\vec{r}) = t_e(\vec{r})$ and $\psi_e(\vec{r}) = P_t(\vec{r})$ has been assumed for the second and third conjugate images to account for interference effects, $o_{2est}(\vec{r}) = \mathbf{FT^{-1}}\{T_{est}^2(\vec{q})\}$.

Thus, in order to determine defocus, beam separation, and the spherical aberration coefficient directly from the region of interest it is suggested that:

1. The STEM sideband off-axis hologram $I(\vec{q})$ is recorded under the optimum imaging conditions, providing the minimum interference of the conjugate images.

2. The exit plane wave function is restored from the expression Eq. (10.24) using estimated values for defocus and C_S.

3. The location of the atomic columns is determined from the amplitude/phase of the $H_1(\vec{r})$ distribution.

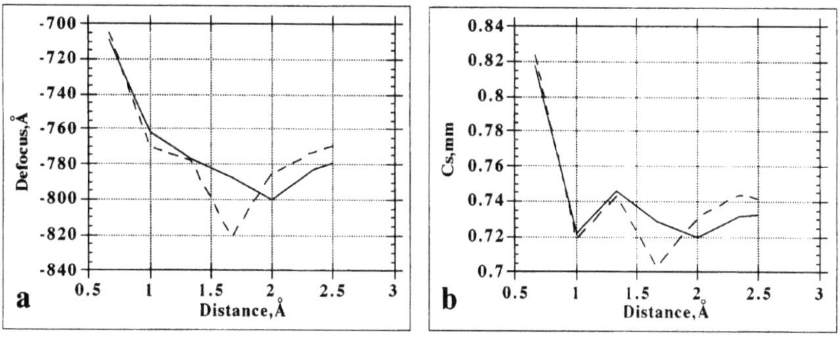

Figure 10.12. Effect of the crystal thickness and the finite spatial resolution of a hologram on the accuracy of defocus (a) and spherical aberration coefficient (b) refinement in 300 kV holography. True values $\Delta z_{tr} = -800$ Å, $C_{str} = 0.72$ mm. Refinement error is presented as a function of a distance s from the section R-R to location of atomic columns (see Fig. 10.8). The spatial resolution of a hologram $d = 1.6$ Å. Crystal thickness $t = 20$ Å(solid line), 40 Å(dashed line).

4. The section R-R of the distribution $H_1(\vec{r})$ is considered at some distance s from the columns. The refined values of defocus, C_S, and the beam separation a, correspond to the minimum of the $J(R)$.

5. The restoration of the exit plane wave function is performed using the new imaging parameters. Refined locations of the atomic columns are determined, since the exit plane wave function is now affected less by the errors in the determination of the imaging parameters.

6. Steps 4 and 5 are repeated until the difference between subsequent values of the refined imaging conditions does not exceed a certain limit.

Parameters affecting the accuracy of imaging conditions determination include the distance s, the resolution of a hologram $d = 1/q_0$, crystal thickness, and the inaccuracy of atomic position determination. The effect of the first three factors is concerned with the deviation of the wave function considered at the points R-R from the approximation Eq. (10.23). The error in atomic position determination indicates that one should find an optimum interval of distances s rather than its single value. Fig. 10.12 presents dependencies of the refined values of defocus and C_S on the distance s as a function of crystal thickness and the resolution of a hologram. The 300 kV model hologram was calculated for optimum imaging conditions ($\Delta z = -800$ Å, $C_S = 0.72$ mm) providing the minimum interference of conjugate images. The interval 1.08 Å $< s <$ 1.68 Å was found as optimum if the resolution of a hologram is at least $d = 1.6$ Å. The interval width corresponds to the presently achievable accuracy $\Delta_r = \pm 0.25$ Å of atomic positions determination. Refinement errors $\Delta z = \Delta z_{est} - \Delta z$ and $\Delta_c = C_{S\,est} - C_S$ increase as the resolution of a hologram d decreases. They were maintained in the range $|\Delta z| = 14$ Å to 20 Å, $|\Delta_c| = (1.6 \text{ to } 2.0)10^{-2}$ mm for 1.2 Å $\leq d \leq$ 1.6 Å and the crystal thickness $t \leq 40$ Å. The beam separation was refined with a sufficiently high accuracy $\Delta_a < 0.01$ Å.

The crystal thickness $t = 40$ Å implies a limit set by the dynamical scattering effects. It corresponds to an atomic column which consists of 20 uranium atoms with a maximum phase shift at the atomic site $\alpha_{max} = \max_r(\sigma V(\vec{r})t) = 5$. The maximum phase shift can be considered as the structure-independent criterion: its high value indicates that dynamical scattering imposes no serious restriction on the refinement method, and so the procedure can be applied to most experimental situations met in high resolution imaging. The requirement that only one row of atomic columns produces a contribution to the hologram along the considered line R-R is satisfied in most structures observed in the low-zone orientation since the appropriate values of s_{opt} are comparable with interatomic distances. The line R-R presents a simplest way of choosing a region from which imaging conditions are extracted; if interatomic distances in the structure are small and do not allow construction of a line with a sufficient number of pixels, the latter can be chosen in a special way. For example, only channels in the structure are analyzed. Finally, it is clear from Eqs. (10.23)-(10.26) that axial astigmatism can be also incorporated in the refinement procedure.

We conclude that the suggested method allows one to considerably improve the accuracy with which imaging conditions can be determined. However, it does not provide a knowledge of crystal thickness, which is important for successful application of STEM sideband off-axis holography. The development of appropriate methods should become the subject of future research. Since the method implies certain considerations to the exit plane wave function $\psi_e(\vec{r})$ (but not to the hologram!) it can be applied with minor modifications to any form of electron holography used for derivation of $\psi_e(\vec{r})$, for example, to TEM off-axis sideband holography.[153]

9. Conclusions

Considerable progress has been made in advancing the method off-axis STEM holography, especially in its application to the study of magnetic materials. The experimental realization of the full wave reconstruction with atomic spatial resolution awaits experimental realization, despite the fact that theoretical formulation and the ways for aberration correction are reasonably well understood. Its problems are concerned with finding a compromise between the sample stability and the required reconstruction accuracy, as well as handling very large amounts of data. If these problems are solved the method can offer significant resolution enhancement. The application of special detectors has proved to be effective for fast on-line display of both the amplitude and phase of the hologram. This is very important for initial evaluation of the object. Its application has been demonstrated with some limited spatial resolution both to biological molecules and thin crystals. It is clear that the method offers much promise and that efforts developing it further can bring new and exciting results.

Acknowledgments

M.A.G. is most grateful to Prof. J.M. Cowley for his kind encouragement and helpful discussions. J.S. appreciates the help and discussions with Prof. H. Lichte, Prof. K.-H. Herrmann and Prof. F. Lenz as well as the great teamwork with his friend Dr. Tommy Leuthner. Dr. J.K. Weiss is acknowledged for sharing his information on the ASU acquisition system. We are grateful to Dr. M. Mankos, Prof. M.R. Scheinfein and Prof. J.M. Cowley for Figs. 10.6 and 10.7. The project on the theory of off-axis STEM holography was supported by NSF grant DMR-8810238.

FOCUS VARIATION ELECTRON HOLOGRAPHY

Dirk Van Dyck and Marc Op de Beeck

University of Antwerp (RUCA-EMAT), B-2020 Antwerp, Belgium

1. Introduction

The idea of improving the resolution in HRTEM from a focal series of HRTEM images of weakly scattering objects by computer-aided compensation of the lens aberrations for the reconstruction of the complex wave function information at the level of the specimen is already quite old.[226,383,393] More recently, a more advanced method for wave function restoration, applicable to general and thicker objects, has been proposed.[456] But only very recently has the technical means become available both to capture the images in a quantitative way by using CCD cameras and to process them on fast computer systems. Moreover, during the last few years field emission gun (FEG) TEMs have become available, having a lower energy spread and a dramatically higher brightness than standard thermionic emitters. Therefore, the illuminating electron beam is much more spatially coherent, yielding an effective information limit ρ_i for the FEG-TEM that is far beyond that for a LaB_6–TEM, and that is almost exclusively controlled by the focal spread ($\rho_i = \sqrt{\pi\,\lambda\,\Delta/2}$), as described, e.g., in Ref. 460. This makes the idea of pushing the interpretable resolution down to the information limit using image processing techniques much more appealing.[458]

We believe that reconstruction techniques have not only become technically feasible, but even indispensable, since most of the high resolution information beyond the point resolution of the instrument is highly delocalized within the image, so that direct interpretation cannot be done without a reliable reconstruction method. Another aspect which is sometimes overlooked is that a reconstruction method may lead toward a direct quantitative interpretation of the HRTEM information in terms of the crystal structure, which avoids cumbersome simulation techniques.

The final goal of reconstruction methods is to obtain the structure of the object projected along the beam direction. However, the relation between the wave function and the projected structure is not always straightforward. In a sense, the dynamical scattering of the electrons in the object has to be reversed. In terms of the classical plane wave description for electron scattering, this is a tedious problem. Another problem is that, in the reconstruction procedure, the exact values of defocus, spherical aberration and astigmatism are not known with sufficient accuracy, so that the retrieved object function is still convoluted with a residual impulse response function. The fine correction of these effects requires prior knowledge of the object itself and can thus only be performed in the final structure retrieval step. Hence we need a simple and robust, albeit approximate, analytical expression for the wave function in terms of the projected structure which holds for realistic crystal thicknesses and which can be used for the final fitting of the object structure. For this purpose we developed the single column channelling theory. There is need for a simple intuitive theory that is valid for larger crystal thicknesses. In our view, a channelling theory fullfills this need.

2. Image formation in a TEM

The image formation process in HRTEM can be split in two distinct parts. In a first step, the incoming electron wave is modulated due to diffraction at the atoms in the specimen. In the case of a zone-axis orientation, this yields a regular grating of atom columns. For very thin crystals, this results primarily in a phase shift which is larger close to the atom cores, while the amplitude nearly remains unchanged. The specimen is then considered to be a "pure phase object." For thicker specimens, both amplitude and phase will be changed and the information about the crystal potential will be distributed over both. The resulting wave function at the exit plane of the specimen is called "the object wave function." It can be shown that the detail in the object wave function, even for thicker specimens, is directly related to the projected column potentials.[462]

In the second step of the image formation process, the phase and amplitude of the object wave function are convoluted with a complex point spread function which accounts for the aberrations of the objective lens. Moreover, only the intensity of the resulting "image wave function" can be recorded, while the phase information is lost. Figure 11.1 represents the CTF for a typical 200 kV LaB_6–TEM and a FEG-TEM at Scherzer focus. For the FEG-TEM it clearly shows the large gap between the point resolution ρ_S (the first zero-crossing of the CTF) and the information resolution ρ_i. Beyond ρ_S the wave aberration function oscillates rapidly. This strongly hampers the direct interpretation of the recorded HRTEM images, even for very thin specimens. Each high resolution image can therefore be considered as a highly blurred representation of the atomic structure of the specimen. The interference between the different diffracted beams can be varied by altering an electron-optical parameter which is reflected in a change of the applied microscope transfer function. The most common method is defocusing, often called "through-focusing." We will take advantage of the knowledge of the variation of the lens aberration function with the defocus settings to find an optimum object wave function which is compatible with the recorded image.

3. Least squares wave function reconstruction

Throughout this chapter, we will describe the exit wave function as well as the image intensity as a function of their reciprocal vectors $\vec{q}$ rather than their real space vectors $\vec{r}$, since this allows for a more convenient description of the effect of the transfer functions. The principle of reconstructing the specimen exit face wave function $\Psi(\vec{q})$ out of a focal series of HRTEM images $I_n^{\exp}(\vec{q})$ with $(n = 1, \ldots, N)$ can then be formulated within a "maximum likelihood" framework.[226] This involves an ordinary least squares formulation under the assumption of Gaussian noise characteristics

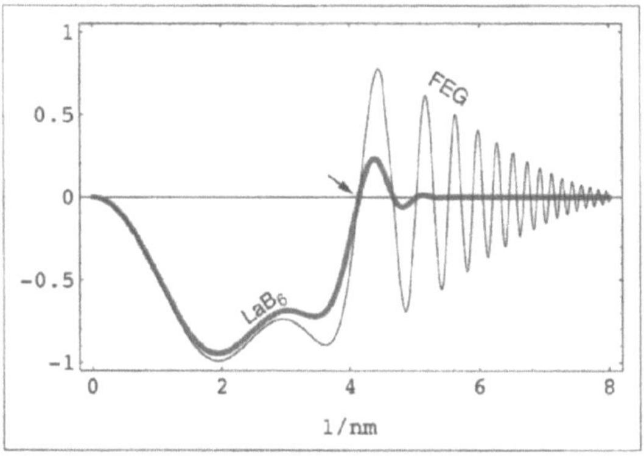

Figure 11.1. Contrast transfer function for a typical LaB_6 (thick line) and a FEG-TEM (thin line) at 200 keV. Both exhibit the same point resolution (i.e., the first zero-crossing; see arrow), but the FEG microscope allows information transfer up to higher frequencies.

$$S = \sum_{n=1}^{N} \iint |M_n(\vec{q})|^2 \, \mathrm{d}\vec{q} \tag{11.1}$$

with the image difference $M_n(\vec{q})$ defined as

$$M_n(\vec{q}) = I_n^{\exp}(\vec{q}) - I_n(\Psi(\vec{q})) - N_n(\vec{q})\,\delta(\vec{q}) \tag{11.2}$$

where $N_n(\vec{q})\,\delta(\vec{q})$ represents a background constant ("fog" level) which might be different for every recorded image. Minimizing S with respect to $\Psi(\vec{q})$ and $\Psi^*(-\vec{q})$ leads to the following equation pair representing weighted correlations between the object wave function $\Psi(\vec{q})$ and the difference image $M_n(\vec{q})$

$$\begin{aligned} 0 &= \sum_{n=1}^{N} \int \Psi(\vec{p}) \cdot T_n(\vec{p}, \vec{q}) \cdot M_n(\vec{q} - \vec{p}) \, \mathrm{d}\vec{p} \\ 0 &= \sum_{n=1}^{N} \int \Psi^*(-\vec{p}) \cdot T_n \cdot (\vec{q}, -\vec{p}) \cdot M_n(\vec{q} - \vec{p}) \, \mathrm{d}\vec{p} \end{aligned} \tag{11.3}$$

In this formulation, the function T_n is the so-called Transmission Cross Coefficient[208] which is used to estimate the image intensities $I_n(\Psi(\vec{q}))$ at defocus Δz_n. Equations (11.3) are highly non-linear in the wave function $\Psi(\vec{q})$; therefore a solution in a closed form cannot be found.

This set of least squares requirements forms the basis for both the parabola and the MAximum Likelihood (MAL) methods.[463] In the MAL method, this requirement is used in the form of a Picard operator. It starts from a "proposal" for the wave function Ψ, and checks whether this wave function fulfils the requirements of Eq. (11.3) for all reciprocal q vectors. If not, the difference function is partially fed back in order to get an updated proposal for Ψ. This procedure is repeated until convergence is reached. In the parabola method on the other hand, the requirements Eq. (11.3) are approximated, and only the terms $p = 0$ are retained. We will further denote $\Psi(\vec{q}) = \alpha\,\delta(\vec{q}) + \Phi(\vec{q})$ with $\Phi(0) = 0$. Physically, this is just a separation of the central beam α from the diffracted beams $\Phi(\vec{q})$. Neglecting the non-linear contribution to the image formation, (i.e., the interference between the diffracted beams $\Phi(\vec{q})$ themselves), in a first instance, this procedure directly leads to the filter functions for $\vec{q} = \vec{o}$

$$f_n(\vec{q}) = \frac{T_n(\vec{o},\vec{q}) - [P^*(\vec{q})/Q(-\vec{q})] \cdot T_n(-\vec{q},\vec{o})}{[Q(\vec{q}) \cdot Q(-\vec{q}) - |P(\vec{q})|^2]/Q(-\vec{q}) + \eta(\vec{q})} \tag{11.4}$$

with $\vec{o}$ the zero vector (or origin in Fourier space), where $\eta(\vec{q})$ is the noise-to-signal ratio and the P and Q functions are given by

$$P(\vec{q}) = \sum_{n=1}^{N} T_n(\vec{q},\vec{o}) \cdot T_n(-\vec{q},\vec{o}) \tag{11.5}$$

$$Q(\vec{q}) = \sum_{n=1}^{N} |\, T_n(\vec{q},\vec{o})\,|^2 \tag{11.6}$$

Note that these filter functions are only valid for the determination of the interference between the central and the diffracted beams $\alpha^* \, \Phi(\vec{q})$. The central beam α can be calculated from the normalization requirements. Different methods have to be applied for zero and non-zero average noise contributions.[328]

It is clear that a minimum of two images (which are linearly independent in the mathematical sense) is required to reconstruct the exit wave function since each recorded image only shows the modulus squared of the image wave functions, thus omitting the phase information. Obviously, if we want to reconstruct the amplitude and phase of the object wave function, the number of data points should at least be equal to the number of unknowns. The use of more images will certainly improve the validity of the solution since the restrictions on the object wave function become more stringent.

3.1. Physical meaning of the filter functions

By performing an extra Fourier transform with respect to the reciprocal defocus axis, the filter function Eq. (11.4) can be shown to be a parabolic filter "$\zeta = 0.5\lambda\, q^2$" in the 3-D reciprocal space $(\vec{q},\zeta)$, since in the limit of an infinite defocus range and zero beam divergence, $P(\vec{q})$ will vanish and $Q(\vec{q})$ will reduce to the square of the damping envelope due to chromatic aberration.

The first term in the numerator of Eq. (11.4) basically selects information on the parabola "$\zeta = 0.5\,\lambda\, q^2$", on which the linear wave function $\alpha^* \, \Phi(\vec{q})$ is located. The application of the transfer functions $T_n(\vec{o},\vec{q})$ will introduce phase shifts which can be interpreted as a "back propagation" of the linear information of $\alpha^* \, \Phi(\vec{q})$ in the images at defoci Δz_n toward the exit plane of the specimen. Due to this first term, the conjugate wave function $\alpha \, \Phi^*(-\vec{q})$ on the parabola "$\zeta = 0.5\,\lambda\, q^2$" equally exhibits the same phase shift, yielding a parabola at "$\zeta = \lambda\, q^2$." In a practical application, however, only a limited defocus range can be used. Therefore, this parabola will be convoluted with a kind of spread function along the reciprocal ζ-axis, yielding a finite chance that the conjugate wave function will contribute to the information at $\zeta = 0$. This is corrected for by the second term.

The denominator of Eq. (11.4) basically represents how well the frequencies can be reconstructed for a specific defocus series. For zero beam convergence and in the case of an equidistant defocus series with step Δz, the denominator can thus be rewritten as

$$N\, e^{-(\pi\,\lambda\,\Delta\,q^2)^2} \cdot \left\{ 1 - \left[\frac{1}{N} \cdot \frac{\sin(N\,\pi\,\lambda\,q^2\,\Delta z)}{\sin(\pi\,\lambda\,q^2\,\Delta z)} \right]^2 \right\} + N\,\eta(\vec{q}) \tag{11.7}$$

This clearly shows that the low frequencies are difficult to retrieve. The lowest frequency which can easily be reconstructed is of the order of

$$q_{low} = \frac{1}{\sqrt{\lambda\,\Delta z^{tot}}} \tag{11.8}$$

with Δz^{tot} the total defocus range. In addition, if the defocus series contains a certain periodicity in the defocus step, the denominator rapidly goes to zero for frequencies which are resonant for this period. This will happen for the first time for the g_{max} value which satisfies the condition

$$q_{max} = \frac{1}{\sqrt{\lambda\,\delta_m}} \tag{11.9}$$

with δ_m the largest periodicity along the defocus series. Thus, in case one wants to reconstruct down to the information limit, the largest periodicity δ_m should be $< 0.5\,\pi\,\Delta$.

The combination of both requirements puts some restrictions on the possible defocus values used in the reconstruction experiment. Tests on both simulations and experimental images have proven that a series of 20 images is more than sufficient for the proposed reconstruction procedure, provided an iterative procedure is applied to correct for the inaccuracies of the first linearized guess, using Eq. (11.4).

3.2. Correction for the non-linear contribution

In the case of (quasi-)linear imaging, i.e., in the Weak Phase Object approximation, the second order interference effects $I_n(\Phi(\vec{q}))$ can be omitted, because the non-linear terms $|\Phi(\vec{r})|^2$ (interference between scattered beams) can be neglected compared to $\alpha^*\,\Phi(\vec{r})$ (interference between central beam and one of the scattered beams). In the absence of noise and non-linear terms, Eq. (11.4) is equivalent to the result obtained by Ref. 383 for image reconstruction of a complex object in the regime of linear image formation.

In the general case of non-linear imaging, the term $I_n(\Phi(\vec{q}))$ should definitely not be omitted. This term is clearly correlated with the linear contribution and should therefore not be included in the noise. The linear parabola reconstruction formula, however, may be the ideal starting point for a further iteration procedure in order to account for non-linear effects. For the first iteration, the non-linear terms $I_n(\Phi(\vec{q}))$ are neglected, yielding

$$\alpha^*\,\Phi^0(\vec{q}) = \alpha^*\,\Phi(\vec{q}) + \sum_n F_n(\vec{q}) \cdot I_n(\Phi(\vec{q})) \tag{11.10}$$

where $\alpha^* \Phi(\vec{q})$ is the required final solution and the last term represents the erroneous contribution of the non-linear terms to the first guess. We will now try to find an optimized update for this approximate solution and therefore we write

$$\Psi_{j+1} = \Psi_j + \varphi \tag{11.11}$$

where φ can be considered to be small in comparison with the diffracted beams Φ_j. Again using the ordinary least squares technique we get up to first order in the deviation, which can, in a first order approximation, be reduced to a 2x2 set of linear inhomogeneous equations. These can be solved as

$$\varphi(\vec{q}) = \frac{c(\vec{q}) \cdot a^*(-\vec{q}) - c^*(-\vec{q}) \cdot b(\vec{q})}{a(\vec{q}) \cdot a^*(-\vec{q}) - b(\vec{q}) \cdot b^*(-\vec{q})} \tag{11.12}$$

where the coefficients are defined as

$$\begin{aligned} a(\vec{q}) &= \sum_n \int |\, \Psi_j(\vec{p}) \cdot T_n(\vec{p}, \vec{q}) \,|^2 \, \mathrm{d}\vec{p} \\ b(\vec{q}) &= \sum_n \int \Psi_j(\vec{p}) \cdot \Psi_j^*(-\vec{p}) \cdot T_n(\vec{p}, \vec{q}) \cdot T_n(-\vec{p}, -\vec{q}) \, \mathrm{d}\vec{p} \\ c(\vec{q}) &= \sum_n \int \Psi_j(\vec{p}) \cdot T_n(\vec{p}, \vec{q}) \cdot M_{n,j}(\vec{q} - \vec{p}) \, \mathrm{d}\vec{p} \end{aligned} \tag{11.13}$$

This general approach has been suggested in Ref. 52 as the Self Consistent MAL procedure, and can be approximated to the parabola equivalent for $q \neq 0$

$$\alpha^* \Phi_{j+1}(\vec{q}) = \alpha^* \Phi_j(\vec{q}) + \sum_n F_n(\vec{q}) \cdot [I_n^{\exp}(\vec{q}) - I_n(\Psi_j(\vec{q}))] \tag{11.14}$$

with $I_n(\Psi_j(\vec{q}))$ the theoretically estimated intensity for the reconstructed wave function in the j-th iteration step. It is clear that the reconstruction procedure should try to minimize the second term. This implicitly means that the parabola iterator is incomplete: the filter-function $F_n(\vec{q})$ will only select information on a parabola, which, in the limit of zero beam convergence and an infinite defocus interval, is a two dimensional section in the three dimensional reciprocal space. The iteration procedure, as described in Eq. 11.14, in fact only compensates for the non-linear contribution on this 2D section. Therefore, not all information in the defocus series is optimally used, which makes the parabola method less robust, i.e., more sensitive to noise than the MAL method, in which all data in the 3D reciprocal space is used.

But precisely the fact that only the data on the parabolas is selected allows us to speed up the calculations enormously since both linear and non-linear terms on this parabola feel exactly the same damping envelope due to chromatic aberration. Therefore, the time-consuming convolutions in the Transmission Cross Coefficient can be totally avoided, since the spatial damping envelopes $E_{\alpha,n}(\vec{q}, \vec{o})$ can be described coherently for a FEG-TEM. We thus first calculate the coherent image intensity including the spatial coherence with the modified transfer functions

$$t_n(\vec{q}, \vec{o}) = e^{-i\chi(\vec{q})} \cdot E_{\alpha,n}(\vec{q}, \vec{o}) \tag{11.15}$$

The effect of the chromatic aberration on the parabolas can then be introduced by convoluting this intensity with the chromatic spread function. Even this step can be avoided, if the specific form of the filter functions is taken into account. In the MAL approach, the chromatic aberration envelope has to be calculated correctly in the complete field of the 3D reciprocal space. However, this damping envelope can equally well be introduced by weight averaging the coherent image intensities at different foci. It has been proven that averaging over only a few (typically 10) equidistantly spaced images yields a very good approximation. In this way, the calculations can be speeded up by several orders of magnitude.[52]

3.3. General strategy in practical applications

Due to specimen drift and the finite recording time, the different experimental images of the defocus series may not be well aligned with respect to one another. This is reflected in a lateral image shift $\delta\vec{r}_n$ across the field of view. A coupled solution for the correction of the defocus Δz_n and lateral misalignment ($\delta\vec{r}_n$), assuming a first order approximation and only small deviations from the provided starting values, is given in Ref. 226. In our experience, this procedure works well for sub-pixel lateral alignment, but requires a few iteration steps for the correction of an initial misalignment of more than one-to-two pixels. Using a cross-correlation function (CCF), the exact misalignment parameter can be found in only one iteration step.

Therefore, we proceed in the following way:

1. The images are aligned as well as possible using the CCF. As pointed out by Frank,[112] the CCF becomes less and less accurate for increasing focus differences between the images due to contrast reversals etc. In our case, the defocus distance between the different images is of the order of the defocal spread (i.e., 4 nm), so that contrast reversals are unlikely to occur. In the case of a very thin specimen, more accurate results can be obtained with the improved CCF of Saxton,[384] which is applicable for even larger defocus distances. But in most general cases, where the object cannot be considered to be thin, only the mutual misalignment between two adjacent images can be measured accurately. Therefore, one has to be aware of the fact that there might still be a substantial alignment error between the first and the last images of the defocus series.

2. The (roughly) aligned focal series is used to calculate the first estimate of the wave function $\alpha^* \Phi(q)$ using Eq. (11.4).

3. The contribution of the background "fog" level $N_n(q)\,\delta(q)$ is estimated in combination with the contribution of the central beam $\alpha\,\delta(q)$, yielding the wave function $\Psi(\vec{q})$.

4. $\Psi(\vec{q})$ and $N_n(q)\,\delta(q)$ are used to calculate an estimate of all the different experimental images of the series. The estimated and experimental intensities are cross-correlated in order to refine the lateral alignment. Note that in this alignment step no contrast reversals occur since the experimental image is correlated with the estimated image for the same defocus.

5. The squared differences between the experimental and the estimated images are calculated to be able to judge the convergence.

6. We recalculate the estimate of the wave function $\alpha^* \Phi(q)$, including the non-linear corrections.

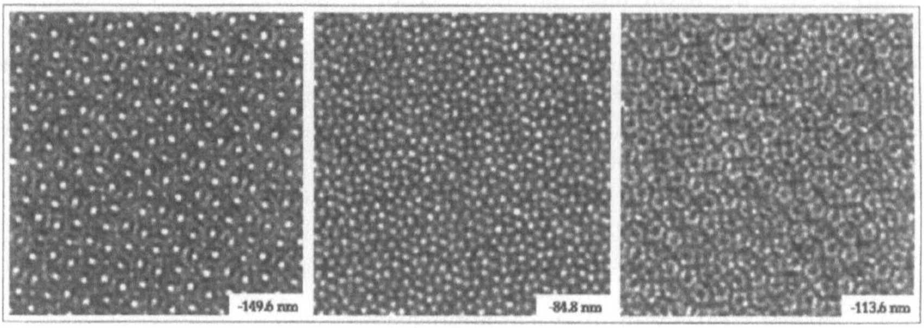

Figure 11.2. Selected images from a through focus series of $Ba_2NaNb_5O_{15}$.

7. We then repeat steps 3) to 6) until convergence is obtained for both the position alignment and the Least Squares criterion.

Experiments have show that perfect alignment and convergence can be achieved in about four iterations.

4. Experimental results

The results we want to present here have been obtained with a Philips CM20 FEG-SuperTwin. This microscope operates at 200 kV and has a point resolution of 2.4 Å and a theoretical information limit of the order of 1.2 Å. A series of 20 images was recorded digitally using a 1024^2 pixel Slow Scan CCD camera. The equidistant defocus step was -72 Å. The microscope and the CCD camera were driven by a TVIPS computer running an extension of the TCL image processing software.[455] Special TCL procedures were developed which enable fully automatic recording of the focal series through the remote control of the CM-microscope. The total exposure time for one micrograph was of the order of 1 second. The whole focal series of 20 images required about one minute and utilized about 10 MB of disk space. Flat field and dark field corrections were performed according to Ref. 79. Electron-optical parameters such as spherical aberration C_S and the defocus increment have been measured by the method described in Ref. 51. The parabola computations themselves were performed on an IBM-590, and required a total reconstruction time of a few minutes. Figure 11.2 shows a through focus series for $Ba_2NaNb_5O_{15}$. Figure 11.3 shows the reconstructed exit wave. The heavy atoms are revealed in the amplitude while the light atoms are found in the phase. From this exit wave, the positions of the atomic columns can be determined. However, in order to determine the structure accurately, one has to invert the dynamical scattering in the object. For this, we will use a simple channelling theory.

5. Interpreting the reconstructed object wave

It is well known that, when a crystal is viewed along a zone axis, i.e., parallel to the atom columns, the high resolution images often show a one-to-one correspondence with the configuration of columns, provided the distance between the columns is large enough and the resolution of the instrument is sufficient. From this, it can be suggested that, for a crystal viewed along a zone axis with sufficient separation between the columns, the wave function at the exit face depends primarily on the projected structure, i.e.,

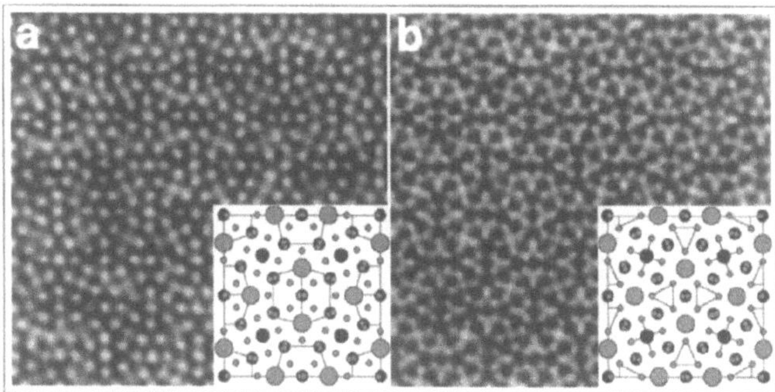

Figure 11.3. Reconstructed amplitude (a) and phase (b) of the exit face wave function.

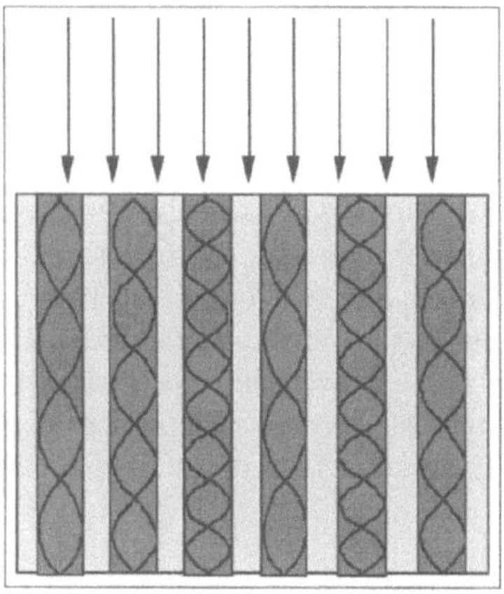

Figure 11.4. Schematical representation of electron channelling. The importance of channelling for interpreting high resolution images has often been ignored or underestimated, probably because for historical reasons, dynamical electron diffraction is often described in reciprocal space. However, most of the high resolution images of crystals are taken in a zone axis orientation, in which the projected structure is the simplest, and in which the number of diffracted beams are the largest. We therefore believe that a simple real-space channelling theory yields a much more useful and intuitive, albeit approximate, description of the dynamical diffraction, and provides an intuitive interpretation of high resolution images, even for thicker objects.

on the type of atom columns. Hence, the classical picture of electrons traversing the crystal as plane-like waves in the directions of the Bragg beams, which stems from the X-ray diffraction picture and upon which most of the simulation programs are based, is in fact misleading. The physical reason for this "local" dynamical diffraction is the channelling of the electrons along the atom columns parallel to the beam direction. Due to the positive electrostatic potential of the atoms, a column acts as a guide or channel for the electron[208,221,274,328] within which the electron can scatter dynamically without leaving the column as displayed in Fig. 11.4. It has been proposed[52] that this so-called atom column approximation can be exploited to speed up the dynamical diffraction calculations by assembling the wave function at the exit face using parts that have been calculated for each atom column separately.

If we assume that the fast electron in the direction of propagation (z-axis) behaves as a classical particle with velocity $v = \hbar k/m$, we can consider the z-axis as a time axis with

$$t = m\, z/\hbar\, k \tag{11.16}$$

Hence we can start from the time-dependent Schrödinger equation

$$-\frac{\hbar}{i}\frac{\partial}{\partial t}\Psi(\vec{r},t) = \mathbf{H}\,\Psi(\vec{r},t) \tag{11.17}$$

with

$$\mathbf{H} = -\frac{\hbar^2}{2m}\Delta_{\vec{r}} - e\,U(\vec{r},t) \tag{11.18}$$

with $U(\vec{r},t)$ the electrostatic crystal potential, m and k the relativistic electron mass and wavelength, and $\Delta_{\vec{r}}$ the Laplacian operator acting in the plane $(\vec{r})$ perpendicular to z. Using Eq. (11.16) we then have

$$\frac{\partial}{\partial z}\Psi(\vec{r},z) = \frac{i}{4\pi\,k}\cdot[\Delta_{\vec{r}} + V(\vec{r},z)]\cdot\Psi(\vec{r},z) \tag{11.19}$$

with

$$V(\vec{r},z) = \frac{2\,m\,e}{\hbar^2}\cdot U(\vec{r},z) \tag{11.20}$$

This is the well-known high energy equation in real space which can also be derived from the stationary Schrödinger equation in the forward scattering approximation. If the periodicity of the crystal along the incident beam direction is not too large so that no higher order Laue zones (HOLZ) contribute to the diffraction, the projection approximation is valid; i.e., the fast electron only sees an "averaged" potential along the beam direction. In that case, $V(\vec{r},z)$ in Eq. (11.20) becomes independent of z. Then the solution of Eq. (11.19) can be expanded in eigenfunctions of the Hamiltonian

$$\Psi(\vec{r},z) = \sum_n c_n\cdot\phi_n(\vec{r})\cdot e^{-i\pi\,k\,z\,E_n/E_0} \tag{11.21}$$

with $\mathbf{H}$ given by Eq. (11.18) and the incident electron energy

$$E_0 = \frac{\hbar^2\,k^2}{2\,m} \tag{11.22}$$

where $\vec{k}$ is the electron wave vector. For $E_n < 0$ the states are bound to the columns. We now rewrite Eq. (11.21) as

$$\Psi(\vec{r},z) = \sum_n c_n\,\phi_n(\vec{r})\cdot\left[\left(1 - \frac{i\pi\,k\,z\,E_n}{E_0}\right) + \left(e^{-i\pi\,k\,z\,E_n/E_0} + \frac{i\pi\,k\,z\,E_n}{E_0} - 1\right)\right] \tag{11.23}$$

The coefficients c_n are determined from the boundary condition

$$\Psi(\vec{r},0) = \sum_n c_n\cdot\phi_n(\vec{r}) \tag{11.24}$$

from which

$$c_n = \int \phi_n^*\cdot\Psi(\vec{r},0)\,\mathrm{d}\vec{r} \tag{11.25}$$

In case of plane wave incidence, one thus gets

$$\sum_n c_n \cdot \phi_n(\vec{r}) = 1 \tag{11.26}$$

and from Eqs. (11.18) and (11.21)

$$\sum_n c_n \cdot \phi_n(\vec{r}) \cdot E_n = \mathbf{H}\Psi(\vec{r}, 0) = \mathbf{H} \cdot 1 = -e\,U(\vec{r}) \tag{11.27}$$

Now Eq. (11.23) becomes

$$\begin{aligned}\psi(\vec{r}, z) = 1 &+ \frac{i\pi\, k\, z\, e\, U(\vec{r})}{E_0} \\ &+ \sum_n c_n \cdot \phi_n(\vec{r}) \cdot \left(e^{-i\pi\, k\, z\, E_n/E_0} + \frac{i\pi\, k\, z\, E_n}{E_0} - 1\right)\end{aligned} \tag{11.28}$$

The first two terms yield the well-known weak phase object approximation. In the third term only those energy states will appear in the summation for which

$$|E_n| \geq \frac{E_0}{k\, z} \tag{11.29}$$

In case the object is very thin, so that no state obeys Eq. (11.29), the weak phase object approximation is valid. For a thicker object, only bound states will appear with very deep energy levels, which are localized near the column cores. Furthermore, a two-dimensional projected column potential has only very few deep states, and when the overlap between adjacent columns is small only the radial symmetric states will be excited. In practice, for most types of atom columns, only one state appears, which can be compared with the 1S state of an atom.

In the case of an isolated column, taking the origin in the center of the column, we then have

$$\begin{aligned}\psi(\vec{r}, z) = 1 &+ i\pi\, k\, z\, e\, U(\vec{r})/E_0 \\ &+ c \cdot \phi(\vec{r}) \cdot \left(e^{-i\pi\, k\, z\, E/E_0} + \frac{i\pi\, k\, z\, E}{E_0} - 1\right)\end{aligned} \tag{11.30}$$

A very interesting consequence of this description is that, since the state ϕ is very localised at the atom cores, the wave function for the total crystal can be expressed as a superposition of the individual column functions

$$\begin{aligned}\psi(\vec{r}, z) = 1 &+ \frac{i\pi\, k\, z\, e\, U(\vec{r})}{E_0} \\ &+ \sum_i c_i \cdot \phi_n(\vec{r} - \vec{r}_i) \cdot \left(e^{-i\pi\, k\, z\, E_i/E_0} + \frac{i\pi\, k\, z\, E_i}{E_0} - 1\right)\end{aligned} \tag{11.31}$$

with

$$U(\vec{r}) = \sum_i U_i(\vec{r} - \vec{r}_i) \tag{11.32}$$

If all the states other than the 1S have very small energies i.e.,

$$|E_n| \ll \frac{E_0}{k\,z} \tag{11.33}$$

then Eq. (11.23) can be simplified to

$$\Psi(\vec{r}, z) = \sum_n c_n \cdot \phi_n(\vec{r}) + \sum_n c_n \cdot \phi_n(\vec{r}) \cdot \left(e^{-i\pi\, k\, z\, E_n/E_0} - 1\right) \tag{11.34}$$

so that Eq. (11.30) becomes

$$\Psi(\vec{r}, z) = 1 + c \cdot \phi(\vec{r}) \cdot \left(e^{-i\pi\, k\, z\, E_i/E_0} - 1\right) \tag{11.35}$$

and Eq. (11.31) becomes

$$\Psi(\vec{r}, z) = 1 + \sum_i c_i \cdot \phi_i\,(\vec{r} - \vec{r}_i) \cdot \left(e^{-i\pi\, k\, z\, E_i/E_0} - 1\right) \tag{11.36}$$

Expressions (11.31) and (11.36) are the basic result of the channelling theory.

The interpretation of Eq. (11.36) is simple. Each column i acts as a channel in which the wave function oscillates periodically with depth. The period is related to the "weight" of the column, i.e., proportional to the atomic number of the atoms in the column and inversely proportional to their distance along the column. The importance of these results lies in the fact that they describe the dynamical diffraction which occurs for larger thicknesses than the usual phase grating approximation, and that they require only the knowledge of one function ϕ_i per column (which can be tabulated similarly to atom scattering factors or potentials). Furthermore, even in the presence of dynamical scattering, the wave function at the exit face still retains a one-to-one relation with the configuration of columns. Hence this description is very useful for interpreting high resolution images and for providing a possible answer to the direct retrieval problem. Equations (11.35 and 11.36) apply to light columns, such as Si(111) or Cu(100) with an accelerating voltage up to about 200 keV. When the atom columns are "heavier" and the accelerating voltage higher (which, due to the relativistic correction, also increases the effective strength of the potential), then Eq. (11.31) has to be used. This is, for example, the case for Au(100).

Figure 11.5 shows the electron density as a function of depth in a Au_4Mn alloy crystal for 200 keV incident electrons. The corners represent the projection of the Mn column. The square in the center represents the four Au columns. The distance between adjacent Au columns is 0.2 nm. The periodicity along the direction of the column is 0.4 nm. From these results it is clear that the electron density in each column fluctuates nearly periodically with depth. For Au this periodicity is about 4 nm and for Mn 13 nm. These periodicities are nearly the same as for isolated columns so that the influence of neighboring columns in this case is still small. The energies of the respective 1S states are respectively about 250 eV and 80 eV.

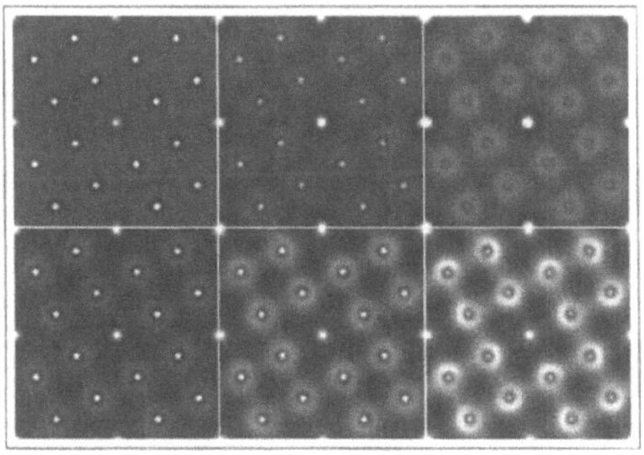

Figure 11.5. Electron density as a function of depth in Au_4Mn (2, 4, 6, 8, 10, 12 nm).

When the atoms are heavy and the accelerating voltage very high (0.5 to 1 MeV), a larger number of states come in and the result becomes more complicated. When the crystal is viewed along a higher index zone axis, the distance between adjacent columns decreases and the weight of the columns also decreases. Hence the bound states broaden, and overlap between adjacent columns starts to occur. This can be incorporated into the theory using perturbation methods. When the overlap between columns is too large, they may be considered as a kind of molecule.[52] The localization can also be improved by using higher voltages. It must be stressed that the derived results are only valid in a perfect zone axis orientation. A slight tilt can destroy the symmetry and excite other, non-symmetric states, so that the results become much more complicated. It is interesting to note that channelling has usually been described in terms of Bloch waves.[221] However, as follows from the foregoing, channelling is not a mere consequence of the periodicity of the crystal but occurs even in an isolated column parallel to the beam direction. In fact, this problem can be treated mathematically by making the column artificially periodical so as to generate a basis of functions (Bloch functions) to expand the wave function. In this view, the Bloch character is only of mathematical importance. This is the case even in a crystal in which the distance between the adjacent columns is sufficient (e.g., 0.2 nm). Bloch wave calculations then yield the same 1S states as found in our simplified treatment. Only when the overlap between columns increases or when the beam is inclined, do the other Bloch states become physically important.

6. Validity of the channelling theory

In order to test the validity of the theory, we compared the theoretical results with simulations using a real space multi-slice program. In the case of one single (not too heavy) column of type i, centered at the origin, Eq. (11.20) gives

$$\Psi(\vec{r}, z) - 1 = 2\,\phi_i(\vec{r}) \cdot \sin(0.5\pi\, k\, z\, E_i/E_0) \cdot e^{-0.5\pi\, i\, k\, z\, (1+E_i/E_0)} \tag{11.37}$$

i.e.,

- the amplitude decreases with increasing distance r from the center of the columns with a slope given by $\phi(\vec{r})$, the 1S eigenfunction

- the amplitude oscillates with depth with a periodicity of $4\,E_0/(k\,E_i)$

- the phase increases linearly with depth, starting from $\pi/2$

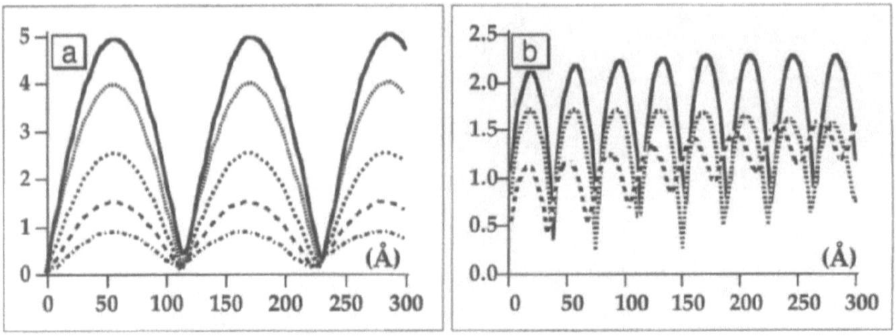

Figure 11.6. Amplitude of $\Psi - 1$ for an isolated Cu (a) and Au (b) column (200 keV).

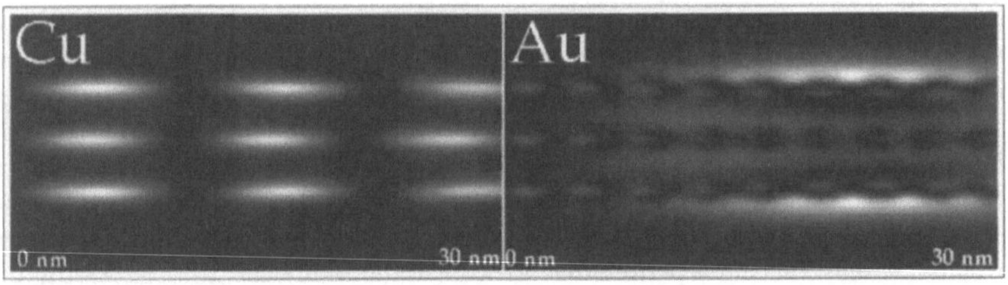

Figure 11.7. $|\Psi - 1|^2$ for an artificial unit cell of (5 x 3 x 4 Å) consisting of three Cu (or Au) columns with inter-column distance of 1 Å.

Figure 11.6a shows the amplitude $(\Psi - 1)$ for one isolated Cu column, with periodicity 3.8 Å along the ordered axis. These results are in complete agreement with Eq. (11.37). Figure 11.6b shows the amplitude for an isolated heavier column of Au with repeat distance of 4.08 Å along the column. Here the oscillation of the amplitude is superimposed onto a linearly increasing function. This is in agreement with the improved expression Eq. (11.30). In order to demonstrate the validity of the theory when the columns are very close we performed a simulation for an artificial unit cell (5 x 3 x 4 Å) consisting of three Cu (or Au) columns with an artificial distance of 1 Å (Fig. 11.7). It is clear that the periodicity, as expected for an isolated column, holds for Cu (Au) up to a thickness of 30 nm (10 nm).

7. Scaling

It is clear from Figs. 11.8 and 11.9 that the eigenfunctions scale with $\sqrt{|E|}$ and the eigenenergy E scales with $Z/d^{5/4}$, where Z is the atomic mass and d the distance between the atoms in the column. This scaling behaviour makes it possible to parametrize the wave function in the form

$$\Psi(\vec{r}, z) \approx 1 + \frac{e^{-i\pi k z E/E_0} - 1}{\sqrt{E}} \cdot \phi_0(\vec{r}\,\sqrt{|E|}\,) \tag{11.38}$$

This expression can still be improved so as to hold for thicker objects and/or heavier columns

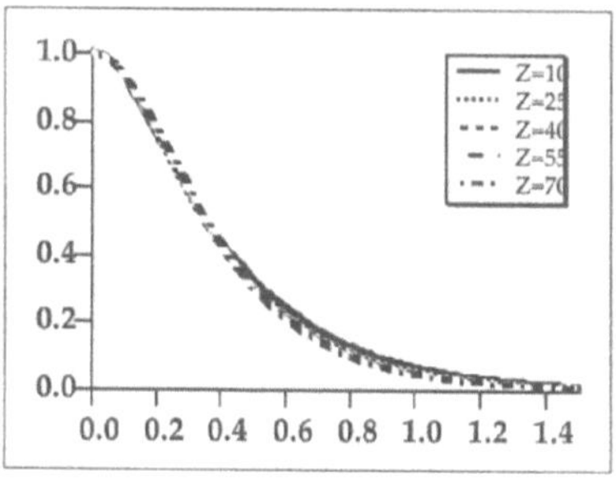

Figure 11.8. The scaled eigenstate $\psi(r)$ as a function of $|E|^{1/2}$ for various mass Z.

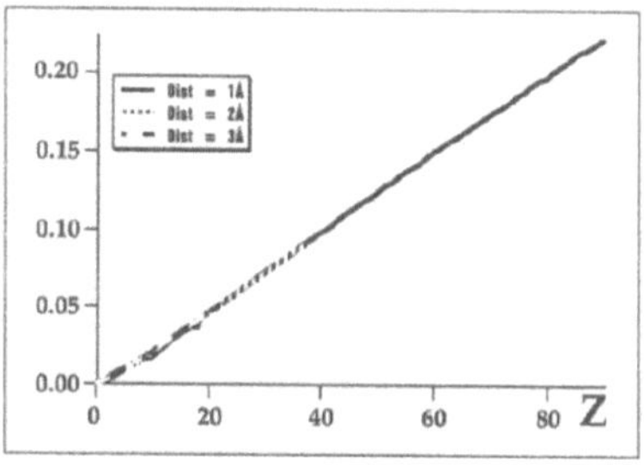

Figure 11.9. The linear scaling of the energy E as a function of $Z/d^{5/4}$.

$$\Psi(\vec{r}, z) \approx 1 + E^{4/5} \cdot V_0(\vec{r}) + \frac{e^{-i\pi\, k\, z\, E/E_0} + i\pi\, k\, z\, E/E_0 - 1}{\sqrt{|E|}} \cdot \phi_0(\vec{r}\,\sqrt{|E|}\,) \tag{11.39}$$

where V_0 and ϕ_0 are universally scaled functions for the projected potential, and the most important channelling eigenstate, i.e., the strongly bound 1S state. The advantage of this result is that the exit wave function is expressed analytically in terms of the parameters (Z,d,z, column position) that describe the crystal structure. This result enables images and diffraction patterns to be calculated in an analytical (and thus very fast) way and also enables the inverse problem of deriving the crystal structure by fitting with the experimentally reconstructed object wave to be solved.

8. Other aspects of channelling

8.1. Channelling and defects

The channelling effect occurs also in the presence of defects such as translation interfaces, twin interfaces, and dislocations, provided the columns parallel to the incident beam are not disrupted. To demonstrate this we take again the Au_4Mn alloy of Fig. 11.5 in which we now introduce translation interfaces (antiphase boundaries) (Fig. 11.10). The electron density, calculated with the periodic continuation method, is shown at the left (thicknesses 8, 10 nm). In the middle, we represent the channelling approximation, i.e., we perform the simulation for individual columns, which are then assembled according to the structure positions. If the channelling along the individual columns were not affected by the interface, the electron density at both sides of the interface would be identical to that of the perfect crystal shifted over the displacement vector of the interface. In order to reveal the deviation from this ideal situation. we subtract this

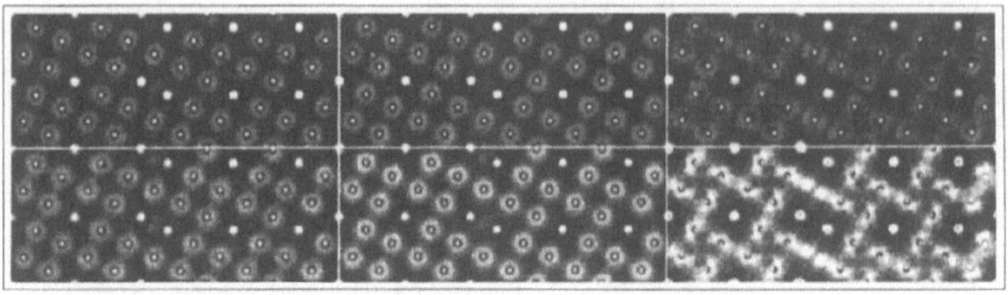

Figure 11.10. Deviation from the ideal channelling at a translation interface in Au_4Mn (8,10 nm). Left: multi-slice calculation of the electron density at the exit face. Center: Channelling approximation. Right: difference image.

"ideal" electron density at both sides and display the difference with increased contrast (right). From this it is clear that the deviation from the ideal channelling condition is very small and is mainly due to very small asymmetric contributions to the image. These can be only be observed at thicknesses at which the gold columns are nearly extinct.

8.2. Diffraction patterns

Fourier transforming the wave function Eq. (11.35) at the exit face of the object yields the wave function in the diffraction plane, which can be written as

$$\psi(\vec{q}, z) = \delta(\vec{q}) + \sum_i F_i(\vec{q}, z) \cdot e^{-2\pi i \vec{q}\vec{r}_i} \tag{11.40}$$

In the case of heavy columns however, we will have to use Eq. (11.31) instead. In a sense, the simple kinematical expression for the diffraction amplitude holds, provided the scattering factor for the atoms is replaced by a dynamical scattering factor for the columns, as obtained in Ref. (457), and is defined by

$$F_i(\vec{q}, z) = c_i \cdot f_i(\vec{q}) \cdot \left(e^{-i\pi\, k\, z\, E_i/E_0} - 1\right) \tag{11.41}$$

with $f_i(\vec{q})$ the Fourier transform of $\phi_i(\vec{r})$. Here $f_i(\vec{q})$ is the kinematical scattering factor including the typical geometry factor, while the last term in Eq. (11.41) accounts for the dynamical corrections on the intensity and phase of the kinematical diffraction spots. It is clear that the dynamical scattering factor varies periodically with depth. This periodicity may be different for different columns.

In case of a mono-atomic crystal, all Φ_i are identical and $\Psi(\vec{q}, z)$ varies perfectly periodically with depth. In a sense, the electrons are periodically transferred from the central beam to the diffracted beams and back. The periodicity of this dynamical oscillation (which can be compared with the Pendellösung effect) is called the *dynamical extinction distance.* It has, for example, been observed in Si(111). An important consequence of Eq. (11.36) is the revelation that the diffraction pattern can still be described by a kinematical type of expression so that existing results and techniques that have been based on the kinematical theory remain valid to some extent for thicker crystals in zone axis orientation. Examples are:

- diffraction at periodical stacking of translation interfaces, twin interfaces and mixed layer compounds

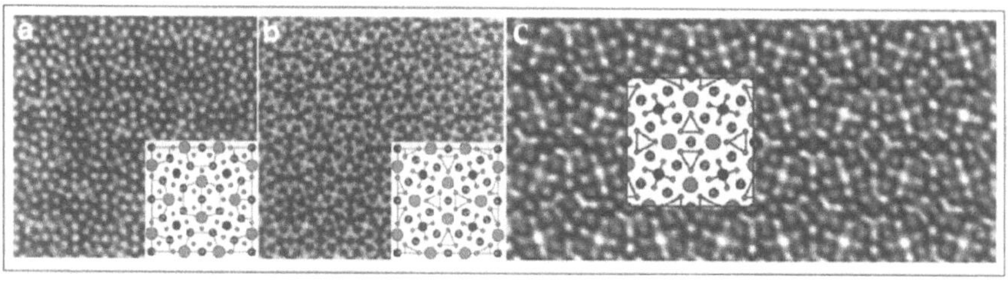

Figure 11.11. Experimentally retrieved amplitude (a), phase (b) and reconstructed structure + structure inset (c) for $Ba_2NaNb_5O_{15}$.

- diffuse scattering from substitutionally ordering alloys with a column structure
- diffraction contrast at defects. In particular, the extinction rule based on the $\vec{q} \cdot \vec{r}$ criterion remains valid if the defect is parallel to the incident beam.

8.3. High resolution images

The wave function in the image plane can be written as the convolution product of the wave function at the exit face of the crystal with the impulse response function $t(\vec{r})$ of the electron microscope

$$\Psi(\vec{r}, z) = 1 + \sum_i \left[c_i \cdot \phi_n(\vec{r} - \vec{r}_i) \cdot \left(e^{-i\pi k z E_i/E_0} - 1\right)\right] \otimes t(\vec{r}) \tag{11.42}$$

If the microscope is operated so the image is close to optimum focus and the beam is in axial mode, the impulse response function is sharply peaked.

If the distance between the columns is larger than the width of the impulse response function $t(\vec{r})$, the overlap between convolution products $\phi_i \otimes t(\vec{r})$ of adjacent sites can be assumed to be small so that each column is imaged separately. The contrast of a particular column varies periodically with thickness. The periodicity can be different for different types of columns. It is interesting to note that the functions ϕ_i as well as $t(\vec{r})$ are symmetrical around the origin, provided the objective aperture is centered around the optical axis. Hence, the image of a column is rotationally symmetric around the position $\vec{r}_i$. of the columns. The intensity at $\vec{r}_i$ is a maximum or a minimum. The positions of the columns can thus be determined from the positions of the intensity extreme.

In case the resolution of the microscope is insufficient to discriminate the individual columns, or the focus is not close to optimum, the overlap between the convolution products of adjacent columns cannot be avoided and the interpretation of the contrast is not straightforward.

9. Results

Figure 11.11 shows the results obtained using the channelling theory to invert the experimentally retrieved exit wave for $Ba_2NaNb_5O_{15}$. All atom columns can be located accurately and their mass can be determined approximately.

10. Conclusions

We have shown how using the focus variaton method allows the object wave function to be reconstructed up to the information limit of the electron microscope. The reconstructed wave function can be interpreted in terms of the projected crystal structure using a simplified channelling theory.

Acknowledgments

This work has been performed within the framework of a Brite-Euram Project (nr 3322). One of us (Marc Op de Beeck) is indebted to the Belgian Fund for Scientific Research (N.F.W.O.) for financial support.

APPLICATIONS OF ELECTRON HOLOGRAPHY

M. Gajdardziska-Josifovska[1] and A.H. Carim[2]

[1]Department of Physics and Laboratory for Surface Studies, University of Wisconsin-Milwaukee, Milwaukee, WI 53201
[2]Department of Materials Science and Engineering, The Pennsylvania State University, University Park, PA 16802

1. Introduction

Electron holography is now at the stage where the emphasis on technique development is slowly shifting towards an emphasis on applications. This is particularly true for electron holography at moderate resolutions, where the phase image gives information about electrostatic and magnetic fields in the sample and in the space around the sample. Such information is not normally available from the intensity images obtained by non-holographic microscopy techniques. This is why most of the published applications of electron holography have been concerned with studies of electromagnetic fields. This book has already devoted Chapter 7 and Chapter 8 to the theoretical and experimental aspects of studying electromagnetic fields by off-axis electron holography. To avoid duplication, in this chapter we will only discuss applications of moderate resolution electron holography to cases when there are no strong electrostatic and/or magnetic fields in the samples. The phase in this case is mainly affected by the mean inner potential and the thickness of the material, with additional contributions from diffraction effects. We will first discuss these three dependencies as a background for the ensuing review of applications of holography to interfaces, surfaces and small particles.

The first background sections of this chapter will discuss the origin of the mean inner potential, its contribution to holograms, its theoretical calculations, and its experimental holographic measurements. For example, the applications of holography to hetero-interfaces relies on the variation of the mean inner potential across the interfaces.

The following background section will discuss the contributions of the projected thickness to the amplitude and phase images reconstructed from a hologram. We will revisit the issue of energy filtering in holography and show how the amplitude image depends on the specimen thickness and the mean free path for inelastic scattering. In cases when the mean inner potential is known, the phase image can give the projected specimen thickness, thus allowing the measurement of steps on surfaces and shapes of three dimensional nanoparticles. In weakly diffracting orientations the thickness dependence can be eliminated from the phase images by combining them with their corresponding amplitude images.

The main advantage of holography over the other electron microscopy imaging techniques is that it gives both the amplitude and the phase of the image wave. The original idea of Gabor[129] was that electron holography will enable enhanced resolution of electron microscopes by removing the effects of the lens aberration from the complex image wave. It has taken several decades of instrumentation and technique development before resolution enhancement by electron holography could, only recently, be demonstrated.[124,212,329] The technique development aspect is still dominant in the area of atomic resolution electron holography and most of the resolution enhancement research is conducted with simple test samples. At this time there are very few applications of resolution enhanced holographic imaging to characterization of unknown samples.

2. Mean inner potential and its effects on phase images

The volume average of the scalar potential of a solid is known as its mean inner potential. This potential is negative and has a value typically between -5 V and -30 V, depending on the composition and structure of the solid. The mean inner potential is given by the zero-order Fourier coefficient (V_0) of the crystal potential for infinitely large perfect crystals, and is usually taken as an ad-hoc zero of the potential. However, V_0 has several important meanings (e.g., see recent review by Spence[417]). Miyake[307] and Ibers[205] have shown that V_0 is proportional to the second moment of the charge density for an atom, and hence to Langevin's diamagnetic susceptibility for a gas. More recently it was shown that V_0 for a crystal depends on sums of dipole and quadrupole moments in the unit cell, and methods were developed for the calculation of V_0 from mean square radii of the atoms ($<r^2>$), or from atomic structure factors for X-ray scattering.[17] The dependence on ($<r^2>$) makes V_0 very sensitive to the redistribution of the outer valence electrons due to bonding: V_0 could be intuitively interpreted as a measure of the effective atomic sizes in a crystal.

In practice, V_0 is important in electron holography, Fresnel imaging of interfaces, reflection high-energy electron diffraction (RHEED), and low-energy electron diffraction (LEED) because it increases the effective energy of the electrons in the solid thereby defining the phase changes and the refraction of the transmitted and diffracted beams. In keeping with the spirit of the book we will limit our discussion to high-energy electrons. Those interested in the effects of the mean inner potential on low-energy electrons can refer to books on LEED (e.g., Van Hove et al.[464] or Pendry[335]). A recent comparison of the effects of the mean inner potential on high-energy and low-energy electrons has been reported by Saldin and Spence.[381]

Refraction effects due to the mean inner potential are usually negligible in transmission electron microscopy and transmission high energy electron diffraction of thin foils. This situation is due to the large difference between the value of the mean inner potential V_0 and the kinetic energy of the incident electrons E ($E \gg V_0$), which results in an electron-optical index of refraction n that is only slightly larger than one[359]

$$n = \frac{K}{k} = 1 + \frac{e\,|V_0|}{E} \cdot \frac{E_0 + E}{2E_0 + E} \tag{12.1}$$

where K is the electron wave vector in the material, k is the wave vector in vacuum, E_0 is the rest energy of an electron and e is the electron charge. Without relativistic effects, Eq. (12.1) reduces to

$$n = 1 + \frac{e\,|V_0|}{2\,E} \tag{12.2}$$

For example, for $E = 100\,\text{keV}$ electrons incident on a crystal with $|V_0| = 15\,\text{V}$, the relativistic and non-relativistic values for the index of refraction are $n = 1.000082$ and $n = 1.000075$ respectively. Hence, detectable refraction effects occur only when the high energy electrons are incident at grazing angles with respect to a sample's surface or an internal interface between two materials. Experimental measurements of V_0 have been obtained from RHEED patterns from surfaces of bulk crystals[141] and from transmission diffraction patterns from electron transparent polyhedral crystals with some of their faces at small angles to the incident electrons.[309] The difference in mean inner potential ΔV_0 between two materials has been measured from refraction effects observed in nanodiffraction from interfaces.[138]

In addition to the change in the propagation direction, the refractive index of Eq. (12.1) also causes a phase shift of an electron wave transmitted through a material, relative to a coherent reference wave which has only traveled through vacuum. For mono-energetic incident electrons with wavelength λ, in the absence of dynamical diffraction effects, this phase difference $\Delta\varphi$ depends only on the projected thickness of the material t and its mean inner potential V_0[359]

$$\Delta\varphi = \frac{2\,\pi}{\lambda}\cdot(n-1)\,t = \frac{2\,\pi\,e}{\lambda\,E}\cdot\frac{E_0+E}{2\,E_0+E}\cdot|V_0|\,t = C_E\,|V_0|\,t \tag{12.3}$$

where C_E is an energy-dependent constant. Due to their small wavelength, high energy electrons undergo substantial phase changes even though the electron-optical index of refraction is only slightly larger than one. These phase changes are also pronounced at normal incidence, when there are no refraction effects.

Equation (12.3) is obtained from the general solution to the relativistic Schroedinger equation (e.g., Aharanov and Bohm[5])

$$\varphi(\vec{r}) = -\frac{2\,\pi\,e}{\lambda\,E}\cdot\frac{E_0+E}{2\,E_0+E}\cdot\int_l V(\vec{r})\,\mathrm{d}s - \frac{2\,\pi\,e}{h}\cdot\int_l \vec{A}(\vec{r})\,\mathrm{d}\vec{s} \tag{12.4}$$

which shows how the phase of an electron wave φ is altered by all scalar V and vector $\vec{A}$ potentials along its path l. In cases when there are no electric and magnetic fields in the material, and when dynamical diffraction effects are weak, this general equation reduces to Eq. (12.3) since the only remaining source for phase change is the volume average of the atomic potential V_0.

From Eqs. (12.3) and (12.4) it follows that the phase image of a material with uniform structure and composition represents a map of its projected thickness $t(x,y)$

$$\varphi(x,y) = C_E\,|V_0|\,t(x,y) \tag{12.5}$$

where V_0 and C_E are multiplicative constants of proportionality. This equation provides a basis for measurement of the mean inner potential for samples with known thickness. Equation (12.5) is kinematical because it ignores phase changes of the transmitted beam introduced by multiple scattering. Fig. 12.1 shows an experimental example for the validity of Eq. (12.5). The phase image in Fig. 12.1a was obtained from a cleaved

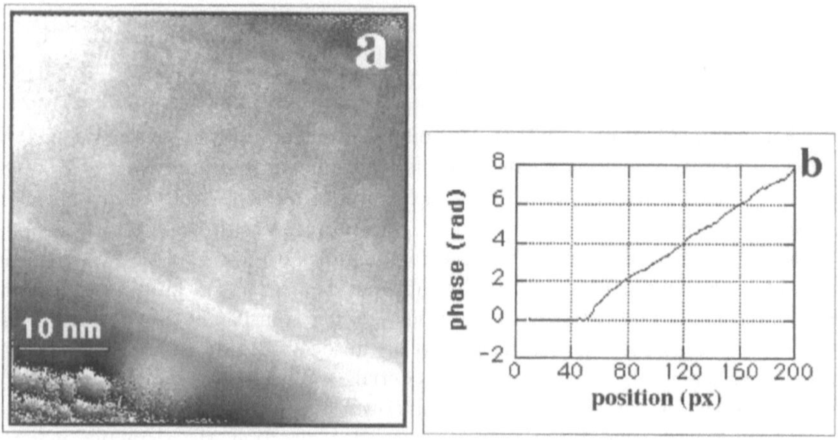

Figure 12.1. Phase image (a) and phase profile (b) from GaAs wedge tilted away from any zone axis and/or systematic row orientation. Note that the phase change increases linearly with thickness, as predicted by the kinematic equation Eq. (12.5).

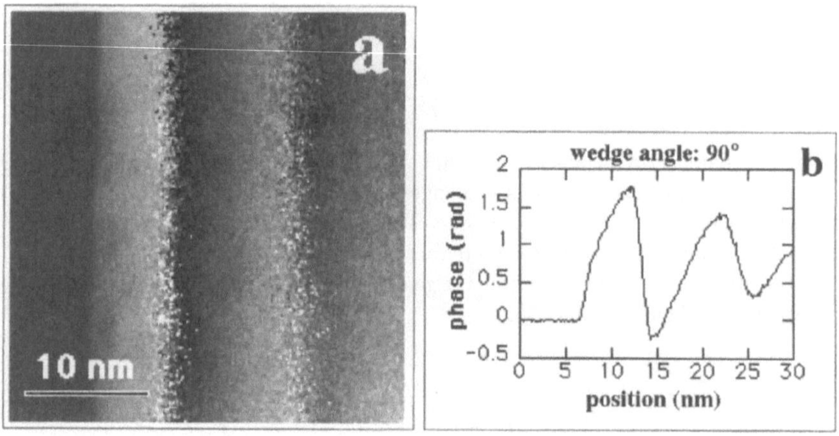

Figure 12.2. Phase image (a) and phase profile (b) from GaAs wedge tilted to a [100] zone axis. Note the strong dynamical diffraction effects which make the phase change nonlinear.

crystal wedge of GaAs which was tilted to a weakly diffracting condition. The phase profile in Fig. 12.1b shows that the phase increases linearly with thickness, as predicted by Eq. (12.5). However, Eq. (12.5) is no longer valid when the crystal is tilted to a strongly diffracting orientation, as shown in Fig. 12.2 for the same GaAs wedge sample tilted to a [100] zone axis. To model this situation one must include the higher order Fourier terms of the periodic crystal potential and allow for multiple scattering between the transmitted and diffracted beams.

In cases when the composition of the sample varies, the phase image maps both the variations in the projected mean inner potential and the projected specimen thickness

$$\varphi(x,y) = C_E\,|V_0(x,y)|\,t(x,y) \tag{12.6}$$

Therefore, both the mean inner potential *and* the thickness of the sample contribute to the phase measured by electron holography. For this reason, we devote a large part of the background discussion to the review of calculations and measurements of the mean inner potential.

3. Theoretical calculations of mean inner potential

The most general expression for the mean inner potential stems from its definition as a volume average of the atomic electrostatic potentials within the material

$$V_0 = \frac{1}{\Omega} \cdot \int\limits_{\Omega} V(\vec{r})\, \mathrm{d}\vec{r} \tag{12.7}$$

where Ω is the volume of the unit cell for a crystalline solid, or the volume of the material for a disordered solid. In principle, V_0 could be calculated from Eq. (12.7) for any material for which $V(\vec{r})$ is known throughout the unit cell, or throughout a given volume. Unfortunately, this information is not generally available and approximations must be made.

The simplest approximation which is often employed to estimate V_0 is to treat the solid as if it were made of an array of *neutral free atoms*. In this non-binding approximation Eq. (12.7) becomes

$$V_0 = \frac{h^2}{2\,\pi\, m_0\, e\, \Omega} \cdot \sum_j f_j(0) \tag{12.8}$$

where Ω is again the volume of the crystal unit cell, and $f_j(0)$ are the atomic scattering amplitudes for forward scattering of electrons. The summing of $f_j(0)$ is performed over the j atoms in the unit cell. The other constants in Eq. (12.8) have their usual meaning. A reader who is interested in the derivation of Eq. (12.8) can refer to the books by Reimer[359] (for the approach based on Bethe's formulation of diffraction combined with the Born approximation), or Cowley[72] and Spence and Zuo[416] (for the approach based on the X-ray scattering factors combined with the Mott formula). Here we will focus on the practical applications of the non-binding approximation and on its limitations.

We will use GaAs to demonstrate how to calculate V_0 using Eq. (12.8). GaAs has a zincblende structure with four gallium and four arsenic atoms per unit cell. The Doyle and Turner[97] atomic scattering amplitudes for ^{31}Ga and ^{33}As at zero scattering angle are $f_{\mathrm{Ga}}(0) = 7.108\,\text{Å}$ and $f_{\mathrm{As}}(0) = 7.320\,\text{Å}$ respectively for electrons of rest mass (i.e., non-relativistic electrons with $m = m_0$). The sum over the unit cell becomes: $\sum f_j(0) = 4\, f_{\mathrm{Ga}}(0) + 4\, f_{\mathrm{As}}(0) = 57.712\,\text{Å}$. When combined with the values for the volume of the unit cell $\Omega = 5.654^3\,\text{Å}^3 = 1.8075 \cdot 10^{-28}\,\mathrm{m}^3$, the Planck constant $h = 6.6262 \cdot 10^{-34}$ Js, electron rest mass $m_0 = 9.1095 \cdot 10^{-31}$ kg, and electron charge $e = 1.6022 \cdot 10^{-19}$ C, the mean inner potential calculated from Eq. (12.8) becomes

$$V_0 = \frac{(6.6262 \cdot 10^{-34}\,\mathrm{Js})^2}{2\,\pi \cdot 9.1095 \cdot 10^{-31}\,\mathrm{kg} \cdot 1.6022 \cdot 10^{-19}\ \mathrm{C}} \cdot \frac{57.712 \cdot 10^{-10}\,\mathrm{m}}{1.8075 \cdot 10^{-28}\,\mathrm{m}^3} = 15.287\,\mathrm{V} \tag{12.9}$$

For the benefit of the reader interested in practical holography, the Doyle-Turner[97] atomic scattering amplitudes for forward scattering of electrons are reproduced in Table 12.1. These calculations by Doyle and Turner are based on the relativistic Hartree-

Fock atomic wave functions. For high energy electrons the atomic scattering amplitudes need to be multiplied by $m/m_0 = [1 - (v/c)^2]^{-1/2}$, where v is the speed of the electron in vacuum and c is the speed of light in vacuum. For example, 100 keV electrons have $m/m_0 = 1.1957$ giving relativistically corrected scattering factors for GaAs as follows: $f_{\mathrm{Ga}}(0) = 8.499$ Å, $f_{\mathrm{As}}(0) = 8.7525$ Å, $\sum f_j(0) = 4 f_{\mathrm{Ga}}(0) + 4 f_{\mathrm{As}}(0) = 69.006$ Å. With these values the mean inner potential calculated from Eq. (12.8) becomes $V_0 = 18.278$ V. This result is one of the two important limitations of the non-binding approximation for calculating the mean inner potential of a solid. From the general definition given by Eq. (12.7), the mean inner potential is a property of the solid, and not of the electrons that are used to probe the solid. This is particularly true for high energy electrons when the exchange and correlation contributions of the incident electrons to the potential of the solid $V(\vec{r})$ are negligible.[416] In this limit the mean inner potential should indeed be a constant which is independent of the energy of the incident electrons. In the limit of low energy electrons, when the exchange and correlation effect can not be ignored, the mean inner potential is expected to vary with the energy of the incident electrons.[335,381,464]

The second important limitation of the non-binding approximation is that it overestimates the values for V_0 because it ignores the redistribution of the valence electrons due to bonding. A correction scheme for this limitation, proposed by Radi,[349] is to treat the values of V_0 obtained from the non-binding approximation as an upper limit of the actual value for the mean inner potential. In that sense it is best to use the scattering amplitudes from Table 12.1 without any relativistic corrections for the mass of the high energy electrons (i.e., take $m/m_0 = 1$) to avoid further increases of the upper limit for V_0. According to Radi,[349] the lower limit to V_0 can be calculated by assuming that the crystal is made of ions at the lattice points with the remaining valence electrons

Z	Element	$f(0)$ [Å]	Z	Element	$f(0)$ [Å]	Z	Element	$f(0)$ [Å]
2	He	0.418	20	Ca	9.913	38	Sr	13.109
3	Li	3.286	21	Sc	9.307	42	Mo	10.260
4	Be	3.052	22	Ti	8.776	47	Ag	8.671
5	B	2.794	23	V	8.305	48	Cd	9.232
6	C	2.509	24	Cr	6.969	49	In	10.434
7	N	2.211	25	Mn	7.506	50	Sn	10.859
8	O	1.983	26	Fe	7.165	51	Sb	10.974
9	F	1.801	27	Co	6.854	53	I	10.905
10	Ne	1.652	28	Ni	6.569	54	Xe	10.794
11	Na	4.778	29	Cu	5.600	55	Cs	16.508
12	Mg	5.207	30	Zn	6.065	56	Ba	18.267
13	Al	5.889	31	Ga	7.108	63	Eu	15.563
14	Si	5.828	32	Ge	7.378	79	Au	10.573
15	P	5.448	33	As	7.320	80	Hg	10.964
16	S	5.161	34	Se	7.205	82	Pb	12.597
17	Cl	4.857	35	Br	7.060	83	Bi	13.096
18	Ar	4.580	36	Kr	6.897	86	Rn	13.492
19	K	8.984	37	Rb	11.778	92	U	19.119

Table 12.1. Atomic scattering amplitudes for forward scattering, $f(0)$, as calculated by Doyle and Turner[97] using the relativistic Hartree-Fock atomic wavefunctions. These values are used in the non-binding approximation, Eq. (12.8), to calculate the mean inner potential V_0 of a solid.

distributed uniformly. The contribution of p perfectly free electrons to V_0 has been given by Bethe[20]

$$V_0 = -\frac{3}{10} \cdot \frac{e p}{4 \pi \epsilon_0 r_0} \tag{12.10}$$

where r_0 is chosen so that a sphere of radius r_0 has the same volume as the crystal atom. The lower limit to V_0 can then be calculated as follows: V_0 of the p-times ionized free atom (i.e., the p outermost electrons of the neutral atom are missing) plus the V_0 of p free electrons from Eq. (12.10). These lower bound values, as calculated by Radi[349] using the binding approximation, are summarized in Tables 12.2 and 12.3 for elemental and compound crystals respectively. Table 12.2 also gives the the upper bound non-binding values of V_0 due to Radi,[349] and the equivalent non-binding values calculated by the authors using the Doyle-Turner[97] (D-T) scattering factors from Table 12.1. Comparison between these two sets of values shows many differences, although they are both obtained from the same non-binding approximation, as given by Eq. (12.8). The reason for the differences is that Radi uses different atomic scattering amplitudes which are based on the Herman and Skillman[181] calculations using the Hartree-Fock-Slater formalism. Therefore, the choice of atomic scattering amplitudes to be used in Eq. (12.8) merits some further discussion.

A number of different tables with atomic scattering amplitudes are commonly available to electron microscopists and holographers. For example, Hirsch et al.[198] give the atomic scattering amplitudes due to Ibers and Vainshtein;[206] the multislice and Bloch wave calculation package EMS by Stadelman[427] uses the Doyle and Turner[97] and the Smith and Burge[401] atomic scattering amplitudes for electrons; older packages often use the x-ray scattering amplitudes combined with the Mott formula. Amongst these standard references from the fifties and the sixties, the Doyle and Turner[97] values are relatively most recent (i.e., 1968). In the nineties, Rez and Rez[364,365] used the latest multi-configuration Dirac-Fock code of Grant et al.[150] to calculate the atomic scattering amplitudes for x-rays. Using the Mott formula they obtained the corresponding scattering amplitudes for electrons and compared them with those of Doyle and Turner[97] to find differences smaller than 0.1% for all elements over the full range of calculated scattering angles. Therefore, we suggest that the Doyle-Turner scattering amplitudes from Table 12.1 should be used whenever the non-binding approximation of Eq. (12.8) is employed to estimate the upper limit to the mean inner potential.

Ross and Stobbs[370] have proposed an empirical rule for compromising between the two extreme estimates of V_0: the upper limit obtained by the non-binding approximation using the Doyle-Turner scattering amplitudes of free neutral atoms, and the lower limit calculated by Radi by taking ions at the lattice sites with all of their valence electrons distributed uniformly in between. By analyzing the ratio between the values of V_0 due to Radi (the average of the upper and the lower limit given in Table 12.2) and those due to the non-binding approximation employing the Doyle and Turner scattering amplitudes, Ross and Stobbs[370] found the following phenomenological dependence

$$\frac{V_0\,(\text{Radi})}{V_0\,(\text{Doyle and Turner})} = 0.0325\, Z/\Omega + 0.6775 \tag{12.11}$$

albeit with a lot of scatter. Ross and Stobbs have proposed a reasonably simple procedure for finding V_0 by first calculating it from the non-binding approximation using Eq. (12.8) with subsequent use of Eq. (12.11) to approximate the effects of

Z	Element	Structure	Lattice parameter a/c (in Å)	ionicity p	Model: Radi, binding V_0 (in V)	Model: Radi, non-binding V_0 (in V)	Model: D-T, non-binding V_0 (in V)
6	C	Dia	3.56	4	15.93	19.75	21.30
14	Si		5.42	4	11.47	12.20	14.02
32	Ge		5.66	4	13.69	13.82	15.59
50	Sn		6.49	4	14.03	13.69	15.22
3	Li	BCC	3.46	1	2.88	6.16	7.60
11	Na		4.30	1	3.30	4.66	5.75
19	K		5.33	1	3.62	4.46	5.68
23	V		3.03	2	15.43	23.93	28.59
24	Cr		2.89	1	17.55	23.84	27.65
26	Fe		2.86	2	17.23	25.03	29.33
37	Rb		5.62	1	4.31	5.08	6.35
41	Nb		3.30	1	19.83	24.38	
42	Mo		3.15	1	22.22	27.27	31.43
55	Cs		6.13	1	4.91	5.58	6.86
56	Ba		5.02	2	9.11	11.40	13.83
73	Ta		3.30	2	24.07	31.08	
74	W		3.16	2	27.12	34.80	
81	Tl		3.88	3	17.26	19.25	
10	Ne	FCC	4.52	0	3.22	3.22	3.43
13	Al		4.04	3	10.00	14.91	17.10
18	Ar		5.43	0	4.96	4.96	5.48
20	Ca		5.58	2	6.75	8.87	10.93
25	Mn		3.70	2	16.38	24.08	28.38
26	Fe		3.63	2	16.95	24.51	28.69
27	Co		3.54	2	17.70	25.28	29.59
28	Ni		3.52	2	17.78	24.90	28.84
29	Cu		3.61	1	15.82	19.98	22.80
36	Kr		5.59	0	6.87	6.87	7.56
38	Sr		6.08	2	7.35	9.09	11.17
45	Rh		3.80	1	23.68	28.55	
46	Pd		3.89	0	22.37	22.37	
47	Ag		4.08	1	18.74	21.87	24.45
54	Xe		6.24	0	7.77	7.77	8.51
77	Ir		3.84	2	30.07	37.24	
78	Pt		3.92	1	28.14	32.14	
79	Au		4.08	1	25.02	28.20	29.81
82	Pb		4.95	4	18.25	19.41	19.89
90	Th		5.08	2	20.91	26.58	
4	Be	HCP	2.28/3.59	2	7.41	16.32	18.08
12	Mg		3.20/5.20	2	6.53	9.24	10.81
21	Sc		3.31/5.27	2	9.99	14.69	17.82
22	Ti		2.95/4.68	2	12.92	19.77	23.83
27	Co		2.51/4.07	2	17.77	25.44	29.56
30	Zn		2.66/4.95	2	13.55	16.64	19.15
39	Y		3.65/5.73	2	11.49	15.22	
40	Zr		3.23/5.14	2	15.49	21.01	
43	Tc		2.73/4.39	2	23.38	31.54	
44	Ru		2.71/4.28	1	24.30	29.55	
48	Cd		2.98/5.62	2	15.92	18.29	20.45
57	La		3.75/6.06	2	14.60	19.51	
58	Ce		3.65/5.96	2	14.32	20.22	
59	Pr		3.99/5.91	2	14.20	19.93	
60	Nd		3.65/5.89	2	14.27	19.90	
64	Gd		3.63/5.78	2	14.81	19.21	
65	Tb		3.60/5.70	2	14.09	19.43	
66	Dy		3.59/5.65	2	14.08	19.37	
67	Ho		3.58/5.62	2	14.10	19.40	
68	Er		3.56/5.59	2	14.12	19.38	
69	Tm		3.54/5.56	2	14.17	19.45	
72	Hf		3.20/5.05	2	19.75	25.45	
75	Re		2.76/4.46	2	29.08	36.95	
76	Os		2.73/4.32	2	30.42	38.26	
81	Tl		3.46/5.53	3	17.59	19.73	

Table 12.2. Calculated values of mean inner potential, V_0, for elemental crystals using binding (Eqs. (12.8) and (12.10)) and non-binding approximations (Eq. (12.8) only). Note the difference in the non-binding values for V_0 resulting from use of different atomic scattering amplitudes.

bonding. Taking again GaAs as an example, the non-binding approximation gives V_0 (Doyle and Turner) = 15.287 V; the mean atomic number per unit volume is $Z/\Omega = (4 \cdot 31 + 4 \cdot 33)/5.654^3\,\text{Å}^3 = 1.416\,\text{Å}^{-3}$, and the ratio calculated from Eq. (12.11) is $V_0\,(\text{Radi})/V_0\,(\text{Doyle-Turner}) = 0.7235$. From this, the empirical value for the mean inner potential of GaAs including bonding becomes $V_0\,(\text{Radi}) = 11.06\,\text{V}$.

The above calculation schemes have the advantage of being simple, and therefore practical. But the values obtained for V_0 can only be used as guidance when performing quantitative holographic experiments. In the absence of better tables with theoretical values of V_0 it would be best to use experimental measurements for V_0 whenever possible.

4. Holographic experimental measurements of mean inner potential

Even though V_0 is a very basic quantity, there are no extensive tables of experimental values, except for a few materials (e.g., Miyake,[307] Reimer[359] and Spence[417]). In addition, the published values for the same material by different research groups often show higher scatter than the reported errors (e.g., values ranging from 12.3 V to 16 V have been reported for V_0 of MgO, as summarized by Yada et al.[502]). We believe that this lack of reliable experimental values for V_0 will be one of the practical obstacles for applications of quantitative holography to materials problems.

Experimental measurements of V_0 have been based on quantification of refraction effects (electron diffraction methods) and phase changes (electron holography methods). Here we will review the electron holographic methods that have been developed over the years, and assemble the results of these measurements in Table 12.4.

Electron interferometry was first used to measure the mean inner potential of films with known thickness by Möllenstedt and Keller.[314] Several researchers followed in this

Compound	Lattice parameter a (in Å)	ionicity p	Model: Radi, binding V_0 (in V)	Model: D-T, non-binding V_0 (in V)
LiF	4.02	+1,-1	13.61	15.00
LiCl	5.14	+1,-1	10.82	11.48
LiBr	5.49	+1,-1	11.40	11.97
LiI	6.00	+1,-1	12.12	12.58
NaF	4.61	+1,-1	11.34	12.86
NaCl	5.63	+1,-1	9.54	10.34
NaBr	5.95	+1,-1	10.06	10.76
NaI	6.46	+1,-1	10.56	11.14
KF	5.33	+1,-1	11.38	13.64
KCl	6.28	+1,-1	9.33	10.70
KBr	6.58	+1,-1	9.57	10.79
KI	7.05	+1,-1	9.86	10.87
RbF	5.63	+1,-1	12.19	14.57
RbCl	6.56	+1,-1	9.78	11.29
RbBr	6.86	+1,-1	9.85	11.18
RbI	7.32	+1,-1	9.95	11.08
AgCl	5.55	+1,-1	14.32	15.18
AgBr	5.76	+1,-1	15.01	15.76
CsF	6.01	+1,-1	13.55	16.15
MgO*	4.21	+2,-2	17.94	18.44
PbS	5.94	+2,-2	16.88	16.26
PbSe	6.14	+2,-2	17.11	16.38
PbTe	6.44	+2,-2	17.74	

Table 12.3. Calculated values of mean inner potential, V_0, for compound crystals with rocksalt structure. Results of binding approximation, due to Radi[349], are given in column four, while results of non-binding approximation, calculated by the first author using the Doyle-Turner[97] scattering factors from Table 12.1, are given in the last column. (*)Radi gives a non-binding value of 16.23 V only for MgO.

path.[35,199,218,224] Amorphous and polycrystalline films of desired nominal thickness have been used, deposited in the shape of strips or disks on electron-transparent substrates (Fig. 12.3a). The phase change was measured directly from the off-axis electron hologram (interferometric measurements), by recording the shift of a holographic fringe at the edge of the film $\Delta\varphi$, the distance between adjacent fringes corresponding to a phase change of 2π (Fig. 12.3b). The reported statistical accuracy in these studies have ranged from 2.5% to 9.5%, although uncertainties in specimen thickness are very likely sources of systematic errors.

Interferometric measurements of V_0 have also been performed for single crys-

Material	V_0 [V]	Method	Reference
Be	7.8 ± 0.4	interferometry of films	Jönsson et al.[218]
C	7.8 ± 0.6	interferometry of films	Keller[224]
Al	13.0 ± 0.4	interferometry of films	Keller[224]
	12.4 ± 1	interferometry of films	Buhl[35]
	11.9 ± 0.7	interferometry of films	Hoffman and Jönsson[199]
Si	9.26 ± 0.05	holography of wedges	Gajdardziska et al.[135,504]
	12.1 ± 1.3	holography of spheres	Wang et al.[488]
	(11.5)	(RHEED)	(Gaukler and Schwarzer[141])
amorph.	11.9 ± 0.9	holography of spheres	Wang et al.[488]
Cu	23.5 ± 0.6	interferometry of films	Keller[224]
	20.1 ± 1.0	interferometry of films	Hoffman and Jönsson[199]
Ge	15.6 ± 0.8	interferometry of films	Hoffman and Jönsson[199]
Ag	20.7 ± 2	interferometry of films	Buhl[35]
	17.0-21.8	interferometry of films	Keller[224]
(W)	23.4	(RHEED)	(Gaukler and Schwarzer[141])
Au	21.1 ± 2	interferometry of films	Buhl[35]
	22.1-27.0	interferometry of films	Keller[224]
MgO	13.01 ± 0.08	holography of cleaved wedge	Gajdardziska et al.[135,504]
	13.2	interferometry of cleaved wedge	Sonier[403]
	13.5	interferometry of smoked wedge	Herring et al.[182]
	13.9	holography of smoke wedge	Herring et al.[182]
	(13.7)	(electron diffraction)	(Sturkey[430])
	(16)	(electron diffraction)	(Honjo and Mihama[200])
	(15.4)	(electron diffraction)	(Moliére and Niehrs[310])
	(13.4)	(electron diffraction)	(Miyake[308])
	(12.3)	(electron diffraction)	(Tomita and Savelli[438])
	(14.4 ± 0.8)	(STEM with sector detector)	(Chapman and Drummond[43])
SiO_2		holography of	
amorph.	10.1 ± 0.6	SiO_2 films on Si spheres	Wang et al.[488]
ZnS	10.2 ± 1	interferometry of thin films	Buhl[35]
GaAs	14.53 ± 0.17	holography of cleaved wedge	Gajdardziska et al.[135,504]
	(13.2)	(RHEED)	(Yamamoto and Spence[141])
PbS	17.19 ± 0.12	holography of cleaved wedge	Gajdardziska et al.[135,504]
Chrysotile	11.5	interferometry of fiber	Miyake et al.[309]
Kohlenstoff	24 ± 5	interferometry of thin films	Möllenstedt and Keller[314]
(GaP)	(12.2)	(RHEED)	(Yamamoto and Spence[141])

Table 12.4. Experimental values for mean inner potential, V_0. Values without brackets were obtained using holographic methods, either by direct interferometric measurements from a hologram, or by true holographic measurements from a reconstructed phase image. Values in brackets were obtained with other techniques, predominantly based on measurements of refraction effects in electron diffraction patterns.

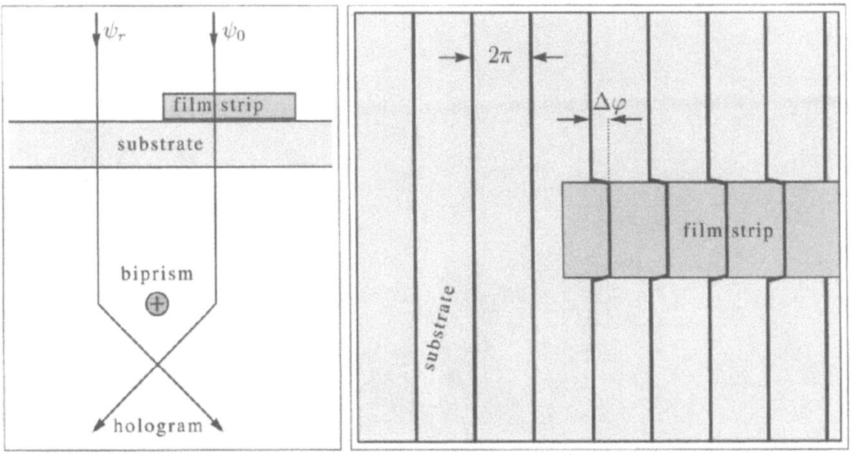

Figure 12.3. Schematic diagram of thin film sample (a) and corresponding hologram (b) used for first interferometric measurements of mean inner potential.[35,199,218,224,314] Note that the sample consists of a uniform substrate film, whose mean inner potential is not measured as both the reference and object waves pass through it, and a film deposited in shape of strips, whose mean inner potential is measured because only the object wave passes through it.

tals by using samples with known shape, such as chrysotile fibers and MgO smoke crystals,[182,403,502] with relative errors between 3.5% and 8%.

Recently Gajdardziska-Josifovska et al.[134,135,504] have developed a method for accurate holographic measurements of V_0 using digital holograms of cleaved crystal wedges of known angle. Fig. 12.4 gives an example of one of these measurements from a Si 109.5° wedge cleaved along {111} crystallographic planes. The off-axis electron hologram in Fig. 12.4a was recorded digitally by a 1024x1024-pixel Gatan 679 Slow-Scan CCD camera mounted on a Philips EM400ST-FEG TEM operated at 100 kV. This microscope is equipped with a thermally assisted tungsten field emitter, and a biprism on one of the selected area aperture positions. The crystal was deliberately tilted away from a zone axis in order to minimize dynamical diffraction effects, as monitored by the disappearance of extinction contours in the hologram. The tilts were further adjusted to minimize the intensity of diffraction spots in the selected area diffraction patterns and to avoid major Kikuchi lines through the 000 disk in the convergent beam diffraction patterns (CBED).

Numerical hologram reconstructions were performed as described in detail in Chapter 6. The sideband, shown in the lower right inset in Fig. 12.4a, was inverse Fourier transformed to calculate the phase image in Fig.12.4b which was contoured with white equiphase lines at $\pi/2$ intervals. This image has been corrected for geometric distortions introduced by the projector lens system by subtracting a reference phase image, reconstructed from a hologram acquired after removal of the specimen from the beam (all the other parameters were kept unchanged). Such correction is made possible by the fixed position of the CCD camera with respect to the microscope imaging system, which in addition to the linear digital recording, is the main advantage of CCD over the photographic detection technique for accurate measurement of the mean inner potential by holography.

The phase profile in Fig. 12.4c was obtained from the phase image by averaging in the direction parallel to the equiphase lines. The averaging improves the signal-to-noise

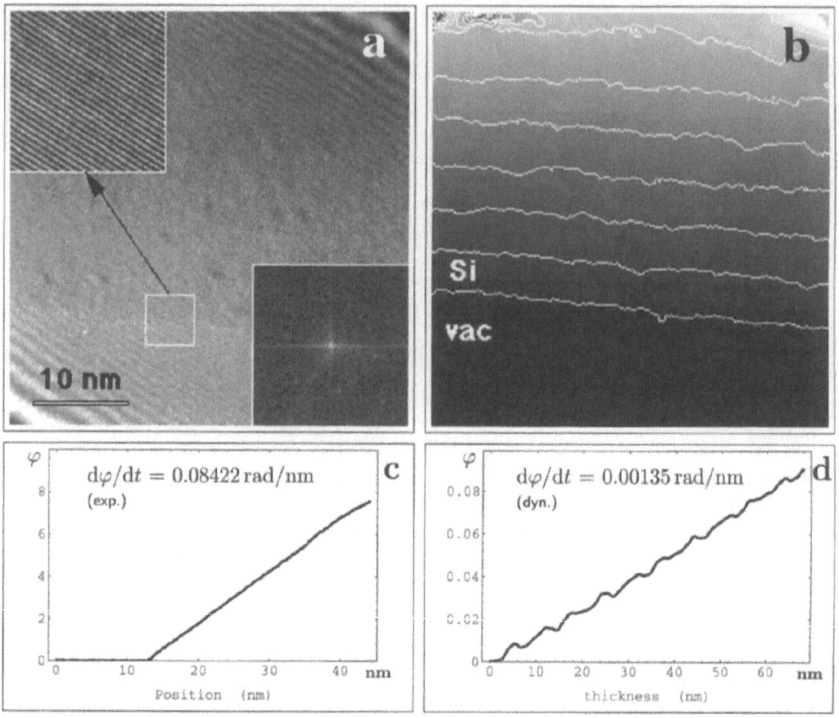

Figure 12.4. Measurement of mean inner potential V_0 using crystal wedge method:[135,504] a) Digital hologram of cleaved Si wedge with sideband (lower inset) and magnified holographic fringes from area with outlined specimen edge (upper inset); b) Phase image in same orientation and magnification as hologram with superimposed equiphase lines; c) Experimental averaged phase profile extracted from phase image in b); d) Calculated dynamical correction to phase of transmitted beam.

ratio and smears out the surface details, thus allowing measurement of the bulk inner potential. The experimental phase profile shows a high level of flatness in the vacuum and a linear increase in phase in the tilted crystal wedge. Rather than using the actual thickness of the wedge at a certain point to calculate V_0 from Eq. (12.5), it is more accurate to measure V_0 from the slope of the phase vs. position profile $\mathrm{d}\Delta\varphi/\mathrm{d}x$

$$|V_0| = \frac{1}{C_E} \cdot \left| \frac{\mathrm{d}\Delta\varphi/\mathrm{d}x}{\mathrm{d}t/\mathrm{d}x} \right| \tag{12.12}$$

where change of the projected thickness with position, $\mathrm{d}t/\mathrm{d}x$, depends upon the electron-optical magnification of the microscope M, the wedge angle θ, and the wedge orientation with respect to the incident beam given by the angles α and β as sketched in Fig. 12.5

$$\frac{\mathrm{d}t}{\mathrm{d}x} = M \cdot \frac{\tan(\theta/2 - \alpha) + \tan(\theta/2 + \alpha)}{\cos\beta} \tag{12.13}$$

It should be noted that the approach given by Eq. (12.12) and Eq. (12.13) removes any contributions from possible contamination overlayers with uniform thickness, such

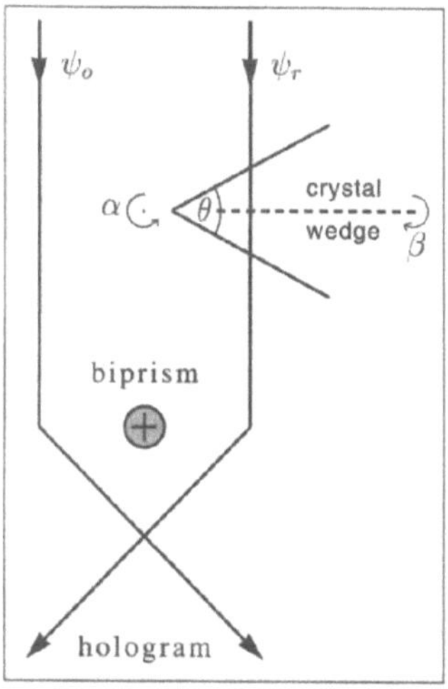

Figure 12.5. Schematic diagram of holographic setup for measurement of mean inner potential using cleaved crystal wedges of known angle θ, tilted away from a zone axis around two perpendicular axes by angles α and β respectively.

as the native oxide often observed on Si surfaces exposed to air, because such overlayers have $dt/dx = 0$. By calibrating the magnification M in the phase image, measuring the exact tilt angles α and β of the crystal wedge from CBED, and calculating the constant C_E from CBED measurements of the energy of the incident electrons, the value of $V_0 = C_E^{-1}\,(d\varphi/dt)_{exp}$ is obtained in the kinematical approximation with an accuracy of 1%. The results of these measurements are given in Table 12.5 for four different crystal wedges. Detailed studies of the errors show that the main uncertainties in this method come from the specimen imperfections and from the accuracy of the phase measurement and the magnification calibration.

Theoretical Bloch wave calculations of the dynamical contributions to the phase of the transmitted beam $(d\varphi/dt)_{dyn}$ were performed to evaluate the validity of the kinematical approximation.[135,504] A calculated dynamical phase profile, which excludes the changes due to V_0, is shown in Fig. 12.4d for Si tilted to the same orientation as in the experimental hologram. The dynamical correction for this case is about 1.6%. The same dynamical calculations were performed for MgO, GaAs and PbS, and the results are given in Table 12.5. The above study has demonstrated that the kinematical approximation gives values for V_0 that are lower than the actual values by less than 5% for suitably tilted single crystals.

crystal	wedge angle θ	tilt angles α ; β	$(d\varphi/dt)_{exp}$ [rad/nm]	$(V_0)_{kin}$ [V]	$(d\varphi/dt)_{dyn}$ [rad/nm]	V_0 [V]
Si	109.5°	5.99°; 2.24°	0.08422	9.11 ± 0.5%	0.001350	9.26 ± 0.5%
MgO	90°	22.12°; 2.60°	0.11816	12.78 ± 0.6%	0.002137	13.01 ± 0.6%
GaAs	90°	4,15°; 1.95°	0.13064	14.13 ± 1.2%	0.003728	14.53 ± 1.2%
PbS	90°	4.07°; 3.21°	0.15118	16.35 ± 0.7%	0.007130	17.19 ± 0.7%

Table 12.5. Mean inner potentials V_0 for four crystals obtained by adding calculated phase change with thickness due to dynamical diffraction $(d\varphi/dt)_{dyn}$ to phase slope $(d\varphi/dt)_{exp}$ measured from electron holograms. Kinematical approximation applied to same experimental data results in lower $(V_0)_{kin}$ values. Wedge angles and tilt from symetrical wedge orientation around axis parallel (α) and perpendicular (β) to wedge edge are included (see schematic diagram in Fig. 12.5).

The results of Table 12.5 are very encouraging for the application of the simple kinematic equations to interpretation of phase images from extended single crystals, such as in studies of crystal surfaces. However, the importance of the crystal tilt must never be under-emphasized, especially when studying nanocrystallites whose orientation to a weakly diffracting conditions is hard to control and quantify in a typical holography experiment. In the case of edge-on holographic studies of epitaxial interfaces the tilts can usually be quantified, but the choice of tilts is limited to zone axis or systematic row orientations. For the case of Si crystals tilted such that the incident beam is parallel to the $\{110\}$ crystal planes, it has been shown that the apparent mean inner potential is increased by about 15% compared to the tilts in which the dynamical diffraction effects are minimal.[135,504] This variation is expected to be different for different systematic row orientations, and even greater for zone axis orientations, as was demonstrated for GaAs in Fig. 12.2.

Herring et al.[182] have proposed a variant to the holographic method described above which could be used in the cases when the wedge angle is not known. The method relies on recording two holograms from the same wedge sample at two different crystal tilts. In this manner one could write Eq. (12.5) twice and solve it for the case when both the thickness and the mean inner potential are unknown, while the sample geometry is known exactly. This technique was demonstrated successfully by using MgO smoke crystallites which have well defined cubic shapes.[182] The need to know the shape poses a limitation for the application of this method to realistic ion milled or electrochemically polished samples, when the shape is usually assumed to be a wedge but in practice it has large local variations from this shape.

Wang et al.[488] have recently extended the crystal wedge methodology to a spherical geometry in their holography studies of silicon nanospheres with surface oxide layers. They have found that the contribution of the uniform overlayer to the phase images can be separated from that of the core of the sphere, yielding measurements of the mean inner potential of both the spherical core and the uniform overlayer. For the smaller nanospheres which solidify as amorphous Si they have found a greater reproducibility of the mean inner potential measurements, compared to the larger nanospheres which solidify as crystalline Si. This observation is consistent with the fact that the dynamical diffraction contributions are expected to be larger in the case of the crystalline nanospheres whose tilts can not be controlled easily. Another conclusion of this study is that the mean inner potential of the crystalline Si $(12.1 \pm 1.3\,V)$ is only slightly larger than that of amorphous Si $(11.9 \pm 0.9\,V)$, consistent with the similar first nearest neighbor configurations in these two structures. The difference in V_0 may be expected to be larger between amorphous and crystalline phases of other materials, but such data is not available at this time.

5. Specimen thickness and its effects on phase and amplitude images

The effects of the specimen thickness on the phase images were already demonstrated in the previous sections. In kinematical orientations the dependence on thickness is linear, while in dynamical orientations this dependence becomes highly nonlinear. The linear dependence is very useful for studies of materials with uniform composition, allowing the phase image to be interpreted directly in terms of the projected specimen thickness. When the mean inner potential of the material is known, the phase image becomes a quantitative map of the projected thickness. This property is the basis for the applications of electron holography to study surface topographies and shapes of nanoparticles, provided that the kinematical approximation can be applied in the cases studied. These applications will be reviewed in detail in the later sections.

In cases of samples with non-uniform composition, the thickness dependence of the phase image complicates its interpretation in terms of composition and structure of the material, especially because most specimen preparation methods yield samples of unknown thickness. For example, different thinning rates at defects and at interfaces between different materials add complexity to the application of electron holography to study of interfaces. While, in principle, it is possible to use one of the standard methods for measurement of specimen thickness, such as electron energy-loss spectroscopy (EELS), convergent beam electron diffraction, or thickness fringes, these require a priori knowledge of the specimen composition and/or structure and they become particularly involved if a thickness map is wanted rather than measurements at a few discrete points. For this reason Gajdardziska-Josifovska and McCartney[137] have developed a holographic method for removal of the thickness dependence, based on combining the thickness dependence of the phase image with that of the amplitude image reconstructed from a same hologram. This method relies on the energy-filtering property of off-axis electron holography which allows us to treat the amplitude image as an equivalent to a zero loss image obtained by electron energy loss filtering. In this section we will briefly review the theoretical background of this method.

Off-axis electron holography is based on interference of a reference plane wave which has passed through vacuum and has been tilted by a biprism by $\Delta\vec{k} = \vec{k}_1 - \vec{k}$ ($\vec{k}$ is the wave vector of the incident electrons; the wave is taken to pass on the left of the biprism and is tilted towards the right side)

$$\psi_r(\vec{r}) = A_r\, e^{2\pi i \vec{k}_1 \vec{r}} \tag{12.14}$$

with an object wave which, after elastic scattering through the material, *remains coherent* with the reference wave

$$\psi_o(\vec{r}) = A_o(\vec{r}) \cdot e^{2\pi i \vec{k}_2 \vec{r}} \cdot e^{i\varphi(\vec{r})} \tag{12.15}$$

The object wave is tilted by $-\Delta\vec{k} = \vec{k}_2 - \vec{k}$ and is altered in phase (φ) and amplitude (A_o) by the object.

At this point it is important to note that only those electrons which are scattered elastically by the object satisfy the coherence requirement and contribute to the interference pattern in the hologram.[137,167,297,489] The background under the interference fringes is given by an incoherent sum of four types of wavefunctions: 1) the incoherent (denoted with a subscript "ic") part of the incident illumination which passes through vacuum on the same side of the biprism as the reference (r) wave ($\psi_{\mathrm{ic,r}}$); 2) the incoherent part of the incident illumination which is scattered elastically (e) by the object ($\psi_{\mathrm{ic,o,e}}$); 3) the incoherent part of the incident illumination which is scattered inelastically (ie) by the object ($\psi_{\mathrm{ic,o,e}}$); and 4) the coherent (c) part of the incident illumination which is scattered inelastically by the object ($\psi_{\mathrm{c,o,ie}}$). Fig. 12.6 gives a schematic representation of the different scattering processes. Therefore the intensity of a realistic hologram is given by

$$\begin{aligned} I_{\mathrm{h}}(\vec{r}) &= \psi^2_{\mathrm{ic,r}} + \psi^2_{\mathrm{ic,o,e}} + \psi^2_{\mathrm{ic,o,ie}} + \psi^2_{\mathrm{c,o,ie}} + |\psi_{\mathrm{r}} + \psi_o|^2 \\ &= A^2_{\mathrm{ic,r}} + A^2_{\mathrm{ic,o,e}}(\vec{r}) + A^2_{\mathrm{ic,o,ie}}(\vec{r}) + A^2_{\mathrm{c,o,ie}}(\vec{r}) + A^2_{\mathrm{r}} + A^2_o(\vec{r}) \\ &\quad + 2\, A_{\mathrm{r}}\, A_o(\vec{r}) \cos\left[4\pi\, \Delta\vec{k}\, \vec{r} + \varphi(\vec{r})\right] \\ &= I_{\mathrm{background}}(\vec{r}) + 2\, A_{\mathrm{r}}\, A_o(\vec{r}) \cdot \cos\left[4\pi\, \Delta\vec{k}\, \vec{r} + \varphi(\vec{r})\right] \end{aligned} \tag{12.16}$$

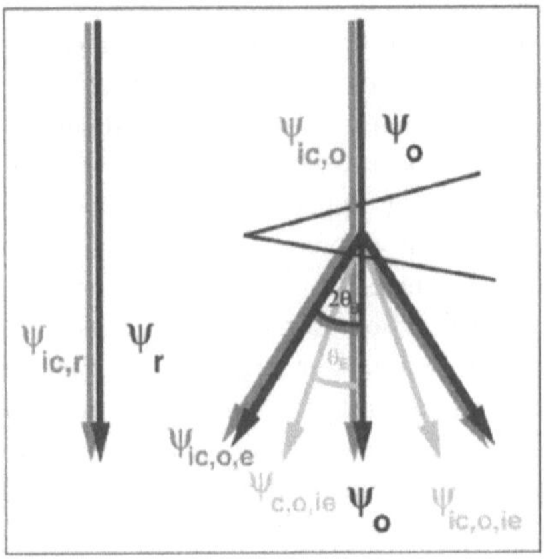

Figure 12.6. Schematic representation of energy filtering in off-axis electron holography: Coherent reference wave ψ_r interferes with coherent object wave ψ_o (scattered elastically by Bragg angles θ_B) to form holographic fringes. Incoherent part of illumination $\psi_{\mathrm{ic,r}}$ which is scattered both elastically $\psi_{\mathrm{ic,o,e}}$ and inelastically $\psi_{\mathrm{ic,o,ie}}$ by object, along with inelastic scattering of originally coherent illumination $\psi_{\mathrm{c,o,ie}}$, form hologram background.

where we have kept the standard shorter notation for the coherent reference wave ($\psi_{\mathrm{r}} = \psi_{\mathrm{c,r}}$) and the coherent object wave due to elastic scattering ($\psi_{\mathrm{o}} = \psi_{\mathrm{c,o,e}}$).

The different contributions to the intensity of a hologram are separated in the reconstruction process. The slowly varying background $I_{\mathrm{background}}(\vec{r})$ contributes to the autocorrelation central part of the Fourier transform of Eq. (12.16), while the holographic fringes given by the cosine term produce the two sidebands centered at $\vec{u} = 2\,\Delta\vec{k}$ and $\vec{u} = -2\,\Delta\vec{k}$ respectively

$$\begin{aligned}\mathbf{FT}\{I_h(\vec{r})\} = \delta(\vec{u}) \otimes \mathbf{FT}\{I_{\mathrm{background}}(\vec{r})\} \\ + \delta(\vec{u} + 2\,\Delta\vec{k}) \otimes \mathbf{FT}\{A_{\mathrm{r}}\, A_o(\vec{r}) \cdot e^{i\varphi(\vec{r})}\} \\ + \delta(\vec{u} - 2\,\Delta\vec{k}) \otimes \mathbf{FT}\{A_{\mathrm{r}}\, A_o(\vec{r}) \cdot e^{-i\varphi(\vec{r})}\}\end{aligned} \tag{12.17}$$

When one of the sidebands is selected, shifted to the center by $2\,\Delta\vec{k}$, and inverse Fourier transformed, the complex image has a modulus equal to $A_r\, A_o(\vec{r})$ and a phase equal to $\varphi(\vec{r})$. Therefore, the amplitude in the reconstructed image is the product of the amplitude of the coherent part of the illumination in vacuum (A_{r}) with the amplitude of only the part of the coherent illumination which was scattered elastically by the specimen (A_o). The size of the selected sideband imposes an effective cut-off angle on the scattering included in the hologram, and only elastic scattering at smaller angles contributes to the reconstructed image.

The energy-filtered amplitude reconstructed from the sideband of a hologram can be normalized using the amplitude measured from that part of the hologram containing vacuum only. McCartney and Gajdardziska-Josifovska[297] have shown that, in the absence of dynamical diffraction effects, the intensity of the vacuum reference wave ($I_r = A_r^2$) can be treated as the total intensity incident on the specimen (I_{total}) and the intensity of the elastically scattered object wave ($I_o = A_o^2$) can be treated as the zero-loss intensity as measured in EELS. The normalized amplitude A_n can then be expressed as an exponential function of the thickness t[100]

$$A_n(x,y) = \frac{A_r\, A_o(x,y)}{A_r^2} = \sqrt{\frac{I_o(x,y)}{I_{total}}} = e^{-t(x,y)/(2\lambda_i)} \tag{12.18}$$

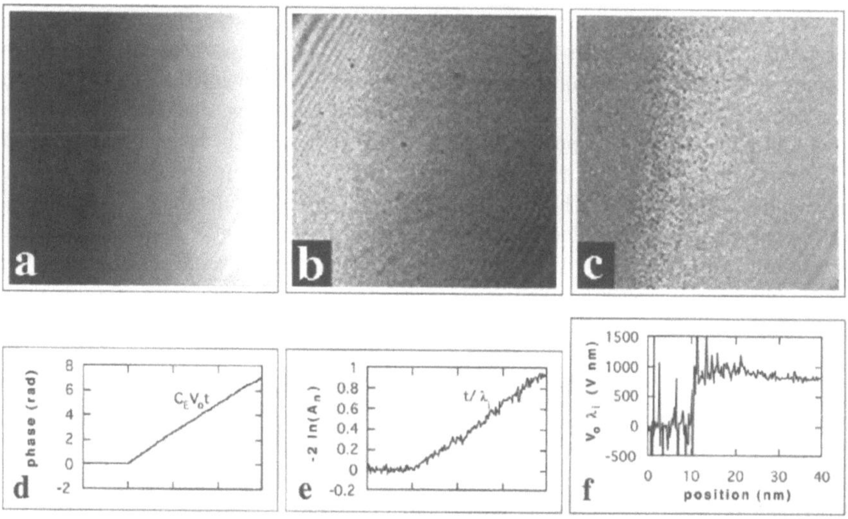

Figure 12.7. Off-axis hologram from Si 109.5° wedge is reconstructed to give energy-filtered phase (a) and amplitude (b) image with contrast dependent on specimen thickness. Phase image is divided by logarithm of normalized amplitude to give ratio image (c) with constant contrast in vacuum and in wedge, dependent solely on composition via the product $V_0\,\lambda_i$. Numerical profiles from images (d-f) allow measurement of mean inner potential (V_0) and mean free path for inelastic scattering (λ_i) when thickness of specimen is known.

where λ_i is the mean free path for inelastic scattering within angles that are smaller than the size of the sideband in Fourier space. For this approximation to hold the elastic scattering to angles larger than the cut-off angle of the sideband aperture must be weak. This requirement is consistent with the requirement that the crystal is tilted to a weakly diffracting orientation.

Equation (12.18) can be used to measure the mean free path for inelastic scattering for samples with known thickness and/or known geometry. Therefore the thin film strips, wedge samples and spherical samples that have been used to measure V_0 from phase images could also be used to measure λ_i from amplitude images. Fig. 12.7 shows an example of such measurements using the same 109.5° Si wedge as was shown in Fig. 12.4. The natural logarithm of the normalized amplitude image (i.e., $-2\ln A_n$) is linearly proportional to the t/λ_i ratio, as can be seen from Fig. 12.7b and Fig. 12.7e. Values for the mean free path obtained from such measurements (i.e., 71 ± 5 nm for MgO and 92 ± 7 nm for Si, see Ref. 297) compare favorably with the experimental values for bulk plasmon mean free paths obtained from EELS. An equivalent method for thin amorphous films has been described and used for measuring λ_i of carbon films by Harscher et al.[167]

By combining Eq. (12.18) with Eq. (12.5), the phase and amplitude images can be used to eliminate the thickness dependence[137]

$$\frac{\varphi(x,y)}{-2\,C_E\,\ln A_n(x,y)} = V_0(x,y)\,\lambda_i(x,y) \tag{12.19}$$

For a homogeneous material Eq. (12.19) predicts constant contrast equal to the product of the mean inner potential and the mean free path. $V_0\,\lambda_i$, both values depending only on the composition of the object. This is illustrated in Fig. 12.7c and 12.7f for the Si wedge. Apart from the noise in the vacuum and the thin regions in the specimen, the ratio image in Fig. 12.7c shows a constant contrast level in the wedge which is different from the contrast level in vacuum. The value of $V_0(x,y)\,\lambda_i(x,y)$ can be read directly from the profile in Fig. 12.7f, demonstrating that Eq. (12.19) is a good first approximation, especially for the thicker regions in the specimen ($t \sim 20\,\text{nm} \rightsquigarrow 120\,\text{nm}$). In vacuum and in the thin specimen regions, the ratio image and its profile are dominated by noise which arises from noise in the amplitude image. The noise is further emphasized in the calculated $V_0(x,y)\,\lambda_i(x,y)$ image because of division by numbers close to zero (i.e.,. $A_n \sim 1$ in vacuum and the thin regions, resulting in $\ln A_n \sim 0$). Better illumination coherence and microscope stability, combined with suitable data processing schemes, would be needed to improve the quality of the amplitude and the $V_0(x,y)\,\lambda_i(x,y)$ images.

For an object with varying composition, Eq. (12.19) gives a $V_0(x,y)\,\lambda_i(x,y)$ image which can be interpreted as a projected composition image. It is important to note that this image is independent of the specimen topography, and could therefore be used for samples with unknown thickness variations. This prediction has been demonstrated experimentally using ion-milled cross-sectional TEM samples from epitaxial CoSi2/Si interfaces.[137] This technique has been used to remove the thickness dependence in the studies of electrostatic fields at dislocations at p-n junctions in Si.[296]

6. Applications of off-axis electron holography to studies of interfaces

The field of interface holography is very new. It is in the stage of development of experimental and theoretical methods. The motivation is kept strong by the prospects for higher spatial resolution, energy filtering, and sensitivity to electric and magnetic fields. The first applications of electron holography have been to homogeneous interfaces, such as p-n junctions, and domain walls in magnetic and ferroelectric material, as reviewed, e.g., in Chapters 7, 8 and 10. In keeping with the scope of this chapter we will limit the discussion to interface studies in non-magnetic materials and in the absence of electrostatic fields at the interface, apart from those that arise from the difference in the mean inner potential across a hetero-interface. In particular, we will review the more recent studies of heterogeneous interfaces and grain boundaries (for practical details about interface holography, such as biprism-interface orientation and hologram reconstruction, the reader may refer to a recent review by Gajdardziska-Josifovska[139]).

6.1. Amorphous and polycrystalline hetero-interfaces: Effects of mean inner potential and specimen thickness

Weiss et al. produced the first published phase image of a hetero-interface, as shown in Fig. 12.8a.[494] The sample was a multilayer of amorphous Si and polycrystalline Mo with a large difference in mean inner potential between the two kinds of layers ($\Delta V_0 = 8 \pm 2\,\text{V}$ as measured from interface refraction effects in nanodiffraction[138]). The phase profile in Fig. 12.8b showed that, in addition to the thickness dependence, the phase advanced more in the Mo layers compared to the Si layers. The asymmetrical width of the Mo/Si and Si/Mo interfaces confirmed previous HREM observations about these multilayers, and showed that holography is sensitive to interface abruptness.[495]

One can not always expect to observe a phase contrast difference across a hetero-interface since the mean inner potential depends both on the composition *and* the structure of the material. Therefore, materials with densely packed light atoms may have mean inner potentials similar to materials with higher atomic numbers and larger

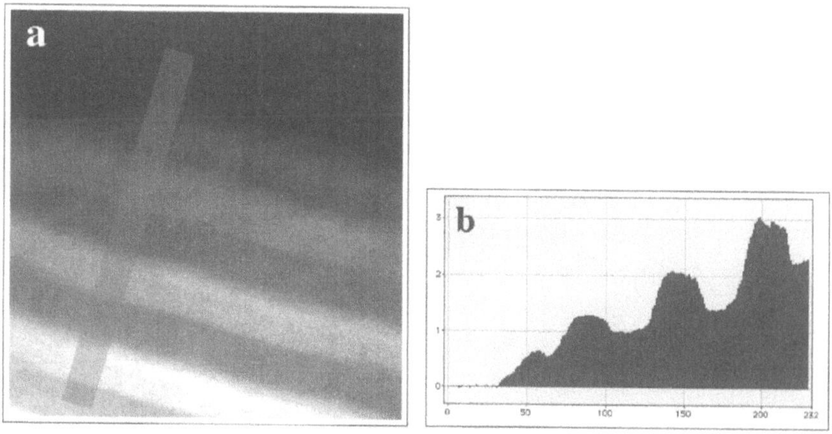

Figure 12.8. a) Phase image from Mo/Si multilayer from first holographic study of heterogeneous interfaces.[495] (Courtesy of J.K. Weiss.) b) Line profile as indicated in a). Phase shift is given in units of π.

interatomic distances. Weiss at al.[495] have demonstrated such lack of phase contrast in holograms from hetero-interfaces between amorphous Si and SiN_x.

In the above studies the interfaces are between amorphous materials and/or polycrystalline materials whose crystallite sizes are much finer than the projected specimen thickness. Such studies can usually ignore the dynamical diffraction contributions to the phase image, unlike the holographic studies of crystalline interfaces in which the dynamical effects must be considered.

6.2. Crystalline hetero-interfaces and grain boundaries: Effects of crystal tilt

The tilt of the specimen is very important for holography from crystals. However, for epitaxial interfaces, the tilt options are usually restricted to zone axis and systematic row orientations for edge-on viewing of the interface. Fig. 12.9 demonstrates the effects of tilt on holograms from a $CoSi_2$/Si(100) interface grown by molecular beam epitaxy.[136] This system is an example of an atomically abrupt crystal/crystal interface with large change in mean inner potential ($\Delta V_0 = 9.1\,\mathrm{V}$ was calculated using the non-binding approximation given by Eq. (12.8)). The amplitude images show similar contrast in the [011] zone axis (Fig. 12.9a) and the systematic (200) row (Fig. 12.9b) orientation, but the contrast between the phase in $CoSi_2$ and Si inverts due to the tilting (Fig. 12.9c and 12.9d). When the incident beam is parallel to the {002} crystal planes (Fig. 12.9d), the phase advances more in $CoSi_2$ compared to Si, consistent with $CoSi_2$ having larger mean inner potential than Si. When the transmitted beam is parallel to the [110] zone (Fig. 12.9c), the dynamical contributions are stronger, causing reversal of the phase contrast from that which is expected from the V_0 dependence solely.

The tilting options are also restricted when amorphous and/or polycrystalline films are grown on single crystal substrates with a surface parallel to a crystallographic plane. The Si/SIPOS interface shown in Fig. 12.10 is an example of such a case.[139] Tilting parallel to the interface changes the contrast in the phase and amplitude images drastically. The amplitude contrast vanishes in the systematic row orientation (Fig. 12.10d), while the phase image from the same hologram displays a phase retardation at the in-

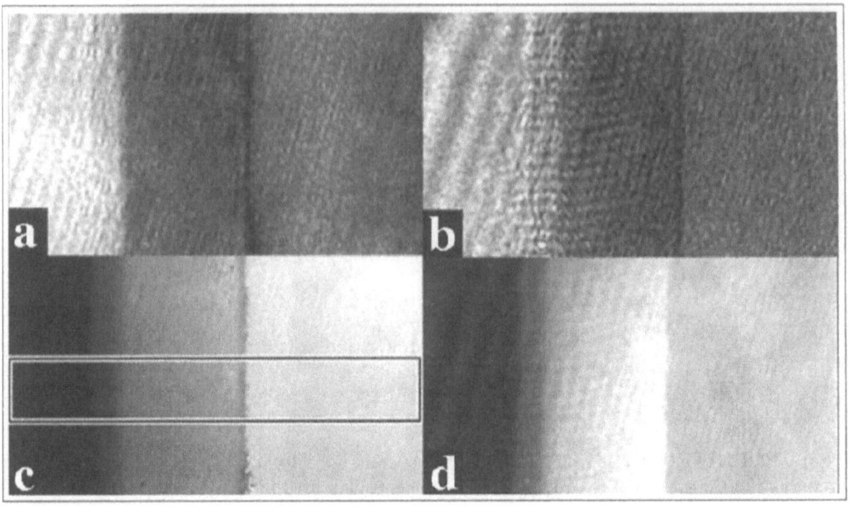

Figure 12.9. Amplitude (a,b) and phase (c,d) images reconstructed from two holograms of $CoSi_2$/Si interface oriented with [011] zone parallel to the beam (a,c) and in (200) systematic row orientation (b,d). Note contrast reversal in phase image upon tilting. Box with length of 30 nm marks region used for profiles shown in Ref. 136.

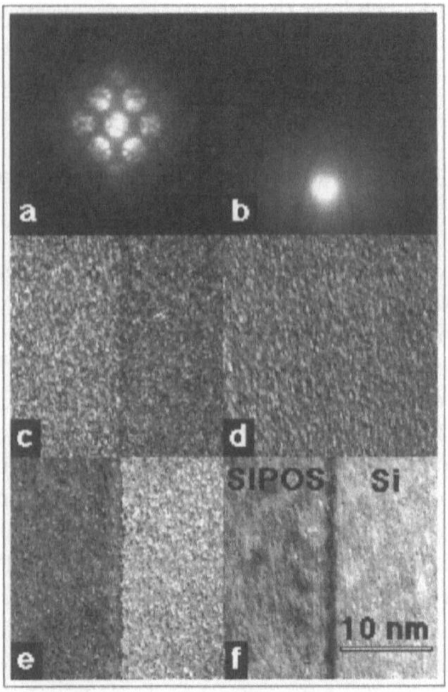

Figure 12.10. Interface between Si(111) and polycrystalline SIPOS at $[0,\bar{1},1]$ zone axis (a,c,e) and at (111) systematic row (b,d,f) orientation. The amplitude (c) and phase (e) images at the zone axis orientation show an abrupt interface, while in the systematic row orientation the amplitude shows no contrast (d) and the phase image (f) reveals a 1 nm wide region with less phase advancement (f).

terface (see dark band in Fig. 12.10f). A similar observation has been reported by Weiss at al.[495] for a Si_3N_4/Si_3N_4 grain boundary which gives a pronounced 12 Å wide band in the phase image while the amplitude image does not show any contrast difference between the two grains and at their interface.

A particularly elegant solution to the problem of dynamical contributions to the phase image has been proposed by Ravikumar et al. and used in their studies of space charge at grain boundaries in strontium titanate electro-ceramics.[356] The idea is to use bicrystals as models of these grain boundaries, in which both crystals have been miscut by the same angle thus forming a symmetrical grain boundary. In this manner the dynamical contributions (i.e., the tilt), as well as the thickness and the mean inner potential contributions can be taken to be equal on both sides of the interface, and the phase change observed at the interface can be interpreted in terms of the space charge at the interface.

6.3. Resolution enhanced interface holography

The complete understanding of the perfect crystal case is needed before holography can be applied to enhanced atomic resolution imaging of interfaces. It can be expected that the full aberration correction will be more involved for holograms of interfaces. Much development is needed in this field, but some of the important rewards for studies of interface structures may be predicted: The superior point resolution of aberration corrected images should enable better initial models of the interface structure, which could then be improved via the established image simulation methods. The comparison of simulated and experimental images should also become more reliable by having an amplitude and a phase image to match against. In addition, the sideband of the hologram is a numerical diffractogram which is sensitive to the crystal tilt.[264] This allows subsequent "nanodiffraction" from any chosen part of the holographic image which should significantly reduce the range of crystal tilts used for simulation of the exit wavefunction.

7. Applications of holography to surfaces

The applications of holography to surfaces rely on the sensitivity of the phase image to thickness, and therefore to the surface topography. Tonomura et al.[444] have demonstrated that single atomic height steps can be observed on a cleaved surface of a molybdenite thin film using transmission off-axis electron holography. Such steps correspond to a thickness change of 6.2 Å, which is one half of the c-axis lattice spacing, and produce a phase shift of $2\pi/50$. Phase amplification was used by these authors to make such a small phase difference visible in a 24 times amplified interference micrograph. Three and five atom high steps were also recorded with this transmission technique.

Detection of single-atom height steps has also been achieved by reflection off-axis electron holography using the Pt(111) surface as a test case.[13,332] The reflection holography technique was also employed to study the strain field around a screw dislocation emerging at a GaAs(110) surface.[333] This technique is a holographic variant of reflection electron microscopy (REM) in which a bulk surface is illuminated at grazing angles of incidence. In a reflection holography experiment the reference wave is formed by electrons which are reflected from an atomically flat part of the surface. The object wave is given by electrons reflected from that part of the surface containing the feature of interest (i.e., a step[13,332] or a dislocation[333]). The objective aperture of the microscope is used to select the specularly reflected beam, and the biprism is used to superimpose the reference and object components of this beam. Phase amplification is not needed in the reflection case because even a single atom-high step introduces a large phase change via the larger geometrical path length difference due to the grazing angle of incidence. The phase shift in the reflection mode is approximately two to three orders of magnitude larger than in transmission holography. For example, a 2 Å high step on GaAs (110), when observed with the (880) specular reflection using electrons accelerated to 100 keV, gives a phase change of 7.6π.[333] Onishi et al. estimated that reflection holography can measure vertical displacements from flatness with precision of ~ 0.10 Å.[333]

A holographic equivalent has also been developed to the high resolution surface profile microscopy by Tanji et al.[436] who used this technique to successfully image the surface potential of a MgO smoke crystal with atomic resolution.

In the above studies the object wave carries the information about the surface feature of interest. However, an "empty hologram" (i.e., an interference pattern that is formed by a biprism when both the reference and the object wave pass through vacuum only) can be used to write a periodic nanowire pattern on a surface, as demonstrated on polymethylmethacrylate (PMMA) thin films in a modified TEM,[325] and on surfaces of PMMA and GaAs bulk samples in a modified SEM.[126] This opens a new venue for applications of electron holography in nanolithography.

While the work to date shows that surface holography can become a powerful technique for the characterization of surfaces, its widespread applications will have to await the wider availability of instruments with field emission sources of electrons and with ultrahigh vacuum at the sample accompanied with appropriate in-situ surface preparation and modification facilities.

8. Applications of holography to fine particles

Since the phase shift is related to the projected thickness of the sample along the beam direction, electron holography provides an opportunity to assess the morphology of fine particles. While shape information in conventional TEM is often inferred from the symmetry and edge faceting observed in images of crystalline particles along low-index zone axes, the conclusions reached are not always accurate for all of the individual crystals. For example, polyhedral and prismatic shapes may not be readily distinguishable. Tilting to multiple zone axes to provide varying projections can be quite difficult, especially for individual nanoscale particles – or may be impractical due to a limited tilt range in the instrument.

In principle, much of the information about three-dimensional shape of an individual particle can be obtained from a single electron hologram. In order to carry out this type of analysis, a number of conditions must be met: 1) there must be no magnetic field associated with the particle; 2) the sample must not develop a charge under the electron beam; 3) the mean inner potential of the sample must be known; and 4) there must be no compositional variation within the field of view (or any compositional variations must be known independently). Even under these conditions, some ambiguity remains. One limitation is that, while the total projected thickness can be calculated, the relative inclinations of the entrance and exit surfaces to the beam are not apparent from the electron hologram. Furthermore, abrupt changes in phase will signal a sharp step in thickness (as may occur at a facet aligned parallel to the beam), but the periodic nature of the phase difference results in an ambiguity of $\pm 2\pi n$ (where n is an integer) in $\Delta\varphi$. From Eq. (12.5), then, the step in thickness will be

$$\Delta t = \frac{\Delta\varphi \pm 2\pi n}{C_E\,|V_0|} \tag{12.20}$$

Examples of these situations are illustrated schematically in Fig. 12.11, where several morphologies (a-d) are shown that will give equivalent phase profiles because their

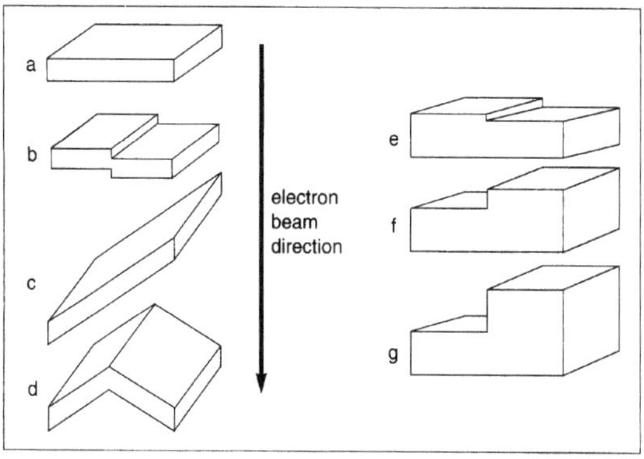

Figure 12.11. The morphologies shown on the left-hand side (a-d), yield equivalent phase profiles because their projected thickness is identical. The morphologies on the right-hand side (e-g) can in some cases be indistinguishable also, as discussed in the text.

projected thickness is identical. On the right-hand side (e-g) morphologies are shown which can not be distinguished despite the difference in their corresponding projected thickness if the thickness jump (from left to right) is: $\Delta t_e = (\Delta\varphi - 2\pi)/(C_E|V_0|)$ for (e), $\Delta t_f = \Delta\varphi/(C_E|V_0|)$ for (f) and $\Delta t_g = (\Delta\varphi + 2\pi)/(C_E|V_0|)$ for (g). Equivalent uncertainties will also occur for polyhedral and many other shapes that are exhibited by fine particles. In most cases uncertainties of this type can be resolved in a number of ways, such as by obtaining a second hologram after slightly tilting the sample.

In practice, useful information about particle shapes can be derived from holograms even if not all of the conditions described above are satisfied. For example, polyhedral particles can readily be distinguished from those that are nearly spherical by changes in slope in the phase profiles, even if the mean inner potential V_0 is not known or the particles are coated with a thin layer of a second phase. Charging problems, which can be substantial for non-conducting materials, appear to be reduced for nanoparticles.[39] Also, if the particles are sufficiently small, they can be regarded as weak phase objects and dynamical effects can be neglected.

Although most morphological studies of fine particles to date have relied on the phase image reconstructed from the hologram, an alternative approach is to obtain a map of thickness $t(x, y)$ divided by λ_i, the mean free path for inelastic scattering, from a normalized amplitude image.[297] As noted previously – see the discussion of Eq. (12.18) – one can take advantage of this relationship to determine λ_i for samples of known geometry, but alternatively the local thickness at all points in the image, and thus the morphology of fine particles, can be evaluated if λ_i is known or can be considered a constant within the material. Nonetheless, use of the phase image may be preferable because the relative noise level is typically lower than in the corresponding amplitude image (see, e.g., Fig. 12.7).

8.1. Particles with known shapes: Proof of principle

A variety of fine particles with known shapes have now been examined in order to demonstrate the effects of thickness variations on phase images and to assist in the development of new techniques. Perhaps the most widely examined material has been MgO, which forms as cubic single crystals with {001} faces when produced by burning magnesium wire or ribbon and collecting the resulting smoke. Characteristic holograms of such cubes viewed at oblique angles clearly show the shifting of interference fringes in the hologram, and the reconstructed phase images exhibit thickness contours consistent with the projected shapes of the cubes.[39,182,219,261,449,452] (The spacing of the thickness contours can be reduced to more closely map the morphology of the particles by phase

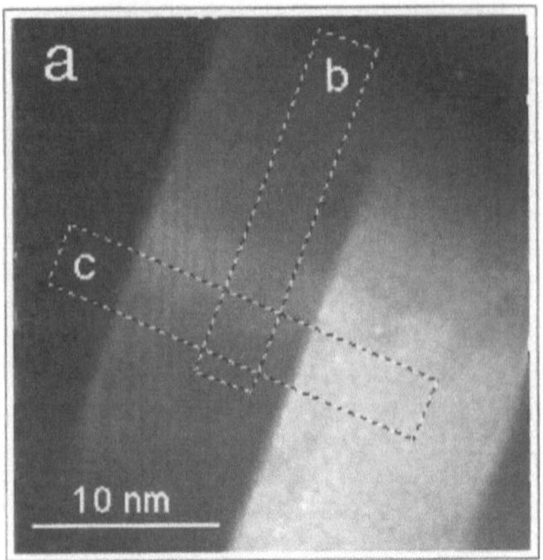

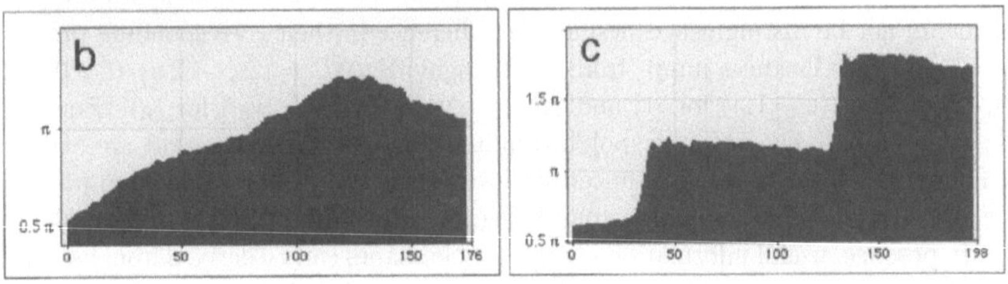

Figure 12.12. (a) Reconstructed phase image of a cubic, nanocrystalline MgO particle aligned with the beam along [110]. The smaller particle is attached to, and overlaps at the right side of the figure with, a larger MgO cube in the same orientation. (b) Phase profile from the edge of the particle in towards the center, displaying a nearly linear increase in phase difference corresponding to the change in projected thickness. (c) A perpendicular phase profile showing the steps in thickness at the (001) edges of the two MgO cubes and the constant thickness within them.

amplification.)

MgO cubes along low-index zone axes such as <011> and <012> present very simple projected thickness profiles, but the dynamical contributions produce phase jumps that are not representative of thickness variations in micron-scale particles. Fortunately, dynamical effects which complicate interpretation of the phase image are diminished for smaller particles. Even for nanoparticles, however, quantitative assessment of specimen thickness is complicated under such imaging conditions by the variation in the effective value of the mean inner potential versus that obtained under kinematical conditions. An example of a phase image for a nanoscale MgO smoke crystal aligned along a <011> axis is shown in Fig. 12.12a. The phase profile along the length of the tilted cube (Fig. 12.12b) does indicate a nearly linear change in projected thickness from the edge to the center of the particle, although there is some fluctuation which may be due to beam damage of the cube surface. The apparent V_0 is approximately 14.5 V – somewhat larger than the valuc (13.01 V) obtained from cleaved wedges oriented so as to minimize dynamical effects.[135,504] Since this particle is attached to a larger one with the same orientation, a perpendicular phase profile shows an abrupt phase shift reflecting the steep (100) face of the smaller crystal, then a constant value as the particle is traversed, and then another jump to a greater $\Delta\varphi$ value as the edge of the second particle is encountered.

A very different type of samplc with known geometry has been used to demonstrate the accuracy of shape information for biological samples, which also exhibit improved contrast in a reconstructed phase image as compared to conventional bright-field TEM images. In this case, an unsupported cylindrical segment of tobacco mosaic virus was examined.[11] The phase profile across the strand was consistent with the projected thickness of the cylinder, and showed a very different intensity distribution than that in a conventional TEM image.

Although electric and magnetic field information are not separable from a single hologram, Tonomura et al.[448] developed an approach to accomplish this end using a pair of holograms. One result is that the shapes of strongly magnetic particles can be evaluated by eliminating the magnetic field information. The technique relies on the appropriate summation and subtraction of phase information from the particle viewed from both sides. Opposite viewing directions do not provide identical information because the magnetic component of $\Delta\varphi$ is reversed in sign, while the thickness contribution is unchanged. The method has been applied to a multiply-twinned particle of cobalt in the approximate shape of a pentagonal dodecahedron (five regular tetrahedra sharing a common edge). Assuming a uniform mean inner potential within the particle, the resulting image exhibits thickness contours without the superposed effect of the magnetic field.

8.2. Particles with unknown or uncertain shapes

While these demonstrations have been useful, much of the interest in electron holography of nanoparticles lies in the ability to derive three-dimensional information which is not otherwise readily accessible. Areas of interest range from the catalytic behavior of noble metals to the consolidation of oxide nanocrystals. With respect to catalysis, Datye et al.[6,8,82,478] have investigated the surface and internal structure of palladium metal particles supported on amorphous silica microspheres. Changes in morphology were observed as a function of pretreatment; most Pd particles contained an internal void after calcination of the precursor and initial H_2 reduction, but were solid when the pretreatment was modified to avoid room-temperature exposure to H_2 and thus the possible intermediate formation of a palladium hydride phase. The porous particles showed linear changes in phase with distance in from the edge of the particle, suggesting surface faceting. The internal features were confirmed to be voids by the phase variation, which was inconsistent with the possibility of an included amorphous core region of another material with a different mean inner potential. Solid particles, on the other hand, were modeled well by phase profiles expected for unfaceted spheres with the exceptions of a few areas where local surface facets may have developed. The extent and orientations of faceting in such materials control the coordinations of the surface atoms, thereby affecting catalytic activity and/or selectivity.

The actual shapes of polyhedral zirconia nanocrystals have similarly been explored by consideration of off-axis electron holograms and the resulting phase images, phase profiles, and phase-amplified thickness maps.[8,39] In this system many of the particles show clear $\{111\}$ and $\{001\}$ facets in high-resolution lattice images, and the projected shapes appear to be consistent with what one might expect for truncated octahedra or cuboctahedra (tetrakaidecahedra). Upon reconstruction of phase images, it was found that some particles did indeed exhibit changes in the slope of the phase shift consistent with the faceting expected for a truncated octahedron. Fig. 12.13 shows high-resolution and reconstructed phase images of such a particle. From the profiles across the phase image, one can see that edge-on facets produced abrupt phase jumps at particle edges, while oblique facets produced linear phase changes correlated to the projected particle thickness. (While the phase image does, in principle, provide all of this information already, the presence of facets and the relative rates of change of projected thickness are sometimes more readily perceived in appropriately selected phase profiles. Averaging over several pixels also serves to reduce noise and can help to emphasize the features of interest.)

Some of the ZrO_2 nanoparticles exhibit shapes that are clearly more complex. Others appear to be cuboctahedral from their projected shapes and the identified facets,

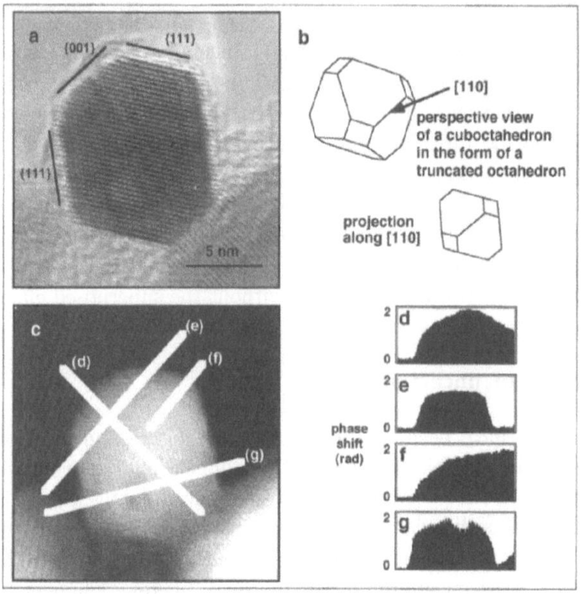

Figure 12.13. (a) High-resolution lattice image of an apparently cuboctahedral zirconia (ZrO_2) particle. (b) Sketch of the corresponding cuboctahedron shape and expected [110] projection. (c) Phase image reconstructed from a hologram of the particle in (a). Line profiles across different sections are shown in (d), (e), (f), and (g). Each starts at the labeled end of the corresponding line in (c). All are consistent with the postulated geometry, with the exception of one area where there is a void in the particle as seen in (g).

but the phase images are not consistent with this inference. At least one such particle viewed along [110] was found to be prismatic, since the phase jumped abruptly at the edges of the crystal but was relatively constant at all points within the particle. As seen in Fig. 12.13, internal voids were also observed in some of these oxide nanoparticles, including both well-faceted polyhedra and particles that appeared to be nearly spherical from their phase profiles.

The shape of a beryllium particle (supported on a carbon film) has also been elucidated by electron holography.[449] In this case, the crystal was much larger (~200 nm) than the Pd and ZrO_2 grains (typically 5-15 nm) discussed above, and so the shape was made apparent by drastically amplifying the phase (x 32). The result was a set of finely-spaced contours representing thickness differences of only 13 Å. Although the authors do not comment on it, the particle clearly exhibits four-fold symmetry in its external form along the viewing axis, and appears to be cuboctahedral in shape – even though elemental Be has a hexagonal close packed crystal structure. The same group has separately demonstrated that such fringes are in fact meaningful by distinguishing surface steps of as little as 6.2 Å in other phase-amplified images,[444] as discussed above.

Supported particles on a film have been investigated by an amplitude-division holography technique, which uses a single-crystal film above the specimen as a beam splitter and thus does not require a biprism in the column.[376] Triangular, epitaxial islands of Au on a substrate of MoS_2 were examined and the phase information was reconstructed. In this case one surface (the side adjacent to the substrate) can be taken as flat, and so the thickness can be mapped directly; the islands appear to be platelets with sloped sidewalls and relatively uniform thickness across the center. This method was also applied to a latex sphere as a test, but the resulting phase map does not appear to directly reflect the projected thickness of a sphere. Others have used holography to examine the electrostatic fields on latex spheres due to charging in the beam,[119] which may play a role in the deviations from the expected phase behavior.

8.3. Contacting particles

Holography can be particularly valuable in determining the conformation of fine particles or fibers that are in contact with one another. For the zirconia nanoparticles described above, part of the interest in faceting arises from the possibility that if particles come together face-to-face, consolidation may be facilitated and/or sintering may be accelerated. A suitable rotation of face-to-face particles that have come together along a common crystallographic facet can even bring grains entirely into registry, turning the biparticle into a single crystal.[351] The extent to which this mechanism contributes to unusually rapid grain growth, defeating the purpose of fabricating such fine powders in the first place, is not yet well understood. Few experimental results using electron holography are available, but occasional ZrO_2 biparticles have been observed.[39] Phase profiles provide information on faceting in both particles at once, revealing their relative rotations in the plane of the interface.

Elastic deformation between contacting bodies is also of interest and can be shown via phase images. Allard et al.[7] have made use of this to examine the local compression along the line of contact between adjacent carbon nanotubes (buckytubes). Phase profiles clearly confirm the hollow cores at the centers of the tubes and are consistent with the model proposed earlier for the manner in which the tubes come into contact.

8.4. Other applications to fine particles

Finally, although this section has focused on the shape determination of fine particles in the absence (or sometimes even in the presence!) of magnetic fields and other contributions to the phase shift, one should note that there are also other types of information on particles that holography can provide. For example, holography of nested fullerenes (encapsulating metal particles) has allowed after-the-fact nanodiffraction to obtain local lattice parameter information.[7] In this case, the spacing of the hologram fringes themselves could be used as a sensitive, precise, and reproducible calibration standard. In other work on buckytubes, the phase image has not only matched well with computed thickness profiles across the hollow tubes, but has also shown sensitivity to defects such as bends in the tube or the closed tube tip.[273] The phase of the electron wave may be sensitive to bonding differences and local strains in such areas, which effectively alter the local value of the mean inner potential. Furthermore, holography may be valuable in detecting and characterizing surface films or layers on particles or fibers of known shape, if there is a sufficient difference in the mean inner potential between the two materials.

9. Conclusions

In conclusion, electron holography is slowly maturing from the stage of technique development to the stage of successful applications to materials problems. Much of the application work to date has been demonstrative, but quite ingenious and gainful. It is laying the framework for the future when a larger number of electron microscopy laboratories will have instruments with coherent sources of electrons, and when electron holography may become a routine technique for the study of solid structures.

ELECTRON HOLOGRAPHY USING DIFFRACTED ELECTRON BEAMS (DBH)

R.A. Herring[1] and G. Pozzi[2]

[1]Microgravity Sciences Program, Canadian Space Agency, St. Hubert, Quebec, Canada J3Y 8Y9
[2]Department of Physics and Instituto Nazionale per la Fisica della Materia, University of Bologna, viale B. Pichat 6/2, I-40126 Bologna, Italy

1. Introduction

In order to relax the stringent beam coherence requirements linked to the standard off-axis scheme of electron holography with the electron biprism, it has been proposed to use single crystal films as amplitude beam splitting devices, either alone[288,342] or in combination with an electron biprism.[286,287,339,340] In particular, it has been conceived to insert the biprism between convergent electron diffracted beams.[341] This method of beam interference was demonstrated by Herring et al.[183,185,187,189] who have shown that good contrast of the interference fringes is possible for a wide range of experimental conditions. The observation plane may be varied from the in-focus diffraction condition, i.e., the Fraunhofer plane, through the Fresnel focus region to the in-focus specimen condition. Also, the beam convergence can be varied from highly convergent to parallel.[190] This method of interferometry was originally referred to as Convergent Beam Electron Diffraction plus an Electron Biprism (CBED&EBI) but then became more generally known as Diffracted Beam Interferometry (DBI). DBI has also been used as a method of holography, now referred to as Diffracted Beam Holography (DBH), in which the phase change at crystal regions such as dislocations, compositional gradients, etc., causes the interference fringes to shift, thus providing quantifyable information on the nature of these features.[184,186,188]

Although a thorough theoretical analysis of this holographic mode is still lacking, the obtained results justify further efforts in this direction. This chapter aims at presenting the state of the art of DBH, showing how the basic ideas underlying DBI can be used to gain insight into the capabilities of the corresponding holographic mode.

2. Basic theoretical considerations

In the DBI set-up (Fig. 13.1) the electron beam coming from the source impinges on the single crystal specimen after having traversed the prefield of the objective lens,

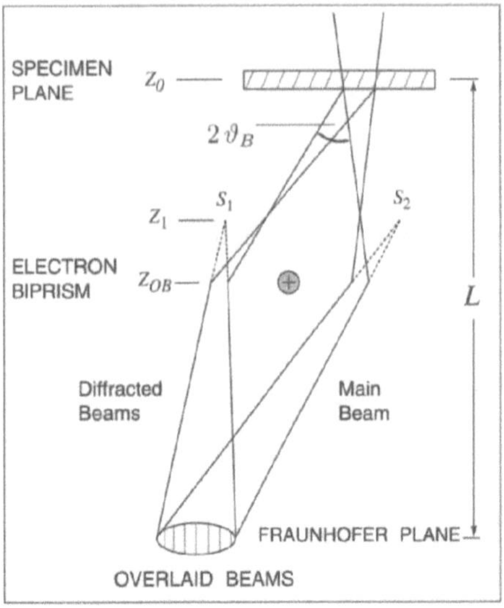

Figure 13.1. A simplified electron ray diagram showing the interference of the main beam with a diffracted beam in the objective back focal plane at the Fraunhofer plane, where the symbols are defined in the text.

which acts as an additional condenser lens. The beam is diffracted by the single crystal. Each diffracted beam travels first through the post-specimen field of the objective lens (i.e., the true imaging lens) and then through the field-free space until it reaches the electron biprism. The biprism is usually inserted in the position of the selected area aperture holder which is fitted in the first intermediate image space. In the Hitachi HF-2000 microscopes, an additional electron biprism can be inserted between the first and second intermediate lenses. For this latter case, the beam impinges on the biprism after having been magnified by the first intermediate lens. In our set-up, the electron biprism is inserted between the diffracted beams in such a way that it does not intercept any part of them but only deflects by opposite angles the beams traveling on opposite sides of the biprism. If this deflection is suitably chosen, the disks of different diffracted beams can be overlapped below the biprism. These propagate through the microscope to the observation plane, where the image is recorded and where, under suitable electron optical conditions, interference fringes can be observed in the overlapping regions.[189]

Calling z the optic axis, and with z_0 the coordinate of the specimen plane, then z_1 and z_{OB} are the planes conjugate to the electron source and electron biprism respectively, whereas the Fraunhofer plane is at a very large (theoretically infinite) distance L from the above planes (Fig. 13.1). S1 and S2 are the apparent focused probe positions (i.e., virtual sources) corresponding to the transmitted and diffracted beam respectively. Note that the order in which these planes are located along the optic axis in the object space depends on the lens excitation and may not reflect the physical order with which the various elements are actually located in the microscope, i.e., source, specimen, biprism, etc. For instance, we can excite the objective lens in order to have the biprism effectively above the specimen plane or, as in the set up shown in Fig. 13.1, the objective lens excitation can be used to place the source plane z_1 below the specimen, near the biprism plane. The actual, physical ray path, however, should be used when the propagation of the electron beam from the source to the detector is investigated.

The image wavefunction in the DBI mode is derived by considering the spatial frequency representation of the wavefunction which is related to the real space representation by a Fourier transform.[189] Let us recall the main results of this analysis. If the electrons are emitted by a δ-like point source at $\vec{r}_1$ in the source plane (referred to the object space, i.e., in the plane intersecting the optical axis at z_1), then we have an ideal spherical illuminating wave, whose complex wave function at a large distance from $\vec{r}_1$ (i.e., in Fourier space) is given by

$$\psi_\delta(\vec{k}) = e^{-2\pi i \vec{k}\,\vec{r}_1} \tag{13.1}$$

Assuming ideal imaging in the condenser, and a circular aperture centered on the axis given by Eq. (13.2),

$$B(\vec{k}) = \begin{cases} 1 & \text{for} \quad |k| \leq K_A \\ 0 & \text{for} \quad |k| > K_A \end{cases} \tag{13.2}$$

then the wave entering the objective lens prefield is given by

$$S(\vec{k}) = B(\vec{k}) \cdot e^{-2\pi i \vec{k}\,\vec{r}_1} \tag{13.3}$$

If we allow for the presence of a tilting stage inserted between the condenser and the objective lens, its net effect, provided $\vec{r}_1$ is unchanged, can be simply described by a rigid shift by $\vec{K}_T$ of the beam amplitude.

The effect of the aberrations of the objective prefield, including its spherical aberration constant C_S^A, and the defocusing factor $\Delta z = z_0 - z_1$, can be accounted for by multiplying (in $\vec{k}$-space) the tilted wave function by the standard exponential factor so that the wave function impinging on the specimen $\psi_{ill}(\vec{k})$ is given (in $\vec{k}$-space) by

$$\psi_{ill}(\vec{k}) = S(\vec{k} - \vec{K}_T) \cdot e^{-i\chi_A(\vec{k})} \tag{13.4}$$

where, for the isoplanatic aberrations defocus and spherical aberration (see page 38),

$$\chi_A(\vec{k}) = \frac{\pi}{2} C_S^A \lambda^3 k^4 + \pi \lambda \Delta z\, k^2 \tag{13.5}$$

By considering the specimen to be an ideal single crystal. it can be described by the transmission function $T_{cr}(\vec{r})$

$$T_{cr}(\vec{r}) = \sum_i A_i \cdot e^{2\pi i \vec{g}_i \vec{r}} \tag{13.6}$$

where A_i is the complex amplitude relative to the diffracted beam of reciprocal wave vector $\vec{g}_i$. It can be easily ascertained* that its effect on the wave function is to generate the following output

$$\psi_0(\vec{q}) = \sum_i A_i \cdot S(\vec{q} - \vec{K}_T - \vec{g}_i) \cdot e^{-i\chi_A(\vec{q} - \vec{g}_i)} \tag{13.7}$$

*i.e., by multiplying Eq. 13.6 with the inverse Fourier transform of Eq. 13.4, and performing a Fourier transform on the result i.e., an array of replicas of the incoming beam, each displaced by the reciprocal vector $\vec{g}_i$. In propagating from the specimen to the biprism, these beams first suffer the spherical aberration C_S^B of the objective lens post-specimen field. They then propagate from the first intermediate plane to the biprism plane, so that the corresponding defocusing factor is given by $\Delta z_B = z_{OB} - z_O$, the biprism defocus. Both phase factors can be grouped in the term $\chi_B(\vec{q})$.

The transmission function of the biprism, $T_B(\vec{r})$, is given in real space by

$$D_R(\vec{r}) \cdot e^{2\pi i \alpha_B \vec{b}\cdot(\vec{r}-\vec{r}_B)/\lambda} + D_L(\vec{r}) \cdot e^{-2\pi i \alpha_B \vec{b}\cdot(\vec{r}-\vec{r}_B)/\lambda} = T_B(\vec{r}) \tag{13.8}$$

where $\vec{r}_B$ is the coordinate of a generic point on the biprism axis, α_B is the angular deflection of the biprism, directly proportional to the applied voltage, and $\vec{b}$ is a unit vector perpendicular to the biprism axis. The two functions D_R and D_L represent the transmitted amplitude of the two half-planes to the right (R) and left (L) of the biprism. This yields a transmission function of the form given in Eq. (13.6). Now the important point is that electron optical conditions should be set up so that the diffracted beams in the biprism plane do not strike the biprism wire, but pass either to the left or to the right. Then the net effect of the biprism is again equivalent to a rigid displacement in the spatial frequency plane of $\pm\vec{K}_B = \pm\alpha_B \vec{b}/\lambda$, plus an additional phase factor $\exp[\pm 2\pi i \vec{K}_B \vec{r}_B]$.

More precisely, the image wave function is the sum of two contributions, one involving the beams passing to the left of the wire, index i_L

$$\psi_L = \sum_{i_L} A_{i_L} \cdot e^{2\pi i \vec{K}_B \vec{r}_B} \cdot S(\vec{q} + \vec{K}_B - \vec{K}_T - \vec{g}_{i_L}) \cdot e^{-i\chi_A(\vec{q}+\vec{K}_B-\vec{g}_{i_L})} \cdot e^{-i\chi_B(\vec{q}+\vec{K}_B)} \tag{13.9}$$

and the other beams passing to the right, index i_R

$$\psi_R = \sum_{i_R} A_{i_R} \cdot e^{-2\pi i \vec{K}_B \vec{r}_B} \cdot S(\vec{q} - \vec{K}_B - \vec{K}_T - \vec{g}_{i_R}) \cdot e^{-i\chi_A(\vec{q}-\vec{K}_B-\vec{g}_{i_R})} \cdot e^{-i\chi_B(\vec{q}-\vec{K}_B)} \tag{13.10}$$

It should be noted that, as observations are carried out in the diffraction or Fraunhofer mode, the observed image intensity is given simply by

$$I = |\psi_R + \psi_L|^2 \tag{13.11}$$

3. Analysis of the interference phenomena

We first investigate the conditions under which two discs overlap, and then the main features of the interference fringes observed in the overlapping region. By considering two generic diffracted beams, one traveling to the left of the biprism and the other to the right, of reciprocal wave vectors $\vec{g}_L$ and $\vec{g}_R$ respectively, the image wave function across them is given by

$$\begin{aligned}\psi_{LR}(\vec{k}) = {} & A_L \cdot e^{2\pi i \vec{K}_B \vec{r}_B} \cdot S(\vec{q} + \vec{K}_B - \vec{K}_T - \vec{g}_L) \cdot e^{-i\chi_A(\vec{q}+\vec{K}_B-\vec{g}_L)} \cdot e^{-i\chi_B(\vec{q}+\vec{K}_B)} + \\ & A_R \cdot e^{-2\pi i \vec{K}_B \vec{r}_B} \cdot S(\vec{q} - \vec{K}_B - \vec{K}_T - \vec{g}_R) \cdot e^{-i\chi_A(\vec{q}-\vec{K}_B)-\vec{g}_R} \cdot e^{-i\chi_B(\vec{q}-\vec{K}_B)}\end{aligned} \tag{13.12}$$

The analysis of the resulting intensity distribution, given by the absolute square of the above expression, looks very complicated, owing to the number of terms involved. Nonetheless, it can be ascertained at once that, because of the form of the amplitude factors in S, perfect overlapping is achieved when

$$\vec{q} + \vec{K}_B - \vec{K}_T - \vec{g}_L = \vec{q} - \vec{K}_B - \vec{K}_T - \vec{g}_R \tag{13.13}$$

i.e., when

$$\vec{g}_L - \vec{g}_R = 2\vec{K}_B \tag{13.14}$$

This means that the biprism should be aligned perpendicular to the line joining the two centers of the diffracted beams, and the deflection angle α_B of the biprism should be equal to the relative Bragg angle $\vartheta_B = \lambda\,|\vec{g}_R - \vec{g}_L|/2$. When this condition is achieved, both the exponential factors containing the position $\vec{r}_1$ of the source Eq. (13.1) and the aberrations χ_A of the prefield are identical in both left and right amplitudes, so that they drop out from the intensity expression. This means that in the case of perfect overlapping, the interference phenomena are not sensitive to the position of the source and can therefore be observed in principle also with extended sources of very poor lateral coherence.

If the simplifying assumption of neglecting the spherical aberration terms in the aberration function is made (i.e., restricting our considerations first to the case of ideal imaging), it is possible to calculate the image intensity without resorting to numerical simulation. Taking into account that non-perfect alignment is the most common condition, by posing

$$\delta\vec{q} = \vec{g}_L - \vec{g}_R - 2\vec{K}_B \tag{13.15}$$

it turns out that the intensity in the overlapping region is given by

$$\begin{aligned} I = {} & |A_L|^2 + |A_R|^2 + 2\,|A_L|\,|A_R| \cdot \\ & \cdot \cos\Big[4\,\pi\,\vec{K}_B\vec{r}_B + 2\,\pi\,\vec{r}_1\,\delta\vec{q} - 4\,\pi\,\lambda\,\Delta z_B\,\vec{q}\,\vec{K}_B + \\ & \qquad 2\,\pi\,i\,\lambda\,\Delta z\,\vec{q}\,\delta\vec{q} - \pi\,\lambda\,\Delta z\,(\vec{g}_L + \vec{g}_R)\,\delta\vec{q} + \beta_R - \beta_L\Big] \end{aligned} \tag{13.16}$$

where β_R and β_L are the intrinsic phases associated with the crystal's diffracted amplitudes A_R and A_L, respectively. In the perfect overlapping case, $\delta\vec{q} = 0$, and the spacing of the interference fringes is given by

$$\Delta = \frac{1}{2\,|K_B|\,\lambda\,\Delta z_B} \tag{13.17}$$

Recalling that the spatial frequency plane is projected onto the final recording plane according to the relation

$$Y = \lambda\,\vec{q}\,L \tag{13.18}$$

where L is the microscope camera length, it turns out that the spacing, ΔY, on the plate is given by

$$\Delta Y = \frac{\lambda L}{2\vartheta_B \Delta z_B} \tag{13.19}$$

This spacing remains almost the same even if $\delta\vec{q} \neq 0$ as in the usual operating conditions the following inequality is satisfied

$$|\Delta z\, \delta\vec{q}\,| \ll |\,\Delta z_B \cdot \vec{K}_B| \tag{13.20}$$

In fact, both defocuses are of the same order of magnitude, but $|\delta\vec{q}| \ll \vec{K}_B$. Moreover, attention should be paid to the terms in the argument of the cosine term in Eq. (13.16), which depends on the positions $\vec{r}_1$ of the source and $\vec{r}_B$ of the biprism, since they become relevant in determining the contrast of the fringes if realistic factors are taken into account. These factors include the finite dimension of the source, i.e., partial lateral coherence, as outlined in the following section.

4. Influence of coherence

Although coherence of the electron beams is not expected to be important for the formation of interference fringes in the exact overlap position, Eq. (13.14), it becomes relevant when the beams are not perfectly aligned (see Eq. (13.15)). The experimental results reported in Ref. 189 show that the fringe contrast diminishes for increasing $\delta\vec{q}$ and disappears at an angular separation of $\vartheta_c = 0.9\,\text{mrad}$, corresponding to a spatial frequency of $0.36\,\text{nm}^{-1}$. A simple theoretical analysis permits correlation of this value to the effective source size. In fact, lateral coherence effects are linked to the finite dimension of the effective electron source size, and their detrimental effects on the contrast of the two beam interference fringes can be taken into account simply by convolving the intensity given by Eq. (13.16) with the function describing the effective source intensity distribution. If we assume a Gaussian distribution[343]

$$J(\vec{r}_1) = \frac{I_s}{\pi R_s^2} \cdot e^{-r_1^2/R_s^2} \tag{13.21}$$

where R_s is the source radius, and I_s the total source current, the intensity of the interference fringes becomes

$$I = |A_L|^2 + |A_R|^2 + 2\,|A_L|\,|A_R| \cdot \cos(\phi) \cdot e^{-\pi^2\,\delta\vec{q}^2\,R_s^2} \tag{13.22}$$

where $\cos(\phi)$ is the cosine phase term given by Eq. (13.16), without the r_1 term. When $A_L = A_R$, the relative visibility is given by

$$C_r = e^{-\pi^2\,(\delta q)^2\,R_s^2} \tag{13.23}$$

which by the Van Cittert-Zernike theorem is the Fourier transform of the intensity distribution, given by Eq. (13.21). Therefore, the vanishing of the fringes as $\delta\vec{q}$ increases, gives directly the source dimension R_s. For a fringe contrast reduction of $1/e$ it results in

$$\pi\,\delta q\,R_s = 1 \tag{13.24}$$

It is easier to detect the spatial frequency at which the contrast falls below the detectability limit, which still gives the correct order of magnitude of the effective source radius (referred to the object space in our coordinate system) from which R_s can be easily calculated, using Eq. (13.24). For our case, we get $R_s = 0.88\,\mathrm{nm}$. Another way of expressing partial coherence is to introduce the coherence parameter of the beam,[343] which is an invariant and is given by the ratio of the coherence distance with the illumination radius. For the reported values, this ratio is ~ 0.1, which is rather low, considering that the microscope is equipped with a field emission source. This discrepancy has also been observed by Zhu, Peng and Cowley,[511] who compared the performances of STEM and TEM instruments equipped with FEGs, and found that the TEM had a broader effective source, probably caused by electrical and mechanical instabilities of the whole instrument.

5. The holography mode

The results of the foregoing sections can be used to obtain insight into the main features of the holographic mode. We assume that, owing to the real specimen structure, the transmission function of each diffracted beam is not described by a single spatial frequency (as assumed for the ideal case, expressed in Eq. 13.6), but by a continuous distribution of spatial frequencies.

Let us consider, for the sake of simplicity, the beam passing on the right side of the biprism to be ideal, i.e., it comprises a single spatial frequency, and the beam passing the left side to be composed of only two spatial frequencies, say $\vec{g}_L$ and $\vec{g}_L + \delta\vec{k}_0$, which then would correspond to the virtual source S_1 effectively splitting into two sources, e.g., S_1 and S_1' in Fig. 13.2, with $\delta\vec{k}_0$, the spatial frequency difference corresponding to the spatial frequency of the phase object. If the perfect overlay condition is realized for interference, i.e., the two beams having spatial frequencies $\vec{g}_L$ and $\vec{g}_R$ are directly overlaid, then Eq. (13.14) can be applied and the disc corresponding to the additional spatial frequency is off-set by $\delta\vec{k}_0$ (Fig. 13.2). This geometry has the following consequences with respect to the process of image formation in the holographic mode.

First, the useful region to be reconstructed, given by the overlapping of the three beams originating from S_1, S_1' and S_2, is reduced with respect to the ideal two beam case, and can even be made absent if $\delta\vec{k}_0$ is too large. Second, to this purely geometrical effect related to the beam overlapping, we should add the effect of the partial beam coherence, whose detrimental influence acts in the same direction. That is, if $\delta\vec{k}_0$ is too large, then the beam is not coherently superimposed to the other two and its information is missing in the holographic reconstruction.

The conclusion is that the phase object's spatial frequency range, approximately given by $\delta\vec{k}_0$, should not be too large. Hence, the resolution of the holographic mode is intrinsically limited by two factors: the available interference width given by the hologram and the partial coherence of the electron beams. This seemingly too pessimistic conclusion can be somewhat mitigated if we consider separate beams only if their spatial frequency range is limited around the main carrier frequency, $\delta\vec{k}_0 < |\vec{g}_L - \vec{g}_R|$.

We can also compare the spatial coherence requirement of the DBH mode with that of the standard image off-axis mode using the biprism alone. In the latter case,

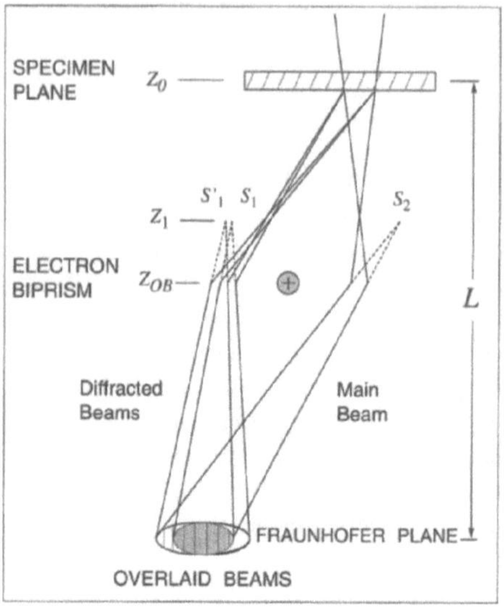

Figure 13.2. A simplified electron ray diagram showing the interference of the main beam with two diffracted beams, representing a very simple object wave function.

in order to achieve complete separation between the beams in the reconstruction, the carrier spatial frequency should be at least three times the highest spatial frequency $\delta\vec{k}_0$ of the specimen. Taking into account that all the spatial frequencies of the specimen should be coherently transferred in the hologram, it turns out that the partial coherence should be high enough to transfer a range of four times the specimen spatial frequency and thus four times that required in the DBH mode, where the same lateral coherence for the standard phase contrast image is needed.

By considering the close analogy of the DBH set-up with point projection microscopy (see Ref. 423 and Spence's chapter on pages 311 ff.) we expect that the wave functions within the diffracted discs are Fresnel images of the transmitted and diffracted beams, with a defocus distance and a projection magnification given by Δz and $L/(\Delta z)$ respectively. If these considerations are confirmed by the theoretical analysis of the set-up, we could conclude that the DBH mode is a Fresnel mode, where the difference between the two overlapping beams is imaged. In this case amplitude and phase information could be obtained as the beam which is used as the reference wave is known and, from the other beam, phase object information can be obtained. In fact, in the DBH case the information about the specimen is contained in the broadened diffraction spots, which are coherent with the main diffracted beam, so that we can recover this information by our holographic methods. This point may be a real advantage of DBH with respect to standard imaging, but this concept still needs to be confirmed by theory and experiments. For this reason it is important to develop a detailed theory of the DBH holography method and exploit its consequences.

Keeping this in mind, two rudimentary methods of DBH holography for measuring the phase of the microscope and for detecting the phase of defects in crystals have been demonstrated recently.[184,186,188,189] One holography method requires the interfering beams to be separated such that a reference wave is taken from the perfect crystal and an object wave is taken from the defect. This method is possible because of the partial coherence of the beams, which continues to produce fringes when they are separated. This method produces two holograms, since, in one part of the micrograph, the information carried by one beam interferes with the unmodulated region of the other

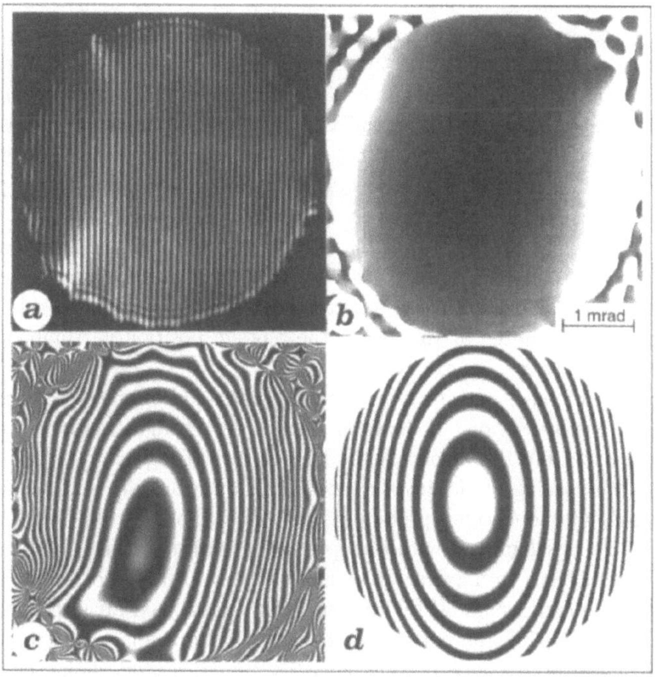

Figure 13.3. (a) An example of an interferogram with the main beam interfering with the (111) beam of GaAs. Reconstructed 2π (b) and $\pi/8$ (c) phase distribution of the interferogram (a). The phase distribution has been flattened around the center of the interferogram. (d) Simulated 16x amplified ($\pi/8$) flattened phase contour map.

beam, and vice versa.[339] Another holography method requires the interfering beams to originate from a thin crystal where kinematical diffraction conditions dominate. The beams are directly overlaid such that one beam (e.g., the 000 beam) interacts weakly or not at all with the defect and thus becomes the reference wave, and the other beam interacts strongly with the defect and therefore becomes the object wave.

6. Experimental method

Diffracted beam interferometry techniques were developed using a Hitachi HF-2000 FEG microscope equipped with a rotatable electron biprism holder. DBI patterns were obtained with this microscope while using many different pole pieces having spherical aberrations, C_S, ranging in value from 0.6 mm to 3.3 mm. The electron biprism (typically $\sim 0.3\,\mu$m $\varnothing$) used an applied voltage ranging from 15 to 270 V to compensate the diffracted beam(s) angle of $2\,\alpha_B$ (Fig. 13.1) with the larger biprism potentials being used with the higher order diffracted beams.

An example of an interferogram is shown Fig. 13.3a It was produced using an electron beam spot size of $\sim$ 3.5 nm, with the main beam interfering with a (111) beam of GaAs. This interferogram shows straight, fine fringes having good contrast. Holograms were produced in the microscope's diffraction mode; however, the biprism must initially be roughly aligned in the imaging mode. In the imaging mode, the image is defocused using the objective lens to show an over- or under-defocused image of the diffracted beams. The rotatable biprism is positioned between the beams to be interfered, perpendicular to the line connecting the beams. The diffraction mode is then chosen and the biprism alignment is readjusted slightly to compensate for the rotation of the electrons along the optical axis. A voltage is then applied to the biprism.

If the beams to be interfered do not intersect, the biprism voltage is increased, and if the biprism cannot be seen in between the diffracted beams once a potential is applied, then a further objective defocus of the image plane is required. That is, in the imaging mode the degree of over- or under-defocussing of the diffracted beams is increased using the objective lens. Also, a smaller condenser aperture can be tried. Once the biprism is correctly placed between the beams, they are merged by increasing the biprism voltage. Often at this point, the biprism position is trimmed slightly so the interfering beams are exactly overlaid.

The phase difference between interfering beams should produce visible interference fringes in the overlap region. If, however, interference fringes cannot be seen in the overlaid beams, the biprism defocus is too large and the fringes are too fine to be seen. To increase the spacing of the interference fringes, the biprism defocus is decreased (making sure that the electron probe does not wander out of the crystal's region of interest due to a misalignment with the optic axis), until the fringes become visible.

It is advantageous to maximize the camera length of the microscope by fully exciting all post-biprism lenses. If the biprism is in the selected area aperture position, this requires maximizing the current of the second intermediate lens, first projector lens and second projector lens. If available, a TV camera system should be used which gives a further $\sim 20\,\mathrm{X}$ magnification, to allow observation of very fine or weak fringes. If the fringes still cannot be seen, then either the beams being interfered have a spatial frequency difference greater than the resolvable detail of the microscope or there is a source of instability in the microscope, biprism or specimen. Holograms obtained by the method outlined above were exposed on film and a negative of the hologram was digitized into 512 x 512 pixels of 256 gray levels and processed by an Apollo DN10000 computer. The hologram was reconstructed digitally using the Fourier transform method of Takeda and Ru.[433]

7. Spherical aberration

The phase between overlapped beams causes the fringe spacing in Fig. 13.3a to become narrower from the left side to the right side of the hologram, although this is somewhat difficult to detect by eye. It is easier to detect in a digital reconstruction of the hologram. To emphasize the curved phase surface, the hologram was reconstructed using a plane reference wave with the phase at the center of the hologram flattened (Fig. 13.3b) and then amplified 16 x, thereby producing $\pi/8$ phase contours. These better show the contours of the phase distribution (Fig. 13.3c). The irregular phase distribution at the edge of the hologram is caused by electrostatic charging of the condenser aperture's edge, due to contamination.

A first attempt to interpret these results has been made by including the spherical aberration in the intensity distribution Eq. (13.22). As the defocus aberration is only responsible for the carrier fringe frequency, the resulting interferogram can be considered a hologram of the fictitious phase object given by

$$\Delta\phi = 0.5\,\pi\,C_S^B\,\lambda^3\left((\vec{K}-\vec{K}_B)^4-(\vec{K}+\vec{K}_B)^4\right) \tag{13.25}$$

This is the phase distribution recovered in the processing of the interferogram, once the phase at the center has been flattened, by subtracting a linear phase term whose coefficient is the gradient of the phase at the center of the diffraction disc. The phase distribution was modeled using Eq. (13.25) assuming $C_S^B = 1.0\,\mathrm{mm}$ and $K_B = g_{111}/2$, which yielded the computer-generated 16x phase amplified image presented in Fig. 13.3d. Figure 13.3c included an experimentally-incorrectable pre-specimen beam tilt of $K_T = 1.4\,g_{111}$. The computer generated image can be seen to reproduce the qualitative features of the experimental phase image quite well.

The numerical simulations carried out so far have also confirmed that these patterns do not depend on the defocus distance, i.e., the carrier fringe spacing. They can be considered as a new form of Ronchigram, so it is hoped that, following Cowley and coworkers,[54,56,64,271] this method of fringe analysis may open interesting perspectives for aberrations analysis and measurement, especially since it can be carried out on-line using a slow scan CCD camera and a desktop computer.

8. Material science applications

Once the electron beam impinges the single crystal specimen, it is diffracted systematically by the interatomic planes having interplanar spacing of d_{hkl} and angular dependence, $2\vartheta_B = l/d_{hkl}$. Each beam contains unique amplitude and phase information of both the crystal and the microscope. For example, for the commonly seen case of a GaAs crystal such as $In_{0.2}\,Ga_{0.8}\,As/GaAs$ containing a strained layer superlattice (SLSL), the main beam contains very little information about the SLSL, whereas the SLSL is clearly seem as lines of bright and dark contrast in the 002 beam. In other beams, information about defects such as dislocations at the SLSL/epi interface is clearly seen in the 004 beam. Thus each beam contains substantial, but different, information about the GaAs material, which the DBH technique is particularly suited to detect.

The interpretation of phase shifts by object features requires both knowledge of the features, as well as knowledge of the imaging conditions. The measurement of the true phase of an object thus requires modeling of the detected phase shifts. Currently, DBH has been shown to be able to "detect" many phase shifts at objects in crystals; two of these are presented below for phase objects which include a threading dislocation in GaAs and a buried amorphous zone in Si. At the time of this writing, these studies are considered to be only qualitative since they rely upon the kinematic approximation for a first, preliminary interpretation. As we do not yet give any equation, say for the phase shift at a dislocation, nor provide a simulation of the phase distribution, we can only state that we have detected phase shifts and have not yet measured the true phase distribution of the phase objects.

To begin with, for a thin TEM specimen, undergoing kinematic diffraction, the amplitude and phase of the crystal is given by the structure factor equation

$$F_{(h,k,l)} = \sum_n A_n \sum_{u,v,w} e^{i\beta_{(h,k,l)}} \tag{13.26}$$

where n is the atom type, A_n are the amplitudes of the different atomic types, β is the phase of each individual beam ($\beta_{(h,k,l)} = [2\pi(hu + kv + lw)]$). $(h,k.l)$ are the beam's Miller indices and u,v,w are the coordinates of the atoms. For thicker TEM specimen, the phase of the (000) beam is given by the specimen's mean inner potential while

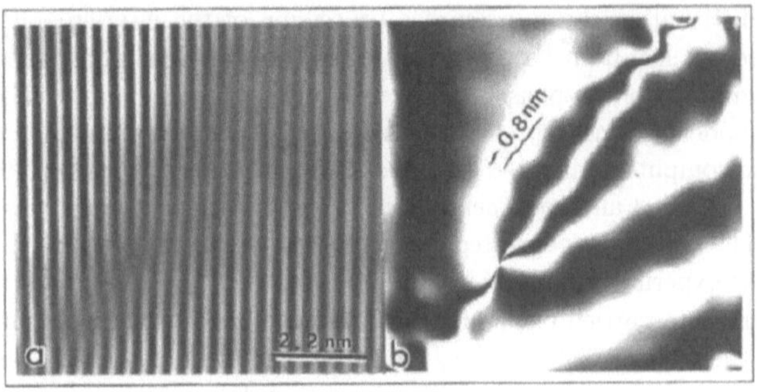

Figure 13.4. Hologram of a threading dislocation in a GaAs epilayer (a), and its reconstructed phase image, amplified 8 times (b).

the phase of each diffracted beam is determined by multi-slice or Bloch wave simulations. For the TEM specimens considered in this chapter we will analyze small phase objects in relatively thin crystals and thus use the above structure factor equation as the basis for the interpretation of the phase of our specimen. The possible changes of phase in crystals, which are measured as phase shifts in the hologram, are due to crystal structures such as compositional gradients, changes of internal strain, electrostatic potentials (e.g., electronic barriers and wells in superlattices, 2-dimensional electron gas at a superlattice interface, unbonded electrons at the cores of dislocations and at grain boundaries) and magnetic fields (e.g., fluxons in superconducting materials). The structure factor Eq. (13.26) is modified by the crystal phases such that,[197]

$$F_{(h,k,l)} = \sum_{n} A_n \sum_{u,v,w} e^{-i\beta_{(h,k,l)}} \sum_{u,v,w} e^{-i\alpha_{(h,k,l)}} \tag{13.27}$$

where, e.g., $\alpha_{(h,k,l)}$ might describe strain which is then given by the atomic displacement $\vec{R}$ and the diffraction vector $\vec{g}_{(h,k,l)}$ according to

$$\alpha_{(h,k,l)} = 2\pi\, \vec{g}_{(h,k,l)}\, \vec{R} \tag{13.28}$$

DBH demonstrates this relationship by detecting the phase shift at a crystal defect which has introduced strain in a material. The phase shifts due to electrostatic and magnetic fields may require some knowledge of the phase object, since their potentials can extend far from their source and thus prevent a good reference beam from being used.[287,340] Magnetic and electrostatic fields have been considered by others in this text and can be incorporated into the phase measurements by DBH with these considerations in mind.

A hologram produced from directly-overlaid, relatively-planar electron beams originating from a GaAs crystal containing a threading dislocation is shown in Fig. 13.4a. Reconstruction of this hologram using a planar reference wave shows the phase distribution (8x phase amplified) undulating along the direction of the dislocation core (Fig. 13.4b). Since the two interfering beams include the (000) beam and the (220) beam, the phase distribution can be represented by the atomic displacement, R, within the crystal Eq. (13.28). The line direction of the dislocation is $\sim 35°$ to the (110) which places it on a [111] habit plane, although it curves to a larger angle (to the right) at the top right hand corner of the Fig. 13.4b. It is interesting to note that the phase undulations along the dislocation core have a period of ~ 0.8 nm and may represent

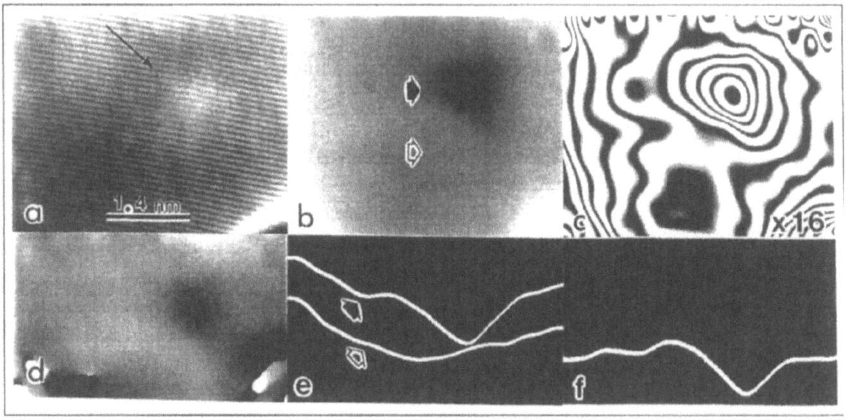

Figure 13.5. Reconstruction of an amorphous zone buried in the Si crystal (dark arrow) using an adjacent region (light arrow) as a reference to determine the absolute maximum phase shift of $\sim 2\pi/5$ where (a) is the hologram, (b) is a 2π phase image, (c) is a $\pi/8$ phase image, (d) is the phase image with the background removed, (e) are the line profiles of the phase passing from side to side in (b), and (f) is the line profile of the phase passing through the amorphous zone in (d), which gives the absolute phase shift of the amorphous zone.

the strain field generated by the dislocation at the As-As or Ga-Ga atomic positions along the dislocation core in the $\langle 110 \rangle$ on the $\{111\}$.

A hologram produced from separated, convergent beams, where the reference beam (the 000) originates from a perfect Si single crystal, and the other beam (the 004) originates from a region which contains a small, buried amorphous zone (produced by high-energy (2 MeV) As-ion implantation), is shown in Fig. 13.5a. Reconstruction of this hologram using a plane reference wave shows that the phase distribution of the amorphous zone lies on a curved phase surface due the use of convergent beams (Figs. 13.5b and 13.5c). Since a good reference hologram representing the curved phase surface was not available, a similar region beside the amorphous zone was used as a reference wave (arrowed region in Fig. 13.5b). The similarity of the phase distribution of these two regions can be seen in the intensity scans of their curved phase surfaces (Fig. 13.5e); their subtraction produced the phase image in Fig. 13.5d. This image has a phase distribution as scanned in Fig. 13.5f, which now allows for an approximation of the absolute phase shift of the amorphous zone. This small amorphous zone, ~ 2 nm in diameter, has a maximum phase shift of $\sim 2\pi/5$, with respect to its surrounding phase, which is equivalent to $\sim 5\,\%$ strain in the crystal. As well, it can be seen in Fig. 13.5f that the wave passing through the amorphous zone is phase shifted forward, i.e., it has been accelerated with respect to the wave passing through the crystal, which means that the amorphous region is more dense than the crystal. The greater density of the amorphous zone and the overall shape of its phase distribution implies that the amorphous region is comprised of a net interstitial concentration, rather than vacancies, and does not reveal a vacancy-rich center as expected from the Brinkman amorphous zone model.

9. Interesting perspectives

DBH is a new method of holography which, because of its flexible methodology, shows potential for being able to measure optical properties of the microscope, as well as physical properties of crystals which are unavailable by other characterization methods. It can be performed using a wide range of beam convergence angles and focus planes, and can cause any two diffracted beams within the information envelop to interfere. As shown on pages 304 ff., C_{SB} can be obtained by modeling the phase distribution of two interfering, exactly overlaid, convergent beams. If the two interfering convergent beams are not directly overlaid, then their phase distribution in the hologram can be expressed as,

$$\Delta\phi = 0.5\,\pi\,\lambda^3 \cdot \Big\{ C_S^A \cdot \Big[\Big((\vec{q} - \vec{K}_B)^4 - (\vec{q} - \vec{K}_B - \delta\vec{q})^4 \Big) \Big] + C_S^B \cdot \Big[\Big((\vec{q} - \vec{K}_B)^4 - (\vec{q} + \vec{K}_B)^4 \Big) \Big] \Big\} \tag{13.29}$$

where $\delta\vec{q}$ is the deviation from the exact overlaid position of the beams Eq. (13.15). It can be seen that if C_{SB} has been measured for the exact overlaid position then C_{SA} should also be able to be measured for a deviated position of $\delta\vec{q}$. Difficulties arise if the partial coherence of the beam is not large enough to allow for a sufficiently large $\delta\vec{q}$ for an accurate measurement of C_{SA}, as was the case for the TEM-FEG used previously[189] which had a measured ratio of the coherence distance with illumination radius of only ~ 0.1 (pages 300 ff.). However, TEMs with FEGs are expected to improve which may enable these measurements to be made in the future. As well, if these experiments are carried out in a STEM which has a higher beam coherence,[511] considerably larger $\delta\vec{q}$ from the exact overlaid position can be used, so that aberration effects due to the prefield-objective lens spherical aberration, C_{SA}, may be measured. Thus DBH may enable the separation and measurement of both spherical aberration components, C_{SA} and C_{SB}. These measurements could enhance our understanding of the focusing behavior of the electro-magnetic lenses which could then provide information for improved objective lens pole piece designs.

Consider two interfering "planar" beams which are generated by a perfect crystal and are directly overlaid, so that $\delta\vec{q} = 0$, then Eq. (13.16), which describes the intensity of the fringes in the hologram, can be expressed as

$$I(\vec{q}) = |A_L|^2 + |A_R|^2 + 2\,|A_L\,A_R| \cdot \cos\Big[4\,\pi\,\vec{r}_B\,\vec{K}_B - \chi_B\,(\vec{q} + \vec{g}_L) + \chi_B\,(\vec{q} - \vec{g}_R) + (\beta_L - \beta_R) \Big] \tag{13.30}$$

For this case, $\vec{q} = a$ constant, and we are left with five phase terms to be considered: the phase due to the biprism, $\vec{K}_B$, the specimen phase for each beam, $\beta_{L,R}$, and the phase of the transfer function of the post-field objective lens, χ_B, for each beam. Since $\vec{q} = a$ constant for interfering planar beams, there is only one phase value across the hologram, i.e., there are no phase shifts. This electron optical condition opens up interesting possibilities for measuring the inherent phase of the crystal since, if $\vec{g}_L = -\vec{g}_R$, the phase due to the transfer function of the objective lens cancels out, leaving only the phases due to the biprism and the specimen. Furthermore, if the biprism is accurately aligned to the optic axis, only the phase of the specimen $\beta_L - \beta_R$ remains, allowing the crystal's inherent structure factors to be directly measured. Because the phase is dependent on the crystal's structure factors, for ordered materials which produce superlattice diffraction, the phase from a given atomic feature (such as a given atomic plane) can be isolated. For example, GaAs produces ±(1/2, 1/2, 1/2) superlattice diffraction which, when they are interfered (i.e., the (1/2, 1/2, 1/2) and (-1/2, -1/2, -1/2) beams), causes the phase due to the Ga atomic planes to cancel, leaving the phase due to the As atomic planes (see Appendix, page 310). This capability of DBH may enable phase objects which are restricted to certain atomic planes in ordered alloys to be better characterized. Thus the above examples illustrate that the various experimental set-ups for DBH will ultimately determine the approach for extracting the hologram's phase information.

10. Conclusions

A description of the method of diffracted beam holography, DBH, has been presented, and a theoretical model for DBH, applied to the Fraunhofer imaging condition, has been provided for a defined set of electron optical conditions. This model agrees well with experimental phase data measurements when compared with the spherical aberration conditions of the microscope. We realize the limitations of this model for interpreting other types of phase objects which are small and easily identifiable but have their phase distribution delocalized due to the out-of-focus imaging conditions. These include dislocations in GaAs and the buried amorphous zone in Si, in which phase distributions have been discussed only qualitatively. Therefore a more comprehensive model must be derived for the Fresnel imaging conditions and tested against the experimental data. Further extension of the DBH method to relatively large phase objects in thick crystals must consider the contributions of the mean inner potential and the phase of the individual diffracted beams of the host crystal, as well as those diffracted beams from the phase object.

Two general methods of DBH have been presented. These two methods interfere the beams in such a way that the beams are either directly overlaid or slightly separated, and these configurations define the method used to interpret the holograms. One method is able to use off-axis interference, because the beams continue to interfere when they are not directly overlaid, due to beam coherence. The perfect crystal produces the reference wave and the object wave originates from a small phase object in the crystal. The other method uses directly overlaid beams where the reference wave originates from a beam which is unaffected by the phase object, and the object wave originates from a beam which interacts strongly with the phase object. As discussed on pages 300 ff., the latter method should not require a coherent electron source, and should allow interferometry and holography studies to be performed with microscopes which do not have a FEG source.

In general, electron holography has a distinct advantage over other conventional material characterization methods such as diffraction contrast imaging, because it allows measurement of the phase of defects and other phase objects such as small electrostatic and magnetic fields. We have shown that Diffracted Beam Holography is a promising technique for material science research.

Acknowledgments

The authors wish to thank A. Tonomura, T. Tanji, J. Ru and T. Hirayama for their support during the Tonomura Electron Wavefront Project, Exploratory Research for Advanced Technology, Japan Research and Development Corporation. One of us (G.P.) also acknowledges S. Patuelli for his skilful technical assistance.

11. Appendix

Short discussion on the cancellation of Ga structure factor phase in GaAs by the interference of the $\pm(1/2, 1/2, 1/2)$ superlattice diffracted beams.

Atomic positions for GaAs, (u,v,w):

Ga	0, 0, 0	0.5, 0, 0.5	0.5, 0.5, 0	0, 0.5, 0.5
As	0.25, 0.25, 0.25	0.75, 0.25, 0.75	0.75, 0.75, 0.25	0.25, 0.75, 0.75

One obtains as structure factor equations:

$$F_{\mathrm{Ga}_{hkl}} = f_{\mathrm{Ga}} \cdot \left\{e^{-2\pi i\,0} + e^{-2\pi i\,(0.5h+0.5l)} + e^{-2\pi i\,(0.5h+0.5k)} + e^{-2\pi i\,(0.5k+0.5l)}\right\}$$

$$\begin{aligned} F_{\mathrm{As}_{hkl}} = f_{\mathrm{As}} \cdot \{ & e^{-2\pi i\,(0.25h+0.25k+0.25l)} + e^{-2\pi i\,(0.75h+0.25k+0.75l)} \\ & + e^{-2\pi i\,(0.75h+0.75k+0.25l)} + e^{-2\pi i\,(0.25h+0.75k+0.75l)} \} \end{aligned}$$

The structure factors for 111-type hkl diffracted beams are defined as

$$F_{111} = F_{\mathrm{Ga}_{111}} + F_{\mathrm{As}_{111}} = 4\,f_{\mathrm{Ga}} + 4\,i\,f_{\mathrm{As}}$$

$$F_{-1-1-1} = F_{\mathrm{Ga}_{-1-1-1}} + F_{\mathrm{As}_{-1-1-1}} = 4\,f_{\mathrm{Ga}} - 4\,i\,f_{\mathrm{As}}$$

$$F_{111} + F_{-1-1-1} = 8\,f_{\mathrm{Ga}}$$

Cancellation of the As structure factor phase appears to be possible, however, a problem exists. The origin can be As or Ga, therefore the solution is not unique.

Solution: For As planes oriented upwards (i.e., giving the 2X4 surface reconstruction and a rigid crystal structure) $\pm 1/2$ order superlattice diffractions yields,

$$F_{1/2,1/2,1/2} = -2\,f_{\mathrm{Ga}} + (1.4 + 1.4i)\,f_{\mathrm{As}}$$

$$F_{-1/2,-1/2,-1/2} = 2f_{\mathrm{Ga}} + (1.4 - 1.4i)\,f_{\mathrm{As}}$$

$$F_{1/2,1/2,1/2} + F_{-1/2,-1/2,-1/2} = 2.8\,f_{\mathrm{As}}$$

therefore a unique solution exists for the separation of the Ga and As phases where the Ga structure factor phase is cancelled by the interference of the $\pm 1/2$ order superlattice beams.

ELECTRON HOLOGRAPHY AT LOW ENERGY

J.C.H. Spence and J.M. Cowley

Department of Physics and Astronomy, Arizona State University, Tempe, AZ 85287-1504

1. Introduction

Shadow images, magnified by projection from a small electron source, were amongst the first images to be obtained by electron microscopes.[24] In the same year (1939) point-projection shadow images of carbon grids were also reported at a few hundred volts using a field-emission electron tip.[320] The authors comment on the absence of aberrations in their lensless system, the removal of which has proved a major obstacle for conventional electron holography at higher voltages. Fig. 14.1(a) shows the optical arrangement for point-projection microscopy (PPM). A tungsten field emitter S is placed at a distance z_1 (a few hundred nanometers) from a thin sample T. and an image recorded on a channel plate D about $z_2 = 10\,\mathrm{cm}$ away. The magnification is approximately $M = z_2/z_1 \approx 10^6$, and the resolution in the image is about equal to the source size. Fig. 14.1(b) emphasises that the resulting image is equivalent to a TEM image with defocus z_1 (under certain conditions). The conducting sample, grounded. acts as the anode. Since field emitters of atomic dimensions can be obtained,[107] atomic resolution should be possible. However, at the very low voltages (about 100 V) needed to sustain field emission with $z_1 = 100\,\mathrm{nm}$, electron penetration in carbon is only about 0.8 nm, and most samples are completely opaque. Then the main features seen in the images are Fresnel fringes around edges, and occasionally interference fringes between waves passing through different pinholes. This arrangement constitutes exactly Gabor's original proposal for electron holography, and also corresponds to the in-line STEM Fresnel holography case described in Chapter 2, Section 3 (with sub-nanometer aberration coefficients appropriate to the PPM virtual field-emission source). For STEM, the probe-forming lens shown in Fig. 2.6(a) is used. A low voltage PPM instrument was built at NIST in 1968[302] and several more have been built more recently[23,108,318,418] using STM-based construction methods.

Two aspects of PPM deserve attention at present. First. the prospects for using the greatly increased penetration of electron beams which occurs at even lower voltages, and secondly the question of radiation damage at very low voltages. We therefore summarize below relevant work on the mean free path of electrons in solids, image reconstruction methods, nanotips, instrumentation, and damage mechanisms at low voltage. The first results from operation of the instrument in the reflection mode are

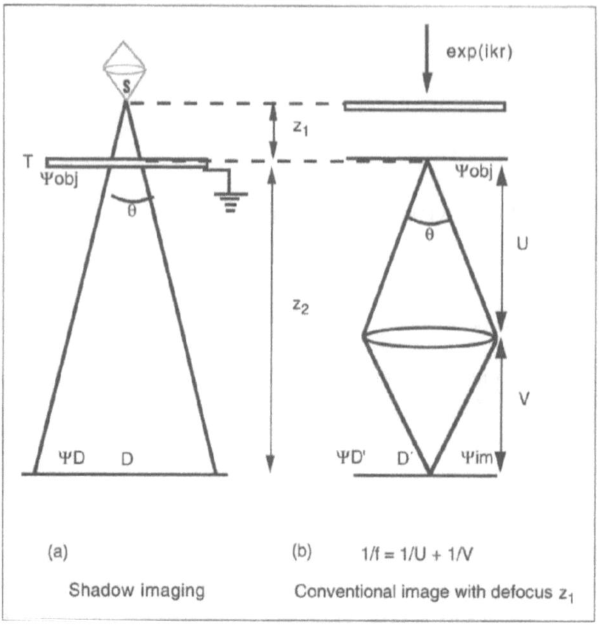

Figure 14.1. Comparison of shadow imaging (a) with conventional imaging (b). Identical out-of-focus images of masks may be formed by arrangements a) and b) at D and D'.

also discussed.

In a typical point-projection microscope, exposure conditions are about 0.1 seconds with less than 1 nA beam current at 100 V. The image (see Fig. 14.2) is superimposed on a point-projection field electron image of the tip (the tip emission pattern, projected from the virtual source inside the tip). This causes a slowly varying background modulation. Note that only movement of the tip relative to the sample is magnified; as in field-ion microscopy (FIM), movement of the tip does not cause magnified movement of the field-emission pattern from the tip, since no movement of the virtual source relative to the surface of the tip occurs. This insensitivity to vibration explains why atoms were first seen by FIM.

If the tip (and detector) are modeled as coaxial parabolae, the on-axis electric field $E(z)$ is given very approximately by

$$E(z) = \frac{U_A}{z \ln(2\, z_1/r)} \approx \frac{U_A}{6z} \tag{14.1}$$

if $z_1 = 200\,\text{nm}$ and $r = 1\,\text{nm}$, where r is the tip radius and z a coordinate along the axis from the focus. The field varies slowly with z_1 but rapidly with r, and falls off very rapidly away from the tip. The virtual source size $d \approx r/2$.[390] Field emission from tungsten requires about 0.4 V/Å at the surface. In practice, reducing the tip-to-sample distance z_1 increases the magnification (as $1/z_1$), reduces the defocus (see below), and reduces the voltage needed for the same tip field and emission current.

Fig. 14.2 shows a low magnification image of a holey carbon film obtained at about 400 V, using an instrument in which the tip motion along the axis is controlled by a

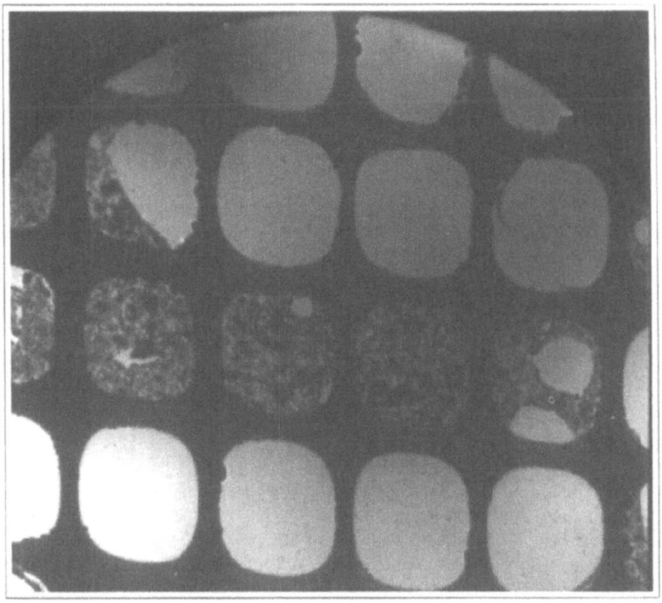

Figure 14.2. Low magnification point projection image of a TEM grid partly covered by a holey carbon film. Since the tip is several millimeters from the sample, the voltage is about 400 V.

Burleigh Inch Worm.

2. Electron ranges in solids

The penetration of low voltage electrons through matter may be understood using the extensive calculations and measurements of the inelastic electron mean free path λ_E which have been made in the fields of photoemission spectroscopy, and high resolution electron energy loss spectroscopy.[396,437] λ_E is defined such that the intensity of the elastic portion of the electron wavefield decays on entering a solid as

$$I(z) = I_0\, e^{-z/\lambda_E} = I_0\, e^{-2\sigma V_0' z} = I_0\, e^{-z N \sigma_t} \tag{14.2}$$

where $\sigma = \pi/\lambda U_A$ (λ the electron wavelength, U_A the beam potential), and V_0' is the imaginary part of the mean inner potential (in V). N is the density of atoms and σ_t is the total inelastic cross section. Measurements and calculations covering the range below 100 V (in which the theory, which is not reviewed here, becomes most difficult) can be found elsewhere.[40,336] The results generally show a minimum of a few Angstroms for λ_E at about 50 eV, below which λ_E rises steeply with the onset of ballistic transport, reaching values of several hundred Angstroms, at a few electron volts. Fig. 14.3 shows experimental measurements of the attenuation length L for low energy electrons in the 7 -1000 eV range.[282] These measurements were obtained using free-standing thin carbon films about 4 nm thick in transmission experiments, using an electron spectrometer to isolate the transmitted elastic scattering. L may be identified with λ_E under the approximation that the detector includes all scattering angles. Thus they are uniquely

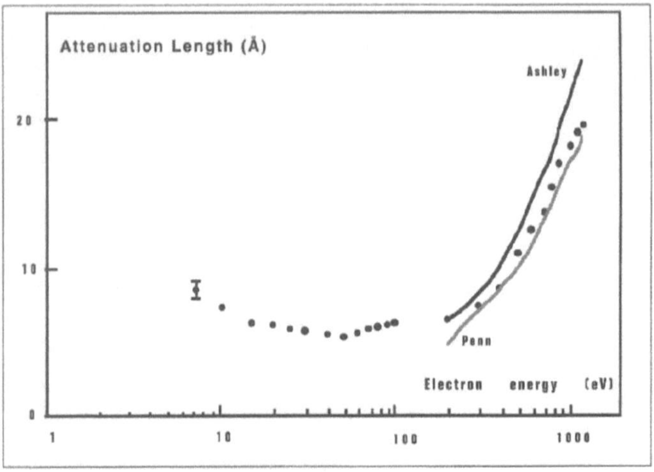

Figure 14.3. Attenuation length for electrons traversing thin carbon films as a function of electron energy, measured using a transmission energy loss spectrometer.[282] Continuous and dashed lines are theoretical predictions. Measurements from other materials rise to 20 nm at 3 eV.

relevant to PPM imaging of organic films. We see that thin films of this material only become reasonably transparent to electrons at energies below about 10 V. At energies around 100 V (where much point-projection work is done), the inelastic mean free path is close to its minimum value. Shedd[398] has found recently that good agreement between simulated and experimental PPM images of holey carbon films can be obtained by treating them as opaque masks containing holes, ignoring completely any partial transmission at edges. The two-dimensional pattern of fringes seen previously[108] were thereby found to be Young's fringes due to the interference between waves passing through three different pinholes, rather than atomic resolution transmission lattice images. The two effects can easily be distinguished if the magnification (and hence the defocus) is accurately known. We therefore assume as an approximation that negligible transmission occurs, and give below the theory of point-projection image formation for masks. Such a theory may be useful for the important problem of determining the shape of molecules, if damage is slight, and if methods can be found for placing them, unsupported, across holes in carbon films. As it is the external shape (conformation, folding) of biomolecules which determines their function, this information may be more important than a knowledge of atomic coordinates.

Our own experimental results suggest that unstained, uncoated purple membrane (thickness 5 nm) is opaque at 100 V, but reasonably transparent at 1 kV, and that significant transmission occurs through samples of the lipid monolayer C_{16} (thickness 2.6 nm) at 160 V. We are not able to determine the fractions of elastic and inelastic scattering in the transmitted beam. At lower voltage, if experimental evidence for significant transmission of elastically scattered electrons is obtained, the theory of transmission low energy electron diffraction (TLEED)[346] would be required, in which case Fourier imaging effects might also be observed.[415]

3. Image formation and reconstruction

We now consider the image-formation process in PPM for mask-like objects. First, we demonstrate that a shadow image of a mask (or a weak phase object) is equivalent to a conventional out-of-focus image, and so discuss focus restoration schemes for masks. Consider a mask illuminated by a spherical-wave originating from a source at distance z_1 from the mask. The transmission function for the sample is $p(x)$, but, for masks, we cannot assume the weak phase object approximation. Choose Cartesian coordinates with z along the beam path. Let the spherical wave incident on the sample be represented in the parabolic approximation by $t_{z_1}(x) = \exp(-i\pi\, x^2/z_1\lambda)$ (with Fourier transform $T_{z_1}(w) = C\exp(i\pi\, z_1\lambda w^2)$, where C is a complex constant) and let the electron wavefunction across the downstream side of the slab be $\psi_i(x)$. Then, with $w := \theta/\lambda$, where θ is the scattering angle,

$$\psi_i(x) = p(x)\, t_{z_1}(x) \tag{14.3}$$

the detected wave function in the far-field will be the Fourier transform

$$\psi_h(w) = \mathbf{FT}\{\psi_i(x)\} = P(w) \otimes T_{z_1}(w) \tag{14.4}$$

where $\otimes$ denotes convolution. Evaluating the convolution in Eq. (14.4), and ignoring unimportant phase factors gives

$$\psi_h(w) = \int P(W) e^{i\pi\, z_1\, \lambda\, W^2} \cdot e^{2\pi i\, z_1\, \lambda\, w\, W}\, \mathrm{d}W \tag{14.5}$$

Now consider $z_1\, \lambda\, w\, W = z_1\lambda\, (\theta/\lambda)\, W = z_1\, \lambda\, (X/z_2\, \lambda)\, W = (X/M)\, W$, where X is the spatial coordinate on the detector and $M = z_2/z_1$ is the magnification of the shadow image. Hence

$$\psi_h(X = z_2\, \lambda\, w) = p\,(X/M) \otimes t_{z_1}(X/M) \tag{14.6}$$

This important result establishes that for masks (or for any thin sample for which a transmission function can be defined), the shadow image consists of an 'ideal' image which is out of focus by the tip-to-sample distance z_1. Equation (14.6) is identical to the expression for an out-of-focus high resolution transmission electron microscope (HREM) image in the absence of lens aberrations or apertures. However, the point-projection image has been magnified by $M = z_2/z_1$ without the use of lenses. No assumption of periodicity has been made for the sample. An in-focus image can only be obtained using Eq. (14.6) if $z_1 = 0$, in which case M is infinite and, if a physical emitter is used, field emission then becomes impossible. The magnification and the focus setting of the PPM are not independent – for given z_2 both are fixed by z_1. For a crystalline sample for which the transmission diffraction orders overlap, the above analysis does not apply in the presence of multiple scattering, since then a transmission function cannot be used, as discussed elsewhere.[414]

It follows from Eq. (14.6) that existing algorithms discussed elsewhere in this book for focus correction (under plane-wave illumination) can be used to reconstruct low voltage in-line holograms (formed with spherical-wave illumination). For masks, Fresnel edge fringes and Young's fringes between nearby holes are the main contrast features, and the aim of reconstruction is to remove these fringes, thus revealing the shape of the holes. The problem of focus correction (or hologram reconstruction) in the in-line geometry has been discussed in Chapter 2, Section 3, for transmission objects in the presence of lens aberrations. The first solution to the unavoidable twin image problem which arises in this geometry may be that of B.J. Thompson and coworkers,[93] who developed the optical method of in-line Fraunhofer holography for small particles (strictly small Fresnel numbers). Solutions to the twin image problem based on recording images at different defoci or lateral tip positions have been described in Chapter 2, Section 3, and might also be applied to PPM. Shadow images of masks have recently become important in the field of semiconductor lithography and contact printing, where the inverse problem arises. For finer line-widths and sharper edges one requires the edge transmission function whose defocused shadow image is most abrupt.

For a mask we define the transmission function

$$p(x) = 1 - b(x) = \begin{cases} 1 & \text{within a hole, and} \\ 0 & \text{within opaque regions} \end{cases} \tag{14.7}$$

Then, assuming the equivalence between point-projection shadow images and conventional images (Eq. (14.6)), the wavefunction on a plane distance z_1 downstream from the sample is, for a magnification of unity,

$$\psi_h(x) = p(x) \otimes t_{z_1}(x) \tag{14.8}$$

The recorded hologram intensity is

$$I(x) = \psi_h(x)\,\psi_h^*(x) = 1 - p(x) \otimes t_{z_1}(x) - p^*(x) \otimes t_{z_1}^*(x) + |p(x) \otimes t_{z_1}(x)|^2 \tag{14.9}$$

To reconstruct the object from the hologram, we first Fourier transform, then multiply by $T_{z_1}^*(u)$, and finally perform an inverse transform

$$\begin{aligned} \mathbf{FT}^{-1}\left\{T_{z_1}^* \cdot \mathbf{FT}\{I(x)\}\right\} &= 1 - p(x) - p^*(x) \otimes t_{2z_1}^*(x) + |p(x) \otimes t_{z_1}(x)|^2 \otimes t_{z_1}^*(x) \\ &= 1 - p'(x) \end{aligned} \tag{14.10}$$

The first two terms give a perfect reconstruction of the original transmission function $p(x)$, the real image. The third term gives the complement of the object, out of focus by twice the tip-to-sample distance z_1. If the object (of width d) is small, and z_1 large enough, this virtual twin image will produce a broad and slowly varying background which allows the real image to be isolated. (The requirement for this is $z_1 \gg d^2/\lambda$ (Fraunhofer holography)). The fourth term is a second order term (actually the reconstruction of the hologram of the complementary object) which is small for weak phase objects, but not necessarily small for masks. Physically, this term might be expected to be troublesome in regions well inside the shadow edge, where the 'reference wave' is weak. The ratio of the wave scattered from the edge to the direct wave controls this term, and, for a finite object, this ratio depends on the Fresnel number $N(d) = \pi\, d^2/(\lambda\, z_1)$. Thus reconstruction for masks introduces greater errors than for weakly scattering objects (WPO); however, for the purpose of identifying simple discontinuities at edges, we shall see that the Fraunhofer condition can be relaxed. We have therefore investigated these issues using computational trials on simulated and experimental data.

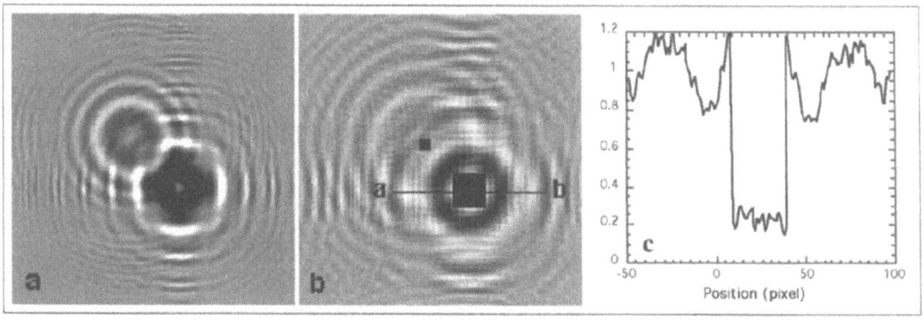

Figure 14.4. Simulated in-line holographic reconstruction of mask. (a) Simulated hologram for z_1 = 60 nm. (b) Real part of the reconstructed wavefunction. (c) Profile of intensity along a-b.

In Fig. 14.4 we illustrate the use of the above algorithm for simulated data. The simulations require periodic boundary conditions (with supercell period L). Interference from neighbouring cells can be avoided if the Fresnel number $N(L) \gg 1$. The reconstruction of two opaque square objects of width 1 nm and 3 nm (Fresnel numbers 0.43 and 3.86) is shown using 100 V electrons. The hologram was simulated using Eq. (14.9) and reconstructed using Eq. (14.10). The shape of the two squares can be clearly seen in Fig. 14.4(b), despite the disturbance from the twin image background, which is least for the smaller square ($N < 1$).

These results suggest that any sharp discontinuity (such as an edge) in a finite object will produce a corresponding sharp discontinuity in the reconstructed image at Gaussian focus if N is not too large. This would allow the shape of objects to be determined with high resolution. We now apply this method to experimental data.

Before attempting reconstruction, the magnification $M = z_2/z_1$ of the images must be known. Five methods have been used: 1. Use of a through-magnification series and an object (such as a grid bar) of known size. This is very inaccurate since errors between successive images compound. 2. Use of parallax-shift of images with lateral tip movement. 3. Analysis of Fresnel edge fringes.[420] 4. Use of Young's fringes observed between two pinholes of spacing D, using $d = z_2 \lambda / D$, where d is the fringe spacing and z_2 the sample-to-screen distance. 5. Trial-and-error values of z_1 may be used in the reconstruction scheme until a sharp image is obtained, as in Fig. 14.4b. We have used methods 4, 5 and 3.

Figure 14.5 shows two experimental point-projection holograms of a holey carbon film obtained at 90 V. Fig. 14.5(b) shows the crossed sets of Young's interference fringes between three pinholes which might be confused with an atomic resolution lattice image. The fringes may be simply interpreted using Eq. (14.6), expressed in two dimensions. Then we see that the fringes are equivalent to near-field Young's fringes, as would be obtained using plane-wave illumination. In the near-field of a mask containing holes, both a shadow image of the holes and interference fringes between different holes are observed, as indicated in Fig. 14.6.

The holograms shown in Fig. 14.5 were also reconstructed using Eq. (14.10) and a program which allows for the change of object pixel size with variation of defocus z_1, which changes the magnification. An approximate value of z_1 was assumed and the corresponding magnification $M = 14\,\text{cm}/z_1$ used. Since the computations are fairly rapid, the effect of varying the trial value of z_1 simulates changing experimental focus near Gaussian focus, and by comparing reconstructed amplitude and phase images, the in-focus image may readily be identified by eye. The focus correction needed to obtain

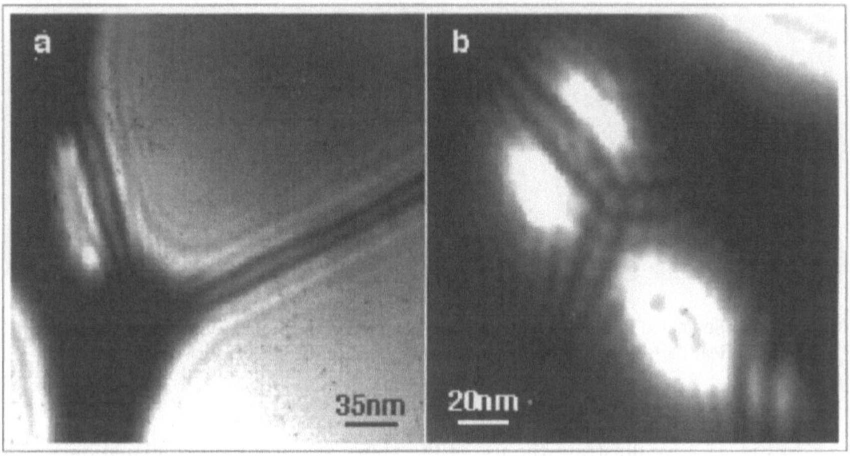

Figure 14.5. Experimental holograms of a holey carbon film obtained at 90 V. (a) Carbon fibers bridging holes. (b) Three pinholes in the same foil.

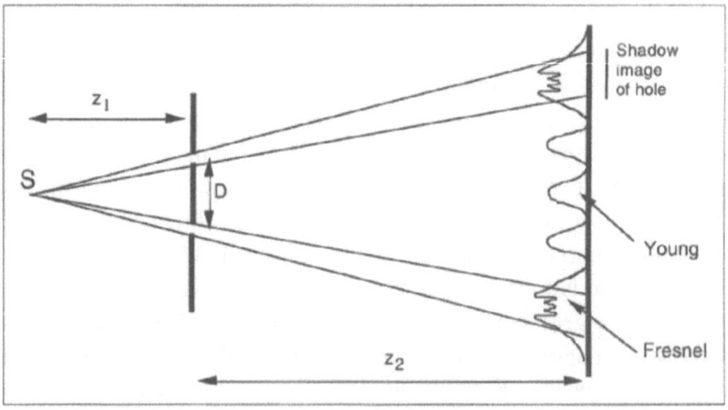

Figure 14.6. The formation of Young's fringes between the shadow images of two pinholes. Fresnel fringes, corresponding to defocus z_1 are seen within each pinhole.

the in-focus image is the experimental tip-to-sample distance.

The shape of the fibers retrieved from the in-focus image reconstructed from Fig. 14.5(a) is shown in Fig. 14.7(a), while Fig. 14.7(b) shows a forward simulation of the hologram based on Fig. 14.7(a), in excellent agreement with the experimental hologram (Fig. 14.5(a)) if a defocus of 1850 nm is used. An integration over the finite source size would improve the fit by washing out the higher order Fresnel fringes. Reconstructions obtained from Fig. 14.5(b) eliminate entirely the Young's fringes at certain focus settings.

The sharpness of the edges in the reconstruction depends on the Fresnel number $N = \pi d^2/\lambda z_1$, which describes the effect of defocus z_1, wavelength λ, and object diameter d. For the edge of an 'infinite' half-plane (N infinite) the reconstruction is noticeably poorer than for a small object. In addition, any phase shift on scattering at edges, or partial transmission, will also affect the reconstruction.

In summary, we find that any sharp discontinuity (such as an edge) in a finite

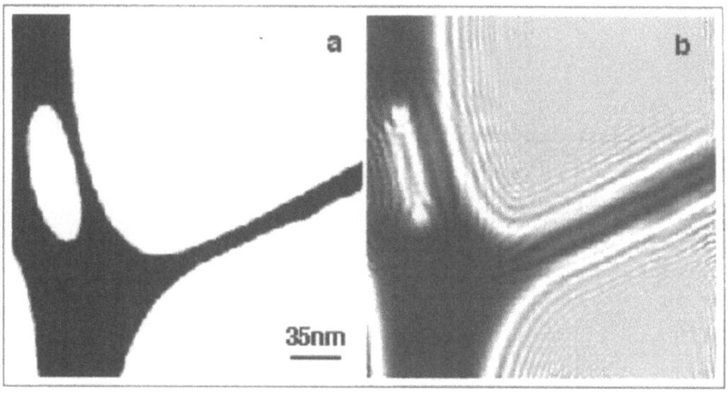

Figure 14.7. Object shape retrieved from experimental image of Fig. 14.5 shown in (a), and (b) simulated hologram with defocus 1850 nm.

object will produce a corresponding sharp discontinuity in the reconstructed image at Gaussian focus if N is not too large. As Fig. 14.4 shows, this focus setting (the tip-to-sample distance) can easily be identified from a focal series of reconstructed images. This discontinuity, identifiable against the background of the twin image and the non-linear terms, easily allows the outline shape of small opaque objects to be identified. Resolution in the reconstructed image will still be governed by the electron source size (which determines the highest order Fresnel fringe, and hence the size of the hologram, since the fringes get progressively finer according to Eq. (14.14)).

4. Nanotips

We now consider briefly the 'nanotips' used to form these images. These may be either naturally occurring asperities (as in STM), or specially prepared with an atomic structure of desired shape. An extensive literature exists in the field of FIM on tip preparation procedures, based on oxygen etching, field evaporation, heating or sputtering.[217] More recently, using FIM imaging to assist in characterization, single-atom and few-atom 'nanotip' field emission tips have been prepared,[107] for which claims of very high brightness and other unique properties have been made, including departures from the Fowler-Nordheim law, 'focusing' effects and unusual emission energy spectra.[21] Theoretical treatments for diffraction of a wavepacket at a laterally confined tunneling barrier have also been given (see Ref. 22 for a review of this work). We now discuss the nanotip properties most relevant to PPM, their aberrations and brightness. Figure 14.8 shows an idealized tip and the curved trajectory of electrons accelerated from the tip. Since most of the potential is dropped within a few hundred nanometers of the tip, the electrons reach their final kinetic energy rapidly. The rearward asymptotic extension of these rays defines the virtual source inside the tip, whose size d depends on the failure of rays leaving at different angles to meet at a point in the Gaussian image plane inside the tip. (This image plane is defined by paraxial rays). Thus the tip acts as a lens of nanometer dimensions, whose aberration coefficients can therefore be expected to be also of nanometer dimensions. Aberration coefficients for such a nanotip have been computed by modern ray-tracing techniques, together with numerical solutions of Laplace's equation.[390] For a tip with 1 nm radius sitting on a boss with 100 nm radius, we find $C_S = 0.177\,\text{nm}$ and $C_C = 0.142\,\text{nm}$ at 100 V, with a

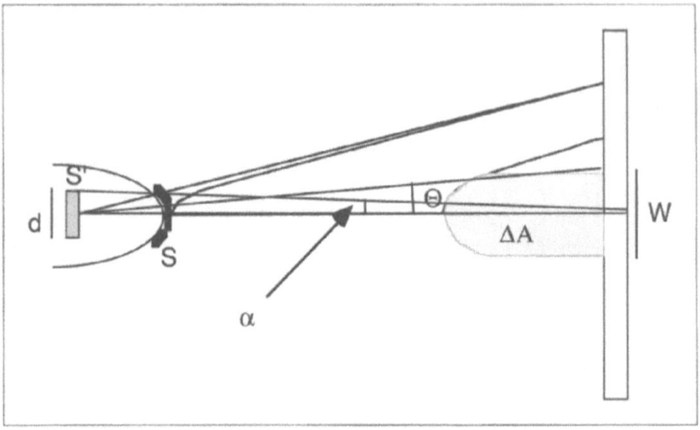

Figure 14.8. Definition of virtual source S', physical source S, solid angle $\Delta\Omega = \pi\alpha^2$, and width of intensity distribution (shown shaded) $\Delta A = \pi\,(W/2)^2$ used to measure brightness. A typical ray is also indicated, with rearward asymptote extending to virtual source.

tip field strength at the surface of the tip of 5 V/nm. The effects of these aberrations on resolution may not be negligible, however, because the virtual source size depends also on emission angle and wavelength. The emission angle q for this tip was found to be 7.4° (half width at e^{-1}).

From Fig. 14.8 the coherence width at the detector is seen to be approximately $X_c = \lambda/a$. It is also useful to define a coherence angle $\gamma = \lambda/d$, equal to the angular range over which emission from the tip is coherent — this is the angle subtended by X_c at the tip. The virtual source diameter d for the 1 nm radius nanotip discussed above is found to be 0.43 nm (i.e., demagnified). For such a source the coherence width on a sample at $z_1 = 100$ nm is 54 nm, while that on a detector at $z_2 = 10$ cm is almost 6 cm. ($\lambda = 0.126$ nm, 100 V). The coherence angle is greater than the emission angle. The coherence width is approximately equal to the width of all the Fresnel edge fringes which can be seen, and hence to the size of the hologram. Because of their high brightness, nanotips may therefore be useful for electron interferometry experiments in which the beam must pass around macroscopic objects smaller that X_c. Measurements of source size have recently been made for a nanotip using a biprism.[319] The longitudinal coherence length $L_c = \lambda\,(E/2\Delta E) = 63$ nm for a nanotip with $\Delta E = 0.1$ eV at 100 V is considerably less than for high voltage field emission guns.

The brightness $\beta = j/\Delta\Omega$ (see Fig. 14.8) of a nanotip has been measured by matching ray-tracing computations and Fowler-Nordheim curves to experimental nanotip intensity distributions.[348] By comparing field ion microscopy and field emission electron images with the ray-tracing results, the effective source size could be obtained. Tungsten (111) single crystal tips were fabricated using Ne sputtering and field evaporation. The average brightness of single atom terminated nanotips was found to be $3.3\cdot10^8$ A cm^{-2} sr^{-1} at 470 V, or $7.7\cdot10^{10}$ A cm^{-2} sr^{-1} when extrapolated to 100 kV. This produces a beam with greater particle flux per unit energy range than obtainable using current synchrotrons. Although this value is almost two orders of magnitude brighter than the values quoted for STEM cold field emission guns, it should be born in mind that the STEM values are based on measurements of the focused probe intensity, which is affected by objective lens aberrations. When corrections are made for this effect, the nanotip brightness is found to be greater than cold field emission by a factor of about 10 at 100 kV. The most valuable aspect of the nanotip for STEM instruments may be the increased emission stability obtained at low extraction voltages, the reduced emission energy spread ΔE, and the increased tolerance to poorer vacuum conditions. It has been speculated (P. Bovey, private communication) that this is due to the reduced energy of back-streaming ions, which, at higher extraction voltages, modify the work-function of the tip.

Source brightness can be measured more directly using the concept of degeneracy $\delta_0 = \beta/F_0$, where F_0 is the upper quantum limit on brightness corresponding to two fermions per cell in phase space (if $\delta_0 = 1$). The degeneracy can also be expressed as

$$\delta_0 = \frac{j}{e} A_c T_c \tag{14.11}$$

This is the mean number of particles per coherence time $T_c = h/\Delta E$ traversing a coherence area $A_c = X_c^2$ (normal to the beam). In practice this means that δ_0 (and hence brightness) can be measured if the beam energy spread and the current density within a patch of Fresnel fringes are known, since the width of these fringes is given approximately by X_c. Note that δ_0 is unity if $j = e/(A_c T_c)$, so that one electron crosses the coherence area per coherence time. Using the expression for the constant F_0, it follows[420] that the experimental brightness can be determined from the wavelength, the current density j and the coherence area at the image plane using

$$\beta = 2j A_c/\lambda^2 \tag{14.12}$$

The effect of source size on resolution may be understood by considering the effect of displacing an ideal emission point transversely by b. This translates the entire image by $M b$, where $M = z_2/z_1$. This imposes a resolution limit on the images, since the complete image intensity must be integrated over the effective source size d, which cannot be smaller than the electron wavelength. This integration takes into account all partial spatial coherence effects. For masks, the resolution of the PPM is most conveniently defined operationally as the width of the finest Fresnel fringe (in the object space), since this is readily shown to be equal to the size of the electron source (including the effects of all instabilities), as we now show. It was first shown by Sommerfeld that a good approximation to the Fresnel integral may be obtained by summing over just two optical paths (rather than infinitely many), the direct ray from source to detector, and that which passes via the edge of the mask. This gives the correct position of maxima and minima, but not the intensity distribution, if path differences $\Delta_0 = 3\lambda/8$ for the edge wave scattered outside the shadow edge and $\lambda/8$ for the edge wave scattered inside are used. Then the lateral position X_n of the nth Fresnel fringe on the screen is

$$X_n = M \cdot \sqrt{2z_1 (n\lambda + \Delta_0)} \tag{14.13}$$

The width of the nth fringe on the screen is

$$\delta X_n = \frac{M^2 z_1 \lambda}{X_n} \tag{14.14}$$

Setting $X_c = X_n/M$ at the sample we find the width of the highest order fringe at the sample is approximately equal to the source size $d = 2z_1 \lambda/X_c$. In practice, however, the contrast of the fringes decreases with order, making measurements of source size and coherence by this method approximate. (See Ref. 420 for a full analysis of Fresnel fringes in PPM. The image magnification, effective source size, transverse coherence width, instrumental resolution, and source brightness are all obtained from an analysis of the fringe spacings and intensity.)

The resolution limit in PPM may be thought of in other ways. By limiting the size of the hologram, the coherence width X_c imposes a diffraction limit on the reconstruction, since the hologram acts rather like a lens (or zone-plate) during reconstruction. Thus the coherence angle γ plays a similar role to the angular limit set by the aperture of the objective lens in HREM (as suggested by θ in Fig. 14.1) – a larger angle is needed for higher resolution d_r. The diameter of the zone plate is approximately equal to the transverse coherence width MX_c at the detector screen. Thus $d_r = 1.6\,\lambda\, z_2/(M\, X_c) = 0.8\lambda/\gamma$ in agreement with the idea that the coherence angle from the source acts as a diffraction-limiting aperture. Higher resolution is obtained by increasing the emission coherence angle $\gamma = \lambda/d$, and this can be achieved by reducing the source size d. A useful result is obtained by noting that, since $X_c = 2z_1\,\lambda/d = \lambda/2\alpha$ (where α is the angle subtended by the source at the sample) and since the resolution is $d_r = 1.6\,\lambda\, z_1/X_c$, we have that $d_r = 3.2\, z_1\, \alpha$. Thus resolution in the reconstruction is improved by using small tip-to-sample distances and the smallest sources. For a thin crystal, we may think again of γ as limiting the 'numerical aperture' of the nanotip lens. Then the finest fringes contributing to a Fourier image will be those with Bragg angle $2\theta_B = \gamma$ for plane spacing d_{hkl}. Using $1/d_{hkl} = 2\theta_B/\lambda$, we again see that the resolution is approximately equal to the source size $d = d_{hkl}$.

The effects on resolution of the spherical aberration of the tip must be considered. Taking the resolution to be very crudely equal to the virtual source diameter ΔC_S, we have, from the definition of spherical aberration,

$$\Delta C_S = 2r_i = 2M(C_{S_3}\sin^3\theta + C_{S_5}\sin^3\theta) \tag{14.15}$$

where θ is measured in the object space (the emitting surface) and, from calculations, $M = 0.531$ at 100 V. For emission at $\theta = 45$ degrees, $\Delta C_S = 0.84$ Å for the 1 nm radius nanotip. It is instructive to compare this with the performance of a magnetic lens for high energy electrons. Here, at 100 kV ($\lambda = 0.037$ nm) one might typically have $C_S =$ 0.5 mm, but an angular aperture set at about $\theta = 10$ mrad. Then $\Delta C_S = 2.5$ Å because of the much smaller angles but larger aberrations. The idea that the aberrated virtual source size is equal to the resolution, however, is not strictly correct, for the reasons given in detail in Section 2.

The effects of a spread of energies ΔE in the beam must be considered. For Fresnel fringes, it was found that energy broadening of the source imposes a chromatic resolution limit such that the highest order fringe which may be observed has order $n_{max} = 2E/\Delta E$. The width of this fringe (referred to the object space) is

$$\Delta C_r = \sqrt{\frac{z_1\,\lambda}{2\,n_{max}}} \tag{14.16}$$

which, by substituting for n_{max}, gives the resolution limit due solely to the energy spread of the beam. For $z_1 = 1000$ nm and $\Delta E = 0.1$ eV we obtain $\Delta C_r = 0.17$ nm at $E = 100$ eV.

In summary, the width of the finest Fresnel fringe from a carbon foil edge is the most useful practical measure of resolution for the point projection microscope. Energy broadening and source size effects may limit resolution to approximately the electron wavelength. They do not impose an absolute limit, however, and, as in near-field and other forms of microscopy, resolution will ultimately be limited by signal-to-noise ratio. The non-linear terms in holographic reconstruction methods also limit resolution. In the presence of multiple scattering, resolution cannot easily be defined. Resolution increases with emission coherence angle, which acts as a diffraction-limiting aperture, as for a lens. This angle $\gamma = \lambda/d$, (where d, the effective source size, is a measure of the resolution) can be measured using the methods outlined above or, better, using an electron biprism. The relatively poor visibility of biprism fringes observed at low voltage[319] may be due to inadequate screening against AC magnetic fields, which has the effect of enlarging the effective source size.

5. Instruments

Several designs for low-voltage point-projection instruments have appeared in the literature. All consist of a tip motion, a sample holder and a detector, for which all recent instruments have used a single-stage channel plate (MCP) followed by a fluorescent screen. These items are supported within an ultra-high vacuum chamber which must be bakeable (at least initially) and supported on a vibration isolation system. A leak valve for sputtering and vacuum gages may be required. Pumps must pump the noble gas used for sputtering. Fine and coarse motion of the tip relative to the sample along x, y, and z is essential to bring a region of interest onto the optic axis, defined by the beam, and to control the magnification and focus. Tip motions have been based on the original piezo tripod scanner used in STM or on a piezo tube scanner for lateral motion and an Inchworm for coarse z-motion.[418] Tip heating (for cleaning) may be provided by resistance heating (in which case the tip support follows the design of the tip assembly in field-emission STEM instruments), or by indirect electron-beam heating, which achieves lower temperatures but allows a higher mechanical resonant frequency for the tip, and does not require thick high-current leads to the tip assembly. If field-ion imaging of the tip is planned, electrical insulation for voltages of at least 10 kV must be provided to the tip. Since resolution is degraded by tip vibration, resistance heaters must be short and rigid, and all the principles of good STM design apply.[44] For a machine supported on a spring system with natural resonance frequency f_{sp}, the attenuation K of vibration amplitude reaching a tip with natural resonance frequency f_{tip} is approximately

$$K = (f_{sp}/f_{tip})^2 = a_{tip}/a_{ext} \tag{14.17}$$

if $f_{sp} \ll f_{tip}$. Here a_{tip} is the vibration amplitude of the tip, while a_{ext} is the amplitude supplied to the frame of the support. The frequencies are proportional to the square root of the relevant spring constant divided by a mass. One makes K as small as possible by making f_{sp} as small as possible (e.g., by using an air-table or elastic suspension system with large mass and small spring constant), and f_{tip} as large as possible (by making the tip as short and rigid as possible). Typical values as $f_{tip} = 2$ kHz, $f_{sp} = 2$ Hz, give $K = 10^{-6}$ or 120 db attenuation. Tips with f_{tip} above 10 kHz are highly desirable. A short mechanical path between tip and sample is required to minimize relative displacements of the tip and sample — a translation of both together is not magnified.

As for STM, the most difficult design problem is the coarse lateral motion of the sample relative to the tip. A design has been described[418] which provides one-dimensional transverse motion of a carriage riding on three sapphire balls, based on

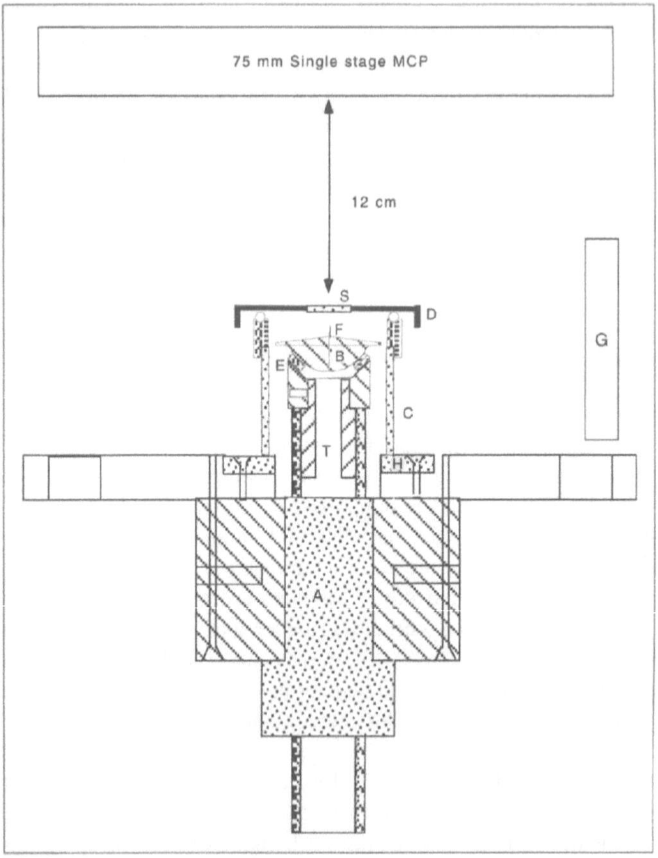

Figure 14.9. Vertical axis low voltage point projection microsope employing eucentric stick-slip goniometer and two-dimensional stick-slip lateral stage motion. For reflection operation the sample S is a semiconductor wafer whose edge lies above the tip.

an inertial stick slip motor. This, however requires careful leveling to avoid consistent downhill motion. The use of a molybdenum plate riding on sapphire balls has been reported to solve this problem. (Unlike STM stick-slip stages, magnets cannot easily be used to stabilize the motion against gravity.) Transverse Inchworm motion has been used; however, jumps in the motion and the loss of a short, rigid mechanical path between tip and sample may limit resolution. The crucial requirement is to be able to control *both* z_1 and the coarse lateral stage motion interactively, in order to center a feature of interest on the optic axis. This is simple once one is within range of the fine lateral motion provided by a tube scanner supporting the tip. High stability, low noise electrical supplies may be required for the scanner, since lateral sensitivities of 50 nm/V are common.

Fig. 14.9 shows a recent vertical axis design[421] which allows convenient tip and sample exchange, together with a stick-slip goniometer[202] for alignment of the beam. The instrument is designed for operation in both the transmission and reflection modes. The tip goniometer allows one to take advantage of natural asperities on a single crystal tungsten tip, which, in our experience (and that of STM workers) can be extremely sharp in the sense of producing very fine Fresnel edge fringes. Since natural asperities are unlikely to be aligned with the [111] axis, the goniometer allows them to be rotated onto the axis. We find that organic samples (which cannot be baked) can be studied at 10^{-9} torr with stable emission. Poorer vacua produce excessive ion background in the MCP. In Fig. 14.9, A is an inchworm and E are three shear-mode piezo stacks which rotate the hemisphere B about two orthogonal axes by the stick-slip mechanism, thus aligning the tip F eucentrically. The stage D (with sample S on a TEM grid) sits on three balls atop a piezo outer tube C whose four electrodes are driven by a ramp, giving lateral fine or (stick-slip) coarse motion of the sample. A pumped load-lock at G allows either the sample to be removed using a pick-up fork, or, subsequently, after extending the inchworm upward, the hemisphere containing the tip to be exchanged by a smaller fork. The previously baked UHV environment within is thus preserved. An internal mu-metal cylinder surrounds the instrument, which sits on a 60 L/S ion pump. Note that the magnetic field from the ion pump magnets is small on the axis. A simple technique for initial chemical etching of W tips has been described which allows optical examination during etching.[303]

The present generation of instruments are capable of providing sub-nanometer resolution images which reveal the external shape of mask-like objects. These may thus be useful for determining the shape of molecules drawn across holes in thin carbon films. In order to obtain images of internal atomic structure, Fig. 14.3 shows one must use either higher or lower voltage.

Higher voltages become possible if the tip field is shielded by a cup-shaped surround, so that a higher voltage is needed for the same tip field. This shield (which is actually a lens) must not obstruct the sample. Lithographic techniques may allow such structures to be formed. Withdrawing the tip and using lenses after the sample to compensate for lost magnification allows higher voltage to be used (up to about 1 kV), but the image then has a very large focus defect. Fig. 14.10(a) shows a PPM image of a cut in a film of unstained, uncoated purple membrane (bacterior-rhodopsin). No changes in the image were observed at the 2 nm resolution level after long exposure.[422,426] However, damage may have saturated in a surface layer within the first millisecond, and samples of this thickness (5 nm) will then still act as masks at 100 V. Figs. 14.10(b) and (c) show simulations based on Eqs. (14.8-14.10). We see that the fringes running normal to the edges of the vertical cut in the film can be explained by roughness at the edges alone, and do not indicate significant transmission of electrons. Because this roughness reflects the periodicity of the crystal, the fringes also show the period of the crystal. By contrast, as shown in Fig. 14.10(d), for C_{16} lipid monolayers (thickness 2.6 nm) significant transmission was obtained at 160 V (perhaps mostly inelastic scattering) and damage was also observed.

6. Radiation damage

We conclude by discussing the possibility of imaging thin organic crystals at voltages between 1 – 10 eV. Figure 14.3 shows that the inelastic mean free path rises to about 20 nm at 3 eV ($\lambda = 0.7$ nm). Preliminary PPM experiments down to 7 eV have been encouraging,[318] but many difficulties remain. The important question is whether radiation damage can be expected to decline at these energies. At high energy, ionization processes follow the Bethe stopping power law, which gives the inelastic ionization cross section as inversely proportional to the electron energy. Thus, if knock-on processes are ignored, damage is minimized by using the highest energies. At low energies

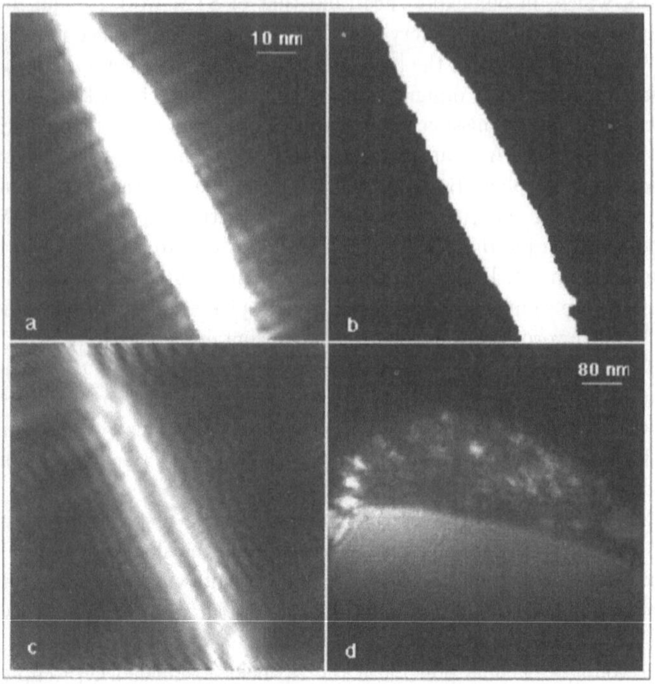

Figure 14.10. (a) Experimental image of purple membrane at 100 V. (b) The mask shape extracted from (a) used to simulate the hologram in Fig. (c) (defocus 550 nm). (d) Experimental image of lipid monolayer obtained at 160 V.[418]

(less than about three times the largest ionization energy) the Bethe law fails, and results depend in detail on the electronic spectrum of excited states for each material. The characteristic increase in λ_E below about 10 eV occurs because fewer excited states become available to the incident electron with decreasing energy. Below about 10 eV neither plasmons nor the damaging carbon inner shell ionization process can be excited, leaving only phonon and valence excitations. Transmission experiments which measure diffraction spot fading as a function of beam energy show a reduction in damage in aromatic (but not aliphatic) materials as the beam energy falls below about three times the carbon inner-shell ionization energy.[203] This suggests that, in these materials, the rarer, more localized inner shell excitation is more damaging than the more frequent valence excitation, which deposits less energy in the sample. Not all ionization events cause damage – the efficiency is believed to be close to one in aliphatics, but much less in the pi-bonded aromatics.

It has been widely accepted that the radiation dose needed to image a single unstained biomolecule at atomic resolution will destroy it.[33] This follows from the Rose criteria for a statistically significant variation in intensity between image pixels[142] and this pessimistic conclusion applies equally to imaging using kilovolt electron beams, neutrons or X-rays of any wavelength.[180] A useful rough figure of merit for radiation damage is the amount of energy G deposited by inelastic events per elastic event. At high energies, the inelastic cross section is about three times the elastic for electron scattering from light elements. The average energy loss at high energies is about 20 eV per event. Thus it has been shown that $G = 3 \cdot 20 = 60\,\mathrm{eV}$ in frozen biological samples for electron microscopes operating below the knock-on threshold.[180] (Note that $G = 80\,\mathrm{keV}$ for 8 keV X-ray imaging, while $G = 400\,\mathrm{keV}$ for soft X-rays ($\sim 400\,\mathrm{eV}$). Soft X-rays deposit less energy, but suffer much stronger inelastic scattering per elastic event.) Extracting the relevant cross sections for low voltages is difficult, but HREELS and LEED experiments have been performed on thin organic films.[149] In the HREELS experiments a monochromatic electron beam of variable energy is reflected from a thin organic film and on a metal substrate, and the reflection energy loss spectrum obtained. Phonon and electronic excitations are resolved in the spectra. For the case of benzene, it has been shown that elastic scattering dominates inelastic over the energy range 1 - 10 eV. This is the reverse of the situation in HREM or X-ray diffraction. In particular, in the range 4 - 8 eV, the mean free paths for phonon and electronic excitation were each found to be about 9 nm, while the elastic mean free path was 1.1 nm. Thus, at 8 eV we would have $G \approx 8/5 = 1.6\,\mathrm{eV}$, compared with $G = 60\,\mathrm{eV}$ for HREM and $G = 400\,\mathrm{keV}$ for soft X-ray imaging, quoted above. Point projection electron images obtained under these conditions could be expected to provide the most information per unit damage. Because electron holography is linear in the amplitude of the scattered wave, cross sections are not strictly relevant, and G provides only a rough guide to damage per unit information extracted. But since the number of unit cells needed to solve a structure at high voltage is much greater than the reduction in G obtained by going to low voltage, it seems likely that both the periodic averaging methods used at high voltage and ice embedding methods will also be needed at very low voltage (1 - 10 eV). In summary, low voltage electrons should provide significantly more information per unit damage than TEM methods, and vastly more information per unit damage than X-ray imaging methods.

Image interpretation will be difficult at very low energies, and the situation may then be similar to scanning tunneling microscopy of organic molecules, where damage may be avoided at large tunneling gaps, but image interpretation is extremely complex. For crystals, it is probably simplest to think of the elastic scattering in this energy range as a Transmission Low Energy Electron Diffraction (TLEED) problem. We have described such a modified LEED theory, and performed calculations for thin gold films, with suitable potential corrections for exchange between beam and crystal electrons.[346] This predicts both transmitted and reflected beams, oscillating in thickness, with the reflected beams approaching constant values for large thickness. As for HEED, cross sections are hardly relevant to Bragg diffraction, and the multiple elastically scattered beam intensities vary strongly with beam orientation, crystal structure and beam energy. For a given crystal in a given orientation, the two-beam extinction distance might be compared with λ_E to compare elastic and inelastic scattering. A more comprehensive theoretical treatment would be based on the density matrix formulation, and include both multiple elastic and multiple inelastic scattering.[98] The kinetic energy of the electron inside the sample is equal to $e(U_A + V_0)$, where V_0 is the mean inner potential, equal to the sum of the Fermi energy and the work function — at low energies V_0 depends on U_A.[381]

Some additional considerations arise at very low voltage. At 3 eV the electron wavelength becomes 0.7 nm; quantum mechanical arguments suggest that this value rather than the source size then limits resolution. For imaging below about 100 eV, channel plate detectors become inefficient, and electrons must be re-accelerated. This requires a lens (since any field with cylindrical symmetry is a lens) whose aberrations must be controlled. By placing this well downstream, the aberrations of the final image may be limited by those of the virtual source in the field emitter. At the very small

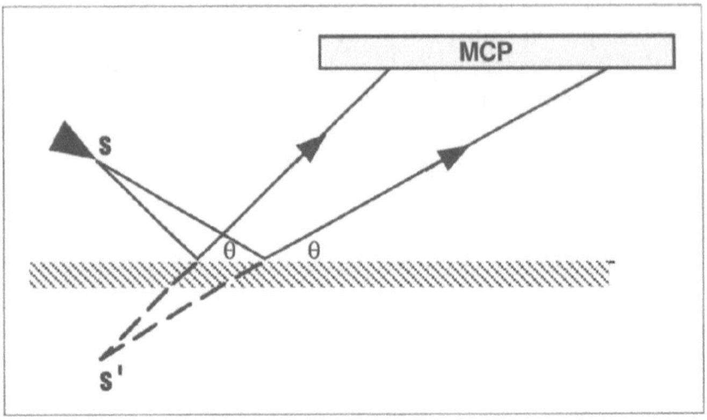

Figure 14.11. Experimental arrangement used to obtain Fig. 14.12. A nanotip field emission source S emits adivergent beam which is reflected from the GaAs (100) surface. A shadow image of opaque objects or surface steps on the mirror surface appears on the detector projected from virtual source S'.

gaps needed to operate at, say 4 eV, the magnification is very large so that z_2 must be small. The electrostatic forces on a self-supporting film become large, and may cause sample movement and damage — preliminary experiments in Tübingen have shown that the unstable tip jump to contact (as in atomic force microscopy) is difficult to control using STM feedback electronics. One also has the problem of making such thin, self-supporting films. As shown in Fig. 14.10, K. Taylor and D. Taylor have been able to provide us with lipid layers about 2 nm thick.[418] Langmuir-Blodgett techniques have also been tried. Polymer fibers have been successfully drawn across holey carbon films for PPM imaging[23] as has DNA for HREM imaging.[127] However, using an imaging energy filter to form the image from elastically scattered electrons alone, thicknesses well below λ_E should be usable.

7. The reflection mode

Useful images might be obtained from low voltage point-projection instruments in the reflection geometry, using bulk crystals with atomically smooth surfaces. The instrument shown in Fig. 14.9 was designed to investigate this possibility, using the tip goniometer to control the beam direction. Two methods have been proposed — one based on the Lloyd's mirror geometry[413] and the other using the specular reflected cone of radiation from a 'point' source (i.e., RHEED with a spherical incident wave). In the second case the crystal surface acts as a mirror, and a virtual source image is formed below the surface, as shown in Fig. 14.11. Both RHEED computations and ray-tracing calculations[424,510] for this case suggested that useful shadow images of surface steps can be obtained if the experimental conditions can be accurately controlled. The method relies on the fact that the electric field falls off rapidly near the tip, so that electron trajectories are relatively unaffected by the orientation of a reflecting surface a few microns or more away. Reflected rays appear to come from a virtual source below the surface, whose distance from the surface defines the defocus. A diffusely scattering object on the surface scatters electrons weakly which interfere with the specular RHEED cone, forming a shadow image (hologram) at infinity. Fig. 14.12 shows the reflection Kikuchi

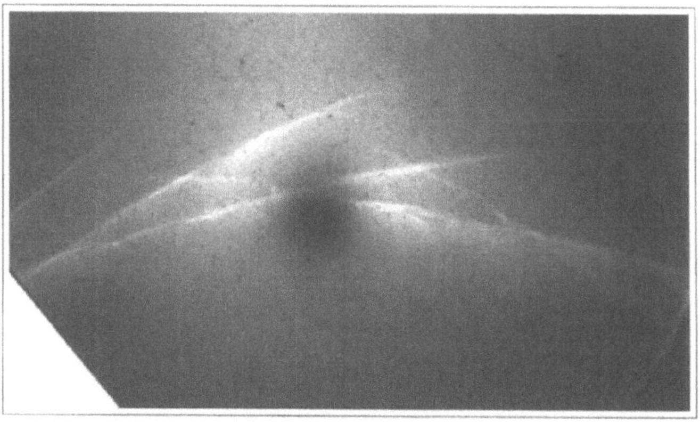

Figure 14.12. Kikuchi lines observed at 450 V and a low angle near the [001] azimuth on the [100] GaAs surface using the arrangement of Fig. 14.11. The black spot is probably due to ions impinging on the nanotip, reducing emission locally.

pattern from a [100] GaAs surface using the geometry of Fig. 14.11. The Kikuchi lines, observed at 450 V and a low angle near the [001] azimuth, are 'attached to the crystal;' however, the direction of the specular cone (containing the image in Fig. 14.13) depends on the beam direction, and so may be separated from the K- lines. (Because the cone angle may exceed the Bragg angle, the K- lines also contain elastic Bragg scattering.) Multiple scattering computations[424,510] showed that for certain orientations, the image may be modulated by the convergent-beam RHEED rocking curve for the specular beam, as in the CBIM method, which uses out-of-focus convergent beam diffraction patterns to generate shadow images superimposed on HOLZ lines. The magnification of the image depends on z_1; however, the angular width of the rocking curve features does not. Because dark-field shadow images are obtained (one in each beam) and projected from different directions, the images give true three-dimensional information. The same object feature (arrowed) can be seen in each image.

A large proportion of the reflected electron scattering is inelastic. These images were obtained by biassing the front element of the channel-plate (MCP) detector to about −400 V, to obtain a crude form of integral energy filtering. However, this results in very inefficient MCP operation, since electrons then enter it with an energy of only a few electron volts. A modified instrument is almost complete containing a filter and re-acceleration electrode which should greatly improve the quality of the images. At high magnification, some image distortion is expected as the sample enters the near-field region of the tip.

8. Conclusions

In summary, the most promising areas of research in this field appear to be:

- The development of imaging energy filters to exclude inelastic scattering. A filter of the imaging retarding Wien type may have advantages.[454] We note that the declining detective quantum efficiency of channel plates below about 100 V has a filtering effect.

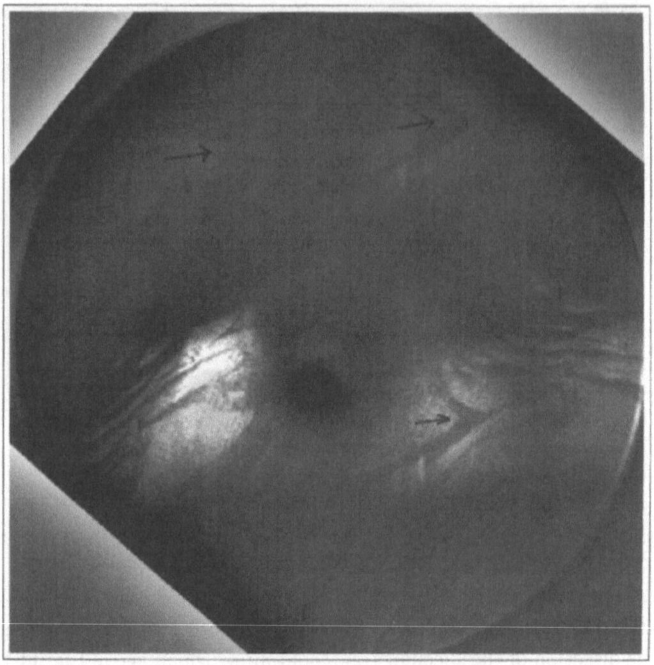

Figure 14.13. Reflection image of cleavage steps obtained by the geometry of Fig. 14.11 at 450 V. Magnification is about 200. The image consists of four identical images recorded in four different Bragg beams, projected from four different points.

- Development of the reflection mode. This could be applied to problems of crystal growth in surface science, and to magnetic and ferroelectric domain imaging, as currently undertaken using LEEM, REM and STM.

- Development of methods for preparing self-supporting molecule assemblies, perhaps drawn across holes in carbon films. The important problem of determining molecular shape for molecules which cannot be crystallized might be tackled in this way with existing instruments.

- Developments in image processing. In order to avoid the restriction to small objects ($N \ll 1$), we are exploring the use of recursive calculation between two defoci, and the imposition of the *a priori* constraint that $q = 1$ or 0 in image processing.

- The use of very low voltages (where the inelastic mean free path increases) should be investigated more fully.

- Development of lithographically formed tip assemblies (mini-lenses) to enable operation at higher voltages, allowing greater penetration, but still below the inner shell ionization energies. For the study of lipid monolayers and Langmuir-Blodgett films, a compromise voltage might be found; however, it is to be expected that the effects of radiation damage will be highly materials dependent.

- Nanotips may also be useful for electron spectroscopy in view of reports of very low energy spreads in the emission distribution.[21]

No general method for imaging individual organic molecules exists, yet the shape of molecules is believed to exert a controlling influence on their function. The overriding problem in molecular imaging is radiation damage, and many proteins, for example, may prove to be impossible to crystallize and so study by alternative crystallographic methods. (Of the 10^5 proteins in the human body, only about 1000 have been crystallized.) For all these reasons an effort aimed at imaging using electrons at very low voltages, say 7 eV, where damage might be expected to be similar to that which occurs in the ultraviolet region of the spectrum, would seem worthwhile. Although there is at present no definitive evidence for the transmission of low energy electrons through free-standing organic films, this is not essential to the problem of determining molecular shape. Three recent developments provide an encouraging note to end on: i) the report of transmission of 8 V ballistic electrons through free-standing films of gold up to 60 nm thick (at very low currents).[301] ii) the finding that the critical dose increases with decreasing voltage in LEED studies below 30 V from certain organic films, the occurrences of resonances in electron-stimulated desorption (the important damage process at these energies), and the use of Cs-coated tips to obtain images at lower voltage, below most of the ionization thresholds,[425] and iii) the recent observation of free-standing PcPS long chain molecules across slots in microstructured silicon by Golzhauser et al.,[147] using the low-voltage point-projection electron microscope.

Acknowledgments

J.C.H.S. is most grateful to Drs. W. Qian, J.M. Zuo, X. Zhang and K. and D. Taylor for their assistance with the low energy electron holography project. Supported by ARO awards DAAG 559810104 and DAAW 049610231.

A PLUS OR MINUS SIGN IN THE FOURIER TRANSFORM?

F. Lenz[1] and J.M. Cowley[2]

[1]Institut für Angewandte Physik, Universität Tübingen, D-72076 Tübingen, Germany
[2]Department of Physics and Astronomy, Arizona State University Tempe, AZ 85287-1504

J. Cowley to editors (September 95)

The sign of the exponential in the Fourier transform is something that we have been concerned with for many years. Of course, there are two conventions that have been used with almost equal frequency but we have been trying to get people to stick to one of them to avoid confusion. In our books and papers (Refs. 65.207, etc.) we have used the convention of the positive sign in the exponential for the forward transform which represents the Fraunhofer diffraction pattern for a real-space object. This is consistent with assuming that a plane wave, going in positive direction in real space is written $\exp[i\,(\omega\,t - \vec{k}\,\vec{r})]$ rather than a minus sign before the i, so that the phase advances with time. This is also consistent with making the Scherzer defocus negative, and also implies that the phase shift in a solid with a positive potential (accelerating the electrons) is negative so that the transmission function is $\exp[-i\,\sigma\,\varphi(x,y)]$. If the other convention is to be used for the Fourier transform exponent sign, then all authors should be advised of all these other implications, which are not immediately obvious. Otherwise, we might find ourselves producing a treatment of positron holography! There is a summary of our conventions and definitions in the appendixes of Refs. 65 and 207.

Lenz to editors (October 1995)

On Complex Notations in Electron Wave Optics

In wave mechanics, a complex wave function $\psi(\vec{r})$ is used to describe the propagation of an electron. It is a solution of Schrödinger's wave equation

$$\frac{1}{2m}\left[\frac{\hbar}{i}\nabla + e\,\vec{A}\,(\vec{r})\right]^2 \psi(\vec{r}) - e\,\Phi(\vec{r})\,\psi(\vec{r}) = E\,\psi(\vec{r}) \tag{15.1}$$

In the special case of field free space ($\vec{A} \equiv 0,\ E + e\,\Phi \equiv eU$) it becomes

$$\Delta\psi(\vec{r}) + k^2\,\psi(\vec{r}) = 0 \tag{15.2}$$

with

$$k^2 := 2\,m\,e\,U/\hbar^2 \tag{15.3}$$

In electron wave optics, Fourier transforms are often used to describe the wave function in the diffraction plane, i.e., the back focal plane of the objective lens. If the object plane $z = 0$ is irradiated by a primary plane wave $\exp(i\,k\,z)$, then the wave function immediately behind the object plane may be written as

$$\psi(x, y, z) = \psi(x, y, 0) \cdot e^{i\,k\,z} \tag{15.4}$$

where the "object function" $\psi(x, y, 0)$ describes the interaction of the primary wave with the object. In the following we shall simply write $\psi(\vec{r}) = \psi(x, y)$ instead of $\psi(x, y, 0)$ for the object funtion, where $\vec{r} = \{x, y\}$ is a two-dimensional vector of position in the object plane. The statement that the wave function in the diffraction plane (the "object spectrum") is equal to the Fourier transform of the object function may be written as

$$\psi_1(\vec{q}) = \iint \psi(\vec{r}) \cdot e^{-2\pi\,i\,\vec{q}\,\vec{r}}\ \mathrm{d}^2\vec{r} \tag{15.5}$$

In Eq. (15.5),

$$\vec{q} := \vec{r}_B/(f\,\lambda) \tag{15.6}$$

is a two-dimensional vector of spatial frequency proportional with the vector of position $\vec{r}_B = \{x_B, y_B\}$ in the diffraction plane. Equation (15.5) may be derived using Kirchhoff's diffraction integral

$$\psi_1(\vec{r}_1) = \frac{1}{4\,\pi} \cdot \iint \left[\frac{e^{i\,k\,r}}{r} \cdot \nabla\psi(\vec{r}) - \psi(\vec{r}) \cdot \nabla\left(\frac{e^{i\,k\,r}}{r}\right)\right]\ \mathrm{d}\vec{S} \tag{15.7}$$

applying the approximation known as "Fraunhofer diffraction." The integration is extended over the object plane, and $r := |\vec{r}_1 - \vec{r}|$. The surface element $\mathrm{d}\vec{S}$ is oriented toward the outside of the volume of integration, i.e.,

$$\mathrm{d}\vec{S} = -\mathrm{d}x\,\mathrm{d}y\,\nabla z \tag{15.8}$$

In conventional transmission electron microscopy, the aberrations of the objective lens are taken into account by replacing the object spectrum $\psi_1(\vec{q})$ by the "image spectrum" $\psi_1(\vec{q})\exp[-i\chi(\vec{q})]$, where the wave aberration $\chi(\vec{q})$ is a measure of the distance between the ideal spherical wave front in image space from the real wave front. For example, if no other aberrations than spherical aberration are present then we have $\chi(\vec{q}) = (\pi/2)\,C_S\,\lambda^3\,q^4 > 0$. This means that the real surface of constant phase, i.e., the real wave front, is, with increasing q, more and more ahead of the ideal spherical wave front. In the special case that $\psi(\vec{r}) = \delta(\vec{r})$, i.e., that the object is formed by one scattering point only, Eq. (15.5) yields $\psi_1 = 1$, i.e., without taking account of any aberration we would have a plane wave $\exp(i\,k\,z)$ in image space. If aberrations are taken into account, we would have $\exp[i\,k\,z - i\chi(\vec{q})]$ instead, i.e., the surfaces of constant phase in image space become $z = k^{-1}\,\chi(\vec{q})+$ const. Thus, the negative sign in the exponent of the phase factor $\exp[-i\chi(\vec{q})]$ is correct, and it should not be replaced by $\exp[+i\chi(\vec{q})]$. It is true that the description of the wave in the image space by $\exp[-i\,k\,z + i\chi(\vec{q})]$ instead of $\exp[i\,k\,z - i\chi(\vec{q})]$ would also yield the correct expression $z = k^{-1}\,\chi(\vec{q})+$ const for the wave fronts, but a wave function $\exp[-i\,k\,z + i\chi(\vec{q})]$ would not be compatible with the basic assumption of wave mechanics that $\vec{p}\,\psi = -i\,\hbar\,\nabla\psi$, since it would yield the eigenvalue $\vec{p} = -\hbar\,k\,\nabla z$ instead of $\vec{p} = \hbar\,k\,\nabla z$. I think we should avoid annotations in which the momentum of the electron has a direction opposite to its velocity.

Another possible approach to the calculation of the phase factor in presence of fields is supplied by the Eikonal approximation

$$\psi(\vec{r}) = a(\vec{r}) \cdot e^{i\,S(\vec{r})/\hbar} \tag{15.9}$$

where $S(\vec{r})$ is Hamilton's characteristic function which, in geometrical–optical approximation, may be written as

$$S = \hbar\,k \cdot \int \mu(\vec{r}, \mathrm{d}\vec{r}/\mathrm{d}s)\ \mathrm{d}s \tag{15.10}$$

where $\mu(\vec{r}, \mathrm{d}\vec{r}/\mathrm{d}s) = \{\,[E + e\,U\,\Phi(\vec{r})]/(e\,U)\,\}^{1/2} - [e/(2\,m\,U)\,]^{1/2} \cdot \vec{A} \cdot \mathrm{d}\vec{r}/\mathrm{d}s$ is Glaser's electron optical index of refraction. Thus we have in the Eikonal approximation

$$\psi(\vec{r}) = a(\vec{r}) \cdot \exp\left[i\,k \int \mu(\vec{r}, \mathrm{d}\vec{r}/\mathrm{d}s)\ \mathrm{d}s\right] \tag{15.11}$$

In the literature on electron wave optics, some authors use instead of the Fourier transform Eq. (15.5)

$$\psi_1(\vec{q}) = \int \psi(\vec{r}) \cdot e^{+2\pi\,i\,\vec{q}\vec{r}}\ \mathrm{d}^2\vec{r} \tag{15.12}$$

and the phase factor $\exp[i\,\chi(\vec{q})]$ instead of $\exp[-i\,\chi(\vec{q})]$. Are these two notations equivalent, or are there any reasons why one of them should be preferred? In the following we shall call the notation using Eq. (15.5) and $\exp[-i\,\chi(\vec{q})]$ "complex notation Reimer" (CNR) and the notation using Eq. (15.12) and $\exp[+i\,\chi(\vec{q})]$ "complex notation Cowley" (CNC).

We shall now derive the Fourier transform Eq. (15.5) from Kirchhoff's formula Eq. (15.6) using CNR and then using CNC.

Complex Notation Reimer

We apply Kirchhoff's formula Eq. (15.6) to the case where the integration is extended over the object plane $z = 0$, and the point P_1 at position $\vec{r}_1$ is assumed to be in a field free image space at a very large axial distance z_1 behind the object plane. We have

$$r = |\vec{r} - \vec{r}_1| = \left[(x-x_1)^2 + (y-y_1)^2 + z_1^2\right]^{1/2}$$
$$= \left(x_1^2 + y_1^2 + z_1^2 - 2\,x\,x_1 - 2\,y\,y_1 + x^2 + y^2\right)^{1/2} \tag{15.13}$$

In the Fraunhofer approximation we neglect the terms $x^2 + y^2$ under the square root on the right hand side of Eq. (15.13), and we introduce the abbreviation $R := (x_1^2 + y_1^2 + z_1^2)^{1/2} \approx z_1$ for the distance of the point P_1 from the origin, i.e., the axis point of the object plane. Thus we obtain

$$r = \left(R^2 - 2\,x\,x_1 - 2\,y\,y_1\right)^{1/2} \approx R - (x\,x_1 + y\,y_1)/R = R - \vec{r}\,\vec{r}_1/R \tag{15.14}$$

For $\nabla\psi(\vec{r})$ we write $\nabla\psi(\vec{r}) \approx i\,k\,\psi(\vec{r})\nabla z$. Further we have for $r \gg k^{-1}$

$$\nabla\left(\frac{e^{ikr}}{r}\right) = \left(\frac{i\,k}{r} - \frac{1}{r^2}\right)\cdot e^{ikr}\cdot\nabla r \approx \frac{i\,k\cdot e^{ikr}}{r}\cdot\nabla|\vec{r} - \vec{r}_1| \tag{15.15}$$

Expressing the area element $\mathrm{d}\vec{S}$ by Eq. (15.7), and with $\nabla r\cdot\nabla z = (z - z_1)/R \approx -1$, we obtain

$$\psi_1(\vec{r}_1) = \frac{k}{2\pi\,i\,R}\cdot e^{ikR}\cdot\int\psi(x,y)\cdot e^{-ik(x\,x_1 + y\,y_1)/R}\;\mathrm{d}x\,\mathrm{d}y \tag{15.16}$$

An ideal objective lens with the focal length f produces in its back focal plane an image of this distant diffraction pattern demagnified by the factor f/R

$$\psi_1(\vec{r}_B) = \frac{k}{2\pi\,i\,f}\cdot e^{ikR}\cdot\int\psi(x,y)\,e^{-ik(x\,x_B + y\,y_B)/f}\;\mathrm{d}x\,\mathrm{d}y$$
$$= \frac{k}{2\pi\,i\,f}\cdot e^{ikR}\cdot\int\psi(\vec{r})\,e^{-ik\vec{r}\,\vec{r}_B/f}\;\mathrm{d}^2\vec{r} \tag{15.17}$$

Elimination of the vector of position $\vec{r}_B$ in the back focal plane by the two-dimensional spatial frequency vector $\vec{q} := \vec{r}_B/(\lambda\,f)$ using Eq. (15.6) yields

$$\psi_1(\vec{q}) = \frac{k}{2\pi\,i\,f}\cdot e^{ikR}\cdot\int\psi(\vec{r})\cdot e^{-2\pi i\vec{r}\vec{q}}\;\mathrm{d}^2\vec{r} \tag{15.18}$$

i.e., an expression which, apart from a constant normalizing factor, coincides with Eq. (15.5). The same relation has been derived by Reimer, using Huygens's principle instead of Kirchhoff's formula.

Complex Notation Cowley

In the same way as Kirchhoff's formula Eq. (15.7) an equivalent formula

$$\psi_1(\vec{r}_1) = \frac{1}{4\pi}\cdot\int\frac{e^{-ikr}}{r}\cdot\nabla\psi(\vec{r}) - \psi(\vec{r})\cdot\nabla\left(\frac{e^{-ikr}}{r}\right)\;\mathrm{d}\vec{S} \tag{15.19}$$

may be derived since $\exp(-i\,k\,r)/r$ is a solution of Eq. (15.2) as well as $\exp(+i\,k\,r)/r$. Equations (15.9) and (15.10) are still valid, but now we have to use

$$\psi(x, y, z) = \psi(x, y, 0) \cdot e^{-i\,k\,z} \tag{15.20}$$

instead of Eq. (15.4). Further we have now $\nabla\psi(\vec{r}) = -i\,k\,\psi\,\nabla z$ and $\nabla\left(e^{-i\,k\,r}/r\right) \approx (-i\,k/r) \cdot e^{-i\,k\,r} \cdot \nabla|\vec{r} - \vec{r}_1|$. Thus we obtain now

$$\psi_1(\vec{r}_1) = \frac{i\,k}{2\,\pi\,R} \cdot e^{-i\,k\,R} \cdot \int \psi(x, y) \cdot e^{i\,k\,(x\,x_1 + y\,y_1)/R}\ \mathrm{d}x\,\mathrm{d}y \tag{15.21}$$

An ideal objective lens with the focal length f produces in its back focal plane an image of this distant diffraction pattern demagnified by the factor f/R

$$\begin{aligned}\psi_1(\vec{r}_B) &= \frac{i\,k}{2\,\pi\,f} \cdot e^{-i\,k\,R} \cdot \int \psi(x, y) \cdot e^{i\,k\,(x\,x_B + y\,y_B)/f}\ \mathrm{d}x\,\mathrm{d}y \\ &= \frac{i\,k}{2\,\pi\,f} \cdot e^{-i\,k\,R} \cdot \int \psi(\vec{r}) \cdot e^{i\,k\,\vec{r}\,\vec{r}_B/f}\ \mathrm{d}^2\vec{r}\end{aligned} \tag{15.22}$$

or

$$\psi_1(\vec{q}) = \frac{i\,k}{2\,\pi\,f} \cdot e^{-i\,k\,R} \cdot \int \psi(\vec{r}) \cdot e^{2\pi\,i\,\vec{r}\,\vec{q}}\ \mathrm{d}^2\vec{r} \tag{15.23}$$

Are both notations equivalent? In Reimer's notation, the object is irradiated by a primary plane wave $\exp(i\,k\,z)$, in Cowley's notation by a plane wave $\exp(-i\,k\,z)$. In Schrödinger's wave mechanics, the momentum operator $\hat{p}$ is defined by $\hat{p}\,\psi(\vec{r}) = -i\,\hbar\,\nabla\psi(\vec{r})$. In Reimer's version, this would lead to $\vec{p} = \hbar\,\vec{k}$, but not so in Cowley's notation. Thus, the same argument which has been used above, against the use of the positive sign in the exponent of the phase factor $\exp[i\chi(\vec{q})]$ speaks against the use of the Fourier transform Eq. (15.12) instead of (15.5).

Schrödinger based his representation of wave mechanics on the operators $\hat{p} = -i\,\hbar\nabla$ and $\hat{H} = i\,\hbar \cdot \partial/\partial t$. One might as well create an anti-Schrödinger notation of wave mechanics by chosing $\hat{p} = i\,\hbar\,\nabla$ and $\hat{H} = -i\,\hbar \cdot \partial/\partial t$, and by replacing all complex expressions in the textbooks by their conjugate complex terms. Since only real quantities such as probability densities, current densities, and the eigenvalues or expectation values of Hermitian operators are observable, this anti-Schrödinger wave mechanics would be equivalent to the original one, as far as the prediction of the values of the observable quantities is concerned. But I think the arbitrary use of once the Schrödinger and once the anti-Schrödinger notation would not yield any new physical results but create unnecessary additional confusion.

Cowley to editors (November 1995)

Many thanks for your note and the copy of the piece by Prof. Lenz. We agree with him, of course, that there are two choices for the signs and that one should be consistent in the choice of one or other of the sets of signs that are used. Most physicists have followed the choice of signs given by Schrödinger in his original papers and this is followed by most quantum mechanics textbooks with the exception of one prominent German author. However, even Schrödinger pointed out that there are two choices, equally valid. Enclosed is a copy of a paper by him, published in 1926, which Peter Goodman found and forwarded to J. Spence a few years ago: so the choice of the opposite convention is not really "anti-Schrödinger."[514] You will see the ± occuring on page 112. Also attached you will find a copy of the section in J. Spence's book which deals with the subject.[411]

Our choice of signs is that which has been adopted universally by crystallographers and officially blessed by the International Union of Crystallography in that it is standard in the International Tables for Crystallography and other publications by or for the Union. Since we are concerned with the structure of matter, we thought it appropriate to follow that convention and avoid any confusion that might arise when comparisons are made with X-ray or other diffraction results.

I know that this is a difficult matter because everyone has their particular preferences for various reasons, but since the deadline for submission of the draft chapters for the holography book is drawing close, it would be a good idea if you, as editors, made the definite decision on what convention we should use and also included a discussion on the conventions in the same way that J. Spence did in his book.[408] If you would like to quote Lenz, that is okay, but maybe you should also quote Schrödinger!

Lenz to editors (November 1995)

Thank you for your letter with the interesting attachments. It appears that the problem is rather old, and that many crystallographers are rather accustomed to what Spence calls the "crystallographic convention." Thus my argumentation that it is in contradiction to Schrödinger's choice of the operators $\vec{p} = [h/(2\pi i)]\nabla$ and $E = -[h/(2\pi i)]\,\partial/\partial t$, i.e., to the "quantum mechanical convention," is known since long but not considered to be important. Thus we cannot avoid some Babylonian language confusion, and I do not think it makes much sense to publish my argumentation once more. I shall certainly not be converted to adopt a special version of Schrödinger's wave mechanics for crystallographers, and I assume that the crystallographers a majority of whom seems to be accustomed to the other convention will not be ready to become converts either. Thus I think we shall have to live with the fact that those authors who are originally physicists will prefer a convention different from that of crystallographers who have learnt the other convention. Perhaps, in books containing contributions from adherents of both conventions, one should make a short note at the beginning of each contribution stating which of the two conventions is used by the author. By the way, who is the "prominent German author" who, according to Cowley, prefers the crystallographic notation even in a textbook on quantum mechanics? Is it Max von Laue? Do you know my article, "Electron Wave Optics?"[253]

Cowley to editors (May 1996)

The "prominent German author(s)" referred to are J.M. Juech and F. Rohrlich.[220]

BIBLIOGRAPHY

REFERENCES

1. G. Ade, Optik, **58** (1981) 321.
2. G. Ade, Optik, **62** (1982) 67-85.
3. G. Ade, R. Lauer, Optik, **88** (1991) 103-108.
4. G. Ade, Advances in Electronics and Electron Physics, **89** (1994) 1-51.
5. Y. Aharonov, D. Bohm, Phys. Rev, **115** (1959) 485.
6. L.F. Allard, E. Völkl, D.S. Kalakkad, A.K. Datye, J. Mater. Sci. **29** (1994) 5612.
7. L.F. Allard, E. Völkl, S. Subramoney, R.S. Ruoff, *Electron Holography*, Eds. A. Tonomura, L.F. Allard, G. Pozzi, D.C. Joy, Y.A. Ono, Elsevier Science B.V., (1995) 219.
8. L.F. Allard, E. Völkl, A. Carim, A.K. Datye, R. Ruoff, Nano. Mater., **7** (1996) 137.
9. L.F. Allard, K. Lang, R. Kincer, E. Völkl, submitted to J. Micros. Res. & Tech. (1998).
10. S. Amelinckx, D. Van Dyck, *Electron Diffraction Techniques*, Ed. J.M. Cowley, Oxford Science Publishers, Vol. 2 (1993).
11. K. Aoyama, G. Lai, Q. Ru, *Electron Holography*, Eds. A. Tonomura, L.F. Allard, G. Pozzi, D.C. Joy, Y.A. Ono, Elsevier Science B.V., (1995) 239.
12. N.W. Ashcroft, N.D. Mermin, *Solid State Physics*, Saunders College, Philadelphia (1976).
13. H. Banzhof, K.-H. Herrmann, H. Lichte, Microscopy Research and Technique, **20** (1992) 457.
14. H. Banzhof, K.-H. Herrmann, Ultramicroscopy, **48** (1993), 475.
15. W. Bayh, Zeitschrift für Physik, **169** (1962) 492-510.
16. V.Y. Bazhenov, M.S. Soskin, M.V. Vasnetsov, J. Mod. Optics, **39** (1992) 985.
17. P. Becker, P. Coppens, Acta Cryst., A46 (1990) 254.
18. C. Beeli, B. Doudin, J.-Ph. Ansermet and P. Stadelmann, J. Magnetism and Magnetic Mat., **164** (1996) 77-90.
19. C. Beeli, B. Doudin, J.-Ph. Ansermet, P. Stadelmann, Ultramicroscopy, **67** (1997) 143-151.
20. H. Bethe, Ann. Phys., **87** (1928) 55.
21. V.T. Binh, S.T. Purcell, N. Garcia, J. Doglioni, Phys. Rev. Letts., **69** (1992) 2527.
22. V.T. Binh, N. Garcia, K. Dransfeld, Manipulation of atoms under high fields and temperatures: Applications, Plenum, New York, (1993) 331.
23. V.T. Binh, Ultramicroscopy, **58** (1995) 307.
24. H. Boersch, Zeitschrift für Techn. Physik, **12** (1939) 346.
25. J.E. Bonevich, H. Kasai, K. Harada, T. Matsuda, T. Yoshida, G. Pozzi, A. Tonomura, Phys. Rev. Lett., **70** (1993) 2952.
26. J.E. Bonevich, K. Harada, H. Kasai, T. Matsuda, T. Joshida, G. Pozzi, A. Tonomura, Phys. Rev. B, **49** (1994) 6800-6807.
27. J.E. Bonevich, K. Harada, H. Kasai, T. Matsuda, T. Joshida, A. Tonomura, Phys. Rev. B, **50** (1994) 567-570.
28. J.E. Bonevich, K. Harada, T. Matsuda, H. Kasai, T. Joshida, G. Pozzi, A. Tonomura, MRS Conf. Proc. No. 332 (Materials Research Society, Pittsburgh, 1994), Eds. M. Sarikaya, H.K. Wickramasinghe, M. Isaacson, p219-224.
29. J.E. Bonevich, D. Cappacci, K. Harada, H. Kasai, T. Matsuda, R. Patti, G. Pozzi, A. Tonomura, Phys. Rev. B, **57** (1998) 1200.
30. M. Born, E. Wolf, *Principles of Optics*, Pergamon Press, (1965).
31. M. Born, E. Wolf, *Principles of Optics*, Pergamon, Oxford, (1980) 505.
32. W.L. Bragg, Zeitschrift für Krist., **70** (1929) 489.
33. J. Breedlove, G. Trammel, Science, **170** (1970) 1310.
34. W. Brünger, M. Klein, Surface Sci., **62** (1977) 317.
35. R. Buhl, Zeitschrift für Physik, **155** (1959) 395-412.
36. B.F. Buxton, P.T. Tremewan, Adv. Cryst., A36 (1980) 304.

37. C. Capiluppi, G. Pozzi, U. Valdrè, Phil. Mag., **26** (1972) 865.
38. A.H. Carim, private communication.
39. A.H. Carim, L.F. Allard, E. Völkl, B.G. Frost, unpublished work.
40. E. Cartier, P. Pfluger, J.-J. Pireaux, M. Vilar, Appl. Phys., A44 (1987) 43-53.
41. J.N. Chapman, G.R. Morrison, J. Mag. Mag. Materials, **35** (1983) 254.
42. J.N. Chapman, J. Phys., D17 (1984) 623.
43. J.N. Chapman, M.J. Drummond, Proc. II Int. Symp. Electron Microsc. and Biophys., Chandigarh, (1986) 193.
44. C. Chen, *Introduction to Scanning Tunnelling Microscopy*, New York, Oxford University Press (1993).
45. J.W. Chen, G. Matteucci, A. Migliori, G. F. Missiroli, E. Nichelatti, G. Pozzi, M. Vanzi, Phys. Rev. A, 40 (1989) 3136.
46. J.W. Chen, T. Hirayama, G. Tai, T. Tanji, A. Tonomura, Optics Letters, **18** (1993) 1887-1889.
47. J. Chen, T. Hirayama, G. Lai, T. Tanji, K. Ishizuka, *Electron Holography*, Eds. A. Tonomura, L.F. Allard, G. Pozzi, D.C. Joy, Y.A. Ono, Elsevier Science B.V., (1995) 81-92.
48. J.R. Clem, J. Low Temperature Phys., **18** (1975) 427.
49. R.J. Collier, L.B. Burckhardt, L.H. Lin, *Optical Holography*, Academic Press, New York, (1971).
50. C. Colliex, B. Jouffrey, M. Kleman, Acta Cryst., A24 (1968) 692.
51. W. Coene, T.J.J. Denteneer, Ultramicroscopy, **38** (1991) 225.
52. W. Coene, G. Janssen, T. Denteneer, M. Op de Beeck, D. Van Dyck, Microscopy Society of America Bulletin, **24** (1994) 472.
53. J.M. Cowley, Appl. Phys. Letters, **15** (1969) 58.
54. J.M. Cowley, Ultramicroscopy, **4** (1979) 435.
55. J.M. Cowley, M. Disko, Ultramicroscopy, **5** (1980) 469.
56. J.M. Cowley, Micron, **11** (1980) 229.
57. J.M. Cowley, D.J. Walker, Ultramicroscopy, **6** (1981) 71.
58. J.M. Cowley, M.A. Osman, P. Humble, Ultramicroscopy, **15** (1984) 311.
59. J.M. Cowley, Progr. in Surface Sci., **21** (1986) 209.
60. J.M. Cowley, *High-Resolution Transmission Electron Microscopy*, Eds., P.R. Buseck, J.M. Cowley, L. Eyring, Oxford Univ. Press, (1988) Chapters 1 and 2.
61. J.M. Cowley, ibid, p20.
62. J.M. Cowley, ibid, p23.
63. J.M. Cowley, Ultramicroscopy, **34** (1990) 293.
64. J.M. Cowley, *Electron Diffraction Techniques*, Vol. 1, Ed. J.M. Cowley, Oxford University Press, (1992) 439.
65. J.M. Cowley, ibid, p567.
66. J.M. Cowley, Ultramicroscopy, **41** (1992) 335.
67. J.M. Cowley, Ultramicroscopy, **49** (1993) 4.
68. J.M. Cowley, Surface Sci., **298** (1993) 336.
69. J.M. Cowley, M.A. Gribelyuk, Micros. Soc. Amer. Bulletin, **24** (1994) 438.
70. J.M. Cowley, Ultramicroscopy, **57** (1995) 327.
71. J.M. Cowley, *Diffraction Physics*, 3rd revised edition, North Holland, (1995).
72. J.M. Cowley, ibid, p83.
73. J.M. Cowley, Applications of Electron Holography, *Handbook of Advanced Materials Testing*, Eds., N.P. Cheremisinoff, P.N. Cheremisinoff, Marcel Dekker, Inc., New York, (1995) 155.
74. J.M. Cowley, J. Electron Microscopy, **45** (1996) 3.
75. J.M. Cowley, M. Mankos, M.R. Scheinfein, Ultramicroscopy, **63** (1996) 133-147.
76. A.V. Crewe et al., Rev. Sci. Instr., **39** (1968) 576.
77. A.V. Crewe, J. Wall, J. Mol. Biol., **48** (1970) 375.
78. A.V. Crewe, *Progress in Optics*, North Holland, Amsterdam, Vol. 11 (1973) 225.
79. I. Daberkow, K.-H. Herrmann, L. Liu, W.D. Rau, Ultramicroscopy, **38** (1991) 215.
80. I. Daberkow, K.-H. Herrmann, L. Liu, W.D. Rau, H. Tietz, Ultramicroscopy, **64** (1996) 35-48.
81. J.C. Dainty, K. Shaw, *Image Science*, Academic Press, New York, (1974) p260.
82. A.K. Datye, D.S. Kalakkad, E. Völkl, L.F. Allard, *Electron Holography*, Eds. A. Tonomura, L.F. Allard, G. Pozzi, D.C. Joy, Y.A. Ono, Elsevier Science B.V., (1995) 199.
83. A.F. de Jong, D. Van Dyck, Ultramicroscopy, **49** (1993) 66.
84. W.J. de Ruijter, M. Gajdardziska-Josifovska, M.R. McCartney, R. Sharma, D.J. Smith, J.K. Weiss, Scanning Micros. Supp., **6** (1992) 347.
85. W.J. de Ruijter, J.K. Weiss, Rev. Sci. Instr., **63** (1992) 4314.

86. W.J. de Ruijter, *Quantitative High-Resolution Electron Microscopy and Holography*. Ph.D. Thesis, The University of Delft, Delft, The Netherlands (1992).
87. W.J. de Ruijter, J.K. Weiss, Ultramicroscopy, **50** (1993) 269-283.
88. W.J. de Ruijter, P.E. Mooney, O.L. Krivanek, Proc. 51st Ann. Meet. MSA, San Francisco Press, San Francisco, (1993) 1062.
89. W.J. de Ruijter, P.E. Mooney, E. Völkl, MSA Bulletin, **24** (1994) 451.
90. W.J. de Ruijter, J. Comp. Assist. Microsc., **6** (1994) 195.
91. W.J. de Ruijter, J.K. Weiss, Ultramicroscopy, **57** (1995) 409.
92. W.J. de Ruijter, Micron, **26** (1995) 247.
93. J.B. DeVelis, G.B. Parrent, B.J. Thompson, J. Opt. Soc. Am., **56** (1966) 423.
94. G.J. Dolan, F. Holtzberg, C. Feild, T.R. Dinger, Phys. Rev. Lett.. **62** no18 (1989) 2184-2187.
95. D.L. Dorset, *Structural Electron Crystallography*, Plenum Press, New York. (1995).
96. K.H. Downing, D. Grano, Ultramicroscopy, **7** (1982) 381.
97. P.A. Doyle, P.S. Turner, Acta Cryst., A24 (1968) 390.
98. S. Dudarev, P. Rez, M. Whelan, Phys. Rev., B51 (1995) 3397.
99. J.A. Eades, Scanning, **16** (2) (Mar-Apr. 1994) 97.
100. R.F. Egerton, *Electron Energy-Loss Spectroscopy in the Electron Microscope*, Plenum, New York, (1986) 127.
101. W. Ehrenberg, R.E. Siday, Proc. Phys. Soc., London, B62 (1949) 8-21.
102. P.M. Epperson, J.V. Sweedler, M.B. Denton, G.R. Sims, T.W. McGurnin, R.S. Aikens. Opt. Eng., **26** (1987) 715.
103. V. Essman, H. Träuble, Phys. Lett., A24 (1967) 526.
104. J. Faget, C. Fert, Cahier de Physique, **83** (1957) 285.
105. U. Fano, Phys. Rev., **72** (1947) 26.
106. P.L. Fejes, Acta Cryst., **A33** (1977) 109.
107. H.W. Fink, Physica Scripta, **38** (1988) 260.
108. H.W. Fink, H. Schmid, H.J. Kreuzer, A. Wierzbicki, Phys. Rev. Letts., **67** (1991) 1543.
109. S. Frabboni, G. Matteucci, G. Pozzi. M. Vanci, Phys. Rev. Lett.. **55** no20 (1985) 2196.
110. S. Frabboni, G. Matteucci, G. Pozzi, Ultramicroscopy, **23** (1987) 29.
111. J. Frank. Optik, **38** No.5 (1973) 519.
112. J. Frank, *Computer Processing of Electron Microscope Images*, (1980) 187.
113. F.J. Franke, K.-H. Herrmann, H. Lichte, 11th Int. Congr. on Electron Microscopy, Kyoto. (1986) 677-678.
114. F.J. Franke, K.-H. Herrmann, H. Lichte, Scanning Microscopy, Supplement. **2** (1988) 59-67.
115. W. Franz, Verhandlungen der Deutschen Physikalischen Gesellschaft. **2** (1939) 65.
116. W. Franz, Physikalische Berichte, **21** (1940) 686.
117. W. Franz, Zeitschrift für Physik, **184** (1965) 85-91.
118. P.P. Freitag, I.G. Trindade, L.V. Melo, J.L. Leal, N. Barradas, J.L. Soares, J. Appl. Phys., **73** (1993) 5527.
119. B.G. Frost, L.F. Allard, E. Völkl, D.C. Joy, *Electron Holography*. Eds. A. Tonomura, L.F. Allard, G. Pozzi, D.C. Joy, Y.A. Ono, Elsevier Science B.V., (1995) 169.
120. B.G. Frost, N.F. van Hulst, E. Lunedei, G. Matteucci, E. Rikkers, Appl. Phys. Lett., **68** (1996) 1865.
121. B.G. Frost, *Problems in the Electron Holographical Analysis of Magnetic Microfields*. Ph.D. Thesis, University of Tübingen, Germany, (1993).
122. J.R. Fryer, C.J. Gilmore, Trans. Amer. Crystallog. Assn., **28** (1992) 57.
123. Q. Fu, H. Lichte, Optik, Suppl. 4, Vol.83, (1989) 27.
124. Q. Fu, H. Lichte. E. Völkl, Phys. Rev. Letters, **67** 17 (1991) 2319-2322.
125. Q. Fu, H. Lichte, Jour. of Microscopy, **179** (1995) 112.
126. S. Fujita, S. Maruno, H. Watanabe, Y. Kusumi, M. Ichikawa, Appl. Phys. Lett., **66** (1995) 2754.
127. Y. Fujiyoshi, N. Uyeda, Ultramicroscopy, **7** (1981) 189.
128. A. Fukuhara, K. Shinagawa, A. Tonomura, H. Fujiwara, Phys. Rev., B27 (1983) 1839-1843.
129. D. Gabor, Nature, **161** (1948) 777.
130. D. Gabor, Proc. Roy. Soc. London, A197 (1949) 454.
131. D. Gabor, Proc. Phys. Soc., B64 (1951) 449.
132. D. Gabor, Rev. Mod. Phys., **28** (1956) 260-276.
133. D. Gabor, Elektrotechnische Zeitschrift - A. **78** (1957) 522.
134. M. Gajdardziska-Josifovska, M.R. McCartney, J.K. Weiss, Proc. 50th Annual EMSA Meeting, Eds. G.W. Bailey, J. Bentley, J.A. Small, San Francisco Press, San Francisco. CA. (1992) 134.

135. M. Gajdardziska-Josifovska, M.R. McCartney, W.J. de Ruijter, D.J. Smith, J.K. Weiss, J.M. Zuo, Ultramicroscopy, **50** (1993) 285.
136. M. Gajdardziska-Josifovska, MSA Bulletin, **24** (1994) 507.
137. M. Gajdardziska-Josifovska, M.R. McCartney, Ultramicroscopy, **53** (1994) 291.
138. M. Gajdardziska-Josifovska, J.K. Weiss, J.M. Cowley, Ultramicroscopy, **58** (1995) 65.
139. M. Gajdardziska-Josifovska, Interface Science, **2** (1995) 425.
140. Gatan Inc., 5933 Coronado Ln, Pleasanton, CA 94588.
141. K.H. Gaukler, R. Schwarzer, Optik, **2** (1971) 215.
142. R. Glaeser, *Introduction to Analytical Electron Micoscopy*, Eds. J. Hren, J. Goldstein, D. Joy, Plenum, New York, (1979) 331.
143. R.W. Glaisher et al., Ultramicroscopy, **27** (1989).
144. S.R. Glanvill, A.F. Moodie, H.J. Whitfield, I.J. Wilson, Aust. J. Phys., **39** (1986) 71.
145. W. Glaser, Hand. Phys., **33** (1956) 330.
146. J.I. Goldstein, D.E. Newbury, P. Echlin, D.C. Joy, C. Fiori, E. Lifshin eds., *Scanning Electron Microscopy and X-Ray Microanalysis*, Plenum Press, New York, (1981) 22.
147. A. Golzhauser, B. Volkel, B. Jager, M. Zharnikov, H. Kreuzer, M. Grunze, J. Vac. Sci. Tech., (1998) in press.
148. M.J. Goringe, J.P. Jakubovics, Phil. Mag., **15** (1967) 393.
149. T. Goulet, V. Pou, J. Jay-Gerin, J. Electr. Spectr. Rel. Phenom., **41** (1986) 157.
150. I.P. Grant, B.J. McKenzie, P.H. Norrington, D.F. Mayers, N.C. Pyper, Comput. Phys. Comm., (1980) 21, 207.
151. M.A. Gribelyuk, J.M. Cowley, Ultramicroscopy, **45** (1992) 103-113.
152. M.A. Gribelyuk, ibid, 115-125.
153. M.A. Gribelyuk, J.M. Cowley, Ultramicroscopy, **50** (1993) 29-40.
154. M. Haider, S. Uhlemann, E. Schwan, B. Kabius, Europ. J. Cell Biol., **74** (7:7) Suppl. 45 (1997).
155. M. Haider, S. Uhlemann, E. Schwan, H. Rose, B. Kabius, K. Urban, Nature **392** (6678) (Apr. 23, 1998) 768.
156. M.E. Haine, J. Dyson, Nature, **166** (1950) 315.
157. M.E. Haine, T. Mulvey, J. Opt. Soc. Amer., **42** (1952) 763.
158. M. Haider, Optik, **99** No.4, (1995) 167.
159. K.-J. Hanszen, Advances in Electronics and Electron Physics, **59** (1982) 1-77.
160. K.-J. Hanszen, R. Lauer, G. Ade, Optik, **63** (1983) 285-303.
161. K.-J. Hanszen, R. Lauer, G. Ade, Ultramicroscopy, **16** (1985) 47.
162. K. Harada, K. Ogai, R. Shimizu, J. Electron Microsc., **39** (1990) 470.
163. K. Harada, R. Shimizu, Journal of Electron Microscopy, **40** (1991) 92-96.
164. K. Harada, T. Matsuda, J. Bonevich, M. Igarashi, S. Kondo, G. Pozzi, U. Kawabe, A. Tonomura, Nature, **360** (1992) 51-53.
165. K. Harada, T. Matsuda, H. Kasai, J.E. Bonevich, T. Joshida, U. Kawabe, A. Tonomura, Phys. Rev. Lett., **71** (1993) 3371-3374.
166. K. Harada, H. Kasai, T. Matsuda, M. Yamasaki, J.E. Bonevich, A. Tonomura, Jpn. J. Appl. Phys., **33** (1994) 2534-2540.
167. A. Harscher, F. Lenz, H. Lichte, 10th European Congress on Electron Microscopy, Granada, Spain, last minute brochure, Eds. L. Megias, M.I. Rodriquez-Garcia, E. Rios, J.M. Arias (1992) 35-36.
168. A. Harscher, G. Lang, H. Lichte, Ultramicroscopy, **58** (1995) 79-86.
169. A. Harscher, H. Lichte, Ultramicroscopy, **64** (1996) 57.
170. F. Hasselbach, Z. Phys., B151 (1988) 443-449.
171. F. Hasselbach, M. Nicklaus, Phys. Rev. A, **48** (1993) 143-151.
172. P.W. Hawkes, E. Kasper, *Principles of Electron Optics*, Vol. 1, Academic Press (1989) 366.
173. P.W. Hawkes, E. Kasper, *Principles of Electron Optics*, Vol. 2, Academic Press (1989) Chap. 43.
174. P.W. Hawkes, E. Kasper, ibid, p908.
175. P.W. Hawkes, E. Kasper, ibid, p914.
176. P.W. Hawkes, E. Kasper, ibid, p988.
177. P.W. Hawkes, E. Kasper, ibid, p1027.
178. E. Heindl, W.-D. Rau, H. Lichte, Ultramicroscopy, **64** (1996) 87-97.
179. K. Heinz, H. Wedler, Surface Rev. Letters, **1** (1994) 319.
180. R. Henderson, Quarterly Reviews of Biophysics, **28** (1995) 171.
181. F. Herman, S. Skillman, *Atomic Structure Calculations*, Prentice Hall, (1963).

182. R.A. Herring, T. Tanji, A. Tonomura, Proc. 50th EMSA Meeting, Eds. G.W. Bailey, J. Bentley, J.A. Small, San Francisco Press, (1992) 990.
183. R.A. Herring, G. Pozzi, T. Tanji, A. Tonomura, Ultramicroscopy, **50** (1993) 94.
184. R.A. Herring, J.E. Bonevich, T. Tanji, A. Tonomura, Mater. Res. Soc. Symp. Proc. Vol 332, Eds. M. Sarikaya, H.K. Wickramasinghe, M. Isaacson, (Boston, MA, 1993) 231.
185. R.A. Herring, G. Pozzi, T. Tanji, A. Tonomura, Proc. 51st Annual Meeting of Microscopy Society of America, Eds. G.W. Bailey, C.L. Rieder, (Cincinnati, OH. 1993) 1056.
186. R.A. Herring, T. Tanji, ibid, p1086.
187. R.A. Herring, G. Pozzi, T. Tanji, A. Tonomura, 13th Int. Congr. on Electron Microscopy, Eds. J.P. Chevalier, F. Glas, P.W. Hawkes (Paris, 1994) 321.
188. R.A. Herring, T. Tanji, ibid, 325.
189. R.A. Herring, G. Pozzi, T. Tanji, A. Tonomura, Ultramicroscopy, **60** (1995) 153.
190. R.A. Herring, Proc. Ann. Micros. Soc. America Conf., Eds. G.W. Bailey, M.H. Ellisman, R.A. Hennigar, N.J. Zaluzec (Kansas City, 1995) 620.
191. K.-H. Herrmann, D. Krahl, H.-P. Rust, O. Ulrichs, Optik, **44** (1976) 393.
192. K.-H. Herrmann, D. Krahl, J. Microscopy, **127** (1982) 17.
193. K.-H. Herrmann, D. Krahl, Adv. Opt. El. Micr., **9** (1984) 1.
194. T. Hibi, S. Takahashi, J. Electronmicros., **12** (1963) 129.
195. T. Hirayama, J. Chen, T. Tanji, A. Tonomura, Ultramicroscopy, **54** (1994) 9.
196. T. Hirayama et al., Jpn. J. Appl. Phys., **34** (1995) 3294.
197. P.B. Hirsch, A. Howie, R.B. Nicholson, D.W. Pashley, M.J. Whelan, *Electron Microscopy of Thin Crystals*, (Butterworths, London, 1965) 163.
198. P. Hirsch, A. Howie, R.B. Nicholson, D.W. Paschley, M.J. Whelan, *Electron Microscopy of Thin Crystals*, Krieger, (1977) 504.
199. H. Hoffmann, C. Jönsson, Z. Phys., **182** (1965) 360.
200. G. Honjo, K. Mihama, J. Phys. Soc. Japan, **9** (1954) 184.
201. W. Hoppe, Acta Cryst., A25 (1969) 495.
202. L. Howald, H. Rudin, H. Guntherodt, Rev. Sci. Instr., **63** (1992) 3909.
203. A. Howie, M. Mohd Muhid, F. Rocca, U. Valdre, Inst. Phys. Conf. Ser., 90 (EMAG 87) (1987) 155.
204. R.P. Huebener, *Magnetic Flux Structures in Superconductors*, (Springer, Berlin, 1979).
205. J.A. Ibers, Acta Cryst., **11** (1958) 178.
206. J.A. Ibers, B.K. Vainshtein, *International Crystallographic Tables*. Vol. III. (Kynoch Press, 1962), Tables 3.3.3A(1) and A(2).
207. *International Tables for Crystallography*, Vol. B, (Kluwer, Academic Publishers Dordrecht, 1993).
208. K. Ishizuka, Ultramicroscopy, **5** (1980) 55.
209. K. Ishizuka, Ultramicroscopy, **52** (1993) 7.
210. K. Ishizuka, Ultramicroscopy, **53** (1994) 297-303.
211. K. Ishizuka, T. Tanji, A. Tonomura, T. Ohno, Y. Murayama, Ultramicrocopy, **53** (1994) 361.
212. K. Ishizuka, T. Tanji, A. Tonomura, T. Ohno, Y. Murayama, Ultramicroscopy, **55** (1994) 197.
213. K. Ishizuka, Ultramicroscopy, **55** (1994) 407.
214. K. Ishizuka, J. Electron Microsc., **44** (1995) 154.
215. S. Isoda, K. Saitoh, S. Moriguchi, T. Kobayashi, Ultramicroscopy, **35** (1991) 329.
216. S. Isoda, K. Saitoh, T. Ogawa, S. Moriguchi, T. Kobayashi, Ultramicroscopy, **41** (1992) 99.
217. R. Jansen, B. Jones, J. Phys., D4 (1971) 118.
218. C. Jönsson, H. Hoffmann, G. Möllenstedt, Phys. Kondens. Mater., **3** (1965) 193.
219. D.C. Joy, Y.-S. Zhang, X. Zhang, T. Hashimoto, R.D. Bunn, L.F. Allard, T.A. Nolan, Ultramicroscopy, **51** (1993) 1-14.
220. J.M. Juech, F. Rohrlich, *The Theory of Photons and Electrons*. Addison Wesley, New York, (1995).
221. K. Kambe, G. Lehmpfuhl, F. Fujimoto, Z. Natur., 29A (1974) 1034.
222. T. Kawasaki, T. Matsuda, J. Endo, A. Tonomura, Jpn. J. Appl. Phys., **29** (1990) L508-10.
223. T. Kawasaki, Qing Xin Ru, T. Matsuda, Y. Bando, A. Tonomura, Jpn. J. Appl. Phys., **30** (1991) L1830-2.
224. M. Keller, Zeitschrift für Physik, **164** (1961) 274-291.
225. W.E. King, G.H. Campbell, Ultramicroscopy, **56** (1994) 46.
226. E.J. Kirkland, B.M. Siegel, Ultramicroscopy, **6** (1981) 169.
227. E.J. Kirkland, Ultramicroscopy, **15** (1984) 151.
228. E.J. Kirkland, B.M. Siegel, N. Uyeda, Y. Fujiyoshi, Ultramicroscopy, **17** (1985) 87.

229. J. Kirz, C. Jacobsen, M. Howells, Quart. Rev. Biophys., **28** (1995) 1.
230. J. Konnert, P. D'Antonio, J.M. Cowley, A. Higgs, H. Ou, Ultramicroscopy, **30** (1989) 371.
231. J. Konnert, P. D'Antonio, Ultramicroscopy, **45** (1992) 281-290.
232. A.J. Koster, A.F. de Jong, Ultramicroscopy, **38** (1991) 225.
233. L. Kou, J. Chen, J. Mod. Optics, **42** (1995) 1171.
234. W. Krakow, M. O'Keefe, *Computer Simulation of Electron Microscope Diffraction and Images*, The Minerals, Metals and Materials Soc., 1989.
235. O.L. Krivanek, Optik, **45** (1976) 97.
236. O.L. Krivanek, *High-Resolution Transmission Electron Microscopy*, Eds., P.R. Buseck, J.M. Cowley, L. Eyring, Oxford Univ. Press, (1988) 528.
237. O.L. Krivanek, G.Y. Fan, Proc. 10th Pfefferkorn Conf., Cambridge, (1992) 105.
238. O.L. Krivanek, P.E. Mooney, Ultramicroscopy, **49** (1993) 95.
239. O.L. Krivanek, G.Y. Fan, Scanning Microsc. Suppl., **6** (1994) 105.
240. O.L. Krivanek, Ultramicroscopy, **55** (1995) 419.
241. P. Kruit, A.H. Buist, M.R. McCartney, M.R. Scheinfein, Proc. 53rd. Ann. Meet. Microscope Society of America (1995) 606.
242. S. Kujawa, D. Krahl, Ultramicroscopy **46** (1992) 395.
243. W. Kunath, W.D. Riecke, Optik, **23** (1965) 322.
244. D.B. Langmuir, Proc. IRE 25 (1937) 977.
245. R. Lauer, Optik, **66** (1984) 159-174.
246. M. Lehmann, F. Lenz, E. Völkl, H. Lichte, 10th European Congress on Electron Microscopy, EUREM92, Granada, Last Minute Brochure (1992) 21-22.
247. M. Lehmann, H. Lichte, Proc. Int. Conf. El. Micr. ICEM 13, Vol. 1, (1994) 293.
248. M. Lehmann, E. Völkl, F. Lenz, Ultramicroscopy, **54** (1994) 335 – 344.
249. M. Lehmann, H. Lichte, *Electron Holography*, Eds. A. Tonomura, L.F. Allard, G. Pozzi, D.C. Joy, Y.A. Ono, Elsevier Science B.V., (1995) 69-79.
250. M. Lehmann, *Numerical Reconstruction of the Aberration-Free Object Wave of Off-Axis Electron Holograms*, Ph.D. Thesis (1997), University of Tübingen, Germany.
251. E.N. Leith, J. Upatnieks, J. Opt. Soc. of Amer., **52** (1962) 1123.
252. F. Lenz, J. Vac. Surf. Phys., **3** (1989) 195.
253. F. Lenz, Vacuum and Surface Analysis, **3** (1989) 195-303.
254. F. Lenz, E. Völkl, Proc. 12th Int. Cong. EM (Seattle, 1990) Vol. 1, 228.
255. T. Leuthner, H. Lichte, K.-H. Herrmann, Phys. Stat. Sol. (a), **116** (1989) 113.
256. J. Li, R. Dunin-Borkowski, M.R. McCartney, D.J. Smith, unpublished results.
257. H. Lichte, G. Möllenstedt, J. Phys. E, Sci. Instr., 12, (1979).
258. H. Lichte, Ultramicroscopy, **20** (1986) 293.
259. H. Lichte, Habilitation, University of Tübingen, Germany, (1987).
260. H. Lichte, K.-H. Herrmann, F. Lenz, Optik, **77** No. 3 (1987) 135-140.
261. H. Lichte, Adv. in Opt. and Electr. Micr., Eds. T. Mulvey, T.R. Sheppard, Acad. Press NY, **12** (1991) 25.
262. H. Lichte, Ultramicroscopy, **38** (1991) 13.
263. H. Lichte, Ultramicroscopy, **47** (1992) 223-230.
264. H. Lichte, E. Völkl, K. Scheerschmidt, Ultramicroscopy, **47** (1992) 231.
265. H. Lichte, Ultramicroscopy, **51** (1993) 15-20.
266. H. Lichte, P. Kessler, F. Lenz, W.D. Rau, Ultramicroscopy, **52** (1993) 575-580.
267. H. Lichte, Philips Electron Optics Bulletin, **130** (1991) 29-35.
268. H. Lichte, *Electron Holography*, Eds. A. Tonomura, L.F. Allard, G. Pozzi, D.C. Joy, Y.A. Ono, Elsevier Science B.V., (1995) 11.
269. H. Lichte, D. Geiger, A. Harscher, E. Heindl, M. Lehmann, D. Malamidis, A. Orchowski, W.-D. Rau, Ultramicroscopy, **64** (1996) 67-77.
270. H. Lichte, Ultramicroscopy, **64** (1996) 79-86.
271. J.A. Lin, J.M. Cowley, Ultramicroscopy, **19** (1986) 31.
272. J.A. Lin, J.M. Cowley, Ultramicroscopy, **19** (1986) 179.
273. X. Lin, V. Ravikumar, R. Rodrigues, N. Wilcox, V.P. Dravid, *Electron Holography*, Eds. A. Tonomura, L.F. Allard, G. Pozzi, D.C. Joy, Y.A. Ono, Elsevier Science B.V., (1995) 209.
274. J. Lindhard, Mat. fys. Medd. Dan. Vidensk. Selsk. 34 (1965).
275. E. Lunedei, G. Matteucci, B.G. Frost, J. Greve, J. Mag. Mag. Mat., 157/158 (1996) 434-435.
276. E. Lunedei, G. Matteucci, Atti XX Congresso di Microscopia Elettronica Rimini 11-14 sett., 95 Suppl., Microscopia Elettronica n.2, 93.
277. M. Mankos, M.R. Scheinfein, J.M. Cowley, J. Appl. Phys., **75** (11) (1994) 7418-7424.

278. M. Mankos, A.A. Higgs, M.R. Scheinfein, J.M. Cowley, Ultramicroscopy. **58** (1995) 87.
279. M. Mankos, J.M. Cowley, M.R. Scheinfein, Phys. Stat. Sol., (a) **154** (1996) 469.
280. M. Mankos, M.R. Scheinfein, J.M. Cowley, *Advances in Imaging And Electron Physics*, Ed. P. Hawkes, 98 (1996) 323.
281. L.D. Marks, D.J. Smith, Nature, **303** (1983) 316.
282. C. Martin, E. Arakawa, T. Callcott, R.J. Warmack, J. Electr. Spectr. Related Phenom., **42** (1987) 171.
283. L. Marton, Phys. Rev., **85** (1952) 1057-1058.
284. L. Marton, J. Arol Simpson, J.A. Suddeth, Rev. Sci. Instr., **25** no.11 (1954) 1099-1104.
285. T. Matsuda, S. Hasegawa, M. Igarashi, T. Kobayashi, M. Naito. H. Kaijama, J. Endo, N. Osakabe, A. Tonomura, Phys. Rev. Lett., **62** (1989) 2519-2522.
286. G. Matteucci, G. Pozzi, Ultramicroscopy, **5** (1980) 219.
287. G. Matteucci, G.F. Missiroli, G. Pozzi, Ultramicroscopy, **7** (1982) 277.
288. G. Matteucci, G.F. Missiroli, G. Pozzi, Ultramicroscopy, **8** (1982) 403.
289. G. Matteucci, G.F. Missiroli, J.W. Chen, G. Pozzi, Appl. Phys. Lett., **52** (1988) 176.
290. G. Matteucci, G.F. Missiroli, E. Nichelatti, A. Migliori, M. Vanzi, G. Pozzi, J. Appl. Phys., **69** (1991) 1835.
291. G. Matteucci, G.F. Missiroli, M. Muccini, G. Pozzi, Ultramicroscopy. **45** (1992) 77-83.
292. G. Matteucci, M. Muccini, U. Hartmann, Appl. Phys. Lett., **62** (1993) 1839.
293. G. Matteucci, M. Muccini, U. Hartmann, Phys. Rev. B, **50** (1994) 6823.
294. G. Matteucci, M. Muccini, Ultramicroscopy, **53** (1994) 19.
295. G. Matteucci, G.F. Missiroli, G. Pozzi, *Advances in Imaging and Electron Physics*. Eds. P.W. Hawkes, Academic Press, **99** (1997) 171-240.
296. M. R. McCartney, D. Smith, R. Hull, J.C. Bean, E. Völkl, B.G. Frost, Appl. Phys. Lett., **65** (1994) 2603.
297. M.R. McCartney, M. Gajdardziska-Josifovska, Ultramicroscopy, **53** (1994) 283.
298. M.R. McCartney, B. Frost, R. Hull, M.R. Scheinfein, D.J. Smith, E. Völkl. *Electron Holography*, Eds. A. Tonomura, L.F. Allard, G. Pozzi, D.C. Joy, Y.A. Ono, Elsevier Science B.V., (1995) 189.
299. M.R. McCartney, P. Kruit, A.H. Buist, M.R. Scheinfein, Ultramicroscopy. **65** (1996) 179.
300. F.F. Medina, G. Pozzi, J. Opt. Soc. Am., A7 (1990) 1027.
301. S. Meepagala and M. Baykul, Phys. Rev., **50** (1994) 13786
302. A.J. Melmed, Applied Physics Letters, **12** (1968) 100.
303. A. Melmed, J. Vac. Sci. Tech., B9 (1991) 601.
304. A. Migliori, G. Pozzi, Ultramicroscopy, **41** (1992) 169.
305. A. Migliori, G. Pozzi, A. Tonomura, Ultramicroscopy, **49** (1993) 87.
306. G.F. Missiroli, G. Pozzi, U. Valdrè, J. Phys. E: Sci. Instr., **14** (1981) 649.
307. S. Miyake, Proc. Phys.-Math. Soc. Japan, **22** (1940) 666.
308. S. Miyake, J. Phys. Soc. Japan, **17** (1962) 124.
309. S. Miyake, K. Fujiwara, K. Suzuki, J. Phys. Soc. Japan, **18** (1963) 1306.
310. K. Moliére, H. Niehrs, Zeitschrift für Physik, **140** (1955) 581.
311. G. Möllenstedt, H. Düker, Naturwissenschaften, **42** (1955) 41.
312. G. Möllenstedt, H. Düker, Zeitschrift für Physik, **145** (1956) 377-397.
313. G. Möllenstedt, Optik, **13** (1956) 209.
314. G. Möllenstedt, M. Keller, Zeitschrift für Physik, **148** (1957) 34.
315. G. Möllenstedt, H. Wahl, Naturwissenschaften, **55** (1968) 340.
316. P.E. Mooney, W.J. de Ruijter, O.L. Krivanek, Proc. **51st** Ann. Meet. MSA. San Francisco Press, San Francisco, (1993) 262.
317. N. Mori, T. Oikawa, T. Katoh, J. Miyahara, Y. Harada, Ultramicroscopy. **25** (1988) 195.
318. R. Morin, A. Gargani, Phys. Rev. B, **48** (1993) 6643.
319. R. Morin, A. Degiovanni, J. Vac. Sci. Tech. B, **13** (1995) 407.
320. G.A. Morton, E.G. Ramberg, Phys. Rev., **56** (1939) 705.
321. R.K. Mueller, Acoustic Holography Survey, Vol. 1 of Advances in Holography. N.H. Farhat, Ed., M. Dekker, Inc., (1975).
322. J. Munch, Optik, **43** (1975) 79.
323. HolograFREE software located at "http://holomac.nist.gov/".
324. M. Nicklaus, F. Hasselbach. Phys.Rev. A, **48** (1993) 152-160.
325. K. Ogai, S. Matsui, Y. Kimura, R. Shimizu, Jpn. J. Appl. Phys., **33** (1993) 5988.
326. M.A. O'Keefe, Ultramicroscopy, **47** (1992) 282.
327. M. Op de Beeck, D. Van Dyck, W. Coene, *Electron Holography*. Eds. A. Tonomura, L.F. Allard, G. Pozzi, D.C. Joy, Y.A. Ono, Elsevier Science B.V., (1995) 307.

328. M. Op de Beeck, D. Van Dyck, W. Coene, Ultramicroscopy, **64** (1996) 167-183.
329. A. Orchowski, W.D. Rau, H. Lichte, Phys. Rev. Lett., **74** No.3 (1995) 399.
330. A. Orchowski, H. Lichte, Ultramicroscopy, **64** (1996) 199-209.
331. N. Osakabe, K. Yoshida, Y. Horiuchi, T. Matsuda, H. Tanabe, T. Okuwaki, J. Endo, H. Fujiwara, A. Tonomura, Appl. Phys. Lett., **42** (1983) 746-8.
332. N. Osakabe, T. Matsuda, J. Endo, A. Tonomura, Jpn. J. Appl. Phys., **27** (1988) L1772.
333. N. Osakabe, J. Endo, T. Matsuda, A. Tonomura, A. Fukuhara, Phys. Rev. Letters, **62** (1989) 2969.
334. M.T. Otten, W.M.J. Coene, Ultramicroscopy, **48** (1993) 77-91.
335. J.B. Pendry, *Low Energy Electron Diffraction*, Academic Press, Boston, (1974) 32, 262.
336. D.R. Penn, Phys. Rev. B, **35** (1987) 482.
337. M. Peshkin, A. Tonomura, *The Aharonov-Bohm Effect*, Springer-Verlag, Berlin, (1989).
338. A.K. Petford-Long, M.B. Stearns, C.-H. Chang, S.R. Nutt, D.G. Stearns, N.M. Ceglio, A.M. Hawryluk, J. Appl. Phys., **61** (1987) 1422.
339. G. Pozzi, Optik, **47** (1977) 105.
340. G. Pozzi, Optik, **63** (1983) 227.
341. G. Pozzi, Optik, **65** 1 (1983) 77.
342. G. Pozzi, Optik, **66** (1983) 91.
343. G. Pozzi, Optik, **77** 2 (1987) 69.
344. G. Pozzi, J. Bonevich, A. Tonomura, Proc. 13th International Congress on Electron Microscopy, Vol. 1 (1994) 309.
345. G. Pozzi, J.E. Bonevich, K. Harada, H. Kasai, T. Matsuda, T. Yoshida, A. Tonomura, *Electron Holography*, Eds. A. Tonomura, L.F. Allard, G. Pozzi, D.C. Joy, Y.A. Ono, Elsevier Science B.V., (1995) 125.
346. W. Qian, J.C.H. Spence, J.M. Zuo, Acta Cryst., A49 (1993) 436.
347. W. Qian, M.R. Scheinfein, J.C.H. Spence, Appl. Phys. Letts., **62** (1993) 315.
348. W. Qian, M. Scheinfein, J.C.H. Spence, J. Appl. Phys., **73** (1993) 7041.
349. G. Radi, Acta Cryst. A, **26** (1970) 41.
350. O. Rang, Optik, **10** (1953) 90-106.
351. J. Rankin, B.W. Sheldon, Mater. Sci. Eng. A, **204** (1995) 48.
352. W.D. Rau, MSA Bulletin, Vol. 25 No. 1, (1994) 466.
353. W.D. Rau, H. Lichte, E. Völkl, U. Weierstall, Journal of Computer-Assisted Microscopy, **3** (1991) 51-63.
354. W.D. Rau, Ph.D. Thesis, University of Tübingen, Germany (1994).
355. W.D. Rau, H. Lichte, K.-H. Herrmann, Optik Supplement 4, **83** (1989) 79.
356. V. Ravikumar, R. P. Rodrigues, V. P. Dravid, Phys. Rev. Lett., **75** (1995) 4063.
357. L. Reimer, *Transmission Electron Microscopy*, Springer, Berlin (1989).
358. L. Reimer, ibid, p.37.
359. L. Reimer, ibid, p.57, 288.
360. L. Reimer, ibid, p.90.
361. L. Reimer, ibid, p.98.
362. L. Reimer, ibid, p.226.
363. L. Reimer, ibid, Section 6.4.
364. D. Rez, P. Rez, Proc. 51st MSA, (1993) 1202.
365. D. Rez, P. Rez, I.P. Grant, Acta Cryst., A50 (1994) 481.
366. J.M. Rodenburg, Ultramicroscopy, **27** (1989) 413.
367. J.M. Rodenburg, B.C. McCallum, P.D. Nellist, Ultramicroscopy, **48** (1993) 304.
368. H. Rose, Ultramicroscopy, **2** (1977) 251.
369. H. Rose, Optik, **85** No.1 (1990) 19.
370. F.M. Ross, W.M. Stobbs, Phil. Mag., **63** (1991) 37.
371. Q. Ru et al., J. Opt. Soc. Am. A, **8** (1991) 1739.
372. Q. Ru, J. Endo, T. Tanji, A. Tonomura, Applied Physics Letters, **59** (1991) 2372-2374.
373. Q. Ru, J. Endo, T. Tanji, A. Tonomura, Optik, **92** (1992) 51-55.
374. Q. Ru, J. Endo, A. Tonomura, Applied Physics Letters, **60** (1992) 2840-2842.
375. Q. Ru, T. Hirayama, J. Endo, A. Tonomura, Jap. J. of Appl. Phys., **31** (1992) 1919-1921.
376. Q. Ru, N. Osakabe, J. Endo, A. Tonomura, Ultramicroscopy, **53** (1994) 1.
377. Q. Ru, G. Lai, K. Aoyama, J. Endo, A. Tonomura, Ultramicroscopy, **55** (1994) 209-220.
378. Q. Ru, *Electron Holography*, Eds. A. Tonomura, L.F. Allard, G. Pozzi, D.C. Joy, Y.A. Ono, Elsevier Science B.V., (1995) 55-68.
379. Q. Ru, ibid, p343.

380. J.C. Russ, *The Image Processing Handbook*, CRC Press, Inc., (1992).
381. D.K. Saldin, J.C.H. Spence, Ultramicroscopy, **55** (1994) 397.
382. O. Saxton, *Computer Techniques for Image Processing in Electron Microscopy*, Academic Press, New York (1978).
383. W.O. Saxton, *Advances in Electronics and Electron Physics*, Academic Press (1978).
384. W.O. Saxton, Journal of Microscopy, **174** (1994) 61-68.
385. W.O. Saxton, Journal of Microscopy, 179 (1995) 201.
386. W.O. Saxton, Ultramicroscopy, **55** (1995) 171.
387. A.L. Schawlow, G.E. Devlin, Phys. Rev., **113** (1959) 120.
388. K. Scheerschmidt, J. Microsc., **190** Ptsl/2 (1998) 238-248.
389. K. Scheerschmidt, *Inverse Problems of Wave Propagation and Diffraction*, Eds. G. Chavent, P.C. Sabatier, Springer-Verlag, Berlin u. Heidelberg 1997, (Lecture notes in physics, Vol. 486) 71-85.
390. M. Scheinfein, W. Qian, J.C.H. Spence, J. Appl. Phys., **73** (1993) 2057.
391. O. Scherzer, Z. f. Phys., **101** (1936) 593.
392. O. Scherzer, J. Appl. Phys., **20** (1949) 20.
393. P. Schiske, Proc EUREM IV (1968) 145.
394. P. Schiske, J. Phys. D; Appl. Phys., **8** (1975) 1372.
395. H. Schmid, Ph.D. Thesis, University of Tübingen, Germany (1985).
396. M. Seah, W. Dench, Surface and interface analysis, **1** (1979) 2.
397. R. Seeliger, Optik, **5** (1949) 409.
398. G. Shedd, J. Vac. Sci. Techn., A12 (1994) 2595.
399. H. Shimoyama, A. Ohshita, S. Maruse, Y. Minamikawa, J. Elect. Micros., **21** No.2 (1972) 119.
400. D.J. Smith, W.O. Saxton, M.A. OKeefe, G.J. Wood, W.M. Stobbs, Ultramicroscopy, **11** (1983) 263.
401. G.H. Smith, R.E. Burge, Acta Cryst., **15** (1962) 182.
402. L. Solymar, D.J. Cooke, *Volume Holography and Volume Gratings*, Academic Press, London, (1981).
403. F. Sonier, J. Microscopie, **12** (1971) 17.
404. L.M. Soroko, *Holography and Coherent Optics*, Plenum Press, New York, (1980).
405. J.C.H. Spence, J.M. Cowley, Optik, **50** (1978) 129.
406. J.C H. Spence, *Scanning Electron Microscopy*, Ed. Om Johari, I.I.T.R.I., Chicago, (1978) 61.
407. J.C.H. Spence, J.M. Zuo, Rev. Sci. Instr., **59** (1988) 2102.
408. J.C.H. Spence, *Experimental High-Resolution Electron Microscopy*, 2nd ed., Oxford Univ. Press, (1988).
409. J.C.H. Spence, ibid, p24.
410. J.C.H. Spence, ibid, p80.
411. J.C.H. Spence, ibid, p155-156.
412. J.C.H. Spence, ibid, p261.
413. J.C.H. Spence, W. Qian, Proc. Micros. Soc. Amer., Ed. G. Bailey, San Francisco Press, San Francisco, (1992) 938.
414. J.C.H. Spence, Optik, **92** (1992) 57.
415. J.C.H. Spence, W. Qian, Phys. Rev. B, **45** (1992) 10271.
416. J.C.H. Spence, J.M. Zuo, *Electron Microdiffraction*, Plenum Press, New York, (1992) 31, 104.
417. J.C.H. Spence, Acta Cryst., A49 (1993) 231.
418. J.C.H. Spence, W. Qian, A. Melmed, Ultramicroscopy, **52** (1993) 473.
419. J.C.H. Spence, J.M. Cowley, J.M. Zuo, Appl. Phys. Letts., **62** (1993) 2446.
420. J.C.H. Spence, W. Qian, M. Silverman, J. Vac. Sci. Tech., A12 (1994) 542.
421. J.C.H. Spence, Proc. 13th Int. Congr. Electr. Micros., (Paris, 1994).
422. J.C.H. Spence, W. Qian, X. Zhang, Ultramicroscopy, **55** (1994) 19.
423. J.C.H. Spence, X. Zhang, W. Qian, *Electron Holography*, Eds. A. Tonomura, L.F. Allard, G. Pozzi, D.C. Joy, Y.A. Ono, Elsevier Science B.V., (1995) 267.
424. J.C.H. Spence, X. Zhang, U. Weierstall, J. Zuo, Surface Rev. Letts., **4** (1996) 577.
425. J. Spence, X. Huang, U. Weierstall, D. Taylor, K. Taylor, *Topics in Electron Microscopy*, ed. P.B. Hirsch, Inst. of Physics, Bristol, (1998), in press.
426. J. Spence, Micron, **28** (1997) 101.
427. P.A. Stadelmann, Ultramicroscopy, **21** (1987) 131.
428. J.A. Stratton, *Electromagnetic Theory*, McGraw-Hill Book Company, (1941) 257-258.
429. W.G. Stroke, M. Halioua, F. Thon, D. Willasch, Optik, **41** (1974) 319.
430. L. Sturkey, Phys. Rev., **73** (1948) 183.

431. J. Sum, *Off-Axis STEM Holography*, Ph.D. Thesis, University of Tübingen, Germany (1995).
432. L.W. Swanson, FEI Company Tecnnical Note No. 1 (1988).
433. M. Takeda, Q. Ru, Applied Optics, **24** (1985) 3068.
434. M. Tanaka, M. Terauchi, K. Tsuda, *Convergent Beam Electron Diffraction*, III, JEOL Ltd., (1994).
435. T. Tanji, K. Ishizuka, *Electron Holography*, Eds. A. Tonomura, L.F. Allard, G. Pozzi, D.C. Joy, Y.A. Ono, Elsevier Science B.V., (1995) 45-54.
436. T. Tanji, K. Urata, K. Ishizuka, Q. Ru, A. Tonomura, Ultramicroscopy, **49** (1993) 259.
437. S. Tanuma, C.J. Powell, D.R. Penn, Surface and interface analysis, 17, (1991) 911.
438. H. Tomita, M. Savelli, C.R. Acad. Sc. Paris, B267 (1968) 580.
439. S.Y. Tong, H. Li, H. Huang, Surface Rev. Letters, **1** (1994), 303.
440. A. Tonomura, T. Matsuda, J. Endo, H. Todokoro, T. Komada, J. Electron Microsc., **28** (1979) 1-11.
441. A. Tonumura, T. Matsude, J. Endo, Jpn. J. Appl. Phys., **18** (1979) 1373.
442. A. Tonomura, T. Matsuda, J. Endo, T. Arii, K. Mihama, Phys. Rev. Lett., **44** (1980) 1430-3.
443. A. Tonomura, T. Matsuda, T. Tanabe, N. Osakabe, J. Endo, J. Fukuhara, A. Shinagawa, H. Fujiwara, Phys. Rev. B, **25** (1982) 6799.
444. A. Tonomura, T. Matsuda, T. Kawasaki, J. Endo, N. Osakabe, Phys. Rev. Lett., **54** (1985) 60.
445. A. Tonomura, N. Osakabe, T. Matsuda, T. Kawasaki, J. Endo, S. Yano, H. Yamada, Phys. Rev. Lett., **56** (1986) 792-5.
446. A. Tonomura, Progress in Optics, Ed. E. Wolf, Elsevier Serv. Pub. Vol. 23 (1986) 18.
447. A. Tonomura, Prog. Opt., **23** (1986) 185.
448. A. Tonomura, T. Matsudo, J. Endo, T. Arii, K. Mihama, Phys. Rev. B, **34** (1986) 3397.
449. A. Tonomura, Rev. Mod. Phys., **59** (1987) 639.
450. A. Tonomura, Adv. Phys., **41** (1992) 59.
451. A. Tonomura, *Electron Holography*, Springer-Verlag, Berlin, (1993).
452. A. Tonomura, ibid, p70-71.
453. *Electron Holography*, Eds. A. Tonomura, L.F. Allard, G. Pozzi, D.C. Joy, Y.A. Ono, Elsevier Science B.V., (1995).
454. K. Tsuno, Rev. Sci. Instr., **64** (1992) 659.
455. "TCL, Technical Command Language," TNO-TUD, TPD, The Netherlands.
456. D. Van Dyck, W. Coene, Optik, 77/3 (1987) 125.
457. D. Van Dyck, J. Danckaert, W. Coene, E. Selderslaghs, D. Broddin, J. Van Landuyt, S. Amelinckx, *Computer Simulation of Electron Microscope Diffraction and Images*, Eds. W. Krakow, M. O'Keefe, (1989).
458. D. Van Dyck, Proc. 12th ICEM (Seattle), Francisco Press Inc., **1** (1990) 26.
459. D. Van Dyck, ibid, p64.
460. D. Van Dyck, A.F. De Jong, Ultramicroscopy, **47** (1992) 266.
461. D. Van Dyck, M. Op de Beeck, *Electron Holography*, Eds. A. Tonomura, L.F. Allard, G. Pozzi, D.C. Joy, Y.A. Ono, Elsevier Science B.V., (1995) 297-306.
462. D. Van Dyck, M. Op de Beeck, Ultramicroscopy, **64** (1996) 99.
463. D. Van Dyck, M. Op de Beeck, W. Coene, Ultramicroscopy, **64** (1996) 167.
464. M.A. Van Hove, W.H. Weinberg, C.-M. Chan, Low-Energy Electron Diffraction (Springer, Berlin, 1986) 132.
465. M. Vanzi, Optik, **58** (1981) 103.
466. L.H. Veneklasen, B.M. Siegel, J. Appl. Phys., **43** (1972) 1600.
467. L.H. Veneklasen, Optik, **36** No.4 (1972) 410.
468. L.H. Veneklasen, Optik, **44** (1975) 447.
469. W.J. Vine, R. Vincent, P. Spellward, J.W. Steeds, Ultramicroscopy, **41** (1992) 423.
470. E. Völkl, H. Lichte, Optik Suppl., 4 Vol.83, (1989) 102.
471. E. Völkl, H. Lichte, Ultramicroscopy, **32** (1990) 177-180.
472. E. Völkl, *High Resolution Electron Holography*, PhD thesis, University of Tübingen, Germany (1991).
473. E. Völkl, L.F. Allard, Proc. **51st** Annual Meeting of the Microscopy Society of America, (1993) 1066.
474. E. Völkl. L.F. Allard, 13th Int. Congr. on Electron Microscopy, (Paris, 1994) 287-288.
475. E. Völkl, F. Lenz, Q. Fu, H. Lichte, Ultramicroscopy, **55** (1994) 75-89.
476. E. Völkl, L.F. Allard, MSA Bulletin, **24** No. 1, (1994) 466-471.
477. E. Völkl, L.F. Allard, Proc. **52nd** Annual Meeting of the Microscopy Society of America, (1994) 918-919.

478. E. Völkl, L.F. Allard, A. Dayte, B. Frost, Ultramicroscopy, **58** (1995) 97-103.
479. E. Völkl, L.F. Allard, B. Frost, Journal of Microscopy, **180** (1995) 39-50.
480. E. Völkl, L.F. Allard, B. Frost and T.A. Nolan, Proc. **53rd** Annual Meeting of the Microscopy Society of America, (1995) 616.
481. E. Völkl, L.F. Allard, B. Frost, *Electron Holography*, Eds. A. Tonomura, L.F. Allard, G. Pozzi, D.C. Joy, Y.A. Ono, Elsevier Science B.V., (1995) 103-116.
482. E. Völkl, L.F. Allard, B. Frost, Scanning Microscopy, Supplement 11, in press.
483. H. Wahl, Z. Naturwissenschaften, 55 (1968) 340.
484. H. Wahl, Optik, **30** (1970) 508-520 and 577-589.
485. H. Wahl, Optik, **39** (1974) 585-588.
486. H. Wahl, Habilitationsschrift, University of Tübingen, Germany, (1975).
487. S.-Y. Wang, J.M. Cowley, Micros. Res. Tech., **30** (1995) 181.
488. Y.C. Wang, T.M. Chou, M. Libera, T.F. Kelly, Appl. Phys. Lett., **70** (1997) 1296.
489. Z.L. Wang, Ultramicroscopy, **52** (1993) 504.
490. U. Weierstall, Masters Thesis, University of Tübingen, Germany, (1989).
491. C.M. Wei, Surface Rev. Letters, **1** (1994), 335.
492. I. Weingärtner et al., Optik, **30** (1969) 318.
493. J.K. Weiss, private communication
494. J.K. Weiss, W.J. de Ruijter, M. Gajdardziska-Josifovska, D.J. Smith, E. Völkl. H. Lichte, Proc. 49th Annual EMSA Meeting, Ed. G.W. Bailey, San Francisco Press. San Francisco, (1991) 674-5.
495. J.K. Weiss, W.J. de Ruijter, M. Gajdardziska-Josifovska, M.R. McCartney, D.J. Smith, Ultramicroscopy, **50** (1993) 301-311.
496. D.B. Williams, C.B. Carter, *Transmission Electron Microscopy I*, Plenum Press, New York, (1996) 71.
497. D.B. Williams, C.B. Carter, ibid, p78.
498. D.B. Williams, C.B. Carter, ibid, p73.
499. D.J. Wohlleben, Appl. Phys., **38** (1967) 3341.
500. D.J. Wohlleben, *Electron Microscopy in Materials Science*, Academic Press, New York, **2** (1971) 712.
501. T.T. Wu, C.N. Yang, Phys. Rev. D, **12** (1975) 3845.
502. K. Yada, K. Shibata, T. Hibi, J. Electron Microscopy, **22** (1973) 223.
503. K. Yagi, *Electron Diffraction Techniques*, Vol. 2, Ed., J.M. Cowley. Oxford Uni. Press, (1993) Chapter 3, 260.
504. N. Yamamoto, J.C.H. Spence, Thin Solid Films, **104** (1983) 43.
505. T. Yoshida, T. Matsuda, A. Tonomura, Proc. 50th Ann. Meet. Electr. Microsc. Soc. Amer., Ed. G.W. Bailey, San Francisco Press, San Francisco, (1992) 68.
506. T. Yoshida, H. Kasai, J. Bonevich, T. Matsuda, A. Tonomura, Proc. of 53rd Ann. Meet. Microsc. Soc. Amer., eds. G.W. Bailey et al., San Francisco Press, San Francisco, (1995) 366.
507. H. Yoshioka, J. Phys. Soc. Jpn., **21** (1966) 948.
508. F. Zemlin, K. Weiss, W. Schiske, W. Kunath, K.-H. Herrmann, Ultramicroscopy, **3** (1978) 49-60.
509. X. Zhang, T. Hashimoto, D.C. Joy, Appl. Phys. Letters, **60** (1992) 784.
510. X. Zhang, J.C.H. Spence, J.M. Zuo, Proc. Micros. Soc. Am. 1995, Ed. G. Bailey, Jones Publishing, New York, (1995) 610.
511. J. Zhu, L.-M. Peng, J.M. Cowley, J. Electron Microscopy Technique, **7** (1987) 177.
512. J.M. Zuo, Ultramicroscopy, **66** (1996) 21.
513. J.M. Zuo, M.R. McCartney, J.C.H. Spence, Ultramicroscopy, **66** (1996) 35.
514. Ann. Phys., (Leipzig) **81** (1926) 109.

Index

MIX
Papier aus verantwortungsvollen Quellen
Paper from responsible sources
FSC® C105338

If you have any concerns about our products,
you can contact us on
ProductSafety@springernature.com

In case Publisher is established outside the EU,
the EU authorized representative is:
Springer Nature Customer Service Center GmbH
Europaplatz 3, 69115 Heidelberg, Germany

Printed by Libri Plureos GmbH
in Hamburg, Germany